DESCRIPTIVE STATISTICS

The Weight of Numbers, the Strength of Data and the Power of Information

Courses and Exercises

By

Aimé Mbobi, PhD, MSc, MEng,

Peggy Thimothée Mauline, MSc and

Gaëlle Mbobi, BSc

Table Of Contents

Dedication

To Adèle, Félicianne, Martin, Raymond, Christine, Monique, Patrice, Erika, Capucine, Aimé Jr, Chaïnna, Naïma, Naëly, Bruno, Achilles, Serge, Hugues, Patrick, Guylain, Irène, Stern, Yasmine, Isahia, Ayden, Victoria et Zion.

Acknowledgments

We would like to take this opportunity to extend our sincere thanks to the individuals who generously invested their time and expertise in reading and reviewing the manuscript of this book. Your valuable feedback, constructive criticism, and encouragement played a pivotal role in shaping the final version of this work. Your insights and suggestions have been invaluable in helping us to improve the quality and clarity of the content.

Your dedication, commitment, and unwavering support have been a source of motivation and inspiration for us throughout this journey. We are deeply grateful for your contributions and honored to have had your assistance in bringing this book to fruition. Thank you for your willingness to share your knowledge and expertise with us. Your efforts and support are deeply appreciated and will not be forgotten.

Special thanks to: Christine Mbobi, Patrice Bellaire, Sophie Bestory-Théodose, Marc Théodose, Mark Cheley, Monique Thimothée, Simon Yomputu, Véronique Delannay-Lassus, Capt Gabriel Mukunda, Erika Mbobi, Capucine Mbobi, A.J. Mbobi, Valérie Pluton, Bienvenu Biyoudi, Brandon Mingo, Denise Lorbel, Gisèle Kyomba, Johann Graham, Blaise Mompere, Henri Savina, Jean-Jacques Bokino, Jean-Robert Bokino, Natasha Khan, Eva Matamba, Katia Sanctussy, Emmanuelle Gustave, Ing. Suresh Guruvayurappan, Ing. Jean-Claude Kazadi, Dr. Javad Jafarian, Dr. Hisham Alasady et Dr. Osama Bazan.

About the Authors

Dr. Aimé Mbobi is a Telecommunications Engineer, holding degrees from the École Nationale Supérieure Mines-Télécom Lille Douai and the University of Lille 1 in France. He also possesses a Specialized Master's Degree in Network Application Engineering from the École Nationale Supérieure des Mines de Paris. Furthermore, he has earned a Ph.D. in Computer Science from the University of Paris 11-Saclay in France.
With over twenty-five years of experience in higher education, Dr. Mbobi is a seasoned Professor. He has taught at the School of Engineering Technology and Applied Science at Centennial College of Applied Arts and Technology, as well as at the Faculty of Applied Arts and Technology at Sheridan College Institute of Technology in Toronto, Canada. He has also delivered courses at the graduate level at the Institut Supérieur des Techniques Appliquées (ISTA) in Kinshasa, DRC, and at the École CentraleSupélec in France. Additionally, he serves as a Visiting Professor at the Faculty of Mathematics and Computer Science at the University of Kinshasa in DRC.
In parallel with his teaching commitments, Dr. Aimé Mbobi currently serves as the Head of the Solution and System Engineering Department at Aviat Networks. Prior to this role, he held the position of Head of Learning Services at Huawei-Canada for a decade. With over thirty years of experience in the ICT industry, Engineer Mbobi has played a pivotal role as a senior executive in several cutting-edge technology companies. Among these entities are Redknee Inc. in Canada, Siemens Nixdorf, France Telecom, Française d'Electronique Recherches et Mathématiques, Telecom Développement in France, and the United Nations.

Peggy Thimothée Mauline holds a Bachelor's degree in Applied Mathematics and a Master's degree in Economic Engineering from the University of the French West Indies and Guiana, Fouillole campus in France. With 20 years of teaching experience, Peggy is a certified Mathematics Professor who has actively participated in various teaching programs and committees throughout her career. She has held positions as a specialized teacher and trainer within a Rased. As a member of the Section for Adapted General and Vocational Education, she played a significant role in the implementation of various Mathematics projects.

Gaëlle Mbobi holds a Bachelor's degree in Chemistry from York University in Canada. With over 8 years of experience, she has honed her expertise in statistical analysis and control, with a focus on enhancing efficiency in managing and reporting key performance indicators (KPIs) across various entities. Thanks to her in-depth and extensive knowledge, as well as her active involvement in implementing processes for data processing, analysis, and interpretation, Gaëlle collaborates with several scientists in this field.

Foreword

This book entitled "*Descriptive Statistics - The Weight of Numbers, the Strength of Data and the Power of Information, Courses and Exercises*" is designed to cater to individuals who frequently employ statistics in their daily lives. The book specifically targets undergraduate and graduate students who have descriptive statistics in their curriculum, including those pursuing Applied Mathematics, Biology, Chemistry, Economics, Management, and other relevant applied sciences. The book's contents are congruent with various academic streams, making it suitable for technical curriculum students, such as College Diploma students. Furthermore, researchers in academic laboratories working on fundamental research or in industrial laboratories undertaking applied research who regularly apply descriptive statistics will benefit from the book's contents. It provides answers to commonly asked questions and clarifies any uncertainties in the subject matter.

The aim of this book is to provide solid knowledge and an easily comprehensible approach to descriptive statistics. Its primary objective is to help readers master the process of statistical analysis, which involves utilizing numerical data, gathering and analyzing information, interpreting outcomes, and making informed decisions based on precise and reliable foundations. The book's strength stems from its philosophy, which emphasizes creating an easily understandable guide that enables readers to learn descriptive statistics with minimal or no assistance. The book comprises eleven chapters, each providing in-depth knowledge and practical exercises to solidify the reader's understanding.

The book is divided into two parts, with the first part comprising nine chapters and the second part containing the final two chapters. The initial nine chapters are devoted to the course's core content, supplemented by relevant exercises to reinforce understanding.

Chapter one serves as an introduction to the four essential stages of the statistical approach: collecting, organizing, analyzing, and interpreting data.

Chapter Two focuses on statistical tables and provides a concise explanation of the three fundamental concepts in statistics: numbers, data, and information. This distinction is crucial for readers to comprehend the critical role of numbers in calculations and operations, the significance of data in expressing what is acknowledged or recognized as such, and the potency of information in interpreting outcomes. The chapter also covers qualitative and quantitative characteristics (or measures) of a statistical series, in addition to numbers, frequencies (or proportions), and cumulative frequencies.

Chapter Three delves into the graphical representation of statistical series, with a thorough discussion of the qualitative and quantitative characteristics. For qualitative characteristics, the chapter illustrates three types of representations: the bar graph, the pie chart, and the triangular representation of statistical series. On the other hand, for quantitative characteristics, the chapter introduces the bar chart for discrete variables and histograms for continuous variables. Furthermore, the chapter provides an in-depth discussion on frequencies and cumulative frequencies.

Chapter Four focuses on the characteristics of central tendency, also known as positional parameters or positional measures. These parameters illustrate the order of magnitude of the elements of the statistical series, including the mode, mean, and median. The mode represents the most frequent value in the statistical series, the mean or average represents the value that each character would have if all the modalities were evenly distributed, and the median represents the middle value of the distribution, i.e., the value that divides the series into two equal parts or the value that splits the population into the same number of individuals above and below it.

Chapter Five deals with characteristics of dispersion or measures of dispersion. This chapter explains how dispersion can be perceived in a statistical distribution, as well as the role of related parameters such as variance, standard deviation, and the coefficient of variation. These three characteristics provide essential information on how the series is dispersed around the mean.

Chapter Six primarily explores the shape characteristics (measures of skewness and kurtosis), which provide a clear idea of the shape of the statistical distribution. The chapter presents various coefficients, such as the YULE's coefficient of skewness, the PEARSON's coefficient of skewness, as well as the FISHER's coefficients of Skewness and Kurtosis. The skewness coefficient informs us about the spread to the left or the right of the distribution curve, making it possible to locate the concentration of weak or strong values in comparison to the central ones. Additionally, the FISHER's coefficient of Kurtosis informs us about the flattening of the distribution curve compared to that of the normal distribution, also known as the Gauss curve.

Chapter Seven delves into the characteristics and measures of concentration, which are used to understand how income or wealth is distributed within a population. This distribution can range from relatively homogeneous to completely unequal. In this chapter, we provide a detailed study of the LORENZ concentration curve and the GINI index, including its interpretation. To determine the GINI index, we present two methods for calculating the areas under the LORENZ curve that provide similar results. The first method is the Trapezoidal rule, which does not require advanced mathematics and involves calculating the total area under the LORENZ curve as the sum of the areas of the elementary trapezoids. The second method, based on advanced mathematical concepts, is numerical quadrature, which involves evaluating the definite integral of the function obtained by interpolating the LORENZ curve in the interval from zero to one hundred. We discuss two methods for interpolating the LORENZ concentration curve: the Newton method and the Lagrange method. Using the numerical quadrature method requires a solid foundation in numerical computation and knowledge of solving definite integrals.

Chapter Eight discusses two-character statistics. We develop important notions such as marginal and conditional statistical distributions, as well as marginal and conditional characteristics. We delve into the definitions of marginal mean and variance, as well as those of conditional mean and variance, and explore the relationship between marginal and conditional characteristics. Additionally, we examine the notion of independence of characteristics. By the end of this chapter, readers will have a solid understanding of these important statistical concepts.

Chapter Nine introduces the concept of linear adjustment, covering both cases where one or two characters are controlled. Several methods are presented, including the empirical method, the method of moving averages, the method of Mayer, and the method of least squares. To aid in understanding the use of linear adjustment, we apply these methods to project future values, using the example of the significant drop in cryptocurrency prices on May 21, 2021.

The chapter concludes with a comprehensive exercise that synthesizes the concepts covered in previous chapters, using demographic and territorial data from a given country. These data produce various statistical series, allowing us to explore nearly all concepts studied thus far. The study is divided into three parts: the first part examines population distribution across different municipalities, the second part investigates the distribution of the entire area of the country into municipalities, and the third part conducts a two-dimensional analysis of both population and area, examining the existence of functional dependence.

The second part of the book contains extensive and in-depth case studies that combine all of the statistical concepts and measures discussed throughout the previous chapters. The first study is a socio-economic analysis, while the second study presents various cases that guide decision-making processes. The last two chapters of this part provide readers with the opportunity to see firsthand how the statistical concepts and measures can be applied in real-world scenarios, making the content even more tangible and relevant.

Chapter ten presents a comprehensive, in-depth, and refined socio-economic analysis of a real-world case study based on the full-time equivalent salaries (FTEQ) of French people working in the private sector in 2018. This study is based on raw data collected from the French National Institute of Statistics and Economic Studies (INSEE). This study is segmented into three parts: analysis of basic statistical decision-making parameters, analysis of parameters related to taxation, and analysis of socio-economic parameters.

The choice of this data is based on its real, complete, and exhaustively inclusive nature, making it representative of the statistical population it characterizes. The study covers all positions, including part-time roles, based on the proportion of their actual volume of work and is obtained from a complete census of all French people working in the private sector in France.

The first part of the study focuses on the analysis of basic statistical decision-making parameters, where all parameters of central tendency, dispersion, shape, and concentration are studied and interpreted with related decisive conclusions.

The second part evaluates the percentages of French people working in the private sector subject to taxes and provides an analysis of parameters related to taxation. The study covers two cases, namely French people not subject to taxes and those subject to taxes, but of which a taxable salary component is subject to the percentage of the last tax bracket, constrained to a tax rate of 45%.

The last part of the study focuses on the socio-economic aspects, drawing comparisons between the data and the benchmarks of the Minimum Interprofessional Growth Wage (SMIC), poverty, wealth, and the middle class. The definitions established by official bodies such as INSEE, the Inequalities Observatory, the Research Center for the Study and Observation of Living Conditions (CREDOC), and the Organization for Economic Cooperation and Development (OECD) were used as a reference for these notions.

Chapter eleven concludes the book with a series of case studies that demonstrate how data can be processed, analyzed, and interpreted to optimize outcomes, such as gain, profit, satisfaction, and well-being while minimizing negative effects such as loss, dissatisfaction, and risk. Four case studies are presented, with the final three sharing similarities in terms of data processing and analysis but differing significantly in terms of interpretation and decision-making.

The first case study involves an international institution seeking to establish a long-term collaboration with a non-profit organization (NPO). Three non-profit organizations were selected and underwent rigorous tests to assess their ability to work with rigor, integrity, impartiality, and transparency, as well as their ability to measure and assess weights without using scales or weighing equipment.

The second case study features a fabric store that is transitioning to an automated fabric-cutting system. Three different cutting machines from various manufacturers were tested and validated to confirm the manufacturers' claims.

The third case study, based on the same data as the previous case, involves a biological laboratory searching for a vaccine jet gun injector that can deliver a specific injection dosage limit. Three solutions from different manufacturers were tested and validated for efficacy.

The fourth case study, also based on the same data as the second case, focuses on a specialized metal foundry seeking an optimized software solution with a response time faster than a given maximum time for carrying out a chain of tasks before cooling and hardening the alloy. Three software developers presented their products, which were tested and validated.

Remarkably, despite sharing identical data and processing methods, the three cases arrived at different decisions, highlighting the importance of data interpretation and the value of comprehensive testing and validation.

We'd like to emphasize that each chapter in this book is replete with appropriate, relevant, and clear illustrative examples to facilitate comprehension of the concepts being studied. Moreover, at the end of each chapter, a series of related exercises is provided, including both exercises relating to the current chapter and Cross-Chapter exercises covering several successive chapters. These exercises aim to reinforce understanding of the material and enable readers to apply the concepts to practical scenarios.

CHAPTER 1

Introduction

This introductory chapter presents the four fundamental steps of a statistical approach, namely, gathering, organizing, analyzing, and interpreting data.

1.1 History

Etymologically the word "Statistics" comes from the Latin: ***status***, which means ''state''. Statistics has been a field of study since ancient times, and it owes its development to the groundbreaking work of prominent mathematicians such as BERNOULLI, LAPLACE, PASCAL, GAUSS, etc. Their contributions helped establish the principles of mathematical statistics, which provide a framework for making informed decisions and drawing conclusions based on data analysis.

The contributions of QUETELET, Cournot, and Engel are noteworthy in the history of statistics. QUETELET conducted pioneering studies on human beings, while Cournot applied statistical methods to the economy. Engel's work on consumption further expanded the use of statistics and facilitated its evolution into a dynamic field.

Initially, statistics were primarily concerned with collecting simple counts and providing the state with demographic, economic, and natural phenomena-related information. However, in contemporary times, statistics is defined as a scientific discipline that collects and analyzes numerical and graphical observations pertaining to large sets of analogous facts such as people, animals, things, concepts, and facts.

The four fundamental stages of descriptive statistics are data collection, organization, analysis, and interpretation of measurable or classifiable observations. These steps are interleaved so that the output from the previous stage serves as the input for the subsequent stage.

From a mathematical standpoint, these stages can be viewed as functions that are composed successively.

Indeed, let:

- f denote the function associated with the operation of collecting raw data.
- g be the function associated with the operation of organizing collected data,
- h denote the function associated with the operation of analyzing organized data,
- t be the function associated with the operation of interpreting analyzed data.

We will have:

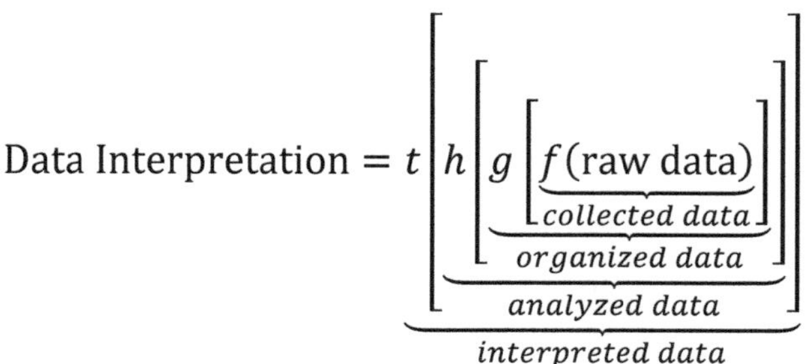

$$\text{Data Interpretation} = \underbrace{t\left[\underbrace{h\left[\underbrace{g\left[\underbrace{f(\text{raw data})}_{\textit{collected data}}\right]}_{\textit{organized data}}\right]}_{\textit{analyzed data}}\right]}_{\textit{interpreted data}}$$

Schematically we will have:

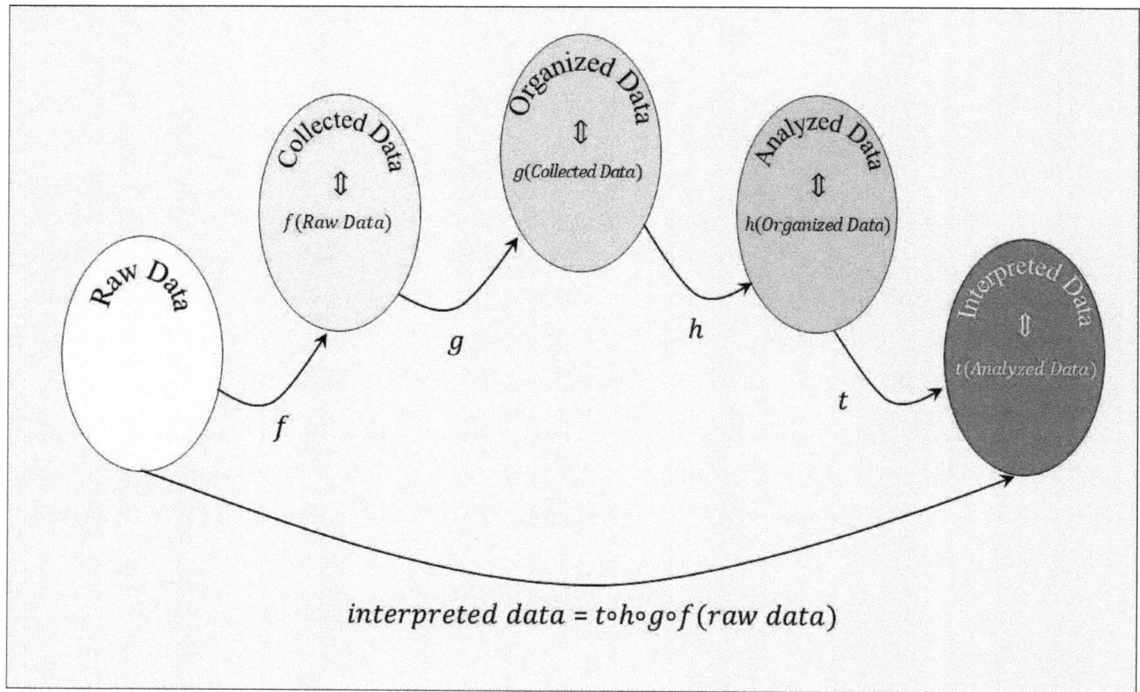

Now, let's examine each of these four steps more closely.

1.2 Data Collection

The statistician plays a crucial role in precisely defining the set of observations that they intend to study. This step is fundamental because it makes it possible to identify the target of the study and to ensure the relevance of the results obtained. This set is known as the ***population***, and each element within it is a ***member***, a ***statistical unit***, or an ***individual***. The word "individual" is derived from the Latin term "indivisus", meaning "indivisible." Individuals can be people, animals, objects, or any other entity that is the subject of study.

Sometimes, for practical reasons, the study is limited to only a portion of the population, which is referred to as a ***sample***. This selection is often made for practical reasons or to reduce the cost of the study. In this case, it is important that the sample is representative of the original population, that is to say, that it faithfully reproduces its characteristics.

From a mathematical perspective, the population can be viewed as a set, which is a collection of elements that share one or more common characteristics. In the context of statistics, the elements of a set can be interpreted as individuals or statistical units that share one or more common characteristics.

Similarly, a sample can be considered a subset of the initial set. This sample must be chosen randomly in order to guarantee the representativeness of the sample in relation to the original population. In addition, the sample size must be sufficient to ensure the reliability of the results obtained. In general, the larger the sample size, the higher the precision of the estimate.

Example 1

Consider the example of a university with 40,000 students. In this case, the statistical population consists of all of these 40,000 students, each of whom is considered a member. The common characteristics can be age, sex, and discipline studied. If we focus on the subset of students at the university who are celebrating their birthdays today, then this subset can be viewed as a sample of the overall population.

1.2.1 Statistical characters

In mathematics, elements of a set are defined by one or more common properties. Similarly, in statistics, population refers to the group of individuals, objects, or events that share a common trait, called the ***character*** of the population.

To make sense of the population, it is necessary to define a ***characteristic*** that is a collection of related attributes or properties that describe a group or a population. This characteristic is known as the ***variable*** or ***characteristic of interest***, and it is the property that we are trying to measure or understand.

Individuals designated as equivalent in terms of characteristics are grouped together in the same category, called the ***modality*** of the character sought.

Analyzing the modalities of the population enables us to understand the characteristics of the population being studied. This involves studying variations in the characteristics and structure of the population and analyzing differences and similarities between different groups of members. This process can highlight the most common obvious characteristics, identify trends and correlations, and draw conclusions about the population.

From a mathematical perspective, a modality is a subset of the population that shares a particular value of the variable. This means that a modality can be represented as a set of individuals, objects, or events that share a common characteristic.

Example 2

1. To better describe the marital status of people in a population, we can classify them into four basic categories: *Single*, *Married*, Divorced, and Widowed. These modalities provide a comprehensive overview of the different types of marital status that exist within the population.

Studied character = Marital Status

Modality1 = Single
Modality2 = Married
Modality3 = Divorced
Modality4 = Widowed

2. To analyze the day of the week on which people in a population are born, we can categorize it into seven modalities: Monday, Tuesday, Wednesday, Thursday, Friday, Saturday, and Sunday. This classification can, for instance, provide a clear overview of the different days on which individuals in the population were born.

Studied character = Day of the week

Modality1 = Monday
Modality2 = Tuesday
Modality3 = Wednesday
Modality4 = Thursday
Modality5 = Friday
Modality6 = Saturday
Modality7 = Sunday

3. When studying the number of days in a month, four main modalities can be identified: 28, 29, 30, and 31 days. This classification is important for planning and organizing various activities based on the number of days available in a given month.

Studied character = Number of days in a month

Modality1 = 28
Modality2 = 29
Modality3 = 30
Modality4 = 31

4. If we look at the number of spouses a woman can have in a non-polyandric country; there are two modalities: 0 and 1.

Studied character = Number of husbands in a non-polyandric country

Modality1 = 0
Modality2 = 1

5. Let's take the example of a company with employees whose minimum legal age is 18, and maximum retirement age is 65. If we consider the duration of employment as the characteristic under study and the stages of seniority that entitle employees to receive a merit medal, we can identify the following modalities (measured in years of seniority):

$$[0\ ;\ 10), [10\ ;\ 20), [20\ ;\ 30), [30\ ;\ 40), [40\ ;\ 47)$$

Studied character = Seniority giving right to a medal of merit in a company

Modality1 = $[0; 10)$
Modality2 = $[10; 20)$
Modality3 = $[20; 30)$
Modality4 = $[30; 40)$
Modality5 = $[40; 47)$

1.2.1.1 Intersection and union of different modalities of the same character

When studying a particular characteristic of a population, we often identify different modalities of that characteristic. Modalities are mutually exclusive and collectively exhaustive categories that represent different values of the characteristic under study. In other words, each statistical unit in the population belongs to one and only one modality of the characteristic being studied.

From a mathematical standpoint, if we consider each modality as a subset of the population, then the intersection of any two modalities must be an empty set, meaning that they do not share any common elements. For example, in **Example 2** mentioned earlier, it's impossible to be both single and married at the same time, just as a day of the week cannot be both Monday and Friday at the same time.

The modalities of a characteristic should enable us to classify all statistical units in the population. From a mathematical perspective, the union of all modalities should be a set that contains all the possible values of the characteristic being studied. This ensures the completeness of the set. For example, in **Example 2**, the set of modalities for the characteristic "day of the week" includes Monday, Tuesday, Wednesday, Thursday, Friday, Saturday, and Sunday, covering all seven days of the week.

There are two types of characteristics or variables: ***qualitative*** and ***quantitative***. Qualitative characteristics, also known as categorical variables, represent attributes or qualities that cannot be measured numerically, such as marital status or day of the week.

Quantitative characteristics, also known as numerical variables, represent measurable quantities, such as the number of days in a month or the number of husbands a woman can have in a non-polyandric country.

Overall, the modalities of a characteristic should be mutually exclusive and collectively exhaustive, enabling us to classify all statistical units in the population. These modalities can be categorized into qualitative and quantitative variables based on the nature of the characteristic being studied.

1.2.1.2 Qualitative characters

Qualitative characters or categorical variables are characters that cannot be measured or identified quantitatively. These variables describe the names or labels of the modalities and are divided into two families: ***the nominal qualitative characters*** and the ***ordinal qualitative characters***.

A qualitative character is nominal when it describes the names or labels of the modalities without referring to any natural order of these modalities. In other words, there is no logical sequence or hierarchy between the different modalities. The mathematical representation of a nominal qualitative character is a set of elements with no inherent order between them.

A qualitative character is ordinal when in addition to describing the name or label of the modalities, it establishes an ordering relationship between these modalities. This means that there is a natural hierarchy or order between the different modalities. The mathematical representation of an ordinal qualitative character includes a real relationship of order between the modalities that is reflexive, antisymmetric, and transitive.

Example 3

- If we take into account the nominal qualitative character "preferred color", the modalities can be *red, blue, green, yellow,* and *orange*. In this case, there is no natural relationship between the different modalities of this character.

- Considering the ordinal qualitative character "level of education", the modalities can be elementary school, middle school, high school, bachelor, master, and doctorate. In this case, there is a natural relationship between the modalities because the " elementary school" level of education is considered lower than the "high school" level of education, and so on up to the "educational level" of doctorate.

In both nominal and ordinal qualitative characters, each category is referred to as an ***item***, and the set of items is called a ***nomenclature***. For convenience, each item is usually assigned an arbitrary number to create a ***code*** for easy reference.

Understanding the differences between nominal and ordinal qualitative characters is important for various fields of study, including psychology, sociology, and economics. By using these qualitative characters, we can categorize and analyze data to make informed decisions and draw meaningful conclusions.

Example 4

If we consider the entire population of the planet as our statistical population, one interesting categorical variable to examine is the "Country of origin of each individual". This is a nominal qualitative variable where each modality is unique and corresponds to all the countries of the planet, forming the "Nomenclature of the countries of the planet".

To represent each country, a code can be created, such as the list of international telephone codes proposed by the International Telecommunications Union or the social identification system used in some countries (e.g., social security number in France or social insurance number in the United States and Canada). Postal codes can also serve as an example of this type of code.

1.2.1.3 Quantitative characters

Quantitative characters are distinct from qualitative characters in that their modalities are measurable. For instance, the number of players on a basketball team, the number of computers in a ministry, the size of students in a class, and the temperature range of a day are all examples of measurable variables. There are two main types of quantitative characters: ***discrete quantitative characters*** or *discrete variables* and ***continuous quantitative characters***.

Discrete quantitative characters are those that can take on any finite numerical value. Mathematically speaking, a quantitative character is considered discrete when its modalities are elements of the set of relative integers, $\mathbb{R}$.

Example 5

The number of children in a family is a discrete quantitative variable, as it can only take integer values (1, 2, 3, etc.) and cannot take intermediate values (such as 1.5 children). For example, one could say that there are x families in a city that have 3 children.

On the other hand, a quantitative character is continuous when it can take all the numerical values of an interval. From the mathematical point of view, a quantitative character is classified as continuous when its modalities are semi-open intervals, i.e., intervals closed on the left and open on the right and whose limits are real numbers.

Example 6

The temperature of a day is a continuous quantitative character because it can take any real value in a temperature interval. We could say, for example, that there were x days in July 2020 when the temperature was between 20 and 35 degrees Celsius, that is, in the semi-open interval [20; 35).

To collect a set of data for analysis, it is recommended to observe the defined population and the trait of interest, which leads to the formation of a statistical series. If all elements within the population have been observed, resulting in a complete set of data, it is referred to as a ***census***. However, if only a portion of the population was observed, such a set of data is known as a **sample**.

Example 7

Let's take **example 1** from **section 1.2**, which concerns the statistical population of all 40,000 students in a university. We can conduct a census by observing all the students. However, if we only focus on the students who are celebrating their birthdays today, we would obtain a sample from the population.

1.2.2 Statistical Variables

We have seen that the common property which can be observed or measured in each of the individuals of a population is called the ***character*** of the population. Moreover, individuals with the same character are placed in the same ***modality*** as the character mentioned.

A ***statistical variable*** is nothing but a character that can take on different values in different modalities. When statistical variables are arranged in order, they form a statistical distribution or table.

As we have previously discussed, the observable or measurable trait in each member of a population is referred to as the character of the population. Furthermore, individuals possessing the same trait are grouped together in a specific modality. A statistical variable refers to this character, which can take on various values across different modalities. When these variables are arranged in a particular order, they create a statistical distribution or table.

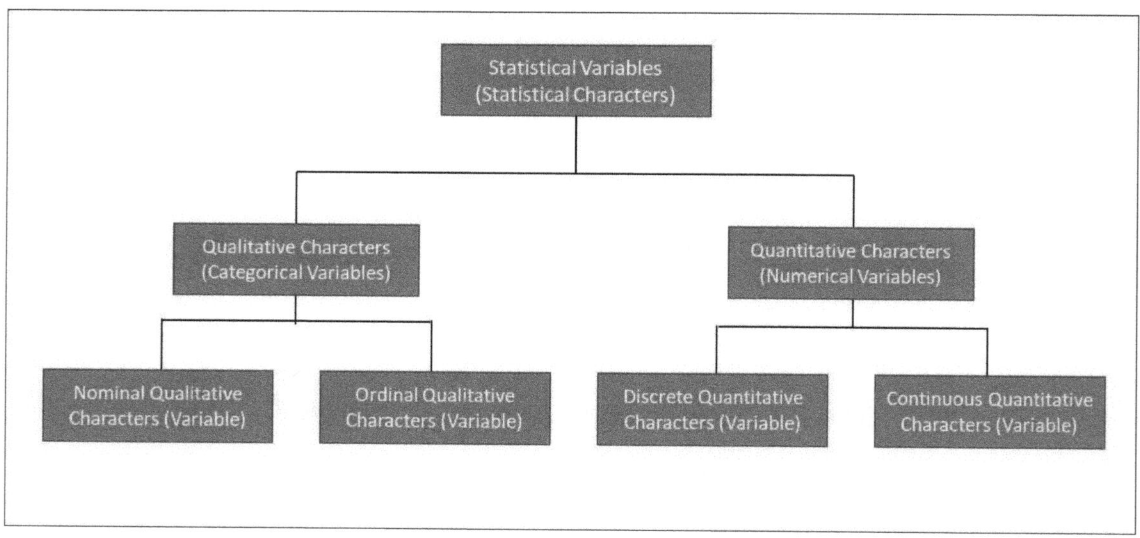

1.3 Data Organization

After conducting observations, the resulting data often comes in a large, unrefined format. To make sense of this data and draw meaningful conclusions, it is crucial to ***analyze*** and ***organize*** it effectively. This typically involves grouping similar observations together to form categories or bins, allowing for the creation of statistical tables and graphic representations.

These tables and graphs provide a visual representation of the data and facilitate a better understanding of trends and patterns that may be present. They allow researchers to see the distribution of the data and identify outliers or other significant characteristics that may be present. Additionally, these representations make it easier to convey findings to others, as visual aids are often more accessible and easier to understand than raw data alone.

Data analysis is a critical step in the research process, as it allows researchers to draw meaningful insights and conclusions from their observations and to make informed decisions.

1.4 Data Analysis

After observing tables and graphical representations, we can gain a general understanding of the behavior of a population with respect to the trait being studied. However, it is often necessary to distill this information into a few key numerical values, known as characteristics, in order to more effectively summarize the trends observed. There are several types of characteristics that are commonly used in data analysis:

Firstly, there are characteristics of central tendency or position that help to describe the order of magnitude of the observations in the dataset. These include measures such as the mean (average), mode, and median, which provide an indication of the typical value within the dataset. The mean is the sum of all values divided by the total number of values. It gives a general idea of the central tendency of the distribution. The mode is the value that appears most frequently in the distribution. It is often used to describe discrete distributions. Finally, the median is the value that divides the distribution into two equal parts, i.e., 50% of the values are below and 50% are above this value. It is particularly useful when the distribution is skewed or has outliers.

Secondly, there are characteristics of dispersion, which describe how the observations are distributed around the central value. Examples of these characteristics include the interquartile range, variance, standard deviation, coefficient of variation, and moments. These measures help to provide an understanding of the spread of the data and the degree to which it varies from the central tendency.

Thirdly, shape characteristics are used to describe and appreciate the shape of the distribution curve. These include indexes that can indicate the skewness and kurtosis of the data compared to that of a normal distribution. Examples of these indexes include the YULE's coefficient of skewness, the PEARSON's coefficient of skewness, the FISHER's coefficient of skewness, and the FISHER's coefficient of kurtosis.

Finally, there are characteristics of concentration, which describe how the population is distributed across different modalities. These characteristics are measured using indicators such as the medial and the GINI concentration index, which can help to identify any potential inequalities within the population and understand the heterogeneity of the trait being studied.

Overall, these characteristics play a crucial role in data analysis, as they allow researchers to extract meaningful insights from large and complex datasets, identify important trends, and make informed decisions based on their findings.

1.5 Data Interpretation

After the data has been collected, organized, and analyzed, the next step is to interpret the results. This is a critical stage, as the findings obtained from the sample need to be generalized to the entire population. To achieve this, it is necessary to formulate a hypothesis and to check whether the forecasts made have a high or low probability of being realized under this hypothesis. This is where the problem of choosing a good sample or sampling theory arises, as it involves the calculation of probabilities and the need to ensure that the sample is representative of the entire population.

In order to ensure that the findings are reliable and can be generalized to the entire population, it is important to carefully consider the sampling method and use statistical tests to assess the significance of the results obtained. By doing so, we can be more confident in the conclusions we draw and the decisions we make based on the data.

It should be noted that the interpretations made in statistical analysis are not established statistical laws but rather observations of trends and patterns within the data. The purpose of statistical analysis is to provide insights and inform decision-making, but it is important to recognize that there are limitations and uncertainties inherent in any statistical analysis.

Overall, statistical interpretation requires careful consideration of the sampling method, statistical tests, and the limitations of the analysis. By taking a systematic approach and carefully assessing the data, we can make more informed decisions and draw more reliable conclusions.

1.6 Exercises and Solutions

• **Exercise 1:** Identify the type of each character listed in the table below as either qualitative or quantitative. For quantitative characters, provide one or more possible modalities.

	Character	Type of Character	Possible modalities
1	The continent of birth of a person		
2	The salaries of employees in a company		
3	The North American seasons		
4	The weight of the fruit in a basket		
5	The meals of the day		
6	The temperature recorded during the day		
7	The Hexadecimal digits symbols		
8	The manufacturer of French cars,		
9	The living area in a house (in square feet)		
10	The number of pages in a book		
11	The speeds recorded by a radar		
12	The three basic colors		
13	The number of passes made in a soccer match		
14	The types of chords in a song		
15	The number of spelling mistakes in a dictation		
16	The duration of a movie		
17	The number of petals on a flower		
18	The age of an individual		
19	The eye color of an individual		
20	The high school grade of a student		

Solution

	Character	Type of Character	Possible modalities
1	The continent of birth of a person	Qualitative	Africa, America, Asia, Europa, Oceania
2	The salaries of employees in a company	Quantitative	
3	The North American seasons	Qualitative	Spring Summer Autumn Winter
4	The weight of the fruit in a basket	Quantitative	
5	The meals of the day	Qualitative	Breakfast, Dinner, Lunch
6	The temperature recorded during the day	Quantitative	
7	The Hexadecimal digits symbols	Qualitative	0, 1, 2, 3, 4, 5, 6, 7
8	The manufacturer of French cars,	Qualitative	Renault, Peugeot, Citroen
9	The living area in a house (in square feet)	Quantitative	
10	The number of pages in a book	Quantitative	
11	The speeds recorded by a radar	Quantitative	
12	The three basic colors	Qualitative	Red, Green, Blue
13	The number of passes made in a soccer match	Quantitative	
14	The types of chords in a song	Qualitative	F, C, G, Am
15	The number of spelling mistakes in a dictation	Quantitative	
16	The duration of a movie	Quantitative	
17	The number of petals on a flower	Quantitative	
18	The age of an individual	Quantitative	
19	The eye color of an individual	Qualitative	Brown, blue, green
20	The high school grade of a student	Qualitative	Freshman, Sophomore, Junior, Senior

• **Exercise 2:** Identify each of the following populations, whether it is a census or a sample.

	Population	Sample or Census
1	The score of all the matches of the champion league	
2	The score of a randomly selected set of games of different pools of the world cup	
3	The size of every last student who gets out of his class in a school	
4	The length of all the highways in Canada	
5	The degree of a randomly selected group of people in Paris	
6	The process of identifying the entire population of a country	
7	Blood sampling from a few babies born on the same day in a hospital	
8	Random selection of products leaving the factory for weighing	
9	The math test result of all students in a class	
10	The color of each car that passes in a roundabout every 10 minutes	
11	The eye color of random tire babies in a maternity ward	
12	The age of all the film actors who live in a given municipality	
13	The salary of all workers in a company	
14	The list of all the schools in a municipality where at least one teacher is absent on a certain Friday	
15	The weight of each fish from the fishery with a caudal fin smaller than the dorsal fin	
16	Some units of perfume bottles for appreciation	
17	Set of houses that served as a model in an urban development	
18	The count of all animals abandoned in an animal shelter.	
19	Partial enumeration of plant species in a forest	
20	The exhaustive list of all universities in a country	

Solution

	Population	Sample or Census
1	The score of all the matches of the champion league	Census
2	The score of a randomly selected set of games of different pools of the world cup	Sample
3	The size of every last student who gets out of his class in a school	Sample
4	The length of all the highways in Canada	Census
5	The degree of a randomly selected group of people in Paris	Sample
6	The process of identifying the entire population of a country	Census
7	Blood sampling from a few babies born on the same day in a hospital	Sample
8	Random selection of products leaving the factory for weighing	Sample
9	The math test result of all students in a class	Census
10	The color of each car that passes in a roundabout every 10 minutes	Sample
11	The eye color of random tire babies in a maternity ward	Sample
12	The age of all the film actors who live in a given municipality	Census
13	The salary of all workers in a company	Census
14	The list of all the schools in a municipality where at least one teacher is absent on a certain Friday	Sample
15	The weight of each fish from the fishery with a caudal fin smaller than the dorsal fin	Sample
16	Some units of perfume bottles for appreciation	Sample
17	Set of houses that served as a model in an urban development	Sample
18	The count of all animals abandoned in an animal shelter.	Census
19	Partial enumeration of plant species in a forest	Sample
20	The exhaustive list of all universities in a country	Census

CHAPTER 2

Statistical Tables

This chapter deals with statistical tables by introducing the key notions of numbers, data, and information. It also involves the concepts of qualitative and quantitative characteristics of a statistical series, as well as the notions of counts, frequencies (or proportions), and cumulative frequencies, which make it possible to describe and synthesize the data. Understanding these basic concepts is essential to be able to analyze the data and draw relevant conclusions.

2.1 Numbers, Data, And Information

Numbers serve as the fundamental building blocks of data. Their weight plays a crucial role not only in the construction of the data but also in the value attributed to this data. The accuracy and precision of numbers play a critical role in the reliability of the data they form. As for the data itself, it is only as useful as the insights that can be extracted from it through proper analysis and interpretation. The true strength of analyzed and interpreted data lies in its ability to offer valuable and reliable information that can inform strategic decision-making.

However, it's not just the quality of the data and the insights derived from it that matter. The way this information is presented and conveyed within its application environment can greatly impact its effectiveness. Clear and concise communication of the findings, as well as proper visualization of the data, can make all the difference in ensuring that the insights are properly understood and acted upon.

Furthermore, the power of information extends beyond its presentation and analysis. The ultimate impact of the insights on the environment in which they are applied is what truly matters. The ability of data-driven decisions to drive positive change and achieve desired outcomes is what makes them valuable. In essence, the power of information lies not only in the accuracy and quality of the data but also in the effectiveness of its analysis, presentation, and the outcomes it achieves.

2.1.1 Definition of Number

A number is a mathematical concept used for counting, enumerating, or quantifying objects, individuals, events, etc. Numbers play a fundamental role in statistics, as they form the basic unit of statistical data. In the context of statistics, numbers can represent a wide range of variables, including measurements, ratings, scores, etc. By assigning numerical values to variables, statisticians are able to manipulate and analyze the data in a meaningful way, drawing insights and making predictions based on patterns and trends.

Example 8

- Numbers ranging from -3 to 6
- Numbers between 10 and 20
- Decimal numbers such as 18.5
- Very large numbers, such as 1,736,000,000,000, etc.

2.1.2 Definition of Data

The concept of data is crucial in statistics as it refers to the information collected or observed on an object or an individual in a specific context or phenomenon. The data representing this information can be quantitative or qualitative. When they are quantitative, they are expressed using numbers and can take different forms, such as measures, sums, percentages, etc.

These numerical values may be discrete or continuous and can be further analyzed through various statistical techniques to draw conclusions about the population or sample from which the data was collected. Therefore, understanding the nature and characteristics of data is fundamental to the study of statistics as it enables researchers to extract meaningful insights and make informed decisions based on the data.

Example 9

- The length of an object ranging from 10 to 20 meters
- The number of children in a family ranging from 3 to 6
- The price of the product at 18.5 euros
- The annual revenue of a business at $1,736,000,000,000
- The number of days in a month to 28, etc.

2.1.3 Definition of Information

In the field of statistics, information refers to a collection of data that relates to a particular topic or subject. This information is made up of a number of statistical individuals and data associated with them. Knowing that such data is only an expression composed of a number and an object or a concept.

This data is obtained through various sources such as surveys, experiments, or observations. Statistical individuals are the units of observation in a given study, and the data collected about them can be further analyzed to extract meaningful insights.

To make sense of this data, it needs to be properly recorded, classified, and organized in a way that can be easily understood and analyzed. This is where statistical frameworks come in, as they provide a structured approach to organizing data and making sense of it. The choice of framework depends on the nature of the data and the questions being asked.

Statistical analysis involves interpreting the data to extract meaningful insights that become information, which can then be used to draw conclusions or make decisions. This may involve calculating various statistical measures such as central tendency or position measures, dispersion measures, shape measures, and concentration measures, or using more complex methods such as regression analysis, factor analysis, or clustering. The interpretation of statistical information requires a good understanding of statistical methods, as well as the ability to communicate findings effectively to others.

Example 10

- Information 1: There are 7 highways whose width is between 10 and 20 meters.
- Information 2: 12 families have between 3 and 6 children.
- Information 3: 21 electrical workers earn an hourly wage of 18.5 euros.
- Fact 4: In 2019, Canada's GDP was $1,736,000,000,000.

Here is a concrete example that demonstrates how the previous four pieces of information can be organized in a database. This method is needed to store and manage data in a simplified and efficient way.

Item	Number	Wide
Highways	7	10 to 20 meters

Item	Number	Number of kids
Families	12	3 to 6

Item	Number	Wage	Periodicity
Electrician workers	21	18.5 euros	Hourly

Year	Country	GDP
2019	Canada	$1,736,000,000,000

2.2 Qualitative Character

The representation of a statistical series with a single qualitative character comprises the nomenclature on a line or on a column and the number of statistical units of each modality on another line or on another column.

This type of representation makes it possible to synthesize and present data in a structured and clear way in the form of a table. This method consists of grouping the data according to a single qualitative criterion.

The nomenclature is the item used for each modality. It can be a category, a class, or a group and can be represented on a line or on a column. The number of statistical units of each modality can be represented either on a row or in a column.

This representation not only makes it easy to visualize the distribution of the data for each modality but also offers a quick understanding of the characteristics of the population studied. This visualization makes it easier to analyze and compare data by identifying the most consistent or sparse modalities and highlighting trends and variables between different modalities. Thus, this representation effectively helps to understand the structure and characteristics of the population under study, which is essential for quickly disseminating relevant information and making informed decisions.

Example 11

In this example, we present the areas of the countries of South America by providing their respective area in square kilometers. This information may be important for studies of various purposes.

Country	Area in Km^2
Argentina	2766890
Bolivia	1098580
Brazil	9511965
Chile	756950
Colombia	1138910
Ecuador	283560
Falkland Islands	12173
French Guyana	91000
Guyana	214970
Paraguay	406750
Peru	1285220
South Georgia and the South Sandwich Islands	3093
Suriname	163270
Uruguay	176220
Venezuela	912050
TOTAL	18821601

2.3 Quantitative Character

2.3.1 Discrete variables

Let's consider a given statistic variable X that can take on k possible values. To analyze the distribution of this variable, we need to consider the number of times each value x_i (with $i \leq k$) has been observed in the population under study. By ordering the x_i in increasing order, one can easily calculate the difference between the largest and the smallest value that the variable X can take. This measure is called *the range or amplitude of the statistical series.*

The range or amplitude is an important measure because it provides an indication of the dispersion of the values in the statistical series. The larger the span, the more scattered the values of the variable are, while a smaller span indicates that the values are more clustered. This can be useful to assess the variability of variable X in the study population.

Example 12

Let's suppose that an exam was taken by 50 students, and their results were recorded as grades out of 20 points (meaning k=21). The obtained grades are as follows).

10	9	16	11	11	8	7	11	6	13
7	15	12	8	7	9	7	8	13	6
8	13	12	13	18	7	9	12	6	8
10	8	8	10	13	15	5	13	12	6
13	9	3	12	4	13	10	13	12	7

After classification we can obtain a series in the form of a set of couples below:

Marks (x_i)	Frequencies (n_i)
0	0
1	0
2	0
3	1
4	1
5	1
6	4
7	6
8	7
9	4
10	4
11	3
12	6
13	9
14	0
15	2
16	1
17	0
18	1
19	0
20	0
Total	50

with x_i = character's value and n_i = corresponding number.

As an illustration, the distribution of scores for the 50 students can be summarized as follows:

- No student received a score of 0 out of 20.
- One student received a score of 5 out of 20.
- Six students received a score of 7 out of 20.
- Nine students received a score of 13 out of 20.
- Finally, one student received a score of 18 out of 20. etc.

2.3.2 Continuous Variables

Continuous variables can take on a large number of values, which makes it difficult to analyze them individually. Therefore, it is more convenient to group these values into classes, which involves combining values that belong to the same modality.

The selection of the number of classes is subjective, but it should be chosen carefully to achieve the desired level of simplification. Too many classes can lead to an overwhelming amount of information, while too few classes can create a risk of grouping together observations that are actually very different.

To standardize the representation of classes, it is customary to define them as semi-open intervals, where the left endpoint is included, but the right endpoint is not. So, the ith class will be determined by:

$$\begin{cases} \text{Its lower bound: } a_{i-1} \\ \text{Its upper bound: } a_i \\ \text{Its amplitude: } (a_i - a_{i-1}) \\ \text{Its center: } x_i = \dfrac{(a_i + a_{i-1})}{2} \end{cases}$$

It is important to stress that it is sometimes difficult to specify the external classes (the first and the last). In this case, we have to recourse to so-called open classes; "*less than*" for the first class and "*more than*" for the last class.

Example 13

Consider the following table, which shows the turnovers of 100 companies in millions of dollars. This table provides the overall distribution of turnovers across the entire population.

46	114	72	143	137	95	100	109	141	96
105	112	55	108	50	105	57	80	29	101
62	136	115	159	83	152	123	113	26	176
107	120	99	79	107	95	20	75	125	67
136	137	144	121	94	118	60	77	46	129
78	149	146	138	119	63	102	144	151	68
100	103	139	110	75	101	105	97	50	75
113	86	64	110	120	47	84	84	97	78
96	109	124	80	104	140	152	33	178	56
50	136	31	93	115	61	96	120	89	77

Sorting this data in ascending order yields:

20	50	68	80	95	101	109	118	136	144
26	55	72	80	96	102	109	119	136	144
29	56	75	83	96	103	110	120	136	146
31	57	75	84	96	104	110	120	137	149
33	60	75	84	97	105	112	120	137	151
46	61	77	86	97	105	113	121	138	152
46	62	77	89	99	105	113	123	139	152
47	63	78	93	100	107	114	124	140	159
50	64	78	94	100	107	115	125	141	176
50	67	79	95	101	108	115	129	143	178

Since the minimum and maximum values in the list are 20 and 178, respectively, it is reasonable to set the class width to 10, as this will allow for a sufficient level of detail in the frequency distribution.

Classes (x_i)	Frequencies (n_i)
[20;30)	3
[30;40)	2
[40;50)	3
[50;60)	6
[60;70)	7
[70;80)	9
[80;90)	7
[90;100)	10
[100;110)	15
[110;120)	10
[120;130)	8
[130;140)	7
[140;150)	7
[150;160)	4
[160;170)	0
[170;180)	2
Total	100

2.4 Relative Frequency (Proportion)

When studying a statistical series, it is often interesting to know the proportion of data that belongs to each class or category. To do this, we calculate the frequency of each class or modality.

The ***relative frequency*** (or ***proportion***) of the ith class or the ith modality of the characteristic is defined as the ratio of the frequency n_i of observations in this class or modality to the total number n of observations. This measure allows us to determine the proportion of data that belongs to this class or modality. We denote this as:

$$f_i = \frac{n_i}{n}$$

In practice, relative frequency is often expressed as a percentage rather than a decimal value, which allows for a more intuitive reading of the data distribution. Thus, to obtain the percentage frequency of a class or category, it is sufficient to multiply the ratio of the n_i relative frequency of the ith class or category by 100.

$$\sum_{i=1}^{i=k} f_i = \sum_{i=1}^{i=k} \frac{n_i}{n} = \frac{n_1}{n} + \frac{n_2}{n} + \dots\dots\dots + \frac{n_n}{n}$$

$$= \frac{n_1 + n_2 + \dots\dots + n_n}{n}$$

$$= \frac{n}{n}$$

$$= 1$$

This results in the following two properties:

$$0 \leq f_i \leq 1 \quad \text{and} \quad \sum_{i=1}^{i=k} f_i = 1$$

It is crucial to note that the total sum of relative frequencies must be equivalent to 100% since all observations must be represented. Checking the sum of frequencies is essential to ensure the quality of statistical analysis.

If the sum of relative frequencies exceeds or is less than 100%, this may be due to an error in relative frequency calculation, missing data, poorly cleaned or recorded data, or approximate values for each relative frequency. In this case, it is necessary to verify the data and calculations or adjust the most important relative frequencies so that the sum is equal to 100%.

Additionally, it is crucial to consider the relative size of each class or category when interpreting the relative frequency distribution. Classes or categories with higher relative frequencies may indicate a more significant trend or pattern in the data, while those with lower relative frequencies may be less relevant. However, it is also important to consider the context of the data and the objectives of the analysis when interpreting relative frequency distributions, as sometimes even small relative frequencies can be interesting.

Upon reviewing the two examples, **example 12** in **Section 2.3.1** concerning student grades and **example 13** in **Section 2.3.2** regarding company turnovers, it is apparent that in the first table, variables with zero frequency have been removed.

x_i	n_i	$f_i(\%)$
3	1	2
4	1	2
5	1	2
6	4	8
7	6	12
8	7	14
9	4	8
10	4	8
11	3	6
12	6	12
13	9	18
14	0	0
15	2	4
16	1	2
17	0	0
18	1	2
Total	50	100

Relative frequencies of the statistical series of student marks

x_i	n_i	$f_i(\%)$
[20;30)	3	3.00
[30;40)	2	2.00
[40;50)	3	3.00
[50;60)	6	6.00
[60;70)	7	7.00
[70;80)	9	9.00
[80;90)	7	7.00
[90;100)	10	10.00
[100;110)	15	15.00
[110;120)	10	10.00
[120;130)	8	8.00
[130;140)	7	7.00
[140;150)	7	7.00
[150;160)	4	4.00
[160;170)	0	0.00
[170;180)	2	2.00
Total	100	100.00

Relative frequencies of the company turnover

2.5 Frequencies and Cumulative Proportions

Once the statistical series is presented in the form of a table, it may be useful to display additional information for each modality of the studied character. This information includes the sum of the frequencies or relative frequencies of the modality and all those that precede it, known as *increasing cumulative frequencies or proportions*. Similarly, the sum of the frequencies or relative frequencies of the modality and all those that follow it, known as *decreasing cumulative frequencies or proportions*.

Displaying cumulative frequencies or proportions helps in understanding the distribution of the data, the proportion of the population that falls within a certain range, and in identifying any outliers or patterns in the data.

We denote by:

$n_i \uparrow$ or $f_i \uparrow$ for increasing cumulative frequencies or proportions.

$n_i \downarrow$ or $f_i \downarrow$ for decreasing cumulative frequencies or proportions.

To better understand these concepts, we can consider the example of counting a population and adding or subtracting frequencies. For example, if we are adding frequencies for increasing cumulative frequencies, we start by not knowing the size of the population. However, if we are subtracting frequencies for decreasing cumulative frequencies, we know the size of the population beforehand.

Suppose that P is the population we want to count and classify, composed of N individuals distributed in different modalities ranging from 1 to the maximum f the modalities that we denote max. This arrangement of the population corresponds to its iterative count, whose incremental factor is the number of each modality.

2.5.1 Formalization of increasing cumulative frequencies and proportions

To begin counting, the cumulative counts of all modalities are initialized to zero. At each iteration of the counting process, after all individuals belonging to a modality have been counted, the total number of individuals corresponding to that modality is established as its frequency. The count of all frequencies from the previous modalities, along with the frequency of the current modality, constitutes the cumulative frequency for that modality. More generally, let $n_i \uparrow$ denote the total frequency of the first i modalities that have been counted. This can be formalized for the following three cases:

- For $i = 1$,

$n_1 \uparrow$ represents the frequency of the first modality and is equal to n_1 since no other modalities have been counted yet, hence the following equation:

$$n_1 \uparrow = n_1$$

- For $1 < i < i_{max}$,

$n_i \uparrow$ represents the total frequency of the first i modalities and is obtained by adding the cumulative frequency of the previous modality with the frequency of the current modality, as shown in the following equation:

$$n_i \uparrow = n_{i-1} \uparrow + n_i$$

- For $i = i_{max}$,

$n_{max} \uparrow$ represents the total frequency of all modalities, indicating the end of the enumeration process. At this point, the total count is equal to the total frequency of all modalities, which is the sum of all n_i. This can be expressed using the following equation:

$$n_{max} \uparrow = n_{max-1} \uparrow + n_{max} = N$$

This same logic is also applied to $f_i \uparrow$.

2.5.2 Formalization of decreasing cumulative frequencies and proportions

Similar to the concept of increasing cumulative frequencies and proportions, the concept of decreasing cumulative frequencies is closely linked to the total frequency. To obtain the decreasing cumulative frequency of each modality, we perform an iterative count of the population by distributing the frequencies of the previous modalities. This operation is repeated until the population is exhausted. The total frequency starts from the value N and gradually decreases according to the frequency of the previous modality.

Let $n_i \downarrow$ denote the count of the population size after distributing the frequency of the previous modality, representing the decreasing cumulative size of the x_i modality. This can be formalized for the following three cases:

- For i = 1, $n_i \downarrow$ is equal to the entire population, that is the total value of n_i, because there is no previous modality as this is the first one and the population is not yet distributed. This gives the following equation:

$$n_1 \downarrow = N$$

- For 1< i < i_{max},

  $n_i \downarrow$ denotes the quantity representing the remaining count of the total frequency after distributing the frequencies of the previous modalities. It is obtained by subtracting the frequency of the previous modality from the cumulative frequency of the previous modality, hence the following relationship:

$$n_i \downarrow = n_{i-1} \downarrow - n_{i-1}$$

- For i = i_{max},

  $n_i \downarrow$ represents the count assigned to the last modality. This corresponds to the subtraction of the frequency of the previous modality from the cumulative frequency of the previous modality. This gives the following equation:

$$n_i \downarrow = n_{i-1} \downarrow - n_{i-1}$$

Like in the case of increasing cumulative frequency and proportions, the same logic is also applied to $f_i \downarrow$.

The previous example of the marks obtained by 50 students during an exam gives the following tables:

x_i	n_i	$n_i \uparrow$	$n_i \downarrow$	$f_i(\%)$	$f_i(\%) \uparrow$	$f_i(\%) \downarrow$
3	1	1	50	2.00	2.00	100.00
4	1	2	49	2.00	4.00	98.00
5	1	3	48	2.00	6.00	96.00
6	4	7	47	8.00	14.00	94.00
7	6	13	43	12.00	26.00	86.00
8	7	20	37	14.00	40.00	74.00
9	4	24	30	8.00	48.00	60.00
10	4	28	26	8.00	56.00	52.00
11	3	31	22	6.00	62.00	44.00
12	6	37	19	12.00	74.00	38.00
13	9	46	13	18.00	92.00	26.00
14	0	46	4	0.00	92.00	8.00
15	2	48	4	4.00	96.00	8.00
16	1	49	2	2.00	98.00	4.00
17	0	49	1	0.00	98.00	2.00
18	1	50	1	2.00	100.00	2.00
Total	50			100.00		

As an illustration for increasing cumulative frequencies and proportions, we have:

For $i = 5$,

$$n_5 \uparrow = 3 = 2 + 1$$
$$f_5 \uparrow = 6\% = 4\% + 2\%$$

INTERPRETATION

- 3 students have obtained a mark less than or equal to 5 out of 20.
- 6% of students have obtained a mark less than or equal to 5 sur 20.

Four $i = 10$

$$n_{10} \uparrow = 28 = 24 + 4$$
$$f_{10} \uparrow = 56\% = 48\% + 8\%$$

INTERPRETATION

- 28 students have obtained a mark less than or equal to 10 out of 20.
- 56% of students have obtained a mark less than or equal to 10 out of 20.

For decreasing cumulative numbers, for example, we will have:

For $i = 7$,

$$n_7 \downarrow = 43 = 47 - 4$$
$$f_7 \downarrow = 86\% = 94\% - 8\%$$

INTERPRETATION

- 43 students have obtained a mark greater than or equal to 7 out of 20.
- 86% of students obtained a mark greater than or equal to 7 out of 20.

Four $i = 16$,

$$n_{16} \downarrow = 2 = 4 - 2$$
$$f_{16} \downarrow = 4\% = 8\% - 4\%$$

INTERPRETATION

- 2 students have obtained a mark greater than or equal to 16 out of 20.
- 4% of students obtained a mark greater than or equal to 16 out of 20.

Note

It should be emphasized that in practice, as well as in later chapters, statistical distribution tables are often presented with frequencies already calculated. Therefore, the counting step is assumed to have been performed beforehand. This means that we will have to take care of the calculation of increasing and decreasing cumulative frequencies and relative frequencies.

2.6 Exercises

• **Exercise 3:** The following table represents the balances recorded in the savings accounts of 300 individuals (rounded and expressed to thousands of euros).

4	2	72	43	37	95	10	10	41	96	2	14	33	95	17
1	6	3	7	88	25	9	11	6	2	1	12	22	25	16
10	14	2	10	4	10	12	80	29	10	8	12	87	10	24
5	2	55	66	50	32	57	51	22	8	2	4	1	32	28
62	9	15	59	3	52	92	13	26	76	6	7	2	52	36
11	11	2	66	83	87	23	14	82	9	4	36	3	87	91
10	13	1	79	2	95	20	75	25	67	3	20	3	95	22
21	3	99	32	10	99	5	45	97	1	7	9	1	99	5
36	1	44	21	94	18	60	77	46	29	21	37	9	18	7
76	12	80	45	75	3	23	2	21	12	54	57	2	3	31
78	14	46	38	19	63	10	44	51	68	17	49	8	63	74
60	17	29	25	86	12	34	62	25	4	9	5	5	12	2
10	21	13	11	62	10	10	97	50	75	7	47	7	10	4
21	2	11	3	75	7	67	9	5	1	5	30	3	7	8
11	9	64	10	12	42	84	84	41	78	3	1	6	42	1
7	7	2	9	3	47	9	8	97	76	13	86	6	47	29
8	3	21	11	11	13	24	36	17	16	15	36	32	13	54
96	31	24	80	10	40	52	33	78	56	41	31	17	5	14
5	19	3	12	23	12	21	14	2	3	12	28	2	4	31
50	16	31	93	15	61	96	20	89	77	6	47	55	22	82

a. Process to the counting by sorting these amounts in ascending order.

b. Group this data into 20 classes of 5,000 Euros amplitude.

c. Give the table of increasing and decreasing cumulative frequencies and relative frequencies.

2.7 Solution

a. Sorting these amounts in ascending order (from smallest to largest).

1	2	3	6	9	10	12	16	21	29	36	47	60	76	87
1	2	4	6	9	10	12	16	21	29	36	47	61	76	88
1	2	4	6	9	10	12	17	22	29	37	49	62	77	89
1	2	4	7	9	10	12	17	22	29	37	50	62	77	91
1	2	4	7	9	10	13	17	22	30	38	50	62	78	92
1	2	4	7	9	11	13	17	22	31	40	50	63	78	93
1	3	4	7	9	11	13	17	23	31	41	51	63	78	94
1	3	4	7	9	11	13	18	23	31	41	51	64	79	95
1	3	5	7	9	11	13	18	23	31	41	52	66	80	95
1	3	5	7	10	11	13	19	24	31	42	52	66	80	95
2	3	5	7	10	11	14	19	24	32	42	52	67	80	95
2	3	5	7	10	11	14	20	24	32	43	54	67	82	96
2	3	5	7	10	11	14	20	25	32	44	54	68	82	96
2	3	5	8	10	12	14	20	25	32	44	55	72	83	96
2	3	5	8	10	12	14	21	25	33	45	55	74	84	97
2	3	5	8	10	12	14	21	25	33	45	56	75	84	97
2	3	5	8	10	12	15	21	25	34	46	57	75	86	97
2	3	6	8	10	12	15	21	26	36	46	57	75	86	99
2	3	6	8	10	12	15	21	28	36	47	59	75	87	99
2	3	6	9	10	12	16	21	28	36	47	60	76	87	99

b. Let group this data into 20 classes of 5,000 euros amplitude

Classes	n_i
[0;5000)	48
[5000;10000)	41
[10000;15000)	47
[15000;20000)	15
[20000;25000)	21
[25000;30000)	12
[30000;35000)	13
[35000;40000)	8
[40000;45000)	9
[45000;50000)	9
[50000; 55000)	10
[55000;60000)	6
[60000;65000)	9
[65000;70000)	5
[70000;75000)	2
[75000;80000)	13
[80000;85000)	8
[85000;90000)	7
[90000;95000)	4
[95000;100000)	13
Total	300

c. The table of increasing and decreasing cumulative frequencies and relative frequencies is:

x_i	n_i	$n_i \uparrow$	$n_i \downarrow$	$f_i(\%)$	$f_i(\%)\uparrow$	$f_i(\%)\downarrow$
[0;5000)	48	48	300	16.00	16.00	100.00
[5000;10000)	41	89	252	13.67	29.67	84.00
[10000;15000)	47	136	211	15.67	45.33	70.33
[15000;20000)	15	151	164	5.00	50.33	54.67
[20000;25000)	21	172	149	7.00	57.33	49.67
[25000;30000)	12	184	128	4.00	61.33	42.67
[30000;35000)	13	197	116	4.33	65.67	38.67
[35000;40000)	8	205	103	2.67	68.33	34.33
[40000;45000)	9	214	95	3.00	71.33	31.67
[45000;50000)	9	223	86	3.00	74.33	28.67
[50000; 55000)	10	233	77	3.33	77.67	25.67
[55000;60000)	6	239	67	2.00	79.67	22.33
[60000;65000)	9	248	61	3.00	82.67	20.33
[65000;70000)	5	253	52	1.67	84.33	17.33
[70000;75000)	2	255	47	0.67	85.00	15.67
[75000;80000)	13	268	45	4.33	89.33	15.00
[80000;85000)	8	276	32	2.67	92.00	10.67
[85000;90000)	7	283	24	2.33	94.33	8.00
[90000;95000)	4	287	17	1.33	95.67	5.67
[95000;100000)	13	300	13	4.33	100.00	4.33
Total	300			100.00		

INTERPRETATION OF THE RESULT

Based on the data presented in the table above, we can make the following observations:

- 151 savers have savings below 20,000 euros, representing 50.33% of the total savers.
- 149 savers have savings greater than 20,000 euros, representing 49.67% of the total savers.

It is worth noting that the 151 savers with savings below 20,000 euros, representing 50.33% of the total, were selected from the column of increasing cumulative frequencies and proportions. On the other hand, the 149 savers with savings greater than 20,000 euros, representing 49.67% of the total, were selected from the column of decreasing cumulative frequencies and proportions.

Furthermore, when we consider the condition of savings greater or less than 20,000 euros, we can see that the sum of frequencies and percentages of savers in both categories adds up to the total frequency of 300 individuals, which represents 100% of the population under consideration.

CHAPTER 3

Graphical Representations of Statistical Series

This chapter delves into the graphic representation of statistical series, building upon the concepts of quantitative and qualitative characteristics introduced in the previous chapter. The discussion on qualitative characteristics includes three types of representations, namely the bar graph, the pie chart, and the triangular representation. On the other hand, the discussion on quantitative characteristics focuses on two types of graphs: the bar graph for discrete variables and the histogram for continuous variables. Additionally, the paragraph provides an in-depth explanation of frequencies and cumulative frequencies.

3.1 Graphical Representations

A statistical series is a set of data presented in the form of a table that contains all the available information about a population or a sample. To better understand the characteristics of this data, it is often useful to represent it graphically. Graphical representation allows for quick visualization of trends and variations in the series.

One of the commonly used methods for graphically representing a statistical series is proportional area representation. This method involves representing each value of the characteristic by a portion of the plane whose area is proportional to the number or frequency of that value. Thus, more frequent values will be represented by larger areas than less frequent values.

Proportional area representation offers many advantages, including allowing for quick visualization of the most important values. It also enables quick identification of significant differences between different values, as well as variations in the series over time or space.

3.2 Qualitative Character

Qualitative data can be represented visually using various methods, including bar graphs and pie charts. Bar graphs are useful for comparing the frequencies or relative frequencies of modalities, while pie charts are effective for displaying the proportional breakdown of categories within a whole.

3.2.1 Bar Graph

Bar graphs are a popular method for visually representing data, particularly for categorical variables. Each category is represented by a rectangular bar whose height is proportional to the frequency or relative frequency of that category. The bars all have the same width, providing a consistent base for comparison between modalities. The x-axis of the graph typically displays the modalities being represented, while the y-axis displays the frequency or relative frequency of each modality.

The choice of this method is often linked to its simplicity and ease of interpretation. Indeed, the height of each rectangle is easily comparable, which makes it possible to quickly visualize the variations of the series. In addition, the rectangles are arranged in such a way as to avoid overlapping, which makes them easier to read.

However, it should be noted that the bar chart representation is less suitable for large statistical series. In these cases, the number of rectangles can become very large, which makes reading more difficult. In these situations, it may be better to use other methods, such as pie charts or bar charts.

Example 14

Let's take the example of the distribution of the area of the departments of Ile de France, as presented in the following table:

Department	Area (Km^2)
Paris	105.4
Seine-Saint-Denis	236
Hauts-de-Seine	176
Yvelines	2284
Seine-et-Marne	5915
Val-de-Marne	245
Essonne	1804
Val-d'Oise	1246
Total	12,011

To create a bar chart representing this statistical series, we need to divide the total land area into as many rectangles as there are departments, with each rectangle having an equal base.

Each rectangle will represent a department, and its height will be proportional to the land area of that department. By comparing the heights of the rectangles, it will be easy to determine the department with the largest land area, as well as the differences in land area between the various departments.

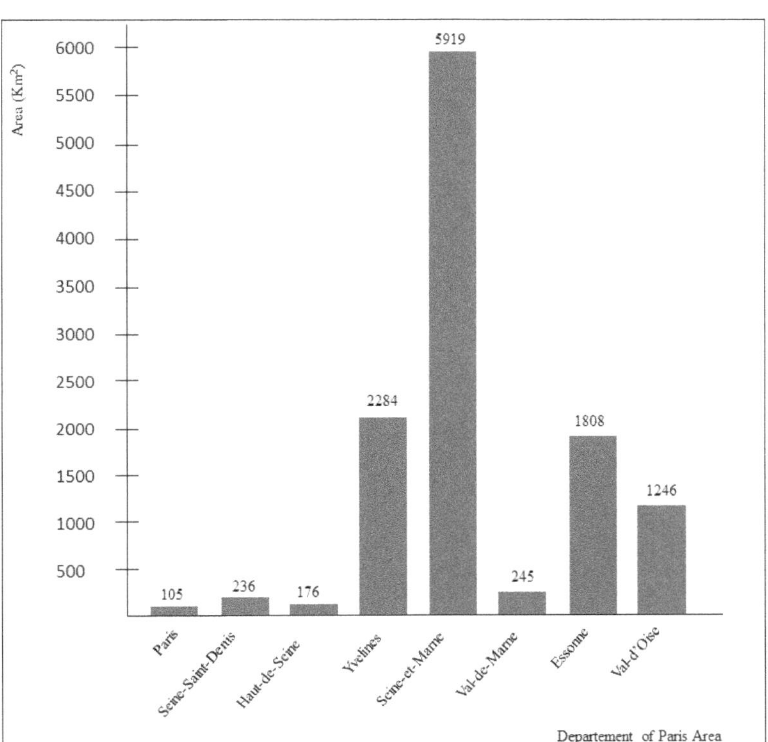

Since there are eight departments in total, we will use eight rectangles arranged side by side, each with the same base. The height of each rectangle will be proportional to the land area of the corresponding department.

This graphical representation allows us to quickly visualize the differences in land area between the departments of Ile-de-France. It is then possible to easily identify the department of Seine-et-Marne with the largest land area, as well as the variations in land area between the different departments.

3.2.2 Pie Chart

The pie chart is widely used in statistics to visualize the distribution of modalities of a qualitative characteristic. This method of graphical representation is particularly useful for circular data, such as percentages or proportions.

In this type of chart, the total population is represented as a circle, and each modality is represented by a circular sector slice. The size of each sector slice is proportional to the frequency or relative frequency of the corresponding modality.

The angle θ_i representing the modality i is proportional to the relative frequency f_i of that modality and is calculated by multiplying that relative frequency by a constant k. To obtain the angle corresponding to each modality, the following formula can be used:

$$\theta_i = kf_i$$

The sum of all angles θ_i must be equal to 360° to cover the entire circle, that is:

$$\boxed{\sum_{i=1}^{i=p} \theta_i = \sum_{i=1}^{i=p} 360° f_i = 360° \sum_{i=1}^{i=p} f_i = 360°}$$

$$\text{because} \quad \sum_{i=1}^{i=p} f_i = 1$$

In practice, the respective proportions are multiplied by 360° to obtain the opening of the corresponding angle.

Example 15

Referring back to **example 14** from **section 3.2.1**, which deals with the distribution of surface area in the Paris region.

Department	Area (Km^2)	Frequency (f_i)	Angle in ° (f_i x 360 °)
Paris	105.4	0.88%	0,88% x 360° = 3.2°
Seine-Saint-Denis	236	1.96%	1,96% x 360° = 7.1°
Hauts-de-Seine	176	1.47%	1,47% x 360° = 5.3°
Yvelines	2284	19.02%	19,02% x 360° = 68.5°
Seine-et-Marne	5915	49.24%	49,24% x 360° = 177.3°
Val-de-Marne	245	2.04%	2,04% x 360° = 7.3°
Essonne	1804	15.02%	15,02% x 360° = 54.1°
Val-d'Oise	1246	10.37%	10,37% x 360° = 37,2°
Total	12,011	100.00%	360°

The resulting pie chart representing the distribution of surface areas of the departments in Île-de-France is shown below:

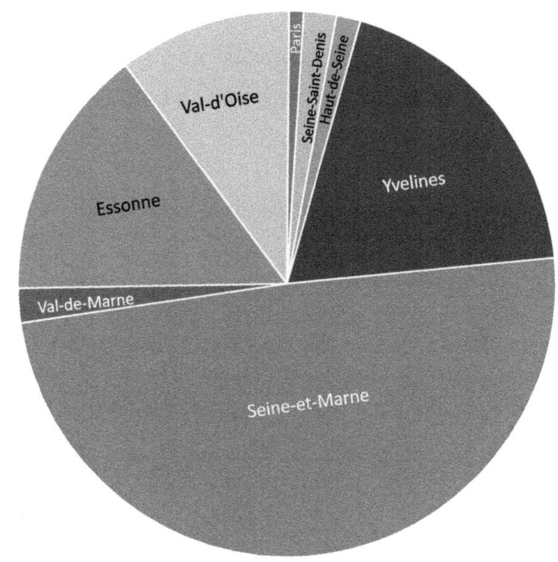

INTERPRETATION

In the pie chart representing the distribution of the surface area of the departments in the Île-de-France region, the opening angle of each circular sector slice is proportional to the frequency of each department. For example:

- For the department of Paris, whose frequency is 0.88%, the opening angle is equal to 0.88% multiplied by 360 degrees, or 3.2 degrees.
- For the department of Seine-Saint-Denis, whose frequency is 1.96%, the opening angle is equal to 1.96% multiplied by 360 degrees, or 7.1 degrees.

3.2.2.1 Comparing Two Series Using Pie Charts

When comparing two statistical series using their pie charts, there are a few assumptions that we must make and a general approach that we can follow.

First postulate: we assume that the weight of a variable in the pie chart is represented by the area it occupies.

Second postulate: if we are comparing two statistical series of variables with the same modalities and using the same unit of measurement, we can determine their relative proportions by comparing the areas occupied by their respective pie charts.

Third postulate: If we have two circles with radii R_1 and R_2, where R_1 is twice the size of R_2 (i.e., $R_1 = 2R_2$), the area of the larger circle (C_1) will be four times greater than the area of the smaller circle (C_2). This can be expressed mathematically as:

$$\underbrace{\pi(R_1)^2}_{Area\ of\ C_1} = \pi(2R_2)^2$$

$$= 4\,\underbrace{[\pi(R_2)^2]}_{Area\ of\ C_2}$$

Generalization

Given two circles C_1 and C_2, where the radius of C_1 is n times greater than the radius of C_2, the area of C_1 is n^2 times greater than that of C_2.

This can be demonstrated mathematically by using R_1 and R_2 to represent the radii of C_1 and C_2, respectively, and if $R_1 = R_2$, the area of C_1 can be calculated as:

$$\underbrace{\pi(R_1)^2}_{Area\ of\ C_1} = \pi(nR_2)^2$$

$$= n^2\,\underbrace{[\pi(R_2)^2]}_{Area\ of\ C_2}$$

To compare the representation of two series S_1 and S_2, with totals T_1 and T_2 respectively, where T_1 is greater than T_2, it is recommended to maintain the principle of proportionality of areas.

Specifically, if we denote $\pi(R_1)^2$as the area of the circle representing the total T_1 and $\pi(R_2)^2$as the area of the circle representing the total T_2, then we must maintain the following proportionality of areas:

$$\frac{\pi(R_2)^2}{\pi(R_1)^2} = \frac{T_2}{T_1} \Rightarrow \frac{R_2}{R_1} = \sqrt{\frac{T_2}{T_1}}$$

Example 16

The table below shows the amounts, in millions of dollars, of the products exported by a developing country (DPC) and a developed country (DC), according to their nature.

Goods	Developing Country	Developed Country
Primary	41.20	51.40
Manufactured	12.70	169.10
Others	0.40	3.70
TOTAL	54.30	224.20

In order to create pie charts representing the developing country and the developed country, we need to first calculate the relative frequencies for each product category: Primary, Manufactured, and Others. Based on these frequencies, we can determine the value of the corresponding angle for each category. This will enable us to construct the following table:

Goods	Developing Country				Developed Country			
	Freq.	R.Freq.	Angle		Freq.	R.Freq.	Angle	
Primary	41.20	75.87	75.87% x 360°=	273°	51.40	22.93	22.93% x 360° =	82.5°
Manufactured	12.70	23.39	23.39% x 360°=	84°	169.10	75.42	75.42% x 360° =	271.5°
Others	0.40	0.74	0.74%x 360°=	3°	3.70	1.65	1.65% x 360° =	6°
TOTAL	54.30	100.00	360°		224.20	100.00	360°	

Expressing the proportionality between DPC and DC

Expressing the proportionality between the Developing Country (DPC) and the Developed Country (DC) requires careful consideration of the representation of their respective pie charts. Although the export volume of the Developed Country is four times greater than that of the Developing Country, representing the circle of the pie chart of the Developed Country with a radius four times greater than that of the Developing Country would be inaccurate. This is because the area of the pie chart of Developed Country would be sixteen times greater than that of the Developing Country, leading to an exaggerated perception of the difference in exports between the two countries.

To calculate the proportionality ratio between the radii of the two circles, we can use the formula:

$$\frac{R_2}{R_1} = \sqrt{\frac{T_2}{T_1}}$$

Where R_1 and R_2 are the radii of the pie charts representing the Developing Country and the Developed Country, respectively, and T_1 and T_2 are the total export volumes of the Developing Country and the Developed Country, respectively.

Applying this formula to the data of our example, we obtain:

$$\frac{R_2}{R_1} = \sqrt{\frac{224.2}{54.3}} = 2.03 \Rightarrow R_2 = 2.03R_1$$

Therefore, the radius of the circle representing the Developed Country should be 2.03 times greater than that of the Developing Country to maintain proportionality. This means that even though the export volume of the Developed Country is four times greater than that of the

Developing Country, the radius of its pie chart is only 2.03 times greater, indicating a more accurate representation of the difference in exports between the two countries.

Note

We use a two-dimensional pie chart without depth to represent the data for two reasons:

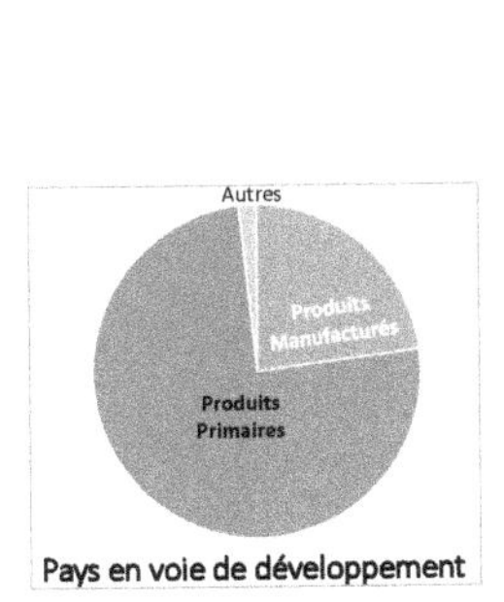

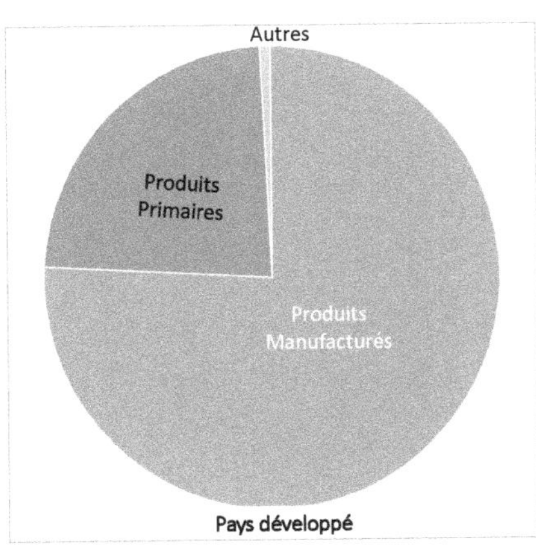

- It is easier to display the proportions on a two-dimensional circle than on a three-dimensional one.
- The visual principle of proportionality of areas is not maintained in a three-dimensional diagram

3.2.3 Triangular Representation

This particular representation is applicable only to characters that possess three modalities. It is a commonly used method to observe variations over time. In this type of representation, one of the noteworthy properties of an equilateral triangle is utilized: ***for any point inside the triangle, the sum of its distances to the three sides remains constant as the point varies***.

Consider the triangle with sides A, B, and C in the diagram below, and the interior points P_1, P_2 and P_3. For each point, the distance is evaluated with respect to each side of the triangle. To simplify this operation, we have chosen to use the length of a solid line located on the dotted segment descending vertically on the corresponding side as a unit of measurement. This value is presented at the foot of the vertical line passing through the point in the figure below. By adding up the distances from each point to the three sides, we get:

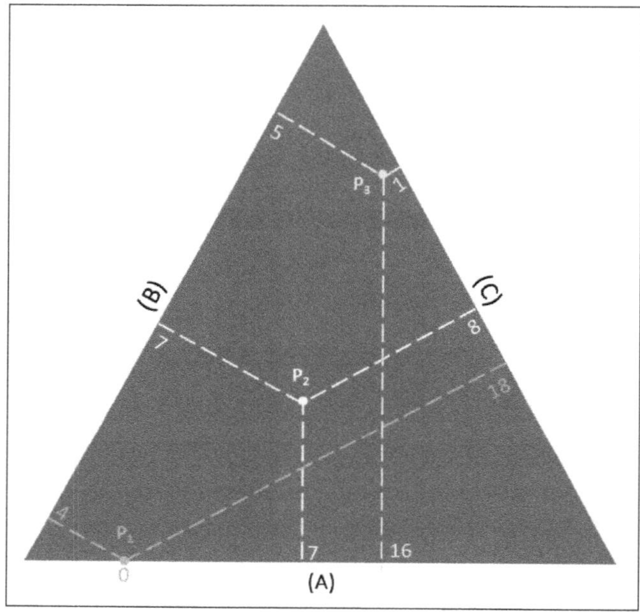

$$P_1 = 0 + 4 + 18 = 22$$

$$P_2 = 7 + 7 + 8 = 22$$

$$P_3 = 1 + 5 + 16 = 22$$

As observed, the sum of distances from a point to the three sides remains constant when the point is moved inside the triangle from P_1 to P_2 and from P_2 to P_3.

Principle

In the figure below, let (A, B, C) represent the triplet denoting the relative frequencies of the three modalities. We can determine the simultaneous relative frequencies of these three modalities using the following steps:

- Draw a line segment on side A that is parallel to side C and passes through the point representing the relative frequency a.

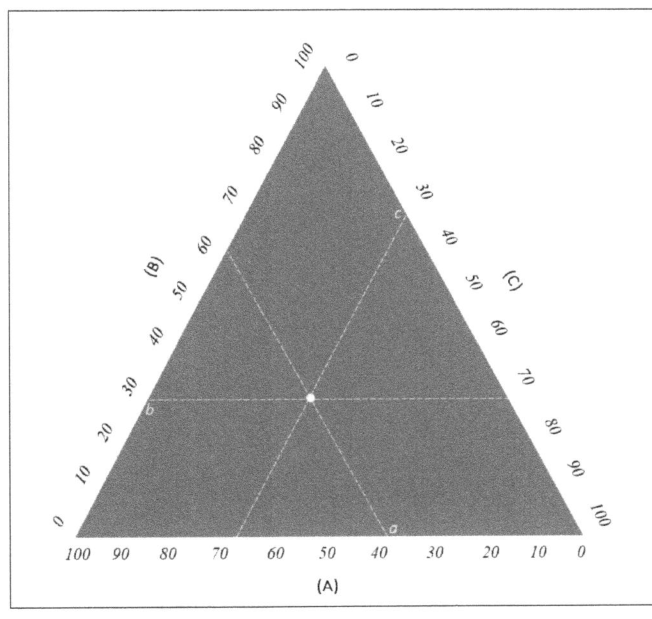

- Draw a line segment on side B that is parallel to side A and passes through the point representing the relative frequency b.
- Draw a line segment on side C that is parallel to side B and passes through the point representing the relative frequency c.
- The point of intersection of these three line segments represents the simultaneous relative frequencies of these three modalities.

Furthermore, the proximity of the point to a vertex indicates the preponderance of its corresponding modality.

Example 17

The table below shows the evolution of the percentage contribution of each economic sector to the UK economy since 1688. The three economic sectors are the primary sector, the secondary sector, and the tertiary sector. The figures indicate the share of each sector in the UK's total national income.

	Primary	Secondary	Tertiary
1688	40	21	39
1770	45	24	31
1801	32	23	45
1841	22	34	44
1901	6	40	54
1907	6	36	58

The data is represented in a triangular form as follows:

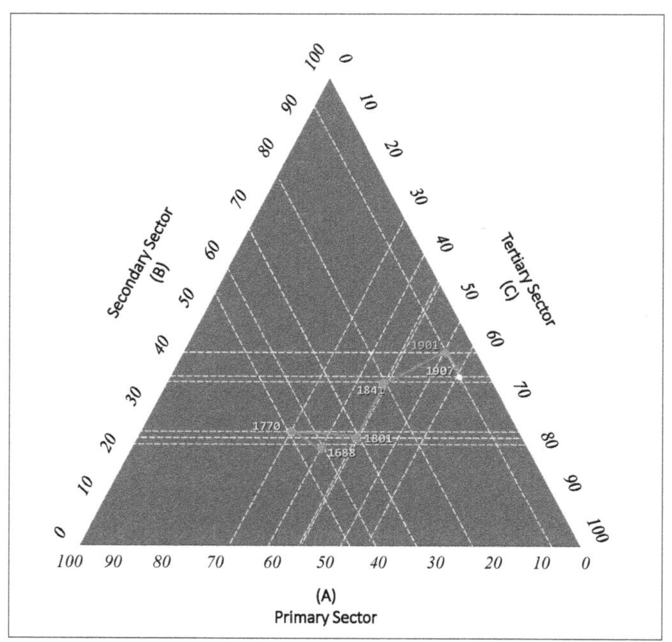

INTERPRETATION OF THE DIAGRAM

An analysis of economic data in the United Kingdom between 1688 and 1907 reveals significant changes in the contributions of different sectors to the national income. It is evident that the primary sector, which includes agriculture, fishing, and mining, experienced a significant decrease, dropping from 40% in 1688 to just 6% in 1901 and 1907. In contrast, the contribution of the tertiary sector, which encompasses services, trade, and finance, increased considerably, rising from 39% in 1688 to 54% in 1901 and 58% in 1907. This reflects a major transformation in the British economy, which shifted from an agricultural and industrial economy to a service-based one.

This transformation was made possible by the Industrial Revolution and urbanization, which led to the emergence of a middle-class demanding services and consumer goods. While the secondary sector (manufacturing, construction, etc.) underwent fluctuations, it played a crucial role in the country's economy, particularly during the Industrial Revolution. Thus, while the dominance of contributions shifted from the primary sector to the tertiary sector, the contributions of the other two sectors cannot be ignored as they were essential for the economic growth and development of the country.

3.3 Quantitative Character

Quantitative characters are represented either by bar graphs or by histograms. The choice of graphical representation for quantitative data depends on several factors, such as the nature of the data, the required precision, and the sample size. Bar charts, also known as bar graphs, are commonly used to represent discrete and categorical data, while histograms are more suitable for continuous data.

3.3.1 Discrete Quantitative Variable

3.3.1.1 Bar Graph

The bar graph is an effective tool for visually representing a discrete statistical variable. It allows for easy comparison of quantities or frequencies of different categories and highlights the differences and similarities between the modalities of the variable. Bars can be differentiated by colors or patterns to facilitate the reading of the graph.

Bar graphs are particularly useful for representing discrete data, such as survey results or sales by product category.

To construct this graph, simply indicate the modalities of the variable on the x-axis and draw a vertical bar, or bar, for each modality. The height of each bar is proportional to the frequency or percentage corresponding to that modality.

Example 18

Let's take **example 12** from **section 2.3.1** on student's marks.

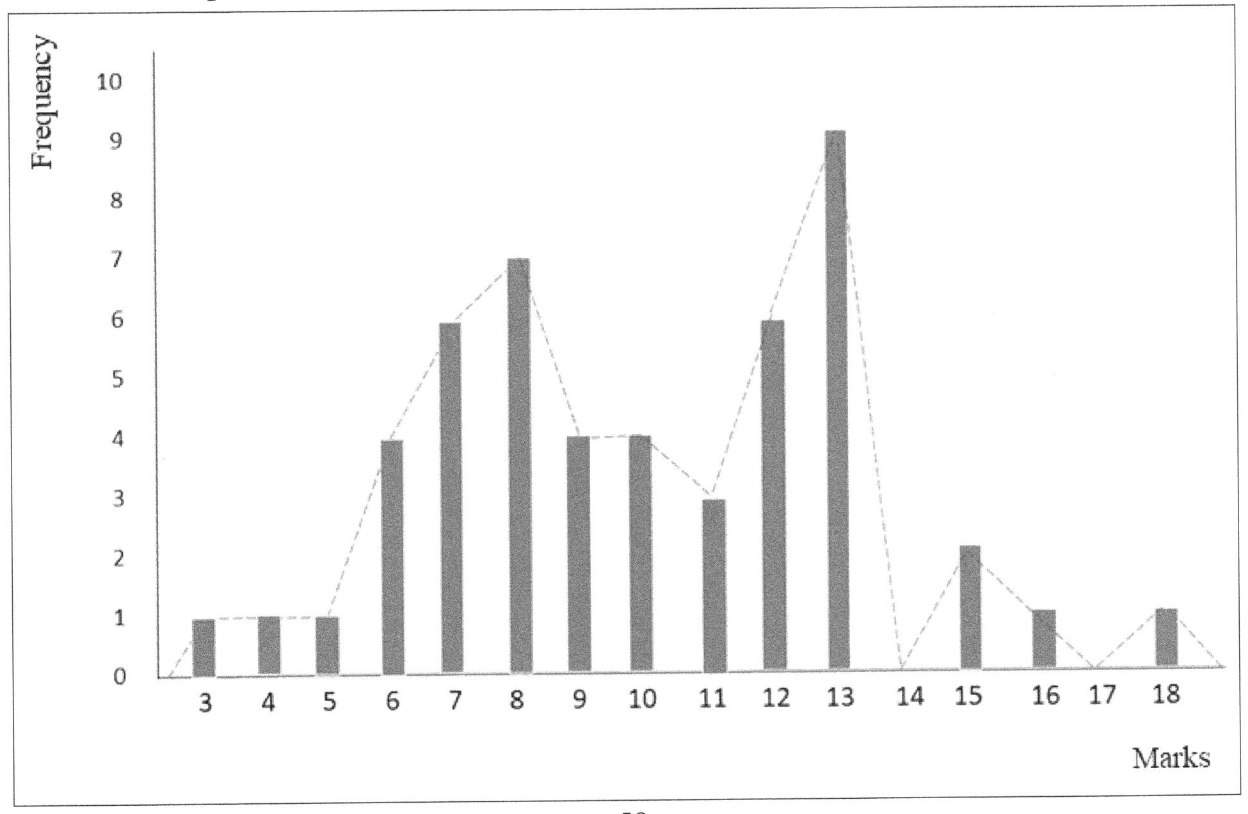

To graphically represent this series of discrete data, we can use a bar graph. The corresponding bar graph allows for easy visualization of the distribution of marks among different categories, with a vertical bar proportional to the frequency or percentage corresponding to each mark.

The general tendency of a distribution is obtained by joining with a line the middle of the upper ends of the frequencies (or proportions) bars with a line.

The resulting polygon is called the ***frequency polygon*** (or the ***proportion polygon***).

3.3.1.2 Distribution function of discrete variables

The ***distribution function***, also known as the ***cumulative function***, serves as a vital tool for graphically representing the cumulative frequencies or relative frequencies of a given statistical variable x. By utilizing the distribution function, one can obtain a comprehensive and concise understanding of the distribution of a statistical variable.

The distribution function of a statistical discrete variable x is defined as the function that associates any real x with the number $F(x)$. This number $F(x)$ represents the proportion of individuals within the population whose character is less than x. By analyzing the distribution function, one can gain valuable insights into the population's characteristics.

The distribution function is graphically represented as a curve. This curve allows for quick visualization of the distribution of the statistical variable by indicating the cumulative proportion of individuals with a variable value that is less than or equal to each possible value of x. Specifically, the distribution function starts at zero for the smallest values of x and reaches one for the largest values of x.

The distribution function is a powerful tool for graphically representing the distribution of a discrete statistical variable and for calculating important measures of this distribution. It is widely used in many fields, such as finance, demographics, psychology, medicine, and more.

It is expressed as follows:

$$F(x_i) \ = n_i \uparrow = n_{i-1} \uparrow + n_i$$

Note:

The distribution function of discrete statistic variables is a step function that is defined for all real values of x. It is important to note that this function is not necessarily increasing, as it is rather constant in each of its intervals.

To better understand this, let us take an interval $[a, b]$ of this distribution function. In this interval, the distribution function is constant for all values of x belonging to $[a, b]$. This constancy means that the proportion of individuals with a value less than or equal to x is the same for all values of x in the interval $[a, b]$. It is this constancy that gives the distribution function its step-like appearance.

Indeed, given an interval $[a, b]$of this distribution function,

$$\forall x_1, x_2 \in [a,b], \text{si } x_2 > x_1, F(x_2) = F(x_1)$$

Example 19

Let's take **example 18** discussed earlier in **section 3.3.1**, which was about the grades of students from the previous section. To better understand the distribution of grades, we will now calculate the cumulative frequencies of these marks.

If

$x = 0$ $F(0) = 0$
$x = 1$ $F(1) = 0$
$x = 2$ $F(2) = 0$
$x = 3$ $F(3) = 0 + 1{=}1$
$x = 4$ $F(3) = 1 + 1 = 2$
$x = 5$ $F(4) = 1 + 1 + 1 = 3$
$x = 6$ $F(6) = 1 + 1 + 1 + 4 = 7$
$x = 7$ $F(7) = 1 + 1 + 1 + 4 + 6 = 13$
$x = 8$ $F(8) = 1 + 1 + 1 + 4 + 6 + 7 = 20$
$x = 9$ $F(9) = 1 + 1 + 1 + 4 + 6 + 7 + 4 = 24$
$x = 10$ $F(10) = 1 + 1 + 1 + 4 + 6 + 7 + 4 + 4 = 28$
$x = 11$ $F(11) = 1 + 1 + 1 + 4 + 6 + 7 + 4 + 4 + 3 = 31$
$x = 12$ $F(12) = 1 + 1 + 1 + 4 + 6 + 7 + 4 + 4 + 3 + 6 = 37$
$x = 13$ $F(13) = 1 + 1 + 1 + 4 + 6 + 7 + 4 + 4 + 3 + 6 + 9 = 46$
$x = 14$ $F(14) = 1 + 1 + 1 + 4 + 6 + 7 + 4 + 4 + 3 + 6 + 9 + 0 = 46$
$x = 15$ $F(15) = 1 + 1 + 1 + 4 + 6 + 7 + 4 + 4 + 3 + 6 + 9 + 0 + 2 = 48$
$x = 16$ $F(16) = 1 + 1 + 1 + 4 + 6 + 7 + 4 + 4 + 3 + 6 + 9 + 0 + 2 + 1 = 49$
$x = 17$ $F(17) = 1 + 1 + 1 + 4 + 6 + 7 + 4 + 4 + 3 + 6 + 9 + 0 + 2 + 1 + 0 = 49$
$x = 18$ $F(18) = 1 + 1 + 1 + 4 + 6 + 7 + 4 + 4 + 3 + 6 + 9 + 0 + 2 + 1 + 0 + 1 = 50$

For this example, we will have the following distribution function:

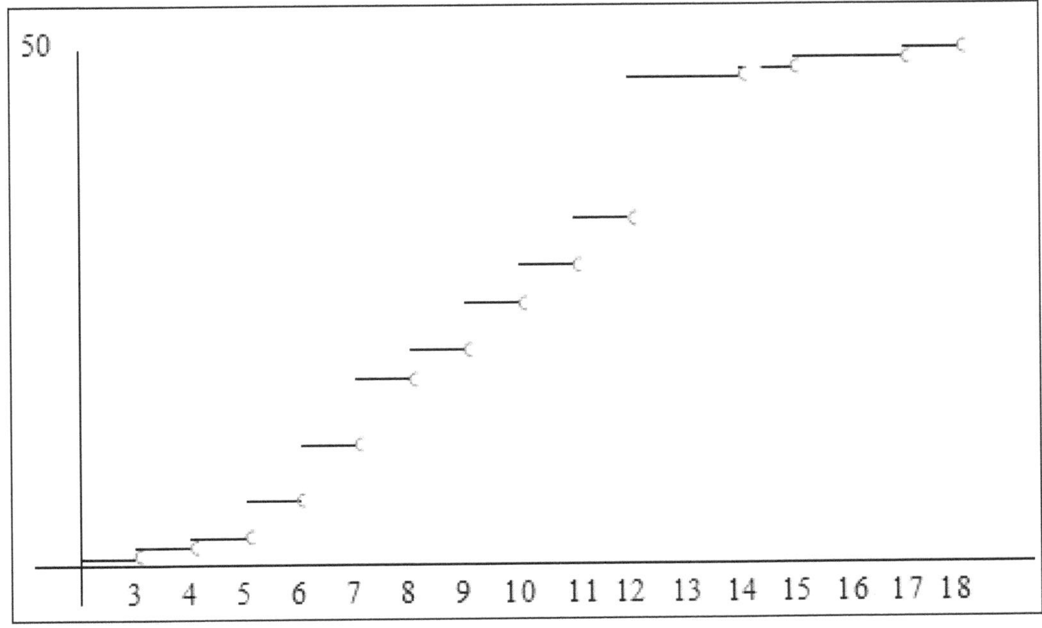

3.3.2 Continuous Quantitative Variable

3.3.2.1 Histogram

Continuous statistical variables can be represented using ***histograms***. In order to create this type of representation, the observations are first grouped into classes. These classes are then displayed on the x-axis in the form of intervals. The length of each interval is determined algebraically and will be represented by a rectangle with a height proportional to the frequency (or the relative frequency) of the corresponding class.

It is important to choose the number of classes appropriately in order to achieve an accurate representation of the data. If there are too few classes, important features of the data may be lost. On the other hand, if there are too many classes, the histogram may become too cluttered and difficult to interpret.

This graphical representation allows for easy visualization of the distribution of data and identification of classes with the highest or lowest frequencies. This helps to highlight the shape of the distribution, such as symmetry, central tendency, and data dispersion. It is also possible to overlay multiple histograms to compare the distribution of different statistical variables.

Example 20

Let's consider the example of table below that gives the statistical distribution of workers in a company grouped by age classes.

Classes	Frequency (n_i)
[20;25)	36
[25;30)	45
[30;35)	27
[35;40)	18
[40;45)	10
[45;50)	8
[50;55)	3
[55;60)	3
Total	150

To create the histogram, we first need to determine the width of each class. The width of each class will be the same and will be calculated by subtracting the lower bound of each class from the upper bound.

For example, the width of the first class $[20,25)$ is 25 - 20 = 5.

We can now plot the histogram. On the x-axis, we will plot the age groups, and on the y-axis, we will plot the frequency of each age group. Each bar will correspond to a class and will have a height proportional to the frequency of that class.

The corresponding histogram is:

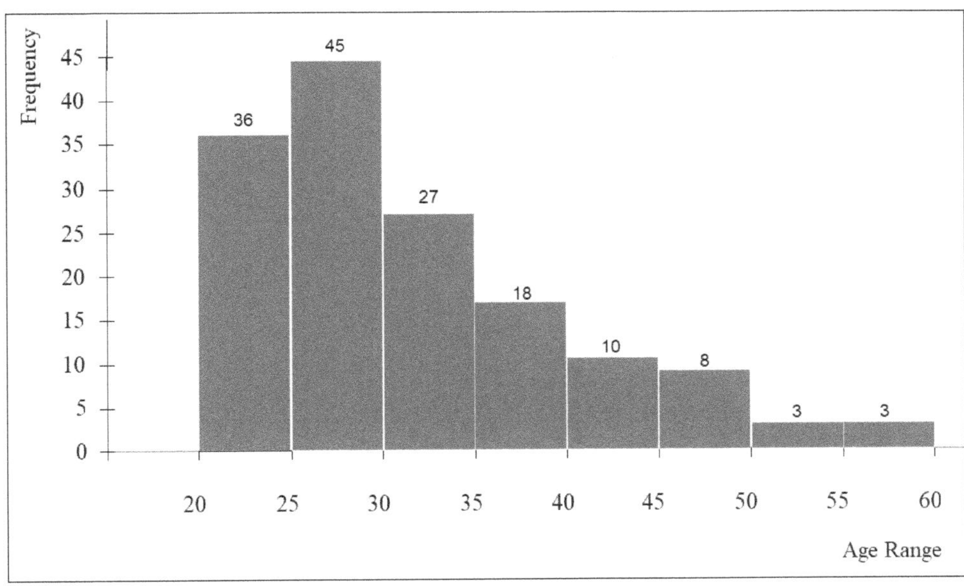

The resulting histogram shows the distribution of ages among the workers in the company, making it easy to identify the class [25,30) as the one with the highest frequency and the classes [50,55) and [55,60) as the ones with the lowest frequency.

3.3.2.2 Histogram in the case of classes with different amplitudes

As mentioned earlier, to properly designate statistical data on a histogram, the y-axis must be proportional to the count or frequency. If the classes have different amplitudes, it is preferable to choose a unit interval of amplitude "a". This normalizes the data for easier comparison.

If the class $[a_i, a_{i+1}[$ has a frequency n_i containing p times the unit interval, which means that the width of the class is equal to the product of p by a, that is, if $a_{i+1} - a_i = pa$, the height of the corresponding rectangle will be:

$$\frac{n_i}{p}$$

Example 21

To illustrate the method of standardizing data on a histogram, consider the example of the following distribution:

Classes	Amplitudes	Frequencies (n_i)	n_i per unit interval
[0;50)	50	100	100
[50;150)	100	140	140:2=70
[150;250)	100	80	80:2=40
[250;300)	50	50	50
[300;450)	150	30	30:3=10
Total		400	

As the amplitude of the classes is not uniform, it is necessary to proportionally reduce the frequency values for each class, taking into account an amplitude interval of $\frac{n_i}{p}$, where p is the ratio between the amplitude of the class and the chosen unit interval.

By applying this method, we obtain the following frequency histogram:

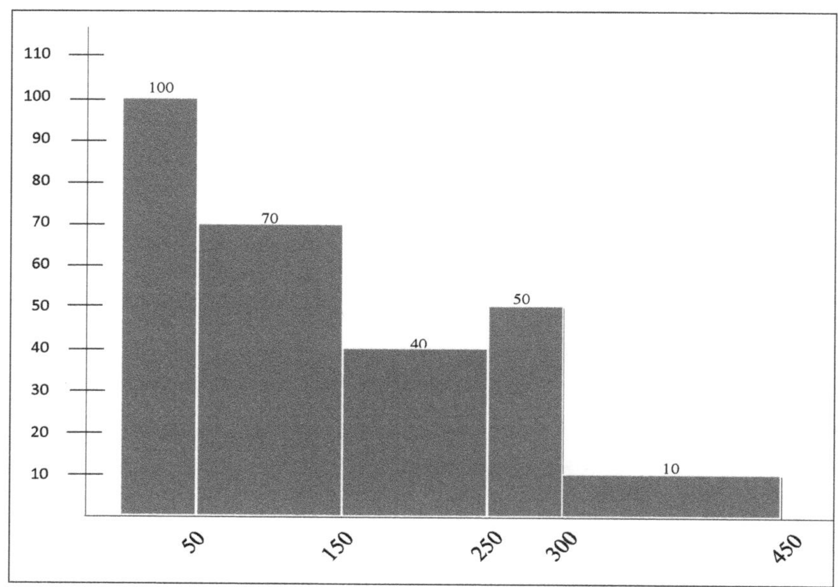

3.3.2.3 Histogram Closure

The polygon obtained connects the vertices of the rectangles forming the histogram and provides an overall description of the distribution shape. However, to obtain a more complete and accurate representation, it is important to "close" the polygon. This operation consists of adding two additional points to the polygon. These points are located to the left and right of the polygon and correspond to the mean values of the fictitious intervals situated to the left and right of the polygon, i.e., values that lie outside the observed range but have a non-zero probability of being observed.

This technique allows for better delimitation of the uncertainty zone and refines the graphical representation of the statistical series distribution. By including these values in the graphical visualization, a more precise representation of the distribution is obtained, which can be very useful for decision-making and data interpretation.

Example 22

Let us revisit **example 20** from section **3.3.2.1**, which concerns the statistical distribution of workers in a company based on their age. Using the previously generated histogram, to obtain a more precise graphical representation of the distribution, we will "close" the polygon of the graph using fictitious intervals.

The initial polygon will connect the vertices of the rectangles forming the histogram. To close it, we add two additional points to the left and right of the polygon. These points correspond to the mean values of the fictitious intervals that have a non-zero probability of being observed.

For example, since the observed age range is from 20 to 60 years old, but there is a non-zero probability that a worker is over 60 years old, we will include the mean value of the fictitious interval to the right of the polygon. Similarly, if a worker is less than 20 years old, we will include the mean value of the fictitious interval to the left of the polygon.

Once these additional points have been added, the polygon is closed, and the graphical representation of the distribution is more precise and complete, as shown in the following figure. This can be particularly useful for better understanding the age distribution of workers in this company and drawing meaningful conclusions for the company and its employees.

When working with data where the class intervals have unequal amplitudes, it is preferable to divide them into equal sub-intervals to obtain a more precise representation of the distribution.

To do so, each class is divided into an equal number of sub-intervals, while respecting the original class limits. Then, the same histogram construction method used for classes of equal amplitude is applied, using the frequencies or counts of each sub-interval.

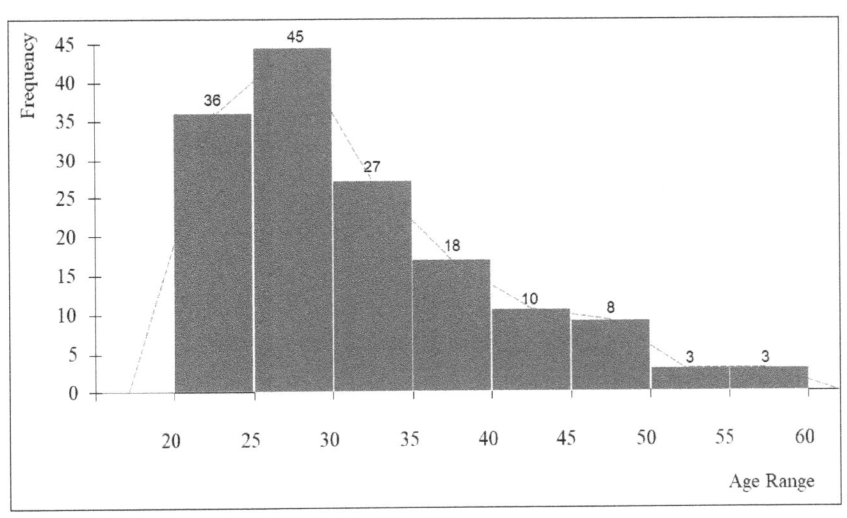

The resulting histogram may have a different shape compared to the one obtained with classes of equal amplitude. This is because the width of the intervals may vary from one sub-interval to another, which can influence the overall shape of the histogram. It is therefore important to consider this particularity when interpreting the results.

However, dividing the classes into sub-intervals allows for a more detailed representation of the distribution and a better visualization of the variations that may exist within each class.

By applying the technique of dividing the histogram into equal sub-intervals to the histogram of **example 21**, a new histogram can be obtained, which will allow for more precise visualization of the distribution.

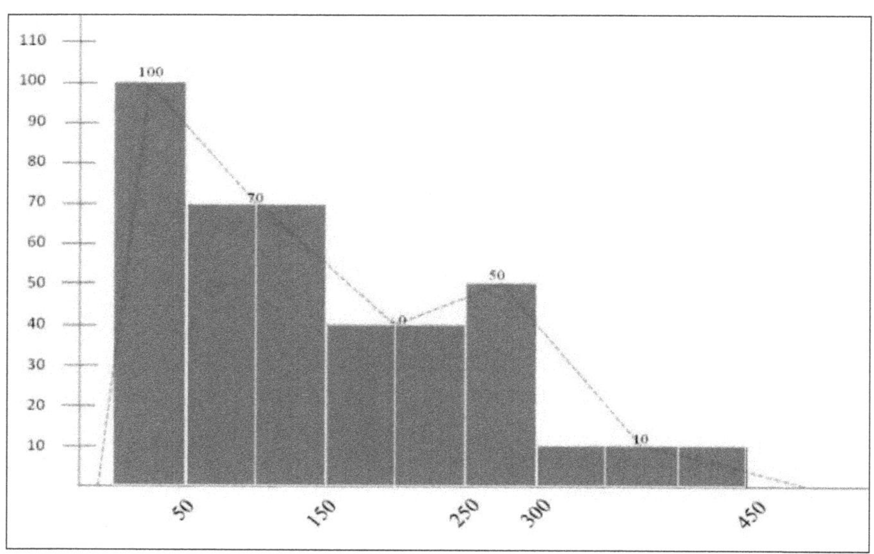

3.2.2.4 Distribution Function of continuous variables

The cumulative distribution function of a continuous statistical variable is defined as the function that associates with any real number *x* the number *F(x)* equivalent to the proportion of individuals in the population whose characteristic is less than or equal to *x*.

$$F(x_i) \;= n_i \uparrow = n_{i-1} \uparrow + n_i$$

As in the case of discrete statistical variables, the cumulative distribution function of a continuous statistical variable is graphically represented as a curve. This curve allows for quick visualization of the distribution of the statistical variable by indicating the cumulative probability of obtaining a value of the variable less than or equal to each possible value of *x*. Specifically, the cumulative distribution function starts at zero for the smallest values of *x* and reaches 1 for the largest value of *x*.

It is important to note that unlike in the case of discrete statistical variables, the cumulative distribution function for a continuous variable is a continuous function and not a step function. However, it retains the same fundamental properties, for example, the fact that the cumulative value increases continuously from zero to one.

If a_i and a_{i+1} respectively represent the lower and upper bounds of a class, the cumulative distribution function *F(x)* that associates any real number *x* with the number of statistical units is linear between the points with coordinates $[a_i, F(a_i)]$ and $[a_{i+1}, F(a_{i+1})]$. This property allows for the graphical representation of the cumulative distribution function of a continuous statistical variable as a curve that is composed of linear segments.

The cumulative distribution function is an increasing function, meaning that for any value of *x*, *F(x)* is greater than or equal to *F(y)* if *x* is greater than or equal to *y*. The curve of the cumulative distribution function starts at zero for the smallest values of x and reaches 1 for the largest values of x, which correspond to the lower and upper bounds of the distribution.

In fact, given an interval $[a, b]$ of this cumulative distribution function,

$$\forall x_1, x_2 \in [a, b], \text{si } x_2 > x_1, F(x_2) > F(x_1)$$

Example 23

Let's take **example 22** discussed earlier in **section 3.3.2.3**, which was about the statistical distribution of workers in a company grouped by age group. To better understand this distribution, we will now calculate the cumulative frequencies of these age groups.

If

$x \in [20\,;\,25)$ $F(x) = 36$

$x \in [25\,;\,30)$ $F(x) = 36 + 45 = 81$

$x \in [30\,;\,35)$ $F(x) = 36 + 45 + 27 = 108$

$x \in [35\,;\,40)$ $F(x) = 36 + 45 + 27 + 18 = 126$

$x \in [40\,;\,45)$ $F(x) = 36 + 45 + 27 + 18 + 10 = 136$

$x \in [45\,;\,50)$ $F(x) = 36 + 45 + 27 + 18 + 10 + 8 = 144$

$x \in [50\,;\,55)$ $F(x) = 36 + 45 + 27 + 18 + 10 + 8 + 3 = 147$

$x \in [55\,;\,60)$ $F(x) = 36 + 45 + 27 + 18 + 10 + 8 + 3 + 3 = 150$

3.3.2.5 Cumulative frequency curves

To better understand statistical data, it is important to understand the two types of cumulative distribution functions, each represented by a different cumulative curve:

- The increasing cumulative distribution function is represented by the curve of increasing cumulative frequencies. This curve connects the points C_i of coordinates $(a_{i+1}; f_i)$, where f_i represents the increasing cumulative frequency for the class $[a_i; a_{i+1}[$ if the increasing cumulative frequency is f_i in the same class.
- The decreasing cumulative distribution function is represented by the curve of decreasing cumulative frequencies. This curve connects the points D_i of coordinates $(a_{i+1}; f_i)$, where f_i represents the decreasing cumulative frequency for the class $[a_i; a_{i+1}[$ if the decreasing cumulative frequency is f_i in the same class.

Example 24

Let's take **example 22** from **section 3.3.2.3**, which concerns the statistical distribution of ages of workers in a company.

Classes	n_i	$f_i(\%)$	$f_i\uparrow$	$f_i\downarrow$
[20;25)	36	24.00	24.00	100.00
[25;30)	45	30.00	54.00	76.00
[30;35)	27	18.00	72.00	46.00
[35;40)	18	12.00	84.00	28.00
[40;45)	10	6.67	90.67	16.00
[45;50)	8	5.33	96.00	9.33
[50;55)	3	2.00	98.00	4.00
[55;60)	3	2.00	100.00	2.00
Total	150	100		

Let's plot the curves of increasing and decreasing cumulative frequencies for the statistical distribution of the workers in this company according to their age classes in order to better understand the distribution of the data.

The observation of the intersection point of the two curves of cumulative relative frequencies is an essential aspect of statistical analysis. It is interesting to note that the coordinates of this intersection point are equal to half of the total population size under study. We will see later that the median, an important measure of central tendency in statistical analysis, can be determined by finding the value of the variable corresponding to this coordinate.

Moreover, this observation remains valid when considering the cumulative frequencies in terms of the absolute population size. In fact, if we plot the cumulative absolute frequencies instead of the cumulative relative frequencies, we can still observe that the intersection point of the two curves has coordinates equal to half of the total population size.

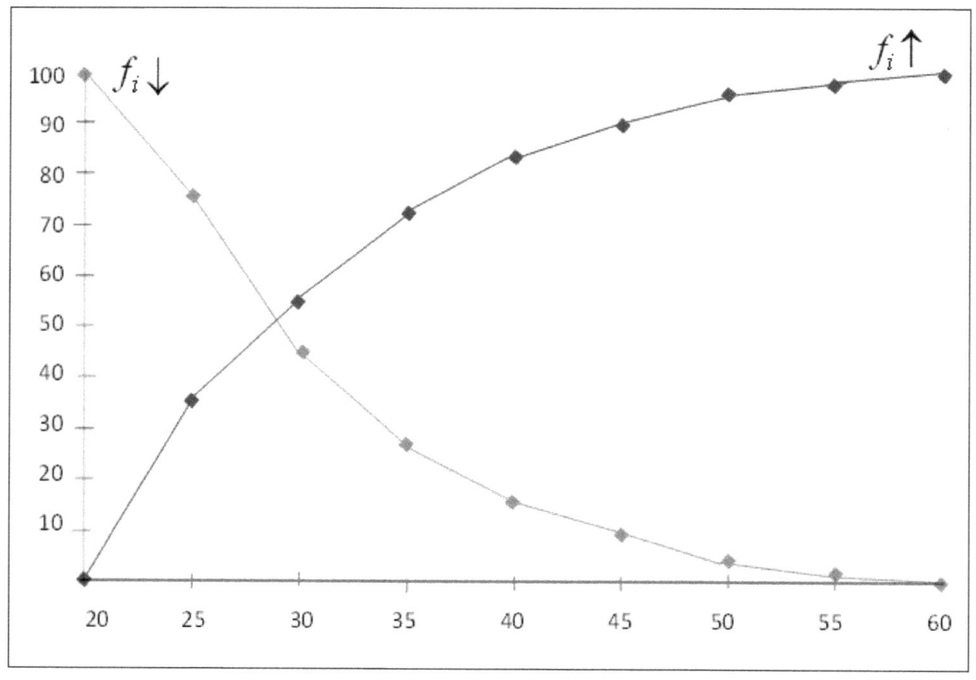

3.4 Exercises and Solutions

3.4.1 Chapter Exercises

• **Exercise 4:** The local administrative authority in a city plans to monitor the number and types of communications in the city. To this end, they have installed a gateway to sort incoming and outgoing phone calls. The collected data resulted in the following distribution by type of device used. The result is the following:

Type of device used	Number of calls
Cell phone	3747
Office landline	218
Domestic landline telephone	48
Computer	545
Public booth	47
Total	4605

Represent this statistical distribution using a pie chart.

Solution

The distribution of phone calls by type of device used is represented with a pie chart:

Type of device used	Number of calls	f_i (%)	Angle
Cell phone	3747	81.37	81.37% x 360° = 292.9°
Office landline	218	4.73	4.73% x 360° = 17°
Domestic landline telephone	48	1.04	1.04% x 360° = 3.8°
Computer	545	11.83	11.83% x 360° = 42.6°
Public booth	47	1.02	1.02% x 360° = 3.7°
Total	4605	100	360°

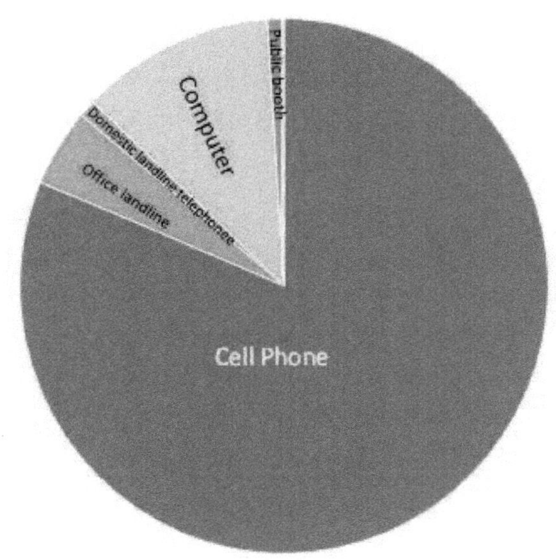

INTERPRETATION OF THE DIAGRAM

This diagram shows that the majority of calls (81.37%) were made from cell phones, followed by calls made from computers (11.8%). Desk and home landline phones represent a small proportion of the total calls, with 4.7% and 1% of calls respectively. Calls made from public phone booths represent the smallest proportion of calls, with only 1% of the total calls.

This statistical distribution allows us to understand how people communicate in the city and which types of devices are most commonly used. It can help administrative authorities make decisions regarding communication infrastructure and urban development. For example, if most calls are made from cell phones, it may be wise to develop mobile phone towers in certain areas of the city to improve coverage and network quality. Similarly, if calls made from public phone booths are few, it may be considered to reduce the number of public phone booths in areas where they are underutilized to save resources.

• **Exercise 5:** The following table represents the population and area of the 10 Canadian provinces in 2016.

Province	Population	Area in Km^2
Alberta	4067175	661848
British Columbia	4648055	944735
Prince Edward Island	142907	5660
Manitoba	1278365	647797
New Brunswick	747101	72908
Nova Scotia	923598	55284
Ontario	13448494	1076395
Quebec	8164361	1542056
Saskatchewan	1098352	651036
Newfoundland and Labrador	519716	405212

a. Represent the distribution of the population:

1. using a pie chart.
2. using a bar graph.

b. Represent the area distribution.

1. using a pie chart.
2. using a bar graph.

Solution

a-1: Distribution of the population using a pie chart:

Province	Population	f_i	Angle
Alberta	4067175	11.61	11.61% x 360° = 41.8°
British Columbia	4648055	13.27	13.27% x 360° = 47.7
Prince Edward Island	142907	0.41	0.41% x 360° = 1.5°
Manitoba	1278365	3.65	3.65% x 360° = 13.1°
New Brunswick	747101	2.13	2.13% x 360° = 7.7°
Nova Scotia	923598	2.64	2.64% x 360° = 9.5°
Ontario	13448494	38.38	38.38% x 360° = 138.2°
Quebec	8164361	23.30	23.30% x 360° = 83.9°
Saskatchewan	1098352	3.13	3.13% x 360° = 11.3°
Newfoundland and Labrador	519716	1.48	1.48% x 360° = 5.3°
Total	35038124	100	360°

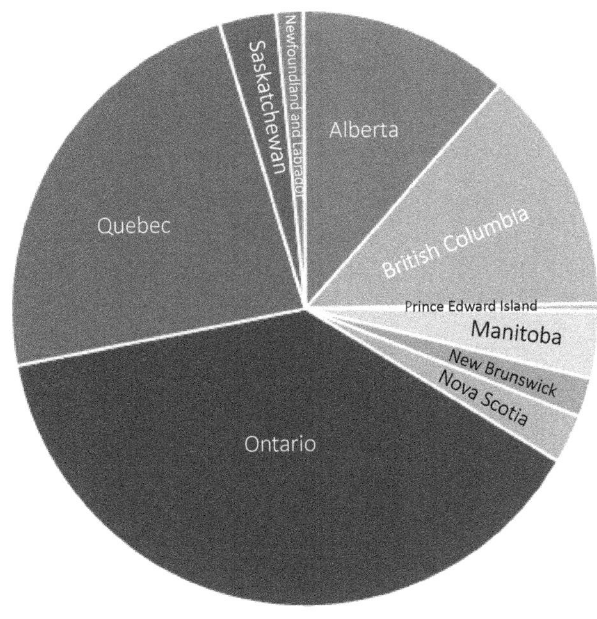

a-2: Distribution of the population using a bar graph:

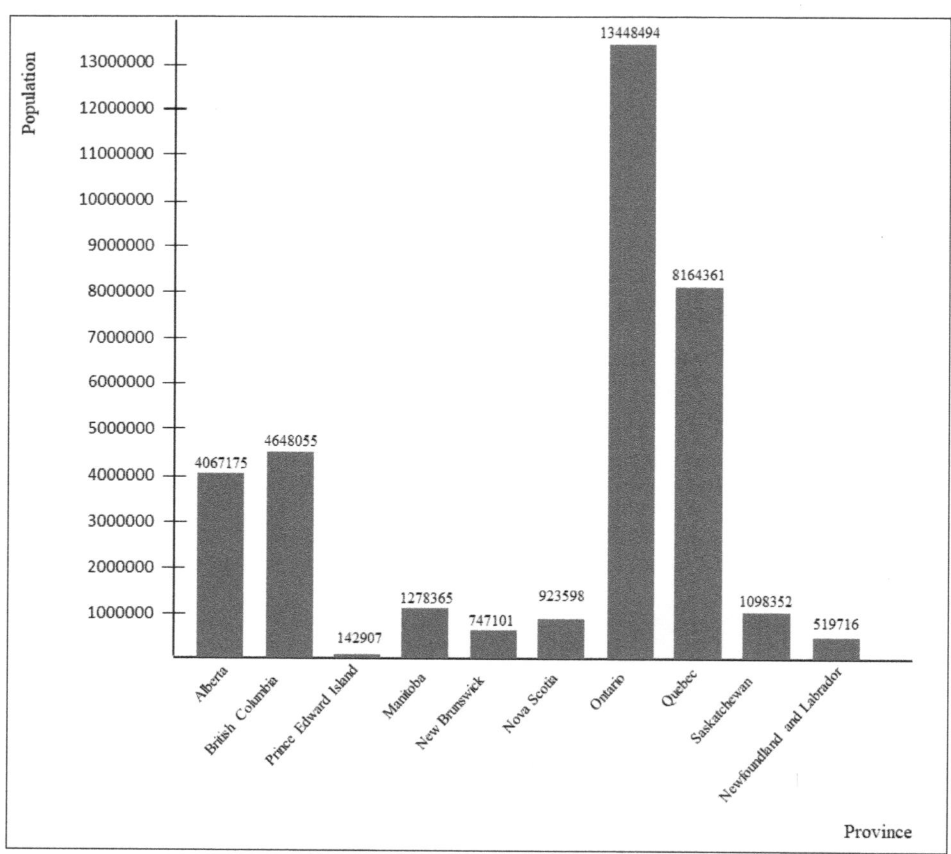

INTERPRETATION OF THE DIAGRAM

These diagrams clearly show the large population disparity between the most populous provinces and the less populous provinces. It can be seen that Ontario has the largest proportion of the population (38.2%), followed by Quebec (23.1%) and British Columbia (12.8%). The three least populated provinces are Prince Edward Island, Newfoundland and Labrador, and Saskatchewan, each representing less than 2% of Canada's total population.

These diagrams demonstrate that the distribution of population in Canada is highly uneven, with a significant concentration of the population in only a few provinces. This information can be important for infrastructure planning, resource allocation, and policy decisions regarding regional development.

b-1: Distribution of the area using a pie chart:

Province	Area	f_i	Angle
Alberta	661848	10.92	10.92% x 360° = 39.3°
British Columbia	944735	15.58	15.58% x 360° = 56.1°
Prince Edward Island	5660	0.09	0.09% x 360° = 0.3°
Manitoba	647797	10.68	10.68% x 360° = 38.5°
New Brunswick	72908	1.20	1.20% x 360° = 4.3°
Nova Scotia	55284	0.91	0.91% x 360° = 3.3°
Ontario	1076395	17.75	17.75% x 360° = 63.9°
Quebec	1542056	25.43	25.43% x 360° = 91.5
Saskatchewan	651036	10.74	10.74% x 360° = 38.7°
Newfoundland and Labrador	405212	6.68	6.68% x 360° = 24.1°
Total	6062931	100	360°

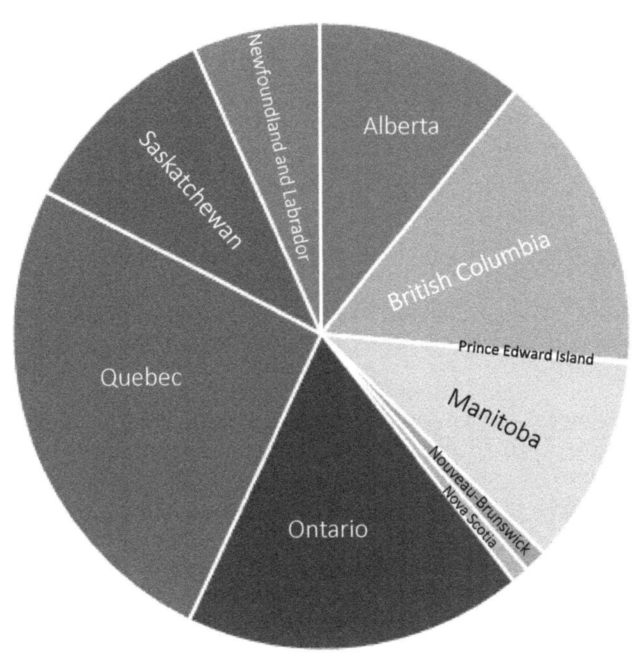

b-2: Distribution of the area using a bar graph:

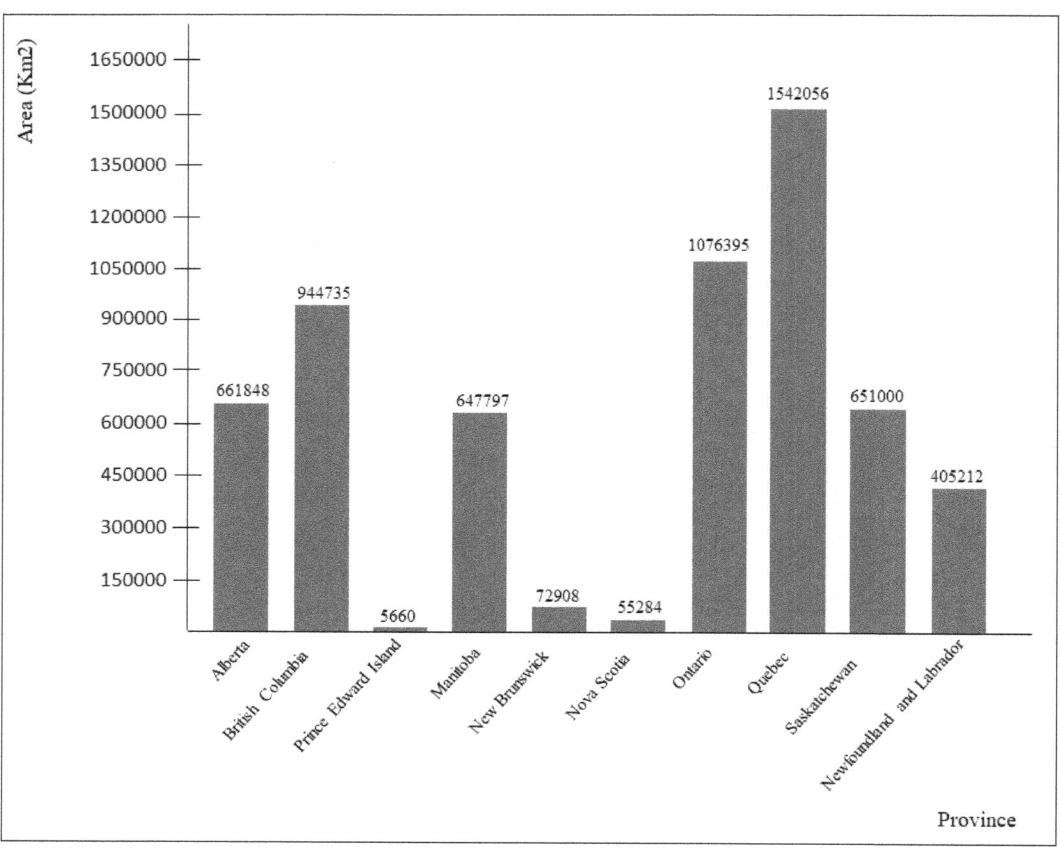

<u>INTERPRETATION OF THE DIAGRAM</u>

These diagrams allow for a quick view of the proportion of each province's area in relation to the total. It can be seen that the largest provinces are Quebec and Ontario, while the smallest provinces are Prince Edward Island, Nova Scotia, and New Brunswick.

By combining the data on population and area of Canadian provinces in 2016, it can be noted that the province of Ontario has the largest population, with 13,448,494 inhabitants, while the province of Prince Edward Island has the smallest population, with only 142,907 inhabitants. Regarding the area, the province of Quebec is the largest, with an area of 1,542,056 km2, while Prince Edward Island is the smallest, with an area of only 5,660 km2.

Therefore, it can be concluded that the population density varies considerably from one province to another. For example, the population of Ontario is very dense, while the population of Prince Edward Island is widely dispersed. Moreover, the geographical distribution of the population is clearly linked to the area of each province. The larger provinces, such as Quebec and Alberta, tend to be less densely populated than the smaller provinces, such as New Brunswick and Nova Scotia.

• **Exercise 6:** The table below shows the import volume of a country in Africa, in millions of US dollars, classified by category:

Category	Import Volume (millions USD)
Consumer Goods	228
Capital Goods	138
Raw Materials	112
Energy	54
Total	532

a: Give its representation in pie charts.

b: Give its representation in bar graphs.

Solution

a: Distribution of import volume using a pie chart:

Category	Imp. Volume	Fréq.	Angle (%)	
Consumer Goods	228	42.86	42.86 x 360° =	154.3°
Capital Goods	138	25.94	25.94 x 360° =	93.4°
Raw Materials	112	21.05	21.05 x 360° =	75.8°
Energy	54	10.15	10.15 x 360° =	36.5°
Total	532	100	360°	

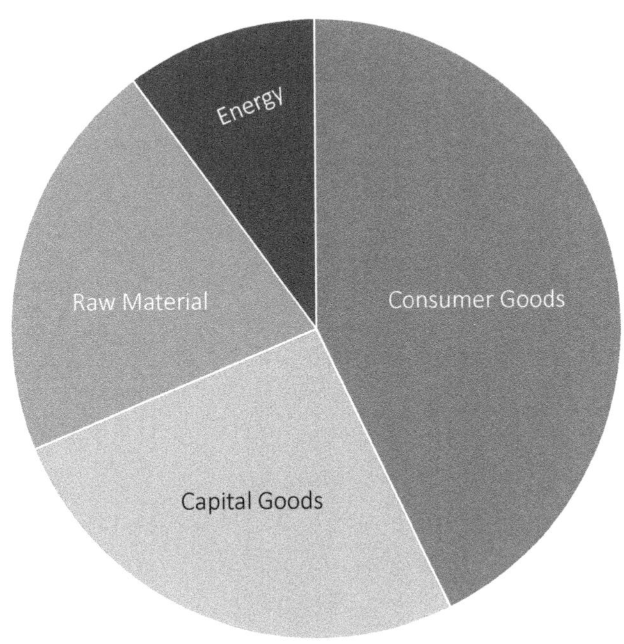

b: Distribution of import volume using bar graph:

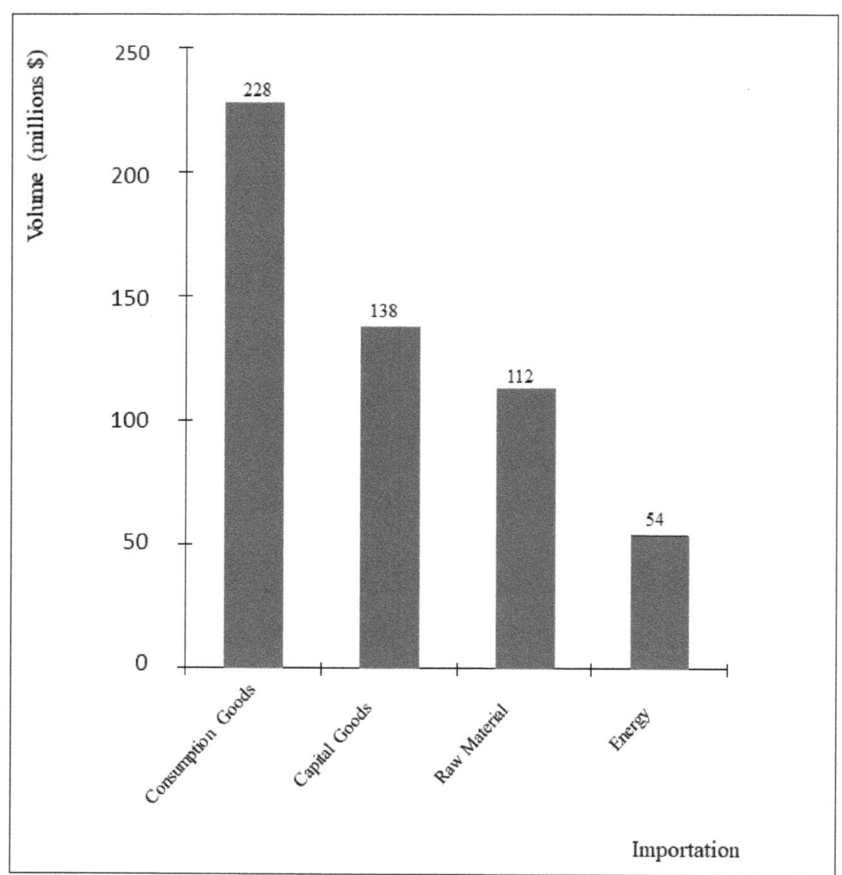

<u>INTERPRETATION OF THE DIAGRAM</u>

The diagrams show that the imports of this African country are dominated by the "Consumer Goods" category, which represents approximately 43% of the total, followed by the "Capital Goods" category with approximately 26%, then the "Raw Materials" category with approximately 21%, and finally the "Energy" category with approximately 10% of the total.

• **Exercise 7:** The following table gives the Gross Domestic Product (GDP) of the 18 French regions in 2018. Give its representation in bar graphs.

Regions	Gross Domestic Product (GDP) (Millions euros)
Île-de-France	726164
Auvergne-Rhône-Alpes	272646
Nouvelle-Aquitaine	176801
Occitanie	173563
Hauts-de-France	166519
Provence-Alpes-Côte d'Azur	166443
Grand Est	160929
Pays de la Loire	117585
Bretagne	98893
Normandie	95064
Bourgogne-Franche-Comté	78367
Centre-Val de Loire	74286
La Réunion	19163
Corse	9443
Guadeloupe	9390
Martinique	8819
Guyane	4164
Mayotte	2449

Solution

The representation in bar charts of the GDP of the 18 French regions in 2018 is:

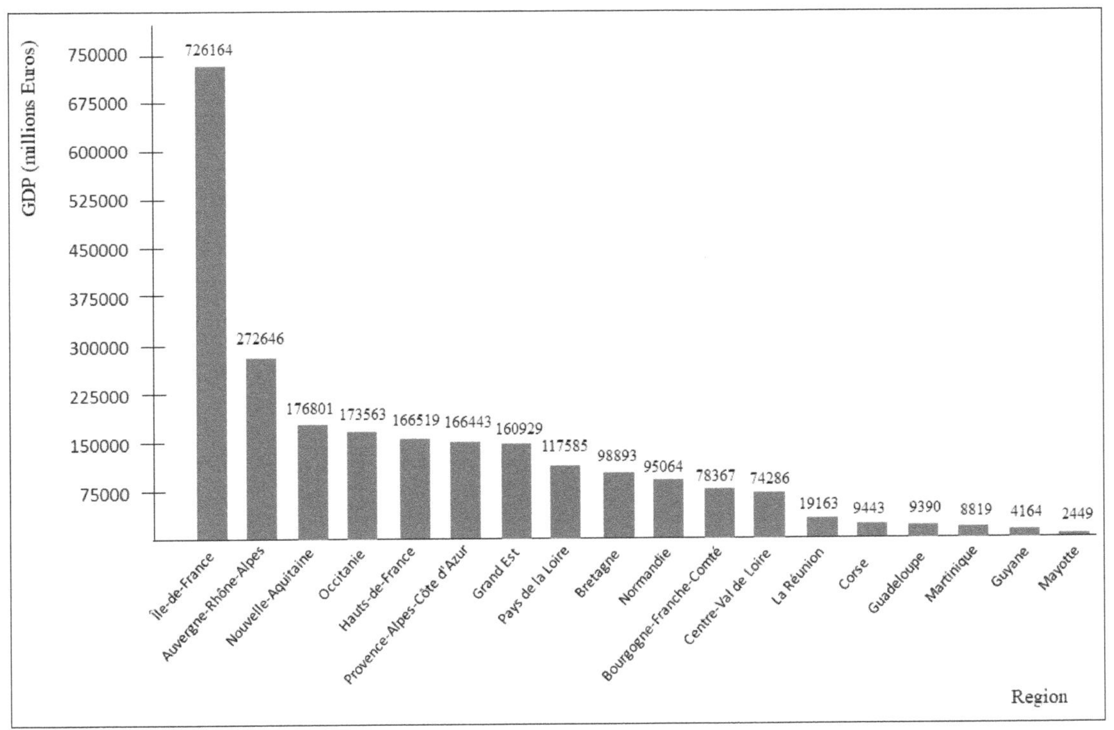

INTERPRETATION OF THE DIAGRAM

The diagram shows that Île-de-France is the region with the highest GDP in 2018, with a GDP of 726 billion euros. It is followed by the Auvergne-Rhône-Alpes region, with a GDP of 272 billion euros, and the Nouvelle-Aquitaine region, with a GDP of 176 billion euros. Overseas regions, such as Reunion, Guadeloupe, Martinique, Guyana and Mayotte as well as the Corsica have the lowest GDPs among the 18 regions.

• **Exercise 8:** For three years diamonds, oil and coffee from an african country were exported as follows:

	2018	2019	2020
Diamonds	593.6	450.8	730.1
Oil	129.9	77.1	122.6
Coffee	122.2	246.9	288.3

Give the triangular graph determining the evolution of the respective shares of each of the three products.

Solution

To construct the triangular graph of the evolution of the shares of the three products, we must highlight the proportions relating to the numbers, hence the following table:

	2018		2019		2020	
Product	n_i	$f_i(\%)$	n_i	$f_i(\%)$	n_i	$f_i(\%)$
Diamond	593.6	70.19	450.8	58.18	730.1	63.99
Oil	129.9	15.36	77.1	9.95	122.6	10.74
Coffe	122.2	14.45	246.9	31.87	288.3	25.27
Total	845.7	100.00	774.8	100.00	1141	100.00

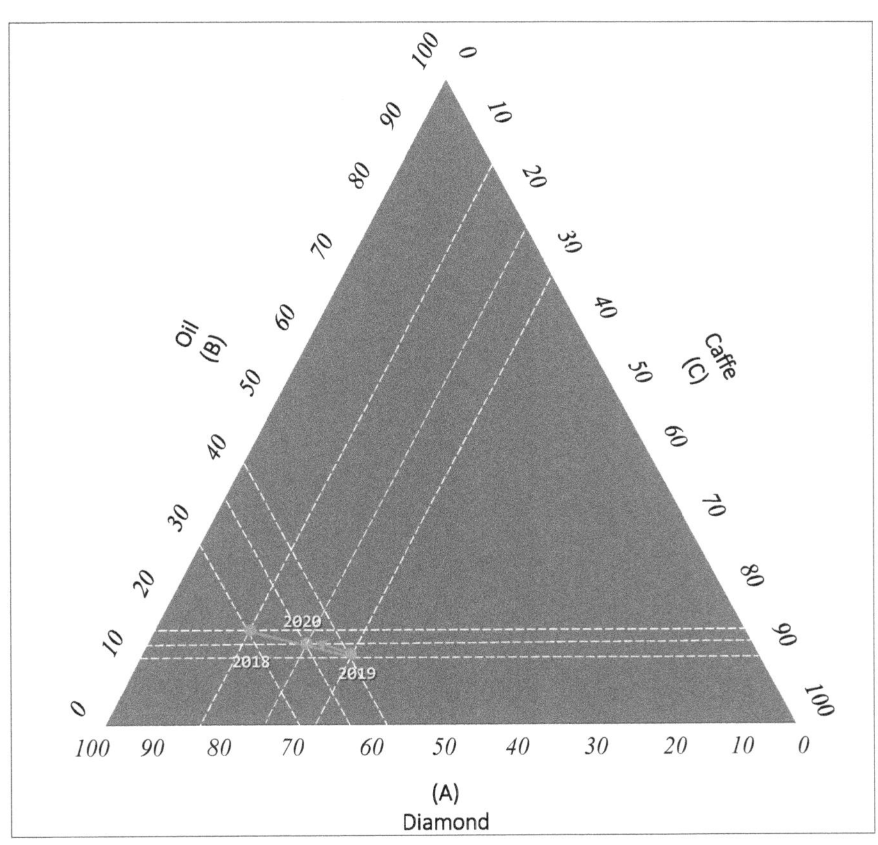

INTERPRETATION OF THE DIAGRAM
Looking at this graph, we can see that the relative share of diamonds has decreased over the years, while the relative share of oil slightly increased in 2019 before decreasing again in 2020. The relative share of coffee has steadily increased over the three years.

• **Exercise 9:** We have the following data, expressed in percentage, concerning the costs of the various production factors of a hospital over three consecutive years: 2018, 2019, and 2020.

Construct a triangular graph to follow the evolution of the respective shares of each of the three major types of factors.

This graph will quickly and clearly visualize the respective proportions of the different types of production factor costs for each year studied.

	2018	2019	2020
Staff	69	65	64
Consumption	24	27	28
Cost of fixed assets	7	8	8
Total	100	100	100

Note: This exercise does not require an additional frequency column because the sum of the numbers for each year is 100, which is equivalent to the maximum relative frequencies.

Solution

The triangular graph to follow the evolution of the respective shares of each of the three major types of factors is:

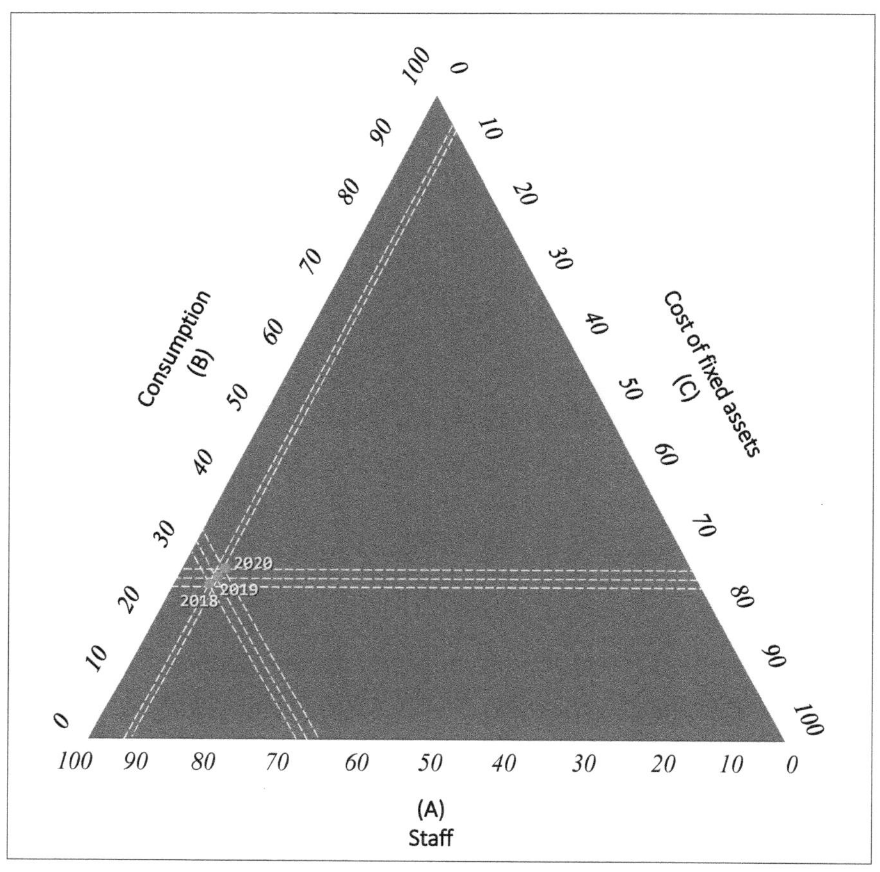

INTERPRETATION OF THE DIAGRAM
Looking at this graph, it can be observed that the cost of the staff represents the largest share of production factor costs for each year studied, with proportions decreasing from 69% in 2018 to 64% in 2020. The consumption costs, which include the costs of goods and services consumed by the hospital, increased from 24% in 2018 to 28% in 2020.

• **Exercise 10:** As part of this exercise, we rolled a die 1,000 times and recorded the number of times each of the six faces appeared. The data is presented in the table below:

Face	1	2	3	4	5	6
Number of appearances	38	144	342	287	164	25

To visualize these statistical data more easily, you are asked to represent them graphically.

It is important to note that the number of appearances of each face are expressed in absolute terms and not in percentages.

Solution

As the data series is of a discrete quantitative nature, we will use a bar chart to represent it. In this type of graph, the length of each bar will be proportional to the frequency of the corresponding modality, allowing for visualization of the distribution of values in the series.

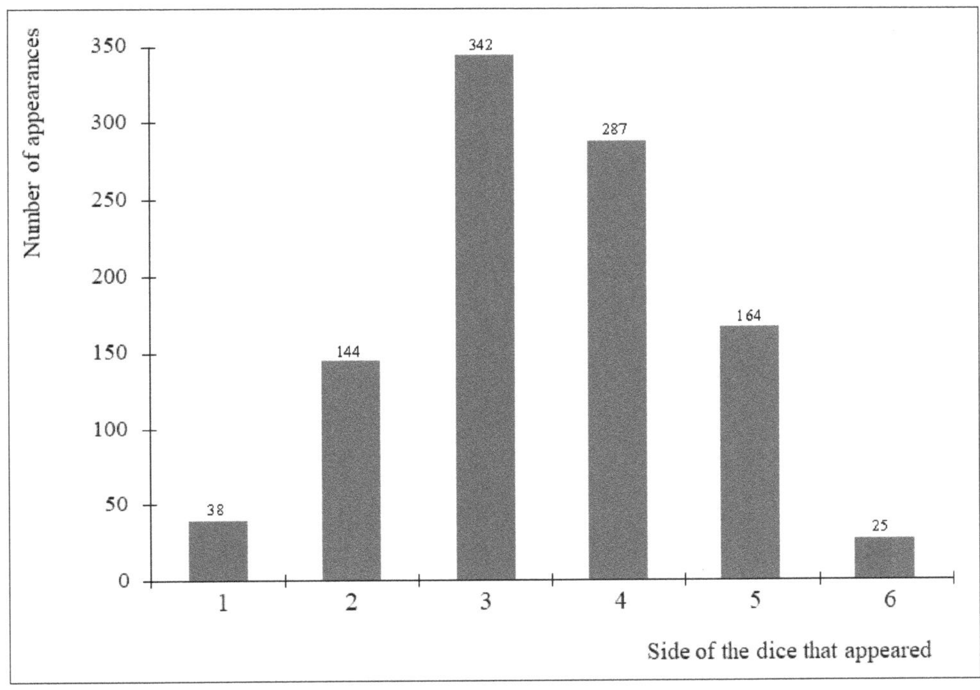

INTERPRETATION OF THE DIAGRAM

The graph representing the results of this die roll allows for quick visualization of the proportions of each face's appearance.

It can be observed that the face that appeared most often is face 3, with a proportion of 34.2%, followed by face 4 with 28.7%, and face 2 with 14.4%. Faces 1, 5, and 6 were the least frequent, with proportions of 3.8%, 16.4%, and 2.5%, respectively.

These results illustrate the probability distribution of possible outcomes for the die roll and allow for verification of whether the die is fair.

If the proportion of appearance of each face is close to 1/6 (or about 16.7%), this suggests that the die is fair. In this case, it can be seen that the proportion of appearance of each face is generally close to this expected value, although some faces are slightly more frequent than others. This could suggest a bias in the dice, which is far from a simple natural random variability of the process.

• **Exercise 11:** The following table shows the production data for a tire factory over a 15-day period. Construct a histogram to represent the distribution of the number of days based on the number of tires produced:

Number of Tires	n_i
[12500;13000)	1
[13000;13500)	2
[13500;14000)	3
[14000;14500)	3
[14500;15000)	3
[15000;15500)	3
	15

Solution

The histogram representing the distribution of the number of days based on the number of tires produced is:

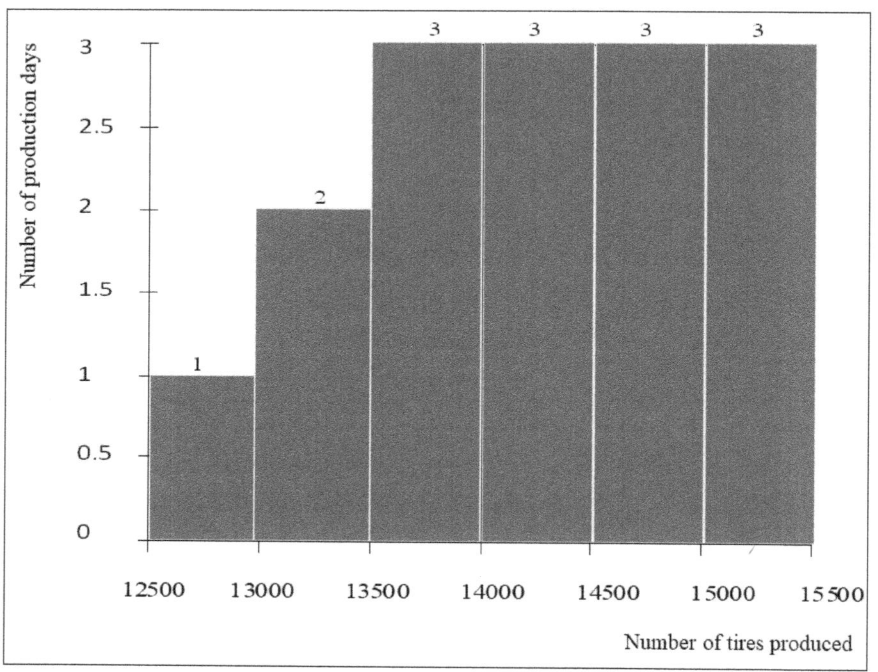

INTERPRETATION OF THE DIAGRAM

The histogram visualizes the distribution of production data of the tire factory over a period of 15 days by grouping the data into classes of 500 units of produced tires.

The majority of the days (9 out of 15) fall within the range of 13500 to 15500 tires produced. This suggests that the factory has a consistent production output within this range and that production is not significantly skewed towards either high or low tire production. The data also indicates that the factory is producing between 12500 to 15500 tires per day, with a relatively uniform distribution within this range.

3.4.2 Cross-Chapter Exercise

• **Exercise 12:** Continuing exercise **3** from section **2.5.** Considering the table below representing the balances recorded in the savings accounts of 300 individuals (in thousands of euros).

Construct its histogram and close the polygon obtained by using the middle of the two fictitious intervals to the left and to the right of the polygon.

Saving Balance	n_i	$n_i \uparrow$	$n_i \downarrow$	f_i	$f_i \uparrow$	$f_i \downarrow$
[0;5000)	48	48	300	16.00	16.00	100.00
[5000;10000)	41	89	252	13.67	29.67	84.00
[10000;15000)	47	136	211	15.67	45.33	70.33
[15000;20000)	15	151	164	5.00	50.33	54.67
[20000;25000)	21	172	149	7.00	57.33	49.67
[25000;30000)	12	184	128	4.00	61.33	42.67
[30000;35000)	13	197	116	4.33	65.67	38.67
[35000;40000)	8	205	103	2.67	68.33	34.33
[40000;45000)	9	214	95	3.00	71.33	31.67
[45000;50000)	9	223	86	3.00	74.33	28.67
[50000; 55000)	10	233	77	3.33	77.67	25.67
[55000;60000)	6	239	67	2.00	79.67	22.33
[60000;65000)	9	248	61	3.00	82.67	20.33
[65000;70000)	5	253	52	1.67	84.33	17.33
[70000;75000)	2	255	47	0.67	85.00	15.67
[75000;80000)	13	268	45	4.33	89.33	15.00
[80000;85000)	8	276	32	2.67	92.00	10.67
[85000;90000)	7	283	24	2.33	94.33	8.00
[90000;95000)	4	287	17	1.33	95.67	5.67
[95000;100000)	13	300	13	4.33	100.00	4.33
Total	300			100.00		

Solution

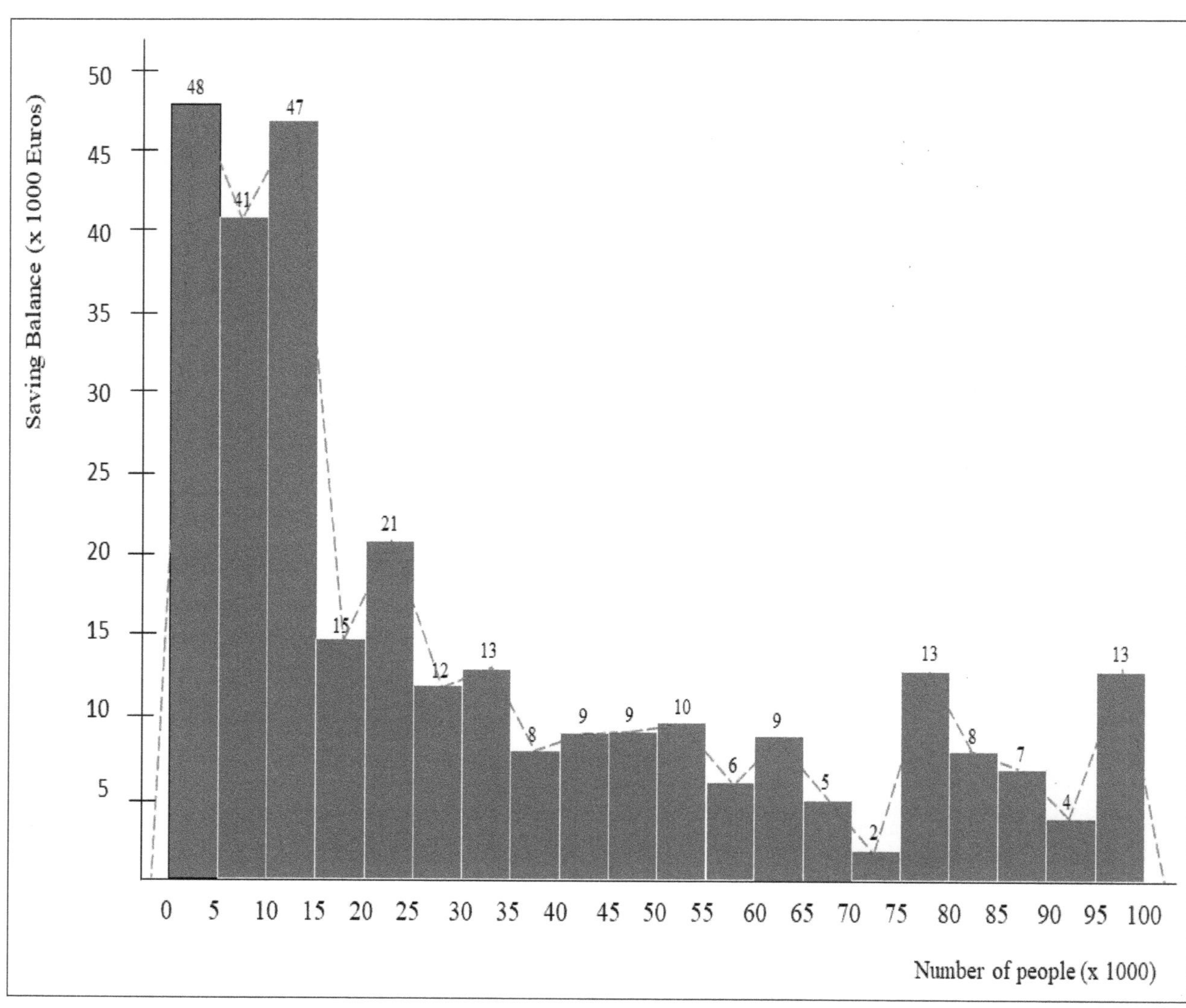

INTERPRETATION OF THE DIAGRAM

We notice that the distribution is asymmetric and heavily concentrated in the lower classes. This suggests that the majority of individuals have relatively modest savings balances. We also note that the frequency decreases rapidly for the upper interval classes, suggesting that there are relatively few individuals with very high balances.

CHAPTER 4

Characteristics of Central Tendency

This chapter explains the characteristics of central tendency, also known as parameters or measures of position, which are used to describe the magnitude order of elements in a statistical series. The three commonly used measures of central tendency, namely mode, mean, and median, are introduced in detail. The mode represents the most frequently occurring value in the statistical series; the mean represents the hypothetical value that each element would have if they were equally distributed, while the median represents the value that sits in the middle of the distribution.

4.1 Introduction

The characteristics of central tendency, also known as parameters or measures of position, are statistical measures that provide valuable information on the order of magnitude of the elements in a statistical series. These measures efficiently summarize information and allow for quick observations on how the series is distributed.

Among these measures, the mode is the most frequent value in the statistical series, i.e., the majority modality of the population. The mean represents the value that each character would have if all modalities were distributed equally. Finally, the median is the value that divides the series into two equal parts, i.e., the central value of the distribution.

These measures of central tendency are important in various fields, including economics, social sciences, healthcare, and more. They can provide valuable insights into the distribution of data and help identify patterns and trends that may not be immediately apparent.

4.2 Mode or Dominant

In a quantitative data set, the ***mode*** or ***dominant*** is the value of the variable that occurs most frequently, i.e., it is the modality of the character by which the frequency has a maximum value.

The mode is useful for describing the shape and identifying the dominant trend of a statistical distribution.

4.2.1 Case of a statistical series with discrete characters

In the case of a discrete data set, the mode is the value of the variable that corresponds to the highest frequency.

Example 25

The distribution of the number of families by the number of children in the family is given in the following table:

Number of children (x_i)	Number of families (n_i)
0	14
1	16
2	12
3	10
4	8
5	5
6	3
7	2
Total	70

In this example the mode is 1

INTERPRETATION OF THE DIAGRAM

Most families have one child.

4.2.2 Case of a statistical series with continuous characters

In the case of a continuous data series, where the data is grouped into classes, the ***modal class*** is the class with the highest frequency or count. The center of this modal class is called the ***mode*** and is used as a measure of central tendency for the data series.

Example 26

Let's take **example 20** from section **3.3.2.1**, which concerns the statistical distribution of workers in a company based on their age:

The modal class is [25 ; 30[.

If, on the other hand, the classes have unequal amplitudes, a method similar to that used for a bar chart is used. For each class, the frequency is expressed per unit interval by calculating the density of each class using the formula:

$$\text{Density} = \frac{\text{Class frequency}}{\text{Class Amplitude}}$$

Classes	Frequencies (n_i)
[20;25)	36
[25;30)	45
[30;35)	27
[35;40)	18
[40;45)	10
[45;50)	8
[50;55)	3
[55;60)	3
Total	150

This method allows for an appropriate representation of continuous data series with classes of unequal amplitudes.

Example 27

Let's take an example of classes of unequal amplitude:

Classes	x_i	Class Amplitude	n_i	Density
[0;50)	25	50	100	100 : 50 = 2
[50;150)	100	100	140	140 : 100 = 1.4
[150;250)	200	100	80	80 : 100 = 0.8
[250;300)	275	50	50	50 : 50 = 1
[300;450)	375	150	30	30 : 150 = 0.2
Total			400	

The modal class is [0 ; 50), and the mode is 25.

Example 28

The distribution of State subsidies budget allocated to companies in an American city (in millions of dollars) is given in the following table:

Subsidies	x_i	Class Amplitude	Nomber of enterprises (n_i)	Density
[0.2 ; 0.6)	0.4	0.4	4	10.0
[0.6 ; 1)	0.8	0.4	5	12.5
[1 ; 2)	1.5	1	9	9.0
[2 ; 3)	2.5	1	7	7.0
[3 ; 5)	4	2	13	6.5
[5 ; 8)	6.5	3	17	5.7
[8 ; 10)	9	2	12	6.0
[10 ; 20)	15	10	21	2.1
[20 ; 30)	25	10	13	1.3
Total			101	

The modal class is [0,6 ; 1) and the mode is $ 0.8 million.

INTERPRETATION OF THE RESULT

The distribution of State subsidies budget allocated to businesses shows that most companies have received transfer subsidies ranging between $ 0.6 and 1 million.

In the case of a ***unimodal*** distribution, there is only one value that appears most frequently in the dataset. However, if there are multiple values with the same maximum frequency, the distribution is considered to be ***multimodal***. If all the values have the same frequency, there is no mode. In a multimodal distribution or a distribution without any mode, the mode cannot be used to represent the center of the distribution.

4.3 Mean and Median

4.3.1 Concepts

When comparing two statistical series ϕ_1 and ϕ_2, the first idea that comes to mind is to compare their means. However, there are situations where the mean does not provide reliable conclusions about the comparison of the two series. Indeed, the mean is strongly influenced by outliers or extreme values, which can distort the results. Therefore, it is important to use other measures of central tendency, such as the median or mode, which are less sensitive to outliers and provide a more representative picture of the series. By using multiple measures of central tendency, it is possible to obtain a more complete picture of the comparison between the two statistical series.

Example 29

Let's consider the example of two series of seven annual grades of two students for the course of Descriptive Statistics (grades out of 200).

Both series have the same mean (120/200):

ϕ_1:	90	130	110	120	150	140	100
ϕ_2:	150	80	70	190	100	130	120

When arranged in ascending order, middle grade of student ϕ_1 is 120/200, while that of student ϕ_2 is 100/200.

ϕ_1:	90	100	110	**120**	130	140	150
ϕ_2:	70	80	90	**100**	140	140	190

Although both series ϕ_1 and ϕ_2 have the same mean, the values in series ϕ_1 appear to be well grouped around their mean, while those in series ϕ_2 appear to be more dispersed around theirs. This can be seen by examining the deviation between the values and their respective mean: the deviations are larger for ϕ_2 than for ϕ_1. Thus, by using the mean as the sole measure of central tendency, one could wrongly conclude that the two series are similar, whereas the dispersion of values in series ϕ_2 suggests some differences with series ϕ_1 as shown in the below diagram:

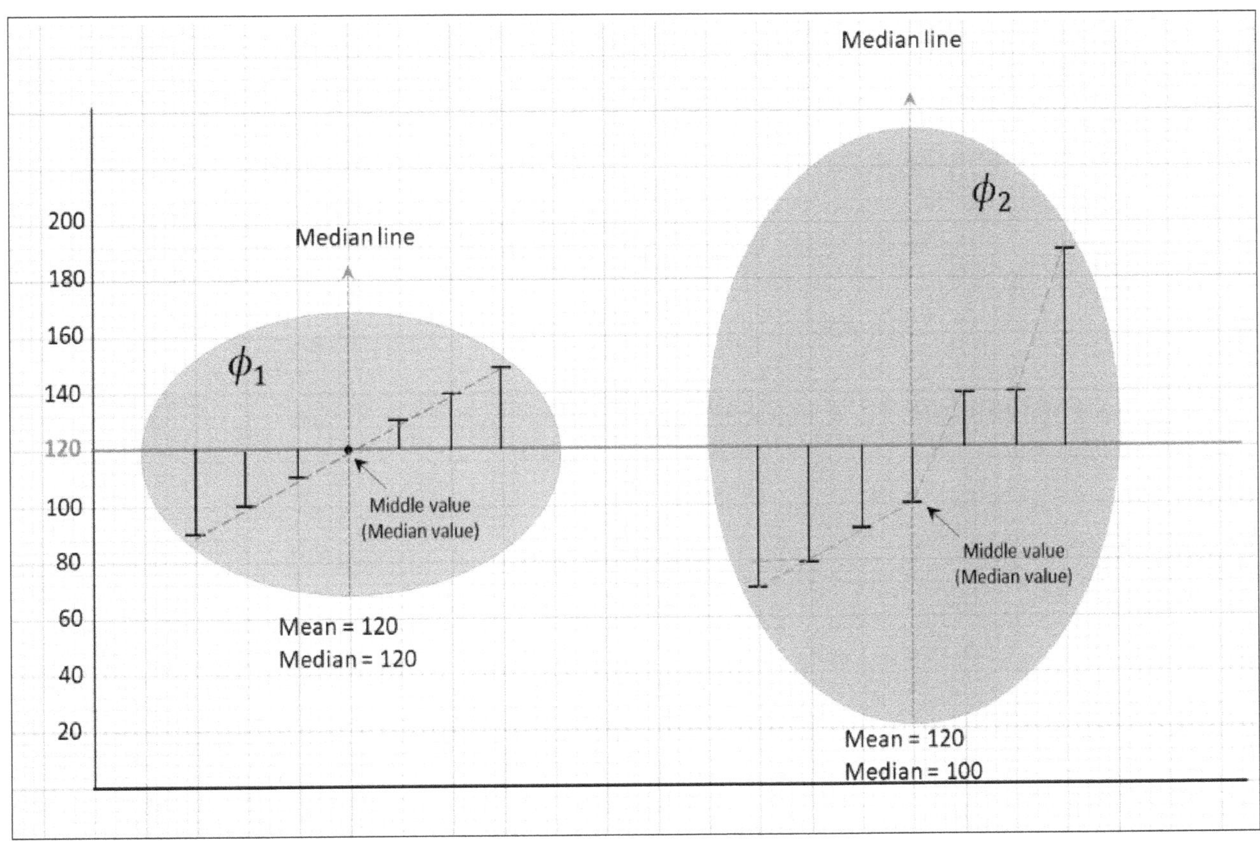

4.3.2 Note on the indivisibility of the modality

The processes of calculating the mean and median involve division operations, while in some cases, the nature of the modality requires it to remain indivisible. In this dilemma, the mean is rounded to the nearest integer, either up or down, depending on whether its decimal part is greater or less than 0.5. For the median, it is rounded to the nearest integer.

Example 30

Let's consider a statistical series where the modality is the number of rivers in a country. In this case, it would be absurd to talk about 12.87 rivers as the mean or median, even if it can be numerically understood. The mean would be rounded to the nearest integer, which is 13 rivers, and for the median, the modality value would be just above 12.87.

4.4 Arithmetic Mean (Average)

The arithmetic mean is a measure of central tendency that allows us to have an idea of the typical value of the data series.

Given *n* observations of numerical data, $n_1, \ldots\ldots\ldots\ldots, n_n$, the ***arithmetic mean*** or simply ***mean*** is defined as the sum of all these observations divided by *n*, and is denoted by $\overline{x}$. Thus, we have:

$$\overline{x} = \frac{n_1 + n_2 + \cdots \ldots \ldots \ldots \ldots .. + n_n}{n}$$

$$\boxed{\overline{x} = \frac{1}{n}\sum_{i=1}^{i=n} n_i}$$

Example 31

Looking back the **example 15** of section 3.2.2 on the distribution of the area of the Paris region:

Department	Area (Km^2)
Paris	105.4
Seine-Saint-Denis	236
Hauts-de-Seine	176
Yvelines	2284
Seine-et-Marne	5915
Val-de-Marne	245
Essonne	1804
Val-d'Oise	1246
Total $= \sum_{i=1}^{i=8} n_i$	12,011

The mean of the surface area in the Parisian region is calculated by adding up the surface areas of all the departments and dividing by the total number of departments. In this case, there are 8 departments, so the mean surface area is:

$$\bar{x} = \frac{1}{n}\sum_{i=1}^{i=8} n_i = \frac{105.4 + 236 + 176 + 2284 + 5915 + 245 + 1804 + 1246}{8}$$

$$= \frac{12{,}011.4}{8}\,\text{Km}^2$$

$$= 1{,}501.43\ \text{Km}^2$$

It is worth noting that the value of the sum $\sum_{i=1}^{i=8} n_i$ could have been directly copied from the table presented above. This would save time and potentially prevent calculation errors when manually retrieving the data.

INTERPRETATION OF THE RESULT

The mean surface area of the Parisian region is approximately 1501.43 square kilometers.

Assuming an equal distribution of the area of the Paris region among the eight departments within it, each of them would have an area equal to that value.

However, as we can see from the data, the surface areas vary greatly between departments, so the mean may not be the most informative measure of central tendency in this case.

4.4.1 Weighted Arithmetic Mean

Weighted arithmetic mean is used when each element of the discrete series has different importance. In this case, weights are assigned to each element, represented by their respective frequencies $n_1, \ldots\ldots\ldots\ldots, n_k$.

The formula for the weighted mean takes these weights into account and is given by:

$$\bar{x} = \frac{x_1 n_1 + x_2 n_2 + \ldots\ldots\ldots\ldots + x_k n_k}{n_1 + n_2 + \ldots\ldots\ldots\ldots n_k}$$

$$= \frac{\sum_{i=1}^{i=k} x_i n_i}{\sum_{i=1}^{i=k} n_i}$$

Given that:

$$f_i = \frac{n_i}{\sum_{i=1}^{i=k} n_i},$$

$$\boxed{\bar{x} = \sum_{i=1}^{i=k} f_i x_i}$$

It is important to note that the weighted mean and the statistical variable have the same unit.

Example 32

Let's take ***example 25*** from ***section 4.3.1***, which concerns the distribution of the number of families based on the number of children in the family, and determine the weighted arithmetic mean of this statistical distribution:

x_i	n_i	$n_i x_i$	f_i	$f_i x_i$
0	14	0	20.00	0.00
1	16	16	22.86	22.86
2	12	24	17.14	34.29
3	10	30	14.29	42.86
4	8	32	11.43	45.71
5	5	25	7.14	35.71
6	3	18	4.29	25.71
7	2	14	2.86	20.00
Total	70	159	100.00	227.1428571

With:

x_i: number of children in the family

n_i: number of families

f_i: frequency

$n_i x_i$: product of the number of children in the family and the number of families

$f_i x_i$: product of the frequency and the number of children in the family

The weighted arithmetic mean of the distribution of the number of families based on the number of children in the family is calculated by adding up all the $n_i x_i$ and dividing by the total number of children.

In this case, there are 7 families, so the weighted arithmetic mean of the number of families based on the number of children in the family is:

$$\bar{x} = \frac{1}{n}\sum_{i=1}^{i=8} n_i x_i = \frac{(0\text{x}14) + (1\text{x}16) + (2\text{x}12) + (3\text{x}10)}{70}$$

$$+ \frac{(4\text{x}8) + (5\text{x}5) + (6\text{x}3) + (7\text{x}2)}{70}$$

$$= \frac{159}{70}$$

$$= 2.27 \text{ children}$$

This same result can be obtained using the frequencies. Indeed, the weighted arithmetic mean of the distribution of the number of families based on the number of children in the family is calculated by adding up all the $f_i x_i$.

In this case, the weighted arithmetic mean of the number of families based on the number of children in the family is:

$$\bar{x} = \sum_{i=1}^{i=8} f_i x_i = \frac{(0\text{x}20) + (1\text{x}22.86) + (2\text{x}17.14) + (3\text{x}14.29)}{100}$$

$$+ \frac{(4\text{x}11.43) + (5\text{x}7.14) + (6\text{x}4.29) + (7\text{x}2.86)}{100}$$

$$= \frac{227.1428}{100}$$

$$= 2.27 \text{ children}$$

INTERPRETATION OF THE RESULT

If every family had an equal number of children, then each family would have 2.27 children.

However, we can notice that the notion of fractional children is not logically conceivable. To simplify, it is possible to round the mean to 2 children per family.

In the case of continuous data that has been grouped into classes, it is common to determine the center of each class denoted by x_i . To calculate the average of these centers, one can use the following formula:

$$\bar{x} = \frac{\sum_{i=1}^{i=k} n_i x_i}{\sum_{i=1}^{i=k} n_i}$$

In this formula, *k* represents the number of classes, n_i represents the number of observations in class *i*, and (a_i, a_{i+1}) represent the boundaries of class *i*. By using this formula, one can easily calculate the average of the centers of each class and obtain a representative measure of the entire dataset.

Example 33

Let's take **example 28** from **section 4.3.2**, which concerns the distribution of subsidy budgets allocated by the government to companies in an American city, expressed in millions of dollars. In this example, x_i represents the amounts of subsidies allocated to each business while n_i represents the number of businesses concerned.

Subventions	x_i	n_i	$x_i n_i$
[0.2 ; 0.6)	0.4	4	1.60
[0.6 ; 1)	0.8	5	4.00
[1 ; 2)	1.5	9	13.50
[2 ; 3)	2.5	7	17.50
[3 ; 5)	4	13	52.00
[5 ; 8)	6.5	17	110.50
[8 ; 10)	9	12	108.00
[10 ; 20)	15	21	315.00
[20 ; 30)	25	13	325.00
Total		101	947.10

To calculate the weighted arithmetic mean, we must divide the sum of the n_i x_i products by the sum of the n_i. Using the data from the table, we obtain:

$$\bar{x} = \frac{1}{n}\sum_{i=1}^{i=9} n_i x_i = \frac{(0.4x4) + (0.8x5) + (1.5x9) + (2.5x7) + (4x13)}{101}$$

$$+ \frac{(6.5x17) + (9x12) + (15x21) + (25x13)}{101}$$

$$= \frac{1.6 + 4 + 13.5 + 17.5 + 52 + 110.5 + 108 + 315 + 325}{101}$$

$$= \frac{947.10}{101}$$

$$= \$\, 9.377227 \text{ million}$$

INTERPRETATION OF THE RESULT

This result means that the weighted arithmetic mean of subsidies granted to businesses in this American city is \$9.337 million. It also means that if the state subsidy budget were evenly distributed among all businesses, each would receive an average of approximately \$9.337 million.

By examining the distribution of subsidies by class, we can also see that most subsidies were allocated to businesses whose subsidy requests were between \$10 million and \$30 million. The companies in this demand range received more subsidies than all other classes combined. Companies that requested less subsidies also received a significant share of total subsidies, while companies requesting over \$30 million received only a small portion of the total subsidies.

These results may be useful for policymakers, financial analysts, or the businesses themselves in understanding the distribution of subsidies and how they can be allocated more effectively in the future.

4.4.2 Properties of The Arithmetic Mean

The arithmetic mean possesses several interesting properties in statistics.
Let $x_1, \ldots\ldots\ldots\ldots, x_k$ be a statistical series, and let $\bar{x}$ denote the mean of this series. The difference between the value x_i and the mean is defined as $x_i - \bar{x}$. The sum of all these differences is equal to zero, which is mathematically expressed by the following formula:

$$\sum_{i=1}^{i=k} (x_i - \bar{x}) = 0$$

Indeed,

$$\sum_{i=1}^{i=k} (x_i - \bar{x}) = \sum_{i=1}^{i=k} x_i - \sum_{i=1}^{i=k} \bar{x}$$
$$= \sum_{i=1}^{i=k} x_i - k\bar{x}$$

Given that:

$$\bar{x} = \frac{1}{k} \sum_{i=1}^{i=k} x_i$$

$$\sum_{i=1}^{i=k} (x_i - \bar{x}) = \sum_{i=1}^{i=k} x_i - k \underbrace{\left(\frac{1}{k}\right) \left[\sum_{i=1}^{i=k} x_i\right]}_{\bar{x}}$$
$$= \sum_{i=1}^{i=k} x_i - \sum_{i=1}^{i=k} x_i$$
$$= 0$$

This property of the arithmetic mean is very important because it allows us to verify if the values of the series are balanced around the mean. If the sum of the differences from the mean is not zero, it means that the values of the series are not symmetrical around the mean and that there is a tendency to deviate from it.

Example 34

Let's take **example 32** of **section 4.4.1**, which concerns the distribution of the number of families based on the number of children in the family, and show that the sum of $x_i - \bar{x}$ is equal to zero:
We know that the average is 2.2714

x_i	n_i	$(x_i - \bar{x})$
0	14	-31.80
1	16	-20.34
2	12	-3.26
3	10	7.29
4	8	13.83
5	5	13.64
6	3	11.19
7	2	9.46
Total	70	0.00

$$\begin{aligned} x_i - \bar{x} &= 14(0 - 2.2714) + 16(1 - 2.2714) + 12(2 - 2.2714) \\ &+10(3 - 2.2714) + 8(4 - 2.2714) + 5(5 - 2.2714) \\ &+3(6 - 2.2714) + 2(7 - 2.2714) \\ &= 0 \end{aligned}$$

INTERPRETATION OF THE RESULT

In this example, the average number of children per family is 2.2714. The above calculation shows the deviations of each x_i value from the mean, multiplied by the corresponding number of observations, which gives a sum of 0. This confirms that the arithmetic mean is a point of balance in the statistical series.

Corollaries

1. The average of the differences from the mean is zero. In other words, if we calculate the sum of the differences of each element from the mean, this sum will be equal to zero. This formula is very useful in many statistical applications:

$$\sum_{i=1}^{i=k} f_i(x_i - \bar{x}) = \frac{1}{n}\sum_{i=1}^{i=k} n_i(x_i - \bar{x}) = 0$$

2. The sum of the squares of the deviations of the observations from the mean is minimal.

The sum of the squares of the differences between each observation and the mean is minimized. This implies that the mean is the most representative value of the set of observations.

To demonstrate this, we use the function $y(q)$, which calculates the squares of the deviations of the observations from a fixed number q. By developing this function, it can be determined that the minimum value of $y(q)$ is reached when q is equal to the mean $\bar{x}$. Thus, the mean is the number that minimizes the sum of the squares of the deviations of the observations.

Indeed, let $y(q)$ be the function of the squares of the deviations of the observations from a fixed number q is given by the following function:

$$y(q) = \sum_{i=1}^{i=k} f_i(x_i - q)^2$$

By developing this function, we have:

$$y(q) = \sum_{i=1}^{i=k} f_i(x_i - q)^2$$

$$= \sum_{i=1}^{i=k} f_i(x_{i^2} - 2qx_i + q^2)$$

$$= \sum_{i=1}^{i=k} f_i x_{i^2} - \sum_{i=1}^{i=k} f_i 2qx_i + \sum_{i=1}^{i=k} f_i q^2$$

Given that

$$\sum_{i=1}^{i=k} f_i = 1 \quad \text{and} \quad \sum_{i=1}^{i=k} f_i x_i = \bar{x}$$

We have

$$y(q) = q^2 - 2q\bar{x} + \sum_{i=1}^{i=k} f_i x_{i^2}$$

By differentiating $y(q)$ with respect to q, we have:

$$\frac{dy}{dq} = 2q - 2\bar{x}$$

This function being quadratic, that is to say of the form $ax^2 + bx + c$ with $a = 1 > 0$, its concavity is therefore turned upwards. It reaches its minimum by the value which cancels out the derivative, that is to say:

$$\frac{dy}{dq} = 0 \Rightarrow q - \bar{x} = 0$$

$$q = \bar{x}$$

Q.E.D

4.4.3 KÖNIG Theorem

The sum of the squares of the deviations of observations from a fixed number q is equal to the sum of the squares of the deviations of observations from the mean plus the square of the deviation of the mean from that fixed number q.

In other words, for any set of observations, the sum of the squares of deviations from a fixed number q is equal to the sum of the squares of deviations from the mean $\bar{x}$ plus the square of the deviation of the mean from q. This relationship is expressed by the function $y(q)$:

$$y(q) = \sum_{i=1}^{i=k} f_i (x_i - q)^2 = \sum_{i=1}^{i=k} f_i (x_i - \bar{x})^2 + (\bar{x} - q)^2$$

By adding and subtracting $\bar{x}$ from the expression in parentheses, we can expand the function $y(q)$ and show that:

$$y(q) = \sum_{i=1}^{i=k} f_i (x_i - q)^2 = \sum_{i=1}^{i=k} f_i (x_i - \bar{x} + \bar{x} - q)^2$$

$$= \sum_{i=1}^{i=k} f_i [(x_i - \bar{x}) + (\bar{x} - q)]^2$$

$$= \sum_{i=1}^{i=k} f_i[(x_i - \bar{x})^2 + 2(x_i - \bar{x})(\bar{x} - q) + (\bar{x} - q)^2]$$

$$= \sum_{i=1}^{i=k} f_i[(x_i - \bar{x})^2 + (\bar{x} - q)^2] + 2\underbrace{\underbrace{\sum_{i=1}^{i=k} f_i(x_i - \bar{x})(\bar{x} - q)}_{\begin{array}{c}\| \\ (\bar{x}-q)\sum_{i=1}^{i=k} f_i(x_i-\bar{x})\end{array}}}_{\begin{array}{c}\| \\ 0\end{array}}$$

Yet, we observe that the sum of the product of the deviations from the mean and the deviation from the mean at q is zero, so we finally obtain:

Thus,

$$y(q) = \sum_{i=1}^{i=k} f_i(x_i - q)^2 = \sum_{i=1}^{i=k} f_i[(x_i - \bar{x})^2 + (\bar{x} - q)^2]$$

Q.F.D

4.4.4 Linearity of the mean operator

The linearity of the mean operator allows for the calculation of the mean of a series of data that has been linearly transformed. To do so, each data point is multiplied by a coefficient, and then another coefficient is added to the set.

In this case, the linear transformation involves subtracting a constant x_0 from each data point and dividing the result by an amplitude a. This produces a new series of data points y_i, which are all centered around zero and have an amplitude equal to 1.

Next, all that is necessary is to calculate the mean of this new series of data points y_i, which is equal to the sum of y_i multiplied by their relative frequency. Using the formula given above, it can be shown that the mean of the original series of data points x_i is equal to the amplitude multiplied by the mean of the transformed series y_i, to which the constant x_0 is added.

Therefore, in statistical exercises, it is common to choose x_0 as the modal value of the series of data points and an as the amplitude of the modalities. This method allows for the centering of data points around the modal value and the normalization of data points for easier comparison.

Indeed, if we put:

$$y_i = \frac{x_i - x_0}{a}$$

Then,

$$\bar{y} = \frac{\bar{x} - x_0}{a}$$

It follows that:

$$\bar{y} = \sum_{i=1}^{p} f_i y_i = \sum_{i=1}^{p} f_i \left(\frac{x_i - x_0}{a}\right)$$

$$= \sum_{i=1}^{p} \left[\left(\frac{f_i x_i}{a}\right) - \left(\frac{f_i x_0}{a}\right)\right]$$

$$= \frac{1}{a}\sum_{i=1}^{p} f_i x_i - \frac{x_0}{a}\sum_{i=1}^{p} f_i$$

$$= \frac{\bar{x}}{a} - \frac{x_0}{a}$$

$$= \frac{\bar{x} - x_0}{a}$$

In the exercises, we will take $x_0 =$ mode and $a =$ amplitude of modalities.

Example 35

Let's revisit **example 26** from **section 4.3.2**, where we have a statistical distribution of workers in a company by age divided into eight age classes. For each age class, we have the number of workers (n_i), the center of each age class (x_i), and the center of each age class weighted by its frequency ($n_i x_i$).

Classes	x_i	n_i	$x_i n_i$
[20;25)	22.5	36	810.00
[25;30)	27.5	45	1237.50
[30;35)	32.5	27	877.50
[35;40)	37.5	18	675.00
[40;45)	42.5	10	425.00
[45;50)	47.5	8	380.00
[50;55)	52.5	3	157.50
[55;60)	57.5	3	172.50
Total		150	4735.00

We will now calculate the mean in the usual way to obtain the average age of the company's workers.

$$\bar{x} = \frac{1}{n}\sum_{i=1}^{i=8} n_i x_i = \frac{(22.5\text{x}36) + (27.5\text{x}45) + (32.5\text{x}27) + (37.5\text{x}18)}{150}$$

$$+ \frac{(42.5\text{x}10) + (47.5\text{x}8) + (52.5\text{x}3) + (57.5\text{x}3)}{150}$$

$$= \frac{4{,}735}{150}$$

$$= 31.566 \text{ years}$$

$$= 31 \text{ years} + 0.566 \text{ x } 12 \text{ months}$$

$$= 31 \text{ years} + 6.792 \text{ months}$$

$$= 31 \text{ years} + 6 \text{ months} + 0.792 \text{ x } 30 \text{ days}$$

$$= 31 \text{ years} + 6 \text{ months} + 23.76 \text{ days}$$

$$= 31 \text{ years} + 6 \text{ months} + 23 \text{ days}$$

INTERPRETATION OF THE RESULT

The average age of the company's workers is 31 years and 6 months, rounded to within 23 days.

This illustrates how the mean is calculated for a statistical distribution of workers in a company by age.

Now let's calculate the mean, using the linearity of the mean operat1or after making a change of variable:

$$y_i = \frac{x_i - 27.5}{5}$$

This table becomes:

Classes	x_i	n_i	$y_i = \frac{x_i - 27.5}{5}$	$y_i n_i$
[20;25)	22.5	36	-1	-36.00
[25;30)	27.5	45	0	0.00
[30;35)	32.5	27	1	27.00
[35;40)	37.5	18	2	36.00
[40;45)	42.5	10	3	30.00
[45;50)	47.5	8	4	32.00
[50;55)	52.5	3	5	15.00
[55;60)	57.5	3	6	18.00
Total		150		122.00

Let's calculate the mean of this distribution with respect to the change of origin and scale:

$$\bar{y} = \frac{1}{n}\sum_{i=1}^{i=8} n_i y_i = \frac{(-1\text{x}36) + (0\text{x}45) + (1\text{x}27) + (2\text{x}18)}{150}$$

$$+ \frac{(3\text{x}10) + (4\text{x}8) + (5\text{x}3) + (6\text{x}3)}{150}$$

$$= \frac{122}{150}$$

$$= 0.813$$

Given that:

$$y_i = \frac{x_i - 27.5}{5}$$

Hence,

$$\bar{y} = \frac{\bar{x} - 27.5}{5}$$

Then,

$$\bar{x} = 5\bar{y} + 27,5$$

$$= 5(0.813) + 27.5$$

$$= 4.065 + 27.5$$

Thus,

$$\overline{x} = 31.56$$
$$= 31 \text{ years } 6 \text{ months and } 23 \text{ days}$$
$$= 31 \text{ years} + 6 \text{ months} + 23 \text{ days}$$

INTERPRETATION OF THE RESULT

The change of variable $y_i = \frac{x_i - 27{,}5}{5}$ has reduced the calculations and facilitated the application of the linearity of the mean operator. Thus, we determined the mean of this statistical distribution using the relationship between the mean of the original variable and that of the transformed variable.

4.5 Other Means

There are different types of averages, whose order is established as follows:

- The harmonic mean is less than the geometric mean.
- The geometric mean is less than the arithmetic mean.
- The arithmetic mean is less than the quadratic mean.

This order is due to the mathematical properties of each type of average. The harmonic mean is used to calculate rates or velocities, the geometric mean is useful for calculating growth rates or rates of return, the arithmetic mean is the most common and used for quantitative data, and the quadratic mean is mainly used in statistics for calculating the standard deviation.

4.5.1 Geometric Mean

The geometric mean is a statistical measure that allows the calculation of the kth root of a product of k numbers, each being multiplied by an associated frequency. It is defined by the following formula:

$$G = \sqrt[n]{x_1{}^{n_1}.x_2{}^{n_2}, \ldots\ldots\ldots\ldots, x_k{}^{n_k}}$$

where $n_1, \ldots\ldots\ldots\ldots, n_k$ are the respective frequencies of each value $x_1, \ldots\ldots\ldots\ldots, x_k$ in the sample.

The geometric mean is often used to calculate the average growth of a return-on-investment rate or to measure the change of a quantity over time. It is also useful for calculating compound annual growth rates:

4.5.2 Mean of Order

The rth-order mean is a generalization of classical means such as the arithmetic mean, the geometric mean, etc. It allows to measure the central tendency of a distribution by giving more weight to higher values (if $r > 1$) or lower values (if $0 < r < 1$).

The expression for the rth-order mean is given by:

$$\chi_r = \sum_{i=1}^{i=k} (f_i x_i^r)^{\frac{1}{r}}$$

4.5.2.1 Quadratic Mean

The quadratic mean, also known as the root mean square, is a measure of the spread of data around their arithmetic mean. It is calculated by squaring each value, multiplying this new value by its respective frequency, and then summing all these products. The square root of this sum is then taken to obtain the quadratic mean. It is often used to measure variance in statistics and physics.

The expression for the quadratic mean is given by:

$$Q = \chi_2 = \sqrt{\sum_{i=1}^{i=k} [f_i x_i^2]}$$

This is:

$$Q = \sqrt{\frac{n_1 x_1^2 + n_2 x_2^2 + \dots\dots\dots + n_k x_k^2}{n_1 + n_2 + \dots\dots\dots + n_k}}$$

4.5.2.2 Harmonic Mean

The harmonic mean is a measure of the average which is the reciprocal of the arithmetic mean of the reciprocals of the observations. It is defined by the mean of order (-1) and is denoted by:

$$H = \chi_{-1} = \frac{1}{\sum_{i=1}^{i=k} f_i \frac{1}{x_i}}$$

In other words, the harmonic mean is the number of observations divided by the sum of the reciprocals of the observations. Therefore, we can write:

$$H = \frac{n}{\frac{1}{x_1} + \frac{1}{x_2} + \dots\dots\dots \frac{1}{x_k}}$$

4.6 Median

The median is a measure of central tendency used in statistics to describe the central position of a set of data. To determine the median, it is necessary to first sort the data in ascending order of the studied variable. The ***median***, denoted by (***Me***), is the value of the variable that separates the set of individuals into two subgroups of equal size. In other words, the median is the value of the variable that divides the statistical series into two equal populations.

The median is often preferred over the mean to represent a data set when it is strongly influenced by outliers or extreme values. It is widely used in various fields, such as finance, medicine, and psychology, to describe the distribution of data and its central position.

4.6.1 Discreet Quantitative Character

4.6.1.1 Case of non-Grouped data

To calculate the median of a set of n non-grouped discrete quantitative data, the distribution must first be sorted in either ascending or descending order. Next, the rank of the median should be calculated using the following formula:

$$\boxed{\text{Rank}_{(median)} = \frac{n+1}{2}}$$

If n is odd, the position of the median is an integer. In other words, the median corresponds to the value that lies in the middle of the ordered series.

Example 36

The grades of a student over the course of a semester are given below:

1 3 8 9 13 14 15 16 17

Since this data set contains 9 ($n = 9$) values, the rank of the median is:

$$\frac{n+1}{2} = \frac{9+1}{2} = 5$$

Therefore, the median is the value of the characteristic located at rank 5, which is the number 13.

1 3 8 9 13 14 15 16 17

↑

Me

INTERPRETATION OF THE RESULT

Half of the grades are lower than 13/20, and the other half is higher than this value. In other words, the grade of 13/20 represents the central value of the distribution.

If n is even, the position of the median is a decimal number. In this case, the median is determined by taking the average of the two central values in the dataset. These two values form the median interval, which contains the position of the median.

Example 37

Consider the dataset of the length of 10 electrical cables (expressed in km), given in the following table:

385 363 369 355 390 372 360 380 381 387

To calculate the median, we first need to order the distribution in ascending order:

355 360 363 369 372 380 381 385 387 390

Since the number of data points is even ($n = 10$), the position of the median is a decimal number. The median rank is:

$$\frac{n+1}{2} = \frac{10+1}{2} = 5.5$$

As 5.5 is between the 5th and 6th position in the distribution, we need to find the median interval which is formed by the two values at these positions.
The median interval is, therefore:

[371; 380].

To find the median value, we calculate the average of these two values:

$$\frac{371+380}{2} = 375.5 \text{ Km}$$

The median interval is, therefore, 375.5 Km.

INTERPRETATION OF THE RESULT

This median value allows us to conclude that there are as many cables with a length less than 375.5 km as there are cables with a length greater than this value. In other words, it represents the dividing point of the distribution into two equal parts, each containing 50% of the data.

4.6.1.2 Case of Grouped data

In the case of grouped data, the median is the measure of central tendency that corresponds to the value of the variable for which the cumulative frequency is equal to half of the total frequency (or 50% if we consider the cumulative frequencies in percentage). There are two methods to calculate the median of grouped data. The first is similar to that used in the case of non-grouped data, and the second, is an analytical method based on linear interpolation.

First method:

In this method, the median is associated with the value in the middle of the cumulative frequency in the same way as for the non-grouped data.

The same formula is applied for determining the position of the median previously used in section **4.6.1.1**.

The median is, therefore, the value of the character corresponding to an increasing cumulative frequency; its rank is:

$$\text{Rank}_{(median)} = \frac{n+1}{2}$$

Example 38

This example is that of an odd cumulative frequency.

The meteorological control body of a country with an equatorial climate wanted to study the annual evolution of national rainfall. To do this, they recorded the number of rainy days per week for a year. The table below shows the number of weeks recorded according to the number of rainy days.

Number of rainy days in a week	Number of weeks (Frequency)	Cumulative Frequency
1	7	7
2	14	21
3	6	27
4	5	32
5	12	44
6	9	53
Total	53	

We will now calculate the median to determine the central value of this distribution. The rank of the median is:

$$\frac{n+1}{2} = \frac{53+1}{2} = 27$$

The median is, therefore, the 27th data point in the ordered distribution. To obtain the value of the median, we will add the cumulative frequencies until we reach or exceed the value 27.

The results obtained allow us to observe that:

- The first 7 rainfall measurements correspond to a value of 1, meaning that during the first 7 weeks, there was only one day of rain per week.
- The measurements from the 8th to the 21st rank correspond to a value of 2, indicating that between the 8th and the 21st week, there were two days of rain per week.
- *The measurements from the 22nd to the 27th rank correspond to a value of 3, indicating that between the 22nd and the 27th week, there were three days of rain per week.*
- The measurements from the 28th to the 32nd rank correspond to a value of 4, indicating that between the 28th and the 32nd week, there were four days of rain per week.
- The measurements from the 33rd to the 44th rank correspond to a value of 5, indicating that between the 33rd and the 44th week, there were five days of rain per week.
- The measurements from the 45th to the 53rd rank correspond to a value of 6, indicating that between the 45th and the 53rd week, there were six days of rain per week.

Thus, we obtain that the measurements from the 22nd to the 27th rank correspond to a value of 3, indicating that between the 22nd and the 27th week, there were three days of rain per week. Therefore, the median has a value of 3.

INTERPRETATION OF THE RESULT

The median of this distribution indicates that there are as many weeks with less than three days of rain as there are weeks with more than three days of rain. In other words, half of the weeks of the year had three or fewer days of rain, while the other half had four or more days of rain. This shows that the rainfall in this equatorial country is relatively uniform throughout the year, with approximately one week out of two having three or fewer days of rain and the other week having four or more days of rain.

Understanding the rainfall data of this country can be useful for planning activities based on rainfall and for assessing the risks associated with floods and droughts.

Example 39

This example is that of an even cumulative frequency.

Let's take the previous example, where we studied the national rainfall in an equatorial country and suppose there was a data entry error: the value of the fourth modality was actually 10, not 5, as we had previously noted. In this case, the distribution now has 58 observations.

Number of rainy days in a week	Number of weeks (Frequency)	Cumulative Frequencies
1	7	7
2	14	21
3	6	27
4	10	37
5	12	49
6	9	58
Total	58	

Since the number of data points is even ($n = 58$), the position of the median is a decimal number. The median rank is:

$$\frac{n+1}{2} = \frac{58+1}{2} = 29.5$$

As 29.5 is between the 3rd and 6th position in the distribution, we need to find the median interval which is formed by the two values at these positions.

The median interval is, therefore:

$$[3; 4].$$

To find the median value, we calculate the average of these two values:

$$\frac{3+4}{2} = 3.5 \text{ rainy days in a week}$$

The median interval is, therefore, 3.5 rainy days in a week.

Second Method:

The second method for calculating the median is analytical and based on linear interpolation. To do this, we use the cumulative frequencies (N_i) and cumulative frequencies (F_i) of a given series. This series is presented in the form of a table with the different modalities and their associated cumulative frequencies and frequencies.

Modalities	N_i	F_i
a_1	N_1	F_1
a_2	N_2	F_2
a_3	N_3	F_3
....	....	....
....	....	....
a_{i-1}	N_{i-1}	F_{i-1}
a_i	N_i	F_i
....	....	....
....	....	....
a_p	$N_p=n$	$F_p=100$

The median is determined by finding the value of the characteristic that corresponds either to the cumulative frequency equal to $\frac{n}{2}$ or to the cumulative frequency equal to 50%.

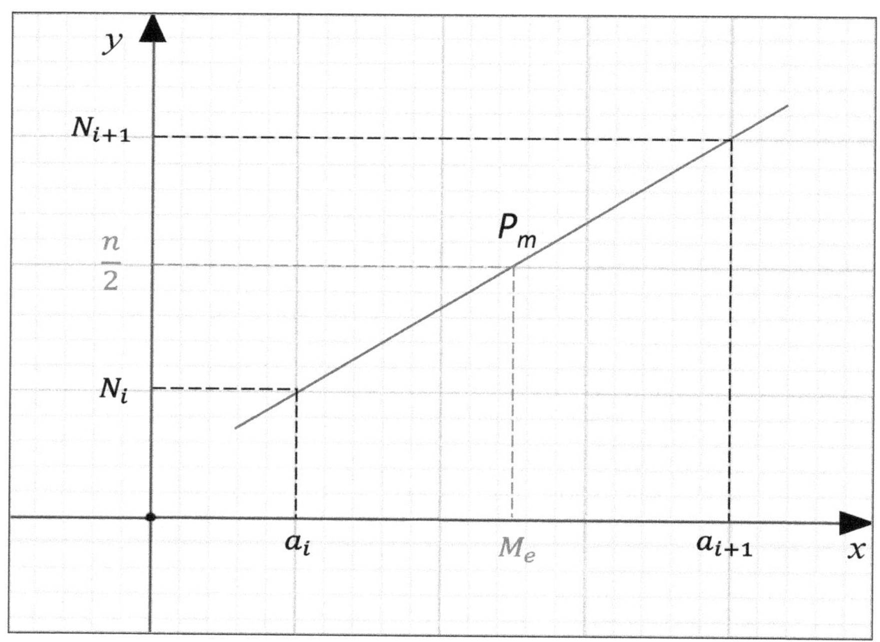

We know that a line passing through two points x and y of coordinates (x_1, y_1) and (x_2, y_2), respectively is represented by the equation:

$$y - y_1 = \frac{y_2 - y_1}{x_2 - x_1}(x - x_1) \dots \dots (1)$$

For the calculation of the median, the abscissas of the two points x_1 and x_2 are replaced by the extreme values of the median interval (a_i, a_{i+1}), and their ordinates are replaced by their respective frequencies (N_i, N_{i+1}) with:

a_i corresponds to x_1
N_i corresponds to y_1
a_{i+1} corresponds to x_2
N_{i+1} corresponds to y_2

By replacing these values in equation (1), we obtain the equation of the line (2):

$$y - N_i = \frac{N_{i+1} - N_i}{a_{i+1} - a_i}(x - a_i) \dots . (2)$$

Consider the point P_m having for ordinate half of the frequencies $\left(\frac{n}{2}\right)$, which corresponds to the abscissa M_e on line (2), and let us express the fact that P_m is on this line, we have:

$$\frac{n}{2} - N_i = \frac{N_{i+1} - N_i}{a_{i+1} - a_i}(M_e - a_i)$$

We deduce the expression of M_e:

$$M_e = a_i + (a_{i+1} - a_i)\frac{\left(\frac{n}{2} - N_i\right)}{(N_{i+1} - N_i)}$$

In the same way, we deduce the expression of the median using the cumulative relative frequencies, replacing the expression $\frac{n}{2}$ by $\frac{100\%}{2} = 50\%$:

$$\boxed{M_e = a_i + (a_{i+1} - a_i)\frac{(50 - F_i)}{(F_{i+1} - F_i)}}$$

Example 40

Let's go back to **example 39**, which was previously discussed in the same section on the national rainfall of an equatorial country. The following data represents the number of rainy days and the corresponding number of weeks:

Number of rainy days	Number of weeks (Frequency)	$f_i(\%)$	$f_i(\%)\uparrow$
1	7	13.21	13.21
2	14	26.42	39.62
3	6	11.32	50.94
4	5	9.43	60.38
5	12	22.64	83.02
6	9	16.98	100.00
Total	53	100.00	

The colored area is the interval that contains the median.

$$
\begin{aligned}
M_e &= a_i + (a_{i+1} - a_i)\frac{(50 - F_i)}{(F_{i+1} - F_i)} \\
&= 3 + (4 - 3)\frac{(50 - 46.55)}{(63.79 - 46.55)} \\
&= 3.2
\end{aligned}
$$

The value of the median is 3.2 rainy days in a week.

4.6.2 Continuous Quantitative Character

For continuous quantitative characteristics, the median is determined in the same way as for discrete quantitative characteristics. It represents the value of the characteristic corresponding to the cumulative frequency equal to half of the total frequency (or 50% if we consider cumulative frequencies taken in percentage).

However, in the case of continuous quantitative characteristics, it is necessary to group the data into classes to calculate the median. To do this, it is necessary to determine the median class, which corresponds to the class containing the median. Then, we use the same formula to calculate the median:

4.6.2.1 Analytical Method for Finding the Median (linear interpolation)

Let the following series be given with increasing cumulative frequencies (N_i) and increasing cumulative relative frequencies (F_i).

Classes	N_i	F_i
$[a_1;a_2)$	N_1	F_1
$[a_2;a_3)$	N_2	F_2
$[a_3;a_4)$	N_3	F_3
....	....	....
....	....	....
$[a_{i-1};a_i)$	N_{i-1}	F_{i-1}
$[a_i;a_{i+1})$	N_i	F_i
....	....	....
....	....	....
$[a_{p-1};a_p)$	$N_p=n$	$F_p=100$

By applying the same logic and calculation as for discrete quantitative characteristics, we can also calculate the median for continuous quantitative characteristics. The only difference is that, in this case, a_i represents the endpoints of the classes:

Using the same formula as for discrete quantitative characteristics, we can determine the expression of the median for continuous quantitative characteristics:

$$M_e = a_i + (a_{i+1} - a_i)\frac{\left(\frac{n}{2} - N_i\right)}{(N_{i+1} - N_i)}$$

In the same way, we deduce the expression of the median using the cumulative frequencies, replacing the expression $\frac{n}{2}$ by $\frac{100\%}{2} = 50\%$

$$\boxed{M_e = a_i + (a_{i+1} - a_i)\frac{(50 - F_i)}{(F_{i+1} - F_i)}}$$

Example 41

Let's take **example 33** from **section 4.4.1,** which shows the distribution of state subsidy budget allocated to companies in an American city (in millions of dollars). In this table, x_i represents the subsidies (in millions) and n_i represents the number of companies:

Subsidies	x_i	n_i	$f_i(\%)$	$f_i(\%)\uparrow$
[0.2 ; 0.6)	0.4	4	3.96	3.96
[0.6 ; 1)	0.8	5	4.95	8.91
[1 ; 2)	1.5	9	8.91	17.82
[2 ; 3)	2.5	7	6.93	24.75
[3 ; 5)	4	13	12.87	37.62
[5 ; 8)	6.5	17	16.83	54.46
[8 ; 10)	9	12	11.88	66.34
[10 ; 20)	15	21	20.79	87.13
[20 ; 30)	25	13	12.87	100.00
Total		101	100.00	

The colored area is the interval that contains the median.

Let's calculate the median:

$$M_e = a_i + (a_{i+1} - a_i)\frac{(50 - F_i)}{(f_{i+1} - F_i)}$$

$$= 5 + (8 - 5)\frac{(50 - 37.62)}{(54.46 - 37.62)}$$

$$= \$\ 7.20$$

Thus, the median of the distribution of subsidies is 7.20 million dollars.

INTERPRETATION OF THE RESULT

The median of the subsidy distribution is 7.20 million dollars. This means that half of the companies received subsidies less than or equal to 7.20 million dollars, while the other half received subsidies higher than this amount.

4.6.2.2 Graphical Method for Finding the Median

The median can be found graphically by finding the point of intersection between the curve of increasing cumulative relative frequencies and that of decreasing cumulative relative frequencies.

Example 42

Let's consider **example 35** from **section 4.4.4,** which deals with workers in a company classified by age and determine the median for this statistical distribution.

Classes	x_i	n_i	$f_i(\%)$	$f_i(\%)\uparrow$	$f_i(\%)\downarrow$
[20;25)	22.5	36	24.00	24.00	100.00
[25;30)	27.5	45	30.00	54.00	76.00
[30;35)	32.5	27	18.00	72.00	46.00
[35;40)	37.5	18	12.00	84.00	28.00
[40;45)	42.5	10	6.67	90.67	16.00
[45;50)	47.5	8	5.33	96.00	9.33
[50;55)	52.5	3	2.00	98.00	4.00
[55;60)	57.5	3	2.00	100.00	2.00
Total		150	100.00		

Analytical determination of the median:

$$M_e = a_i + (a_{i+1} - a_i)\frac{(50 - F_i)}{(F_{i+1} - F_i)}$$

$$= 25 + (30 - 25)\frac{(50 - 24)}{(54 - 24)}$$

$$= 29.33 \text{ years}$$

$$= 29 + 0.33 \text{ x } 12$$

$$= 29 \text{ years and } 4 \text{ months}$$

Graphical determination of the median:

The cumulative increasing frequency curve ($f_i \uparrow$) represents the cumulative percentage of companies that received subsidies less than or equal to a certain value, while the cumulative decreasing frequency curve ($f_i \downarrow$) represents the cumulative percentage of companies that received subsidies higher than this value. The point of intersection of the two curves is the median.

Using the method of projecting the increasing cumulative frequencies, we find that the median is around 29.33 years, which is equal to 29 years and 4 months.

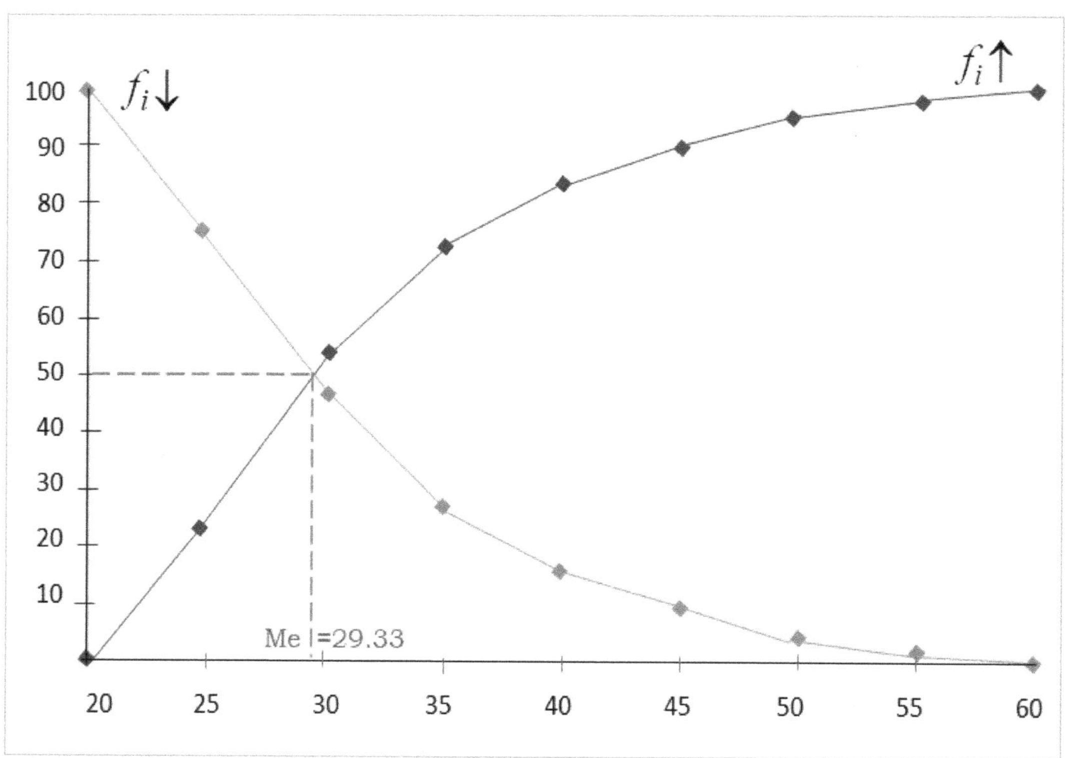

INTERPRETATION OF THE RESULT

Within this company, the number of employees who are at least 29 years and 4 months old is equivalent to the number of employees who are older. This indicates that the median age of the employees falls at 29 years and 4 months, meaning that when arranging the employees in ascending order of age, the middle person would be 29 years and 4 months old.

4.7 The Choice of Mean versus Median in Analytical Contexts

4.7.1 Rule of Thumb for using the Mean and Median

The choice between mean and median is a key issue in statistical analysis. It is important to understand when to use each measure of central tendency in order to obtain an accurate interpretation of the data.

In general, the mean is used to obtain the central tendency when the distribution is almost normal, with very few outliers. However, when the data series exhibits significant disparities, it is recommended not to use the mean as the presence of outliers can influence it, thus altering the interpretation of the data.

On the other hand, the median is used to obtain the central tendency when the distribution is visibly skewed. Thus, when the data series contains outliers, it is recommended to use the median as it is not influenced by extreme values.

Example 43

An insurance company is considering establishing a presence on an island with a population of barely 2,000 engaged in professional activity, including five billionaire entrepreneurs who each pay themselves a monthly salary of around four million dollars.

Before making any decisions, the company prepared a prospective study on the wages of the inhabitants of the island. From the collection of information, the following table results after analysis. What is your interpretation of this data, and what do you conclude from it?

Monthly Salary Bracket	Number of people (n_i)
[0;1000)	380
[1000;2000)	633
[2000;3000)	527
[3000;4000)	151
[4000;5000)	123
[5000;6000)	85
[6000;7000)	58
[7000;8000)	39
[8000;9000)	8
[9000;10000)	6
[10000;4000000)	5
Total	2015

Solution

Solving this exercise requires the columns in the following table:

Classes	x_i	n_i	$n_i x_i$	f_i	$f_i\uparrow$
[0;1000)	500	380	190000	18.86	18.86
[1000;2000)	1500	633	949500	31.41	50.27
[2000;3000)	2500	527	1317500	26.15	76.43
[3000;4000)	3500	151	528500	7.49	83.92
[4000;5000)	4500	123	553500	6.10	90.02
[5000;6000)	5500	85	467500	4.22	94.24
[6000;7000)	6500	58	377000	2.88	97.12
[7000;8000)	7500	39	292500	1.94	99.06
[8000;9000)	8500	8	68000	0.40	99.45
[9000;10000)	9500	6	57000	0.30	99.75
[10000;4000000)	2495000	5	12475000	0.25	100.00
Total		2015	17276000	100.00	

Let's calculate the average:

$$\bar{x} = \frac{1}{n}\sum_{i=1}^{i=11} n_i x_i = \frac{(500\text{x}380) + (1500\text{x}633) + (2500\text{x}527)}{2015}$$

$$+\frac{(3500\text{x}151) + (4500\text{x}123) + (5500\text{x}85)}{2015}$$

$$+\frac{(6500\text{x}58) + (7500\text{x}39) + (8500\text{x}8)}{2015}$$

$$+\frac{(9500\text{x}6) + (2495000\text{x}5)}{2015}$$

$$= \frac{17{,}276{,}000}{2{,}015}$$

$$= 8{,}573.69\ \$$$

Let's calculate the median:

$$M_e = a_i + (a_{i+1} - a_i)\frac{(50 - F_i)}{(F_{i+1} - F_i)}$$

$$= 1{,}000 + (2{,}000 - 1{,}000)\frac{(50 - 18.86)}{(50.27 - 18.86)}$$

$$= \$\ 1{,}991.4$$

INTERPRETATION OF THE RESULTS

The collected data suggests that the average salary on the island is $8,573.69, while the median salary is $1,991.4.

Using the average salary as a measure of financial capacity for insurance calculations would be misleading because half of the working population actually earns less than $1,991.4, which is more than four times less than the estimated average salary. It is evident that the majority of the population would find it difficult to afford insurance that is calculated based on an income estimated to be over four times greater than their actual income.

The presence of five billionaire entrepreneurs with a high monthly salary of four million dollars each greatly influences the average salary, which is likely to give a misleading interpretation of the actual economic situation of the island.

Therefore, it would be more appropriate for the company to base its calculations on the median salary rather than the mean salary. This is because the median salary better reflects the typical income level of the population and allows for a more accurate view of the island's economic situation.

This case illustrates how the mean can be influenced by extreme values and generate a clearly misleading central tendency.

4.7.2 Attempt at Correction

In an attempt to correct the inconsistency of the mean, we decided to eliminate the five outliers from the dataset, and the new distribution is as follows:

Classes	x_i	n_i	$n_i x_i$	f_i	$f_i \uparrow$
[0;1000)	500	380	190000	18.91	18.91
[1000;2000)	1500	633	949500	31.49	50.40
[2000;3000)	2500	527	1317500	26.22	76.62
[3000;4000)	3500	151	528500	7.51	84.13
[4000;5000)	4500	123	553500	6.12	90.25
[5000;6000)	5500	85	467500	4.23	94.48
[6000;7000)	6500	58	377000	2.89	97.36
[7000;8000)	7500	39	292500	1.94	99.30
[8000;9000)	8500	8	68000	0.40	99.70
[9000;10000)	9500	6	57000	0.30	100.00
Total		2010	4801000	100.00	

Let's calculate the new average:

$$\bar{x} = \frac{1}{n}\sum_{i=1}^{i=10} n_i x_i = \frac{(500\text{x}380) + (1500\text{x}633) + (2500\text{x}527)}{2010}$$

$$+\frac{(3500\text{x}151) + (4500\text{x}123) + (5500\text{x}85)}{2010}$$

$$+\frac{(6500\text{x}58) + (7500\text{x}39) + (8500\text{x}8)}{2010}$$

$$+\frac{(9500\text{x}6)}{2010}$$

$$=\frac{4801000}{2010}$$

$$= \$\ 2{,}388.55$$

Let's calculate the new median:

$$M_e = a_i + (a_{i+1} - a_i)\frac{(50 - F_i)}{(F_{i+1} - F_i)}$$

$$= 1{,}000 + (2{,}000 - 1{,}000)\frac{(50 - 18.91)}{(50.40 - 18.91)}$$

$$= \$\ 1{,}987.3$$

INTERPRETATION OF THE RESULTS

The analysis reveals that the average salary has significantly dropped from $8,573.69 to $2,388, while the median salary has slightly changed, going from $1,991.4 to $1,987.30. This indicates that the mean is greatly influenced by outliers, whereas the median is more robust and unaffected by extreme values. As the two central values have significantly converged, either measure can be used interchangeably.

4.8 Exercises and Solutions

4.8.1 Chapter exercises

• **Exercise 13** The table beside presents a table showing, for a given period, the number of people who have recovered from a viral pandemic in a country as well as the time spent in the hospital (in days) before total recovery is confirmed and pronounced by medical professionals.

Time spent in hospital (Classes)	Number of healings (n_i)
[0;5)	12711
[5;20)	3512
[20;40)	13024
[40;60)	52646
[60;70)	65665
[70;100)	136051
Total	283609

a. Plot the histogram of the distribution.

b. Plot the ascending and descending cumulative relative frequency curves of the number of recoveries as a function of time spent in the hospital before total recovery.

c. Determine the modal time for patients in the hospital before total recovery.

d. Calculate the average time spent in the hospital before total recovery.

e. Calculate the median time spent in the hospital before total recovery.

f. How many people spent less than a month in the hospital before total recovery?

Solution

a: Let's plot the histogram of the distribution:

Since the class intervals are not all of the same magnitude, we will consider the smallest interval as the unit interval with an amplitude of 5. This means that the amplitudes of all classes will be divided by 5 to make them comparable.

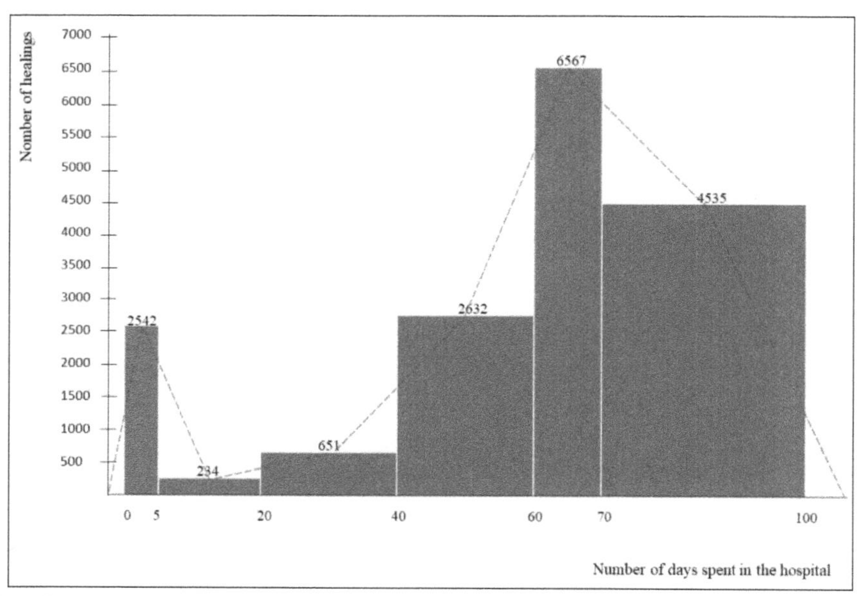

b: The following table shows the data required to solve this exercise:

Classes	x_i	n_i	Density	$n_i x_i$	f_i	$f_i \uparrow$	$f_i \downarrow$
[0;5)	2.5	12711	2542.2	31777.5	4.48	4.48	100.00
[5;20)	12.5	3512	234.1	43900	1.24	5.72	95.52
[20;40)	30	13024	651.2	390720	4.59	10.31	94.28
[40;60)	50	52646	2632.3	2632300	18.56	28.88	89.69
[60;70)	65	65665	6566.5	4268225	23.15	52.03	71.12
[70;100)	85	136051	4535.0	11564335	47.97	100.00	47.97
Total		283609		18931257.5	100.00		

The curves of the ascending and descending cumulative frequencies are as follows:

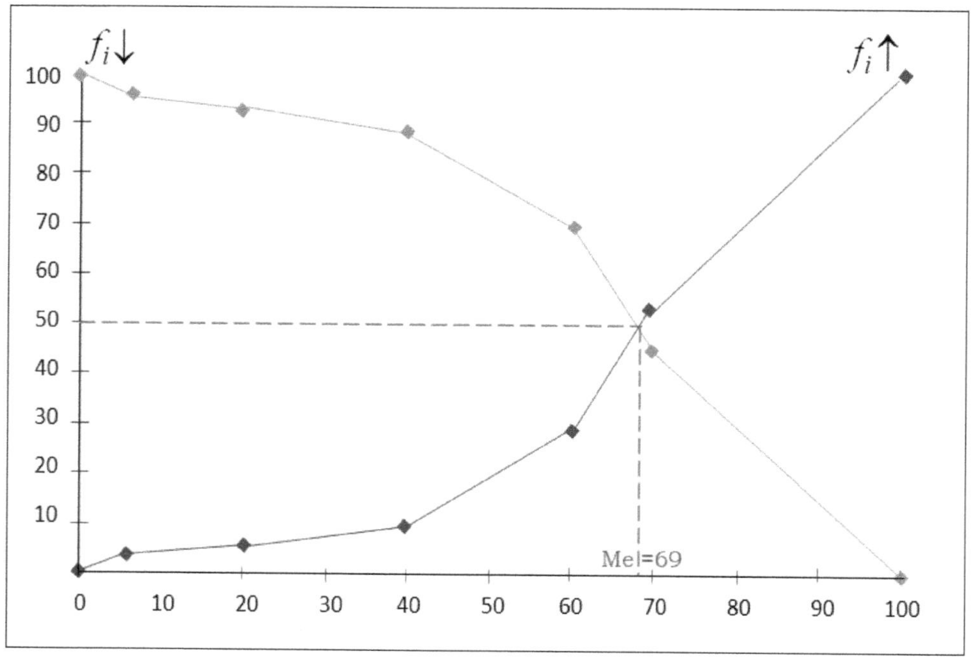

c: The modal class being [70 ;100); thus, the modal time of patients in the hospital before recovery is 85 days.

d: The average time spent in the hospital before full recovery is:

$$\bar{x} = \frac{1}{n}\sum_{i=1}^{i=6} n_i x_i = \frac{18{,}931{,}257.5}{283{,}609}$$

$$= 66.75 \text{ days}$$

$$= 66 \text{ days and } 18 \text{ hours}$$

e: The median time spent in the hospital before full recovery is:

$$M_e = a_i + (a_{i+1} - a_i)\frac{(50 - F_i)}{(F_{i+1} - F_i)}$$

$$= 60 + (70 - 60)\frac{(50 - 28.88)}{(52.03 - 28.88)}$$

$$= 69.12311 \text{ jours}$$

$$= 69 \text{ days, 2hours and 57 minutes}$$

f: Let's calculate the number of people who spent less than a month (30 days) in hospital before full recovery.

First, calculate the percentage of these people, then convert it to a number.

Let $P_{patient}$ be the percentage to find, and we use linear interpolation formula on $(a_i, f_i \uparrow)$:

In order to determine the number of patients who left the hospital less than a month after admission and were completely recovered, we need to follow two steps. First, we need to calculate the percentage of these patients, and then we need to convert this percentage into the actual number of patients.

To do this, we will use the same linear interpolation method as presented above, using the data pair $(a_i, f_i \uparrow)$ as the basis. Let's $P_{patient}$ denote the percentage to be calculated. We will use the same linear interpolation method as presented above, using the corresponding data pair $(a_i, f_i \uparrow)$ as the basis:

$$30 = a_i + (a_{i+1} - a_i)\frac{(P_{patient} - F_i)}{(F_{i+1} - F_i)}$$

$$30 = 20 + (40 - 20)\frac{(P_{patient} - 5.72)}{(10.31 - 5.72)}$$

$$30 - 20 = 20\frac{(P_{patient} - 5.72)}{(10.31 - 5.72)}$$

$$\frac{10}{20} = \frac{(P_{patient} - 5.72)}{(10.31 - 5.72)}$$

$$0.5 = \frac{(P_{patient} - 5.72)}{4.59}$$

$$P_{malade} = 5.72 + 4.59 \text{ x } 0.5$$

We get:

$$P_{patient} = 8.015\%$$

The percentage of people who spent less than a month (30 days) in the hospital before full healing is 8.015%, which represents 22,732 people.

INTERPRETATION OF THE RESULT

c. The majority of patients stayed in the hospital for approximately 85 days (between 70 and 100 days) before recovery.

d. Each patient spent an average of 66 days and 18 hours in the hospital before full recovery.

e. There were as many patients who recovered in less than 69 days, 2 hours, and 57 minutes as there were patients who recovered after that time.

f. Among the recovered individuals, 22,732 left the hospital less than a month (30 days) after admission, which represents 8.015% of the total number of recoveries.

• **Exercise 14** The vehicle fleet of a company consists of a significant number of cars. For 100 of them, the mileage recorded at the time of their maintenance is given in the table beside:

Trip Mileage (in km)	Number of cars
[80000;85000)	5
[85000;90000)	9
[90000;95000)	14
[95000;100000)	18
[100000;105000)	25
[105000;110000)	16
[110000;115000)	7
[115000;120000)	6
Total	100

a. Plot the histogram of the distribution.

b. Plot the cumulative frequency curve of the number of cars based on the mileage covered.

c. Determine the modal mileage.

d. Determine the average mileage.

e. Determine the median mileage.

f. Calculate the percentage of cars with mileage less than 93,000 km at the time of their maintenance.

Solution

a: Histogram of the distribution

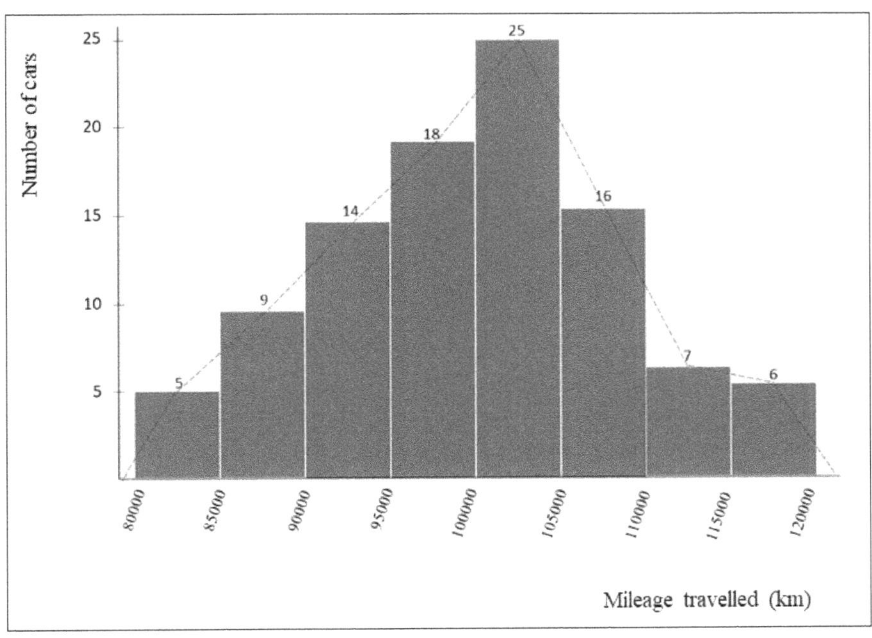

b: The following table shows the data required to solve this exercise:

Classes	x_i	n_i	$n_i x_i$	$f_i \uparrow$	$f_i \downarrow$
[80000;85000)	82500	5	412500	5.00	100.00
[85000;90000)	87500	9	787500	14.00	95.00
[90000;95000)	92500	14	1295000	28.00	86.00
[95000;100000)	97500	18	1755000	46.00	72.00
[100000;105000)	102500	25	2562500	71.00	54.00
[105000;110000)	107500	16	1720000	87.00	29.00
[110000;115000)	112500	7	787500	94.00	13.00
[115000;120000)	117500	6	705000	100.00	6.00
Total		100	10025000		

The curves of the ascending and descending cumulative frequencies are as follows:

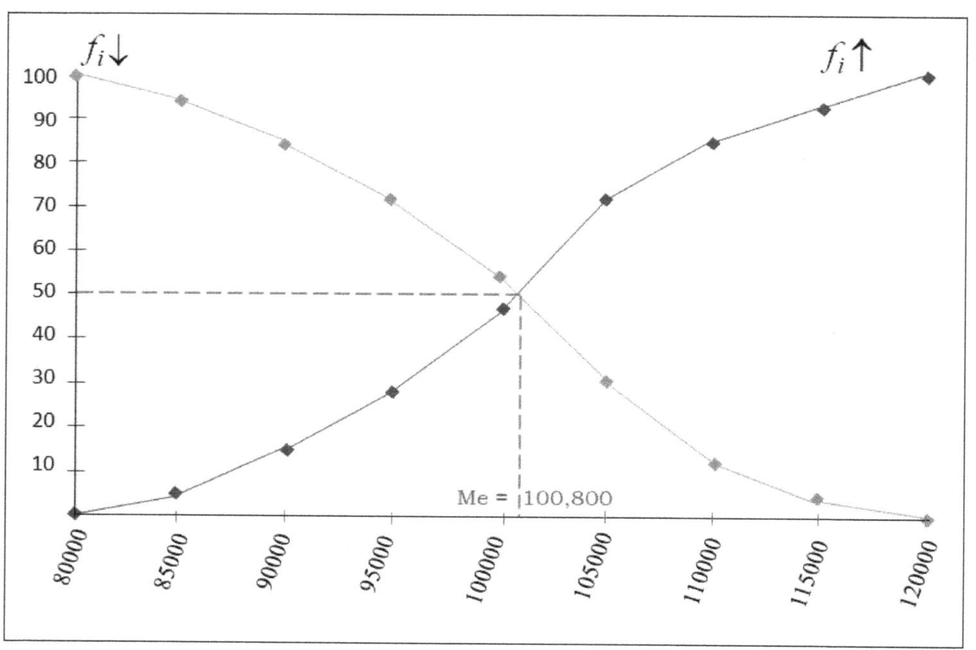

c: The modal mileage is:

$$\text{Mode} = 102{,}500 \text{ Km}$$

d: The average mileage is:

$$\bar{x} = \frac{1}{n}\sum_{i=1}^{i=8} n_i x_i = \frac{10{,}025{,}000}{100}$$

$$= 100{,}250 \text{ Km}$$

e: The median mileage is:

$$M_e = a_i + (a_{i+1} - a_i)\frac{(50 - F_i)}{(F_{i+1} - F_i)}$$

$$= 100{,}000 + (105{,}000 - 100{,}000)\frac{(50 - 46)}{(71 - 46)}$$

$$= 100{,}800 \text{ Km}$$

f: Let's calculate the percentage of cars that have covered a distance of less than 93,000 Km at the time of maintenance.

To do this, let's P_{93000} denote the percentage to be calculated. We will use the same linear interpolation method presented in this chapter, using the corresponding data pair $(a_i, f_i \uparrow)$ as the basis:

$$93{,}000 = a_i + (a_{i+1} - a_i)\frac{(P_{93000} - f_i)}{(f_{i+1} - f_i)}$$

$$93{,}000 = 90{,}000 + (95{,}000 - 90{,}000)\frac{(P_{93000} - 14)}{(28 - 14)}$$

That is:

$$P_{93000} = 22.4\%$$

The percentage of cars having traveled a distance of less than 93,000 km at the time of mechanical service is 22.4%, which represents 22 cars.

INTERPRETATION OF THE RESULT

During the mechanical service, the following observations were made about the cars:

c. Most of the cars had a total mileage of 102,500 km.

d. Assuming equal driving and delivery service, each car should have had 100,250 km at the time of mechanical service.

e. The number of cars with less than 100,800 km was equal to the number of cars with more than 100,800 km.

f. 22.4% of the cars had less than 93,000 km on the odometer at the time of service, while 77.6% had more mileage.

• **Exercise 15** A cyclist performs a route. During 25 Km he drives at a speed $v_1 = 25$ Km/h, during 15 Km he drives at a speed $v_2 = 20$ Km/h, and during 10 Km, he drives at a speed $v_3 = 40$ Km/h.

a. Determine the arithmetic mean of the three speeds.

b. Determine the geometric mean of the three speeds.

c. Determine the harmonic mean of the three speeds.

d. Determine the quadratic mean of the three speeds.

e. Among all these averages, which represents the average speed of the cyclist on the course?

Solution

We have

$$d_1 = 25 \text{ Km} \qquad v_1 = 25 \text{ Km/h}$$

$$d_2 = 15 \text{ Km} \qquad v_2 = 20 \text{ Km/h}$$

$$d_3 = 10 \text{ Km} \qquad v_3 = 40 \text{ Km/h}$$

a: The arithmetic mean of the three speeds is calculated by adding the three speeds and dividing by the number of speeds:

$$v_m = \frac{v_1 + v_2 + v_3}{3}$$

$$= \frac{25 + 20 + 40}{3} = 28.33 \text{ Km/h}$$

b: The geometric mean of the three speeds is calculated by multiplying the three speeds and taking the cubic root of the result:

$$v_g = \sqrt[3]{v_1 \text{ x } v_2 \text{ x } v_3}$$

$$= \sqrt[3]{25\text{x}20\text{x}40}$$

$$= 27.14 \text{ Km/h}$$

c: The harmonic mean of the three speeds is calculated by taking the inverse of the sum of the inverses of the three speeds and then dividing it by the number of speeds:

$$v_h = \frac{n}{\frac{1}{v_1} + \frac{1}{v_2} + \frac{1}{v_3}}$$

$$= \frac{3}{\frac{1}{20} + \frac{1}{25} + \frac{1}{40}}$$

$$= \frac{600}{23}$$

$$=26.087 \text{ Km/h}$$

d: The quadratic mean of the three speeds is calculated by taking the square root of the arithmetic mean of the squares of each speed:

$$v_q = \sqrt{\frac{v_1^2 + v_2^2 + v_3^2}{3}}$$

$$= \sqrt{\frac{20^2 + 25^2 + 40^2}{3}}$$

$$= \sqrt{\frac{2{,}625}{3}}$$

$$= 29.58 \text{ Km/h}$$

e: None of these speeds represent the average speed of the cyclist.

$d_1 = 25\ Km$	$t_1 = 1\ h$	$v_1 = 25\ Km/h$
$d_2 = 15\ Km$	$t_2 = 0.75\ h$	$v_2 = 20\ Km/h$
$d_3 = 10\ Km$	$t_3 = 0.25\ h$	$v_3 = 40\ Km/h$
$d_{total} = 25 + 15 + 10 = 50\ Km/h$	$t_{total} = 1 + 0.75 + 0.25 = 2\ h$	$v_{total} = \frac{50}{2} = 25 \text{ Km/h}$

INTERPRETATION OF THE RESULT

None of these speeds represents the average speed of the cyclist because:

- The arithmetic mean of the three speeds is just the sum of the three speeds divided by three; it cannot then be interpreted as the average speed of the cyclist. We will therefore say that the average of the three speeds is different from the average speed of the cyclist.
- The geometric mean of the three speeds is the cubic root of the product of the three speeds cannot, therefore, be interpreted as the average speed of the cyclist either.
- The harmonic mean of the speeds which is the inverse of the arithmetic mean of the inverses of the speeds can in no way be interpreted as the average speed of the cyclist.
- The quadratic mean of the three speeds, which is the square root of the mean of the squared speeds, cannot in any case be interpreted as the average speed of the cyclist.

From a kinematic point of view, the average speed of the cyclist is the total of the distances covered divided by the total of the times taken to cover them.

In this example, the cyclist's average speed is:

$$v_m = \frac{\sum \text{distances traveled}}{\sum \text{Time to cover the distances}} = \frac{50 \text{ Km}}{2\text{hours}} = 25 \text{ Km/h}$$

Therefore, it is important to carefully consider which type of average is appropriate for a given situation, as each measure has its own unique properties and limitations.

4.8.2 Cross-Chapter Exercises

- **Exercise 16:** Let's continue with exercise **11** from section **3.4.2**, which deals with the production data of a tire factory over a 15-day period.

Number of Tires	n_i
[12500;13000)	1
[13000;13500)	2
[13500;14000)	3
[14000;14500)	3
[14500;15000)	3
[15000;15500)	3
Total	15

a. Plot the ascendant and descendant cumulative relative frequency graphs of the number of days as a function of the number of tires produced.

b. Determine the mode of distribution.

c. Determine the average number of tires produced per day.

d. Determine the median of the distribution.

Solution

b: The following table shows the data required to solve this exercise:

Number of racks	x_i	n_i	$n_i x_i$	f_i	$f_i\uparrow$	$f_i\downarrow$
[12500;13000)	12750	1	12750	6.67	6.67	100.00
[13000;13500)	13250	2	26500	13.33	20.00	93.33
[13500;14000)	13750	3	41250	20.00	40.00	80.00
[14000;14500)	14250	3	42750	20.00	60.00	60.00
[14500;15000)	14750	3	44250	20.00	80.00	40.00
[15000;15500)	15250	3	45750	20.00	100.00	20.00
Total		15	213250	100		

a: The curves of the ascending and descending cumulative frequencies are as follows:

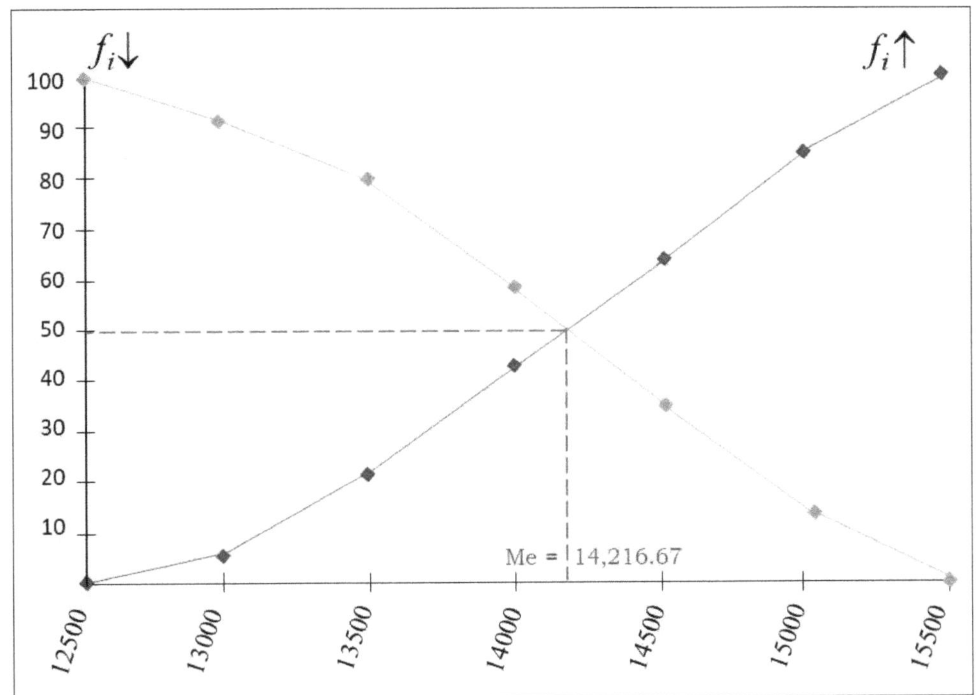

b: This statistical series is multimodal, the class centers 13750, 14250, 14750 and 15250 are modes and classes [13500 ; 14000), [14000 ; 14500), [14500 ; 15000) and [15000 ; 15500) are modal classes.

c: The average number of tires produced per day is:

$$\bar{x} = \frac{1}{n}\sum_{i=1}^{i=6} n_i x_i = \frac{213{,}250}{15}$$

$$= 14{,}216.67 \text{ tires}$$

d: The median production is:

$$M_e = a_i + (a_{i+1} - a_i)\frac{(50 - F_i)}{(F_{i+1} - F_i)}$$

$$= 14{,}000 + (14{,}500 - 14{,}000)\frac{(50 - 40)}{(60 - 40)}$$

$$= 14{,}250 \text{ tires}$$

INTERPRETATION OF THE RESULT

b. This statistical series is multimodal, indicating that it has multiple modes, i.e., multiple values that repeat with a high frequency. The values of daily tire production show that there are several levels of production that occur frequently. This indicates that the tire factory's production can vary significantly from day to day, with production peaks at different values. This could be due to variations in production processes, market demand fluctuations, or other factors.

c. If the daily production were consistent and evenly distributed over the studied period, it would be an average of 14,216.67 tires per day. However, the data shows that the actual production of the factory fluctuates around this value, indicating that there are days with higher or lower production than this theoretical value.

d. There are as many days when the factory produces less than 14,250 tires as there are days when it produces more, indicating significant variability in daily production. This could be due to various factors such as market demand, variations in production processes, or other internal or external factors that influence the factory's performance. There are an equal number of days when the tire factory produces less than 14,250 tires as there are days when it produces more, indicating significant variability in daily production. This means that tire production can fluctuate around the median value of 14,250 tires, with an equal number of days where it is lower or higher than this value.

• **Exercise 17** Let's continue **exercise 12** from **section 3.4.2**, which concerns the savings accounts of 300 individuals (in thousands of euros).

Saving Balance	n_i	$n_i \uparrow$	$n_i \downarrow$	f_i	$f_i \uparrow$	$f_i \downarrow$
[0;5000)	48	48	300	16.00	16.00	100.00
[5000;10000)	41	89	252	13.67	29.67	84.00
[10000;15000)	47	136	211	15.67	45.33	70.33
[15000;20000)	15	151	164	5.00	50.33	54.67
[20000;25000)	21	172	149	7.00	57.33	49.67
[25000;30000)	12	184	128	4.00	61.33	42.67
[30000;35000)	13	197	116	4.33	65.67	38.67
[35000;40000)	8	205	103	2.67	68.33	34.33
[40000;45000)	9	214	95	3.00	71.33	31.67
[45000;50000)	9	223	86	3.00	74.33	28.67
[50000;55000)	10	233	77	3.33	77.67	25.67
[55000;60000)	6	239	67	2.00	79.67	22.33
[60000;65000)	9	248	61	3.00	82.67	20.33
[65000;70000)	5	253	52	1.67	84.33	17.33
[70000;75000)	2	255	47	0.67	85.00	15.67
[75000;80000)	13	268	45	4.33	89.33	15.00
[80000;85000)	8	276	32	2.67	92.00	10.67
[85000;90000)	7	283	24	2.33	94.33	8.00
[90000;95000)	4	287	17	1.33	95.67	5.67
[95000;100000)	13	300	13	4.33	100.00	4.33
Total	300			100.00		

a. Plot the curves of ascending and descending cumulative frequencies of the number of savers according to the amount of savings.

b. Determine the modal saving the distribution.

c. Determine the average amount of savings for each individual.

d. Determine the median amount of the distribution.

Solution

The following table shows the data required to solve this exercise:

Classes	x_i	n_i	$n_i xi$	f_i	$f_i \uparrow$
[0;5000)	2500	48	120000	16.00	16.00
[5000;10000)	7500	41	307500	13.67	29.67
[10000;15000)	12500	47	587500	15.67	45.33
[15000;20000)	17500	15	262500	5.00	50.33
[20000;25000)	22500	21	472500	7.00	57.33
[25000;30000)	27500	12	330000	4.00	61.33
[30000;35000)	32500	13	422500	4.33	65.67
[35000;40000)	37500	8	300000	2.67	68.33
[40000;45000)	42500	9	382500	3.00	71.33
[45000;50000)	47500	9	427500	3.00	74.33
[50000;55000)	52500	10	525000	3.33	77.67
[55000;60000)	57500	6	345000	2.00	79.67
[60000;65000)	62500	9	562500	3.00	82.67
[65000;70000)	67500	5	337500	1.67	84.33
[70000;75000)	72500	2	145000	0.67	85.00
[75000;80000)	77500	13	1007500	4.33	89.33
[80000;85000)	82500	8	660000	2.67	92.00
[85000;90000)	87500	7	612500	2.33	94.33
[90000;95000)	92500	4	370000	1.33	95.67
[95000;100000)	97500	13	1267500	4.33	100.00
Total		300	9445000	100.00	

a: The curves of the ascending and descending cumulative frequencies are as follows:

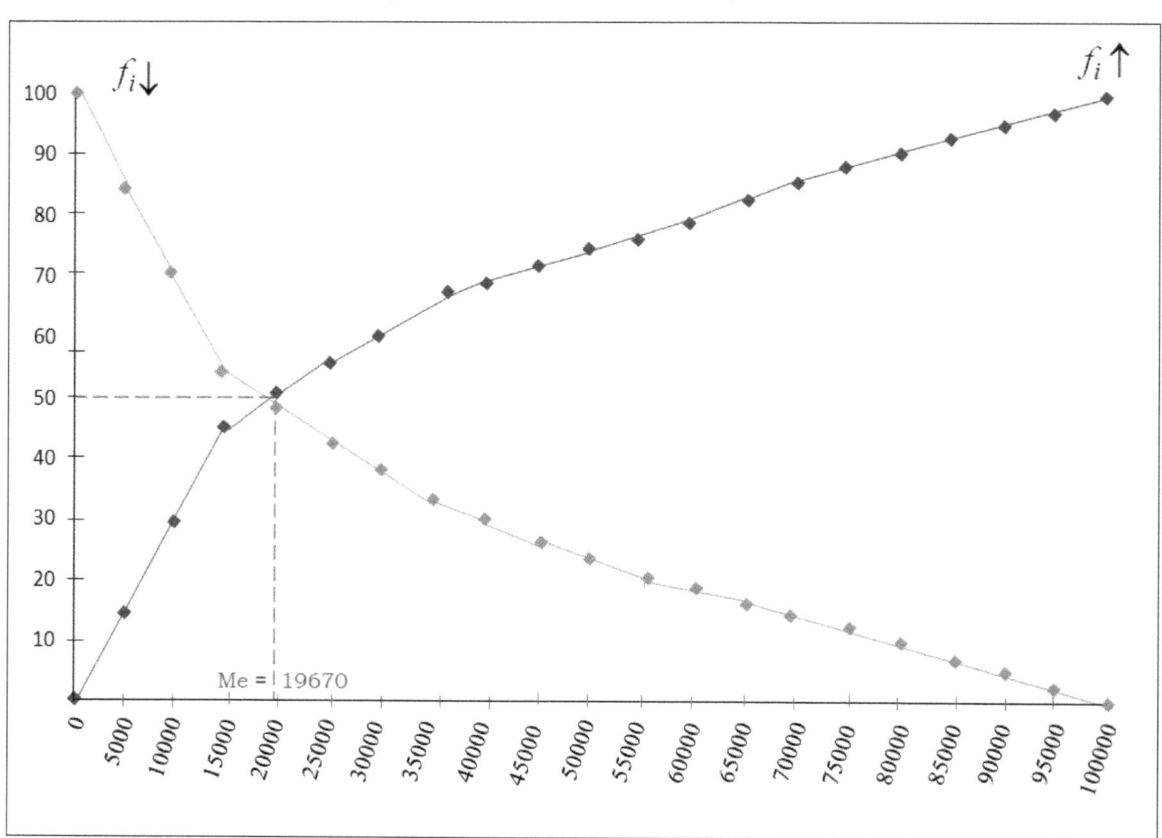

b: The mode of the statistical series is 2500 and the modal class is [0 ; 5000).

c: The average amount of savings for each individual is:

$$\bar{x} = \frac{1}{n}\sum_{i=1}^{i=20} x_i = \frac{9{,}445{,}000}{300}$$

$$= 31.483 \text{ euros}$$

d: The median amount of savings is:

$$M_e = a_i + (a_{i+1} - a_i)\frac{(50 - F_i)}{(F_{i+1} - F_i)}$$

$$= 15{,}000 + (20{,}000 - 15{,}000)\frac{(50 - 45.33)}{(50.33 - 45.33)}$$

$$= 19{,}670 \text{ euros}$$

INTERPRETATION OF THE RESULT

b. The modal value of the savings of individuals in this distribution is 2,500 euros, indicating that most people have around 2,500 euros in their savings account.
c. If the total amount of money saved (9,445,000 euros) were evenly distributed among all the savers, each individual would have an average of 31,483 euros in their savings account.
d. The median of this distribution is 19,670 euros, which means that there are an equal number of people with savings below 19,670 euros as there are people with savings above this value.

CHAPTER 5

Characteristics of Dispersion

This chapter discusses the dispersion characteristics of a statistical distribution and the role of related parameters such as variance, standard deviation, and coefficient of variation in measuring dispersion. These parameters provide valuable information about how spread out a series is around its mean. By understanding dispersion, we can get a better picture of the variability and diversity of data points in a distribution. The variance measures the average distance between each data point and the mean, while the standard deviation provides a measure of the typical distance between a data point and the mean. The coefficient of variation is another useful measure that standardizes the dispersion by expressing the standard deviation as a percentage of the mean. By analyzing dispersion characteristics, we can gain insights into the behavior and variability of a data set, allowing us to make more informed decisions and draw meaningful conclusions.

5.1 Introduction

Describing a statistical series correctly requires more than just using measures of central tendency. It is indeed possible for two series to have the same mode, median, and mean. However, the elements of one series may appear tightly grouped around their mean, while in the other series, the elements may be more dispersed around their mean. The study of this dispersion is the role of dispersion parameters, which define how values deviate from the mean or not.

5.2 Distribution Range

The ***range*** of a statistical series is defined as the difference between the two most extreme values of the variable being studied. It is actually the difference between the maximum value (i.e., the highest observed value) and the minimum value (i.e., the lowest observed value) of the variable. However, its use is limited as these extreme values may be outliers, thus affecting its representativeness.

In the case of a distribution with continuous statistical variables and equal class intervals, the class centers can be used to determine the range as the distance between the two values is preserved.

It is important to note that the range is often used in manufacturing quality control. However, its main weakness lies in the fact that it is not very informative due to the lack of information about the distribution of the variable between these two extreme values.

$$\boxed{\text{Range}(x) = \text{Max}(x) - \text{Min}(x)}$$

Example 44

Let's look back at **example 42** from **section 4.6.2.2** on the workers in a company based on their age in which x_i represent the ages of workers and the n_i represent the number of employees:

Classes	x_i	n_i
[20;25)	22.5	36
[25;30)	27.5	45
[30;35)	32.5	27
[35;40)	37.5	18
[40;45)	42.5	10
[45;50)	47.5	8
[50;55)	52.5	3
[55;60)	57.5	3
Total		150

The range of the distribution is:

$$57.5 - 22.5 = 35 \text{ years}$$

INTERPRETATION OF THE RESULT

The age difference between the youngest employee and the oldest employee in this company is 35 years. This information can be used to provide an indication of the dispersion of data in the age distribution of workers in this company, but it does not provide information about the distribution of ages between these two extreme values. It would be necessary to use other statistical measures to obtain a more comprehensive understanding of the age distribution of workers in this company.

5.3 Quantiles

In an ordered statistical population, meaning arranged in ascending (or descending) order, ***quantiles*** are called discriminating values that divide this population into n ordered subsets of equal size. Depending on the value of n, quantiles are given different names:

- If $n = 2$, they are named median.
- If $n = 4$, they are named quartiles.
- If $n = 5$, they are named quintiles.
- If $n = 10$, they are named deciles.
- If $n = 100$, they are named percentiles.
- etc.

5.3.1 Quartiles and Deciles

When the population is arranged in ascending (or descending) order and subdivided into four equal parts, four ***quartiles*** are obtained, and when subdivided into ten equal parts, ten ***deciles*** are obtained. Each quartile or decile is indexed and identified based on its rank in the subdivision.

Thus, we have the ***first quartile***, denoted as Q_1, as the value of the characteristic such that the count of values less than Q_1 represents the first quarter of the total count, and we call the ***third quartile***, denoted as Q_3, as the value of the characteristic such that the count of values less than Q_3 represents three-quarters of the total count.

Note that the median is indeed the second quartile, allowing us to conclude:

$$Q_1 < Q_2 = M_e < Q_3$$

In an analogous way, we can define the deciles.

Example 45

Considering the same example discussed previously in **section 5.2** on the statistical distribution of the workers of a company based on their age:

Classes	x_i	n_i	$f_i(\%)$	$f_i(\%)\uparrow$	$f_i(\%)\downarrow$
[20;25)	22.5	36	24.00	24.00	100.00
[25;30)	27.5	45	30.00	54.00	76.00
[30;35)	32.5	27	18.00	72.00	46.00
[35;40)	37.5	18	12.00	84.00	28.00
[40;45)	42.5	10	6.67	90.67	16.00
[45;50)	47.5	8	5.33	96.00	9.33
[50;55)	52.5	3	2.00	98.00	4.00
[55;60)	57.5	3	2.00	100.00	2.00
Total		150	100.00		

To calculate quartiles and deciles, we use the linear interpolation formula as presented in the previous chapter, using the data pair $(a_i, f_i \uparrow)$ with the simple difference that the values of the percentages are replaced by those of the quartiles or deciles.

Finding the first quartile (25% of the population):

$$Q_1 = a_i + (a_{i+1} - a_i)\frac{(25 - F_i)}{(F_{i+1} - F_i)}$$

$$= 25 + (30 - 25)\frac{(25 - 24)}{(54 - 24)} = 25.17 \text{ years}$$

$$= 25 \text{ years and 2 months}$$

Finding the third quartile (75% of the population):

$$Q_3 = a_i + (a_{i+1} - a_i)\frac{(75 - F_i)}{(F_{i+1} - F_i)}$$

$$= 35 + (40 - 35)\frac{(75 - 72)}{(84 - 72)} = 36.25 \text{ years}$$

$$= 36 \text{ years and 3 months}$$

Finding the ninth decile (90% of the population):

$$D_9 = a_i + (a_{i+1} - a_i)\frac{(90 - F_i)}{(F_{i+1} - F_i)}$$

$$= 40 + (45 - 40)\frac{(90 - 84)}{(90.67 - 84)} = 44.498 \text{ years}$$

$$= 44 \text{ years, 5 months and 29 days}$$

INTERPRETATION OF THE RESULT

In this company:

- 25% of employees are below 25 years and 2 months, while 75% of employees are above 25 years and 2 months.
- 75% of employees are below 36 years and 3 months, while 25% of employees are above 36 years and 3 months.
- 90% of employees are below 44 years, 5 months, and 29 days, while 10% of employees are above 44 years, 5 months, and 29 days.

Example 46 - Effect of outliers

Let's revisit **example 43** from **section 4.7.1**, which discusses an insurance company's decision to settle on a remote island with a population of only 2,000.

This distribution contains outliers:

Classes	x_i	n_i	$n_i x_i$	f_i	$f_i \uparrow$
[0;1000)	500	380	190000	18.86	18.86
[1000;2000)	1500	633	949500	31.41	50.27
[2000;3000)	2500	527	1317500	26.15	76.43
[3000;4000)	3500	151	528500	7.49	83.92
[4000;5000)	4500	123	553500	6.10	90.02
[5000;6000)	5500	85	467500	4.22	94.24
[6000;7000)	6500	58	377000	2.88	97.12
[7000;8000)	7500	39	292500	1.94	99.06
[8000;9000)	8500	8	68000	0.40	99.45
[9000;10000)	9500	6	57000	0.30	99.75
[10000;4000000)	2495000	5	12475000	0.25	100.00
Total		2015	17276000	100.00	

Let's calculate the first quartile (25% of the population):

$$Q_1 = a_i + (a_{i+1} - a_i)\frac{(25 - F_i)}{(F_{i+1} - F_i)}$$

$$= 1{,}000 + (2{,}000 - 1{,}000)\frac{(25 - 18.86)}{(50.27 - 18.86)}$$

$$= 1{,}195.48\$$$

Let's calculate the third quartile (75% of the population):

$$Q_3 = a_i + (a_{i+1} - a_i)\frac{(75 - F_i)}{(F_{i+1} - F_i)}$$

$$= 2{,}000 + (3{,}000 - 2{,}000)\frac{(75 - 50.27)}{(76.43 - 50.27)}$$

$$= 2{,}945.33\$$$

Let's calculate the ninth decile (90% of the population):

$$D_9 = a_i + (a_{i+1} - a_i)\frac{(90 - F_i)}{(F_{i+1} - F_i)}$$

$$= 4{,}000 + (5{,}000 - 4{,}000)\frac{(90 - 83.92)}{(90.02 - 83.92)}$$

$$= 4{,}996.72\ \$$$

<u>INTERPRETATION OF THE RESULT</u>

It is noteworthy that the gap between the mean calculated in section 4.7, which was $8,573.69, and Q3 evaluated at $2,945.33 is still excessively large. This implies that 75% of the island's workers have salaries lower than half of the mean salary.

In other words, 75% of the island's workers will still not be able to afford this insurance if the calculations are based on this mean.

Furthermore, it is observed that D9 is $4,996.72, indicating that approximately 10% of the island's population has a salary higher than half of the mean salary.

Similarly, if the calculations are based on this mean, 90% of the island's workers will still be unable to afford this insurance.

Now let's evaluate the percentage of people with a salary higher than the average. For this, we will first calculate the percentage of people with salaries below the mean, which we will then subtract from 100%.

The percentage of people with a salary less than $ 8,573.69:

Let $P_{<8573.69}$ be the percentage to find; we use the linear interpolation formula using the data pair $(a_i, f_i \uparrow)$ as proceeded above:

$$8{,}573.69 = a_i + (a_{i+1} - a_i)\frac{(P_{<8573.69} - F_i)}{(F_{i+1} - F_i)}$$

$$8{,}573.69 = 8{,}000 + (9{,}000 - 8{,}000)\frac{(P_{<8573.69} - 99.06)}{(99.45 - 99.06)}$$

$$P_{<8573.69} = 99.28\%$$

The percentage of people earning less than $ 8,573.69 is 99.28%. We deduce that the percentage of people with a salary greater than $8,573.69 is 100% - 99.28% = 0.72%.

INTERPRETATION OF THE RESULT

The average salary with outliers included ($8,573.69) is significantly higher than what the majority of people on the island earn. In fact,

- 99.28% of people on the island earn less than this average.
- Only 0.72% of people have a salary above this average, which represents just 15 individuals out of a sample size of 2015.

The above data interpreted shows the importance of removing outliers from statistical distributions to avoid distorted interpretations and make more accurate conclusions.

After eliminating outliers, we obtain the following distribution:

Classes	x_i	n_i	$n_i x_i$	f_i	$f_i \uparrow$
[0;1000)	500	380	190000	18.91	18.91
[1000;2000)	1500	633	949500	31.49	50.40
[2000;3000)	2500	527	1317500	26.22	76.62
[3000;4000)	3500	151	528500	7.51	84.13
[4000;5000)	4500	123	553500	6.12	90.25
[5000;6000)	5500	85	467500	4.23	94.48
[6000;7000)	6500	58	377000	2.89	97.36
[7000;8000)	7500	39	292500	1.94	99.30
[8000;9000)	8500	8	68000	0.40	99.70
[9000;10000)	9500	6	57000	0.30	100.00
Total		2010	4801000	100.00	

Let's calculate the first quartile (25% of the population):

$$Q_1 = a_i + (a_{i+1} - a_i)\frac{(25 - F_i)}{(F_{i+1} - F_i)}$$

$$= 1{,}000 + (2{,}000 - 1{,}000)\frac{(25 - 18.91)}{(50.40 - 18.91)}$$

$$= 1{,}193.39\ \$$$

Let's calculate the first quartile (75% of the population):

$$Q_3 = a_i + (a_{i+1} - a_i)\frac{(75 - F_i)}{(F_{i+1} - F_i)}$$

$$= 2{,}000 + (3{,}000 - 2{,}000)\frac{(75 - 50.40)}{(76.62 - 50.40)}$$

$$= 2{,}938.21\ \$$$

Let's calculate the first quartile (90% of the population):

$$D_9 = a_i + (a_{i+1} - a_i)\frac{(90 - F_i)}{(F_{i+1} - F_i)}$$
$$= 4{,}000 + (5{,}000 - 4{,}000)\frac{(90 - 84.13)}{(90.25 - 84.13)}$$
$$= 4{,}959.15\ \$$$

INTERPRETATION OF THE RESULT

The previous observation regarding the example with outliers still holds true. This is because quantiles and deciles are not significantly affected by outliers, as they represent specific percentages of the data distribution and are not influenced by extreme values.

Now let's evaluate the percentage of people with a salary higher than the average. For this, we will first calculate the percentage of people with salaries lower than the average, which we will then subtract from 100%.

Percentage of people with a salary less than \$ 2,388.55:

Let $P_{<2388}$ be the percentage to find; we use the linear interpolation formula using the data pair $(a_i, f_i \uparrow)$:

$$2{,}388.5 = a_i + (a_{i+1} - a_i)\frac{(P_{<2388} - F_i)}{(F_{i+1} - F_i)}$$
$$2{,}388.5 = 2{,}000 + (3{,}000 - 2{,}000)\frac{(P_{<2388} - 50.27)}{(76.43 - 50.27)}$$

$$P_{<2388} = 60.43\%$$

The percentage of people earning less than \$ 2,388.55 is 60.43%. As a result, it can be deduced that the percentage of people with a salary greater than \$ 2,388.55 is 100% – 60.43% = 39.57%.

INTERPRETATION OF THE RESULT

The average without outliers is \$2,388.5.
In this island,

- 60.43% of people on the island have a salary lower than this average.
- 39.57% of people have a salary higher than this average, which represents 80 individuals out of 2010.

As mentioned above, it is important to remove outliers from the statistical distribution to avoid biased interpretations.

5.3.2 Interquartile Range

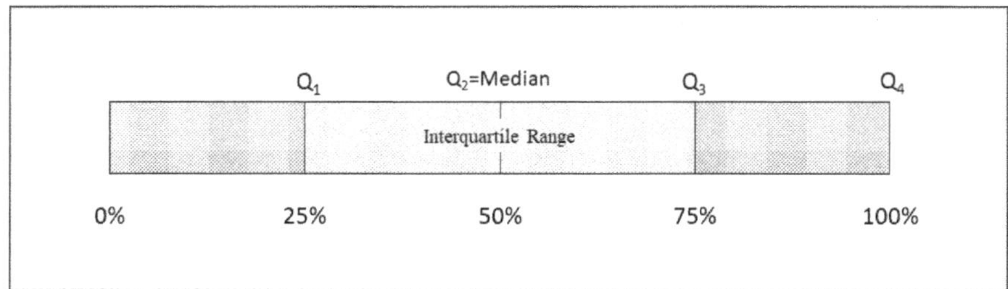

The ***interquartile range***, denoted as $Q_3 - Q_1$, is defined as the difference between the first quartile (Q_1) and the third quartile (Q_3) of a set of data. It provides a measure of the spread of values around the median and gives information about the variability of a data set.

This quantity corresponds to studying a statistical series with a sample size that is half of the initial series and centered around the median. By adopting this parameter, we eliminate the first and last quartiles of the series, excluding any "outlier" values and focusing solely on the central half of the distribution. This allows us to highlight the extent of dispersion within the median half of the series.

Example 47

Let's consider **example 41** from section **4.6.2.1**, which deals with the distribution of the budget of subsidies allocated by the state to companies in an American city, expressed in millions of dollars. In this example, x_i represents the subsidy amounts in millions of dollars, and n_i represents the number of beneficiary businesses.

Subventions	x_i	n_i	f_i	$f_i \uparrow$
[0.2 ; 0.6)	0.4	4	3.96	3.96
[0.6 ; 1)	0.8	5	4.95	8.91
[1 ; 2)	1.5	9	8.91	17.82
[2 ; 3)	2.5	7	6.93	24.75
[3 ; 5)	4	13	12.87	37.62
[5 ; 8)	6.5	17	16.83	54.46
[8 ; 10)	9	12	11.88	66.34
[10 ; 20)	15	21	20.79	87.13
[20 ; 30)	25	13	12.87	100.00
Total		101	100.00	

$$Q_1 = 3 + (5 - 3)\frac{(25 - 24.75)}{(37.62 - 24.75)} = 3.03885$$

$$Q_3 = 10 + (20 - 10)\frac{(75 - 66.34)}{(87.13 - 66.34)} = 14.16546$$

Interquartile range equals:

$$Q_3 - Q_1 = 14.16546 - 3.03885$$
$$= \$\,11.12661 \text{ million}$$

INTERPRETATION OF THE RESULT
The results show that the dispersion of subsidies allocated by the state to businesses in an American city, expressed in millions of dollars, in the median half of the series is significant, with a difference of 11.12661 million dollars between the two furthest apart subsidies. This indicates that the subsidy amounts vary considerably among the beneficiary businesses, which may have an impact on the fairness of the subsidy distribution.

5.4 Variance

In order to fully grasp the concept of variance, we will use a real-life case study as an example.

The data collected by the radar installed at the entrance of the urban area provide important insights into the driving behavior of local drivers. The recorded instantaneous speeds vary greatly, ranging from 30 km/h to 220 km/h, with an average speed of 80 km/h, as shown below:

$$45 - 30 - 220 - 60 - 50 - 40 - 35 - 55 - 65 - 200$$

This information helps to paint a picture of the traffic situation in this urban area.

It is worth noting that the average speed of 80 km/h can serve as an indicator to assess compliance with speed limits and road safety in this urban zone. If the average speed recorded had been significantly higher, it could have indicated a widespread disregard for speed limits, which might require additional measures to ensure the safety of road users. On the other hand, if the average speed was much lower, it could indicate strict adherence to speed limits but also slower and congested traffic, which might require improvements to the road infrastructure.

In order to assess whether drivers generally comply with speed limits or if they tend to drive at extreme speeds (too fast or too slow) compared to the average, the authorities wish to measure the dispersion of this speed data series in relation to the average of 80 km/h. This will provide a better understanding of the distribution and variability of driving behaviors among drivers in this urban area. This can be useful for making informed decisions on road safety and traffic management. By understanding the variability of driving behaviors among drivers, appropriate actions can be taken to raise awareness among drivers, strengthen road safety measures, or improve road infrastructure in this urban area, if needed.

To do this, it is common practice to calculate the differences between each value and the mean and then take their average. However, as we have seen in ***section 4.5.2***, *the sum of the differences from the mean is always zero, regardless of the series considered*, which does not provide us with any useful information. To address this, we can instead consider the differences in absolute value and calculate their average, which will give us the ***mean absolute deviation***.

Example 48

Let's consider the following series of speed measurements of cars with a total distance of 800 km and an average speed of 80 km/h.

											Total	Mean
n_i	45	30	220	60	50	40	35	55	65	200	800	80
$\lvert n_i - 80\rvert$	35	50	140	20	30	40	45	25	15	120	520	52

To assess the dispersion of this series, the mean absolute deviation is calculated, resulting in a value of 52 km/h. It is important to note that absolute values are not ideal for algebraic calculations, as they do not adequately account for large deviations that can increase dispersion.

In practice, to measure the dispersion of a data series, variance is commonly calculated, which is the average of the squared deviations from the mean. Variance is also known as fluctuation and is given by the following formula:

$$V(X) = \frac{\sum_{i=1}^{i=k} n_i(x_i - \bar{x})^2}{\sum_{i=1}^{i=k} n_i} = \sum_{i=1}^{i=k} f_i(x_i - \bar{x})^2$$

$$= \sum_{i=1}^{i=k} f_i x_i^2 - \bar{x}^2$$

Hence

$$\boxed{V(X) = \frac{1}{n}\sum_{i=1}^{i=k} n_i x_i^2 - \bar{x}^2}$$

The higher the variance, the more dispersed the data points are, indicating that they deviate more from their mean.

Example 49

Let's consider **example 45** from **section 5.3.1,** which concerns the statistical distribution of workers in a company based on their age:

Classes	x_i	n_i	$n_i x_i$	$n_i x_i^2$	$(x_i - \bar{x})$	$n_i(x_i - \bar{x})^2$
[20;25)	22.5	36	810.00	18,225.00	-9.07	2958.92
[25;30)	27.5	45	1,237.50	34,031.25	-4.07	743.96
[30;35)	32.5	27	877.50	28,518.75	0.93	23.55
[35;40)	37.5	18	675.00	25,312.50	5.93	633.82
[40;45)	42.5	10	425.00	18,062.50	10.93	1195.52
[45;50)	47.5	8	380.00	18,050.00	15.93	2031.14
[50;55)	52.5	3	157.50	8,268.75	20.93	1314.70
[55;60)	57.5	3	172.50	9,918.75	25.93	2017.72
Total		150	4,735.00	160,387.50	67.47	10,919.33

The mean of this distribution has already been calculated in section **4.4.4**.

$$\bar{x} = 31.566 \text{ ans}$$
$$= 31 \text{ years } 6 \text{ months and } 23 \text{ days}$$

Let's calculate the variance by the first formula:

$$V(x) = \frac{1}{n}\sum_{i=1}^{i=8} n_i x_i^2 - \bar{x}^2 = \frac{[36(22.5)^2] + [45(27.5)^2] + [27(32.5)^2]}{150}$$
$$+ \frac{[18(37.5)^2] + [10(42.5)^2] + [8(47.5)^2]}{150}$$
$$+ \frac{[3(52.5)^2] + [3(57.5)^2]}{150} - (31.566\,)^2$$
$$= \frac{160{,}387.50}{150} - (31.566\,)^2$$
$$= 72.80$$

Let's calculate the variance by the second formula:

$$V(x) = \frac{\sum_{i=1}^{i=8} n_i (x_i - \bar{x})^2}{\sum_{i=1}^{i=8} n_i} = \frac{[36(22.5 - 31.566)^2] + [45(27.5 - 31.566)^2]}{150}$$
$$+ \frac{[27(32.5 - 31.566)^2] + [18(37.5 - 31.566)^2]}{150}$$

$$+\frac{[10(42.5-31.566)^2]+[8(47.5-31.566)^2]}{150}$$

$$+\frac{[3(52.5-31.566)^2]+[3(57.5-31.566)^2]}{150}$$

$$=\frac{10{,}919.33}{150}$$

$$=72.80$$

INTERPRETATION OF THE RESULT

The variance of 72.80 indicates the degree of dispersion of data around the mean. Auch a high variance implies that the ages of the employees are spread over a wide range, with significant deviations from the mean of 31 years, 6 months, and 23 days. This may suggest a significant diversity of ages among the employees in the company, with some employees being younger and others being older than the mean.

Property

Just like the mean, the variance is well-suited for changes in origin and scale. In fact,

$$V(x)=\frac{1}{n}\sum_{i=1}^{i=k}n_i(x_i-\bar{x})^2$$

By setting:

$$y_i=\frac{x_i-x_0}{a}$$

Then

$$\bar{y}=\frac{\bar{x}-x_0}{a}$$

Where x_0 is a constant and *a* is a scaling factor, the variance *V(y)* of the transformed data $y_i$can be expressed in terms of the variance *V(x)* of the original data x_i

Using this transformation, we obtain the following equation:

$$\bar{y}=\frac{\bar{x}-x_0}{a}$$

$$V(x) = \frac{1}{n}\sum_{i=1}^{i=k} n_i(y_i - \bar{y})^2$$

$$= \frac{1}{n}\sum_{i=1}^{i=k} n_i \left[\left(\frac{x_i - x_0}{a}\right) - \left(\frac{\bar{x} - x_0}{a}\right)\right]^2$$

$$= \frac{1}{a^2}\frac{1}{n}\sum_{i=1}^{i=k} n_i(x_i - \bar{x})^2$$

Now, we know that:

$$V(x) = \frac{1}{n}\sum_{i=1}^{i=k} n_i(x_i - \bar{x})^2$$

Thus, we can conclude that:

$$V(x) = a^2 V(y)$$

As was the case for the average, in the exercises x_0 will represent the mode and a will be the amplitude of the modalities.

By applying this property to the previous example, where x_0 represents the mode and a is the amplitude of the modalities, we can calculate the variance of the data series.

Classes	x_i	n_i	$y_i = \frac{x_i - 27.5}{5}$	$n_i y_i$	$n_i y_i^2$
[20;25)	22.5	36	-1	-36	36.00
[25;30)	27.5	45	0	0	0.00
[30;35)	32.5	27	1	27	27.00
[35;40)	37.5	18	2	36	72.00
[40;45)	42.5	10	3	30	90.00
[45;50)	47.5	8	4	32	128.00
[50;55)	52.5	3	5	15	75.00
[55;60)	57.5	3	6	18	108.00
Total		150		122	536.00

This average has already been calculated in section **4.4.4**.

$$\bar{y} = \frac{122}{150} = 0.813$$

Finding of the variance of this distribution with respect to the change of origin and scale is:

$$V(y) = \frac{1}{n}\sum_{i=1}^{i=8} n_i y_i^2 - \bar{y}^2 = \frac{[36(-1)^2] + [45(0)^2] + [27(1)^2]}{150}$$

$$+\frac{[18(2)^2] + [10(31)^2] + [8(4)^2]}{150}$$

$$+\frac{[3(5)^2] + [3(6)^2]}{150} - (0.813\,)^2$$

$$= \frac{536}{150} - (0.813\,)^2$$

$$= 2.911$$

It follows that the variance of this distribution is:

$$V(x) = a^2 V(y) = a^2 \text{x } 2.911$$
$$= 72.8$$

5.5 Standard Deviation (σ)

The standard deviation is defined as a statistical measure that indicates the dispersion or variability of data in a sample relative to their mean, i.e., the difference between each data point and the mean. It is defined as the square root of the variance.

It is expressed by the formula:

$$\sigma_x = \sqrt{\frac{\sum_{i=1}^{i=n} n_i (x_i - \bar{x})^2}{\sum_{i=1}^{i=n} n_i}} = \sqrt{\sum_{i=1}^{i=n} f_i (x_i - \bar{x})^2}$$

$$= \sqrt{\sum_{i=1}^{i=n} f_i x_i^2 - \bar{x}^2}$$

Hence:

$$\sigma_x = \sqrt{\frac{1}{n}\sum_{i=1}^{i=n} n_i x_i^2 - \bar{x}^2}$$

Where n_i represents the number of occurrences of each value in the sample, x_i is the value of each observation, $\bar{x}$ is the mean of the sample and $\sum$ symbolizes the sum of values for i ranging from 1 to n.

The higher the standard deviation, the more dispersed the data points are around their mean, indicating that the values in the sample are further apart from each other. On the other hand, a low standard deviation indicates that the values are more closely clustered around the mean, indicating less dispersion.

The standard deviation can be useful in various fields such as finance, science, economics, and other areas of research and data analysis.

Example 50

Let's consider **example 49** from **section 5.4** which concerns the statistical distribution of workers in a company based on their age:

$$\sigma_x = \sqrt{\frac{1}{n}\sum_{i=1}^{i=n} n_i x_i^2 - \bar{x}^2} = \sqrt{\frac{160{,}387.5}{150} - (31.566)^2}$$

$$= \sqrt{1{,}069.25 - 996.41}$$
$$= 8.53 \text{ years}$$

INTERPRETATION OF THE RESULTS

The square root of the sum of the squares of the deviations of the ages of the employees from the mean age of the company (31 years 6 months and 23 days) is 8.53 years. This value represents the degree of dispersion of the data around the mean.

Example 51

Looking back at **example 47** from **section 5.3.2,** which deals with the distribution of the budget of subsidies allocated by the state to companies in an American city, expressed in millions of dollars. In this example, x_i represents the subsidy amounts in millions of dollars, and n_i represents the number of beneficiary businesses.

Classes	x_i	n_i	$n_i x_i$	$n_i x_i^2$
[0.2 ; 0.6)	0.4	4	1.6	0.64
[0.6 ; 1)	0.8	5	4	3.2
[1 ; 2)	1.5	9	13.5	20.25
[2 ; 3)	2.5	7	17.5	43.75
[3 ; 5)	4	13	52	208
[5 ; 8)	6.5	17	110.5	718.25
[8 ; 10)	9	12	108	972
[10 ; 20)	15	21	315	4725
[20 ;30)	25	13	325	8125
Total		101	947.1	14816.09

The average of this distribution has already been calculated in section **4.4.1**:

$$\bar{x} = 9.37$$

Finding the variance:

$$V(x) = \frac{1}{n}\sum_{i=1}^{i=9} n_i x_i^2 - \bar{x}^2 = \frac{[4(0.4)^2] + [5(0.8)^2] + [9(1.5)^2]}{101}$$

$$+\frac{[7(2.5)^2] + [13(4)^2] + [17(6.5)^2] + [12(9)^2]}{101}$$

$$+\frac{[21(15)^2] + [13(25)^2]}{101} - (9.37\,)^2$$

$$= \frac{14{,}816.09}{101} - (9.37\,)^2$$

$$= 58.76396$$

Finding standard deviation:

$$\sigma = \sqrt{V(x)}$$

$$= \sqrt{58.76396}$$

$$= \$\ 7.66 \text{ millions}$$

INTERPRETATION OF THE RESULTS

The standard deviation of 7.66 million dollars, which is the square root of the sum of the squares of the deviations of the subsidies from the average subsidy of 9.37 million dollars, represents the degree of dispersion of the data. This indicates that the subsidies allocated to the businesses in this American city vary on average by approximately 7.66 million dollars, around the mean of 9.37 million dollars.

5.6 Refining SHEPPARD's Coefficient

We have seen that variance measures the dispersion of data around their mean. However, when using class midpoints to calculate the variance in the case of discrete or grouped data, a slight error may be introduced as the class midpoints do not exactly represent the actual values of the data. The ***SHEPPARD coefficient*** allows for this error to be corrected by adjusting the variance formula. The corrected formula for variance, using the SHEPPARD coefficient, is as follows:

$$\sigma^2_{\text{corrected}} = \sigma^2 - \frac{a^2}{12}$$

Where $\sigma^2_{\text{corrected}}$ represents the corrected variance, σ^2 represents the variance calculated without correction, a represents the class width (i.e., the difference between the upper and lower limits of a class), and 12 is a fixed constant.

This correction allows for a more accurate estimation of the variance by minimizing the error introduced by using class midpoints.

It is important to note that the SHEPPARD coefficient is typically used when the class width is relatively large compared to the dispersion of the data. If the class width is small compared to the dispersion of the data, the effect of correcting with the SHEPPARD coefficient may be negligible.

Example 52

Let's consider **example 50** from **section 5.5,** which concerns the statistical distribution of workers in a company based on their age:

$$\sigma^2_{\text{corrected}} = \sigma^2 - \frac{a^2}{12}$$

$$= (8.532)^2 - \frac{25}{12}$$

$$= 70.7167$$

Thus

$$\sigma_{\text{corrected}} = \sqrt{70.7167}$$
$$= 8.4 \text{ years}$$

5.7 Coefficient of variation

The ***coefficient of variation*** (CV) is a dimensionless quantity that allows for assessing the degree of homogeneity of a statistical distribution. It is defined as the ratio of the standard deviation (σ) to the mean ($|\bar{x}|$), expressed as a percentage:

$$CV = \frac{\sigma}{|\bar{x}|} \text{x} 100$$

The coefficient of variation is used to compare the relative dispersion of data in statistical distributions with different means. The lower the coefficient of variation, the more homogeneous the data is and the less variation there is between the values. Conversely, the higher the coefficient of variation, the more dispersed the data is and the more variation there is between the values.

As for the rules of interpretation based on the value of the coefficient of variation, there is no strict baseline rule. The interpretation of the coefficient of variation (CV) depends on the specific context of the statistical application and the nature of the data being studied. However, in general, a low coefficient of variation indicates a low relative dispersion of data, while a high value of a coefficient of variation indicates a relatively larger relative dispersion of data.

In practice, the CV is often used to assess the stability or reliability of measurements or data, particularly in fields such as scientific research, finance, medicine, economics or quality management etc. Since each industry may have its own specific thresholds for evaluating the quality of a product or process, it is essential to consider the specific context of the statistical application and the nature of the data being studied and to use the coefficient of variation in conjunction with other methods of analysis to obtain a comprehensive understanding of data variability.

Nevertheless, as a general rule, the following baseline rule can be applied, which works when the data and context are not extremely specific.

RULE OF THUMB

$CV = 0\%$	:	the distribution is homogeneous
$0\% < CV < 15\%$	:	the distribution is a little dispersed (quasi-homogeneous)
$15\% \leq CV < 50\%$	:	the distribution is moderately dispersed
$50\% \leq CV < 100\%$	:	the distribution is dispersed
$CV \geq 100\%$	:	the distribution is very dispersed

Example 53

Considering the same **example 52** discussed previously in **section 5.6** on the statistical distribution of the workers of a company according to their age:

The coefficient of variation of this distribution is:

$$CV = \frac{\sigma}{|\bar{x}|} \text{x}100 = \frac{8.53}{31.566} \text{x}100$$

$$= 27\%$$

INTERPRETATION OF THE RESULTS

When observing the coefficient of variation of 27% for the statistical distribution of workers in a company based on their age, it means that this distribution has a relative variation of 27% compared to the mean. A coefficient of variation value below 30% is generally considered moderate, indicating that the distribution is relatively homogeneous, and the values are moderately dispersed around the mean. This may suggest that the ages of the workers are evenly distributed across the different age classes without significant variations.

Example 54

Looking back at **example 51 of section 5.5,** which deals with the distribution of the budget of subsidies allocated by the state to companies in an American city, expressed in millions of dollars.

The coefficient of variation of this distribution is:

$$CV = \frac{\sigma}{|\bar{x}|} \text{x}100 = \frac{7.66}{9.37} \text{x}100$$

$$= 80\%$$

INTERPRETATION OF THE RESULTS

When observing the coefficient of variation of 80% for the distribution of budget allocations in millions of dollars as subsidies from the state to businesses in an American city, it means that this distribution has a high relative variation of 80% compared to the mean. A high coefficient of variation value indicates significant dispersion of subsidy amounts awarded to businesses. This may suggest that the allocated amounts vary considerably among businesses, with some businesses receiving much larger amounts than others.

Example 55 - Effect of outliers

Let's reconsider **example 46** from **section 5.3.1**, which discusses an insurance company's decision to settle on a remote island with a population of barely 2,000 inhabitants.

Let's use the distribution with the outliers.

x_i	n_i	$n_i x_i$	$n_i x_i^2$
500	380	190000	95000000
1500	633	949500	1424250000
2500	527	1317500	3293750000
3500	151	528500	1849750000
4500	123	553500	2490750000
5500	85	467500	2571250000
6500	58	377000	2450500000
7500	39	292500	2193750000
8500	8	68000	578000000
9500	6	57000	541500000
2495000	5	12475000	31125125000000
	2015	17276000	31142613500000

$$\sigma = \sqrt{\frac{1}{n}\sum_{i=1}^{i=n} n_i x_{i2} - \bar{x}^2} = \sqrt{\frac{31,142,613,500,000}{2,015} - (8,573.69)^2}$$

$$= \sqrt{15,455,391,315.13 - 73,508,160.2}$$

$$= 124,023.72 \ \$$$

The coefficient of variation of the distribution is:

$$CV = \frac{\sigma}{|\bar{x}|} x100 = \frac{124,023.72}{8,573.69} x100$$

$$= 1,446\%$$

INTERPRETATION OF THE RESULTS

In this example, the coefficient of variation reaches a high percentage of 1,446%, which is primarily due to the presence of five outlier values in the data series. The corresponding n_i values for these outlier values are relatively low compared to the other values in the distribution, resulting in excessive data dispersion. In other words, these five atypical values have a significant impact on the variation of the statistical distribution, which explains why the coefficient of variation is so high. This indicates that the data series is highly heterogeneous, with extreme values that have a considerable influence on the overall data dispersion.

Now let's use the distribution after the removal of outliers:

x_i	n_i	$n_i x_i$	$n_i x_i^2$
500	380	190000	95000000
1500	633	949500	1424250000
2500	527	1317500	3293750000
3500	151	528500	1849750000
4500	123	553500	2490750000
5500	85	467500	2571250000
6500	58	377000	2450500000
7500	39	292500	2193750000
8500	8	68000	578000000
9500	6	57000	541500000
	2010	4801000	17488500000

$$\sigma_x = \sqrt{\frac{1}{n}\sum_{i=1}^{i=n} n_i x_{i^2} - \bar{x}^2} = \sqrt{\frac{17{,}488{,}500{,}000}{2{,}010} - (2{,}388\ )^2}$$

$$= \sqrt{8{,}700{,}746.26 - 5{,}702{,}544}$$

$$= 1{,}731.53\ \$$$

The coefficient of variation of the distribution is:

$$CV = \frac{\sigma}{|\bar{x}|}\text{x}100 = \frac{1{,}731.53}{2{,}388}\text{x}100$$

$$= 72.5\%$$

INTERPRETATION OF THE RESULTS

The coefficient of variation shows a significant decrease from 1446% to 72.5% after removing outliers. This indicates that the dispersion of the series is now within an acceptable range.

5.8 Normal Distribution

The normal distribution, also known as the Gaussian curve or bell-shaped curve, is a widely used statistical distribution to model many real-world phenomena. It is often used to represent quantitative characteristics such as IQ scores, heart rates, weights, heights, contest results, and many others.

The Gaussian curve has several distinctive characteristics. Firstly, it is symmetric, and its axis of symmetry passes through the mean. This means that the mean, mode, and median of the distribution coincide, making it a perfectly balanced distribution. Furthermore, the Gaussian curve is also referred to as a bell-shaped curve due to its characteristic shape that resembles an inverted bell.

The empirical rules of the normal distribution are also very useful for understanding the distribution of data. According to these rules:

- Approximately 68.27% of observations fall within the interval [mean - standard deviation, mean + standard deviation], which means that the majority of the data is located close to the mean.

- Approximately 95.45% of observations fall within the interval [mean - 2 standard deviations, mean + 2 standard deviations].

- Nearly 99.73% of observations fall within the interval [mean - 3 standard deviations, mean + 3 standard deviations]. This shows how the normal distribution is concentrated around the mean, with fewer values as we move away from the mean.

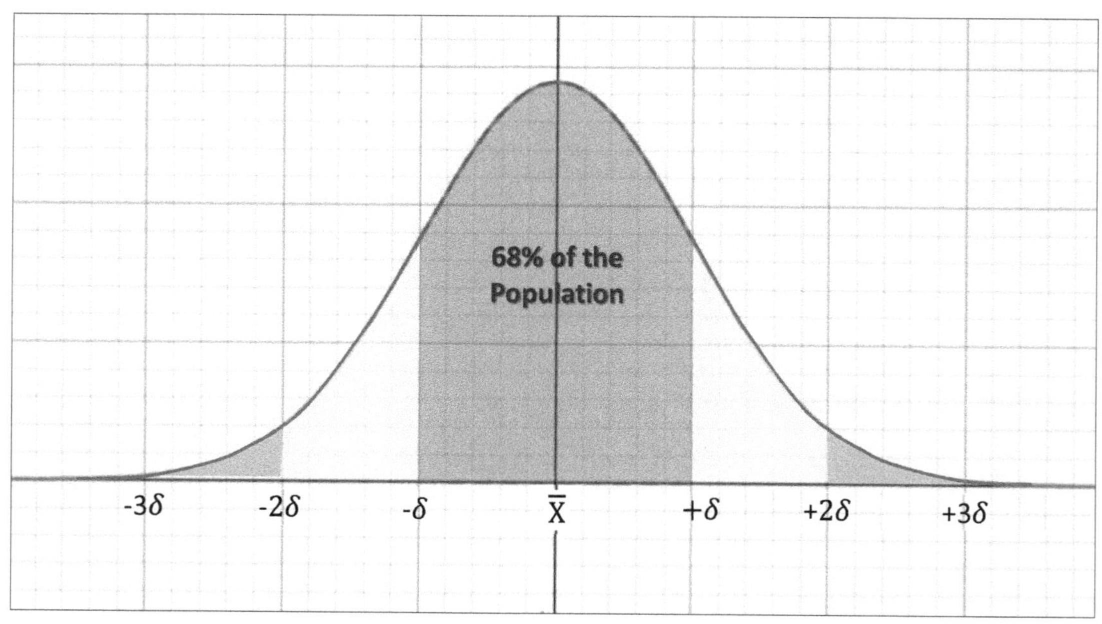

The bell curve shows a strong concentration of values around the mean, then fewer and fewer values at the ends of the series.

Another important aspect of the normal distribution is the use of standard deviations to measure the dispersion of data. The standard deviation is a measure of the dispersion of values around the mean.

A large standard deviation indicates a greater dispersion of data, which can result in a wider (spread out) and less tall (flatter) distribution compared to the bell curve of the normal distribution. In other words, when the standard deviation is large, the data may be spread over a wider range around the mean, resulting in a more spread out and less concentrated distribution.

On the other hand, a small standard deviation indicates a lesser dispersion of data, which can result in a narrower (more concentrated) distribution around the mean, but it does not necessarily guarantee that the distribution will be taller compared to the bell curve of the normal distribution. The height of the distribution depends on several other factors, such as sample size, the shape of the distribution, etc.

Example 56

Let's take **example 53,** as discussed previously in **section 5.7**, regarding the statistical distribution of workers in a company based on their age:

Now, let's see how to calculate the percentage of workers whose age falls between the mean minus the standard deviation and the mean plus the standard deviation.

Classes	x_i	n_i	f_i	$f_i \uparrow$
[20;25)	22.5	36	24.00	24.00
[25;30)	27.5	45	30.00	54.00
[30;35)	32.5	27	18.00	72.00
[35;40)	37.5	18	12.00	84.00
[40;45)	42.5	10	6.67	90.67
[45;50)	47.5	8	5.33	96.00
[50;55)	52.5	3	2.00	98.00
[55;60)	57.5	3	2.00	100.00
Total		150	100.00	

Solution

The values of the mean and the standard deviation have already been calculated in section **5.5**, with respective values of 31.566 and 8.53.

Let's calculate the percentage of workers whose age is between the mean minus the standard deviation and the mean plus the standard deviation.

To calculate the percentage of workers whose age falls between the mean minus the standard deviation and the mean plus the standard deviation, we will use the values we have already obtained. We subtract the standard deviation from the mean to get the lower limit and add the standard deviation to the mean to get the upper limit. Then, we calculate the percentage of workers whose age is below the lower limit, as well as the percentage of workers whose age is above the upper limit. Finally, we will subtract these two values to obtain the percentage of workers whose age falls between the mean minus the standard deviation and the mean plus the standard deviation.

The mean plus or minus the standard deviation is:

$$\bar{x} - \sigma = 31.566 - 8.53$$
$$= 23.036 \text{ years}$$

$$\bar{x} + \sigma = 31.566 + 8.53$$
$$= 40.096 \text{ years}$$

Let's calculate the percentage of workers whose age is less than $\bar{x} - \sigma$, i.e., less than 23.036 years

Let $P_{<\bar{x}-\sigma}$ be the percentage to find; we use the linear interpolation formula using the data pair $(a_i, f_i \uparrow)$:

$$\bar{x} - \sigma = a_i + (a_{i+1} - a_i)\frac{(P_{<\bar{x}-\sigma} - F_i)}{(F_{i+1} - F_i)}$$
$$23.036 = 20 + (25 - 20)\frac{(P_{<\bar{x}-\sigma} - 0)}{(24 - 0)}$$

$$P_{<\bar{x}-\sigma} = 14.57\%$$

The percentage of workers whose age is less than 23.036 years is equal to 14.57%

Let's calculate the percentage of workers whose age is less than $\bar{x} + \sigma$, i.e., less than 40.096 years.

Let $P_{<\bar{x}+\sigma}$ be the percentage to find; we use the linear interpolation formula using the data pair $(a_i, f_i \uparrow)$ as proceeded above:

$$\bar{x} + \sigma = a_i + (a_{i+1} - a_i)\frac{(P_{<\bar{x}+\sigma} - F_i)}{(F_{i+1} - F_i)}$$

$$40.096 = 40 + (45 - 40)\frac{(P_{<\bar{x}+\sigma} - 84)}{(90.67 - 84)}$$

$$P_{<\bar{x}+\sigma} = 84.12\%$$

The percentage of workers whose age is less than 40.096 years is equal to 84.12%

The percentage of workers whose age is between $\bar{x} - \sigma$ (23.036 years) and $\bar{x} + \sigma$ (40.09 years) is:

$$86.016 - 18.24 = 67.78\%$$

Which represents 102 employees out of 150.

INTERPRETATION OF THE RESULTS

69.55% of workers fall within the age range of 23 years minus 13 days $(\bar{x} - \sigma)$ and 40 years 1 month and 4 days $(\bar{x} + \sigma)$. In the interval $[\bar{x} - \sigma ; \bar{x} + \sigma]$, it is observed that this distribution is slightly more populated than the normal distribution, with 69.55% compared to 68.27% of the population.

Example 57

Let's consider **example 54** from section **5.7**, which deals with the distribution of the budget of subsidies allocated by the state to companies in an American city, expressed in millions of dollars. In this example, x_i represents the subsidy amounts in millions of dollars, and n_i represents the number of beneficiary businesses.

Determine the percentage of firms that received state subsidies between the mean minus the standard deviation and the mean plus the standard deviation.

Subventions	x_i	n_i	f_i	$f_i \uparrow$
[0.2 ; 0.6)	0.4	4	3.96	3.96
[0.6 ; 1)	0.8	5	4.95	8.91
[1 ; 2)	1.5	9	8.91	17.82
[2 ; 3)	2.5	7	6.93	24.75
[3 ; 5)	4	13	12.87	37.62
[5 ; 8)	6.5	17	16.83	54.46
[8 ; 10)	9	12	11.88	66.34
[10 ; 20)	15	21	20.79	87.13
[20 ; 30)	25	13	12.87	100.00
Total		101	100.00	

Solution

The mean and standard deviation have already been calculated in **section 5.5.** The mean is $9,377, and the standard deviation is $7.66 million. Now we will calculate the percentage of companies that have received a subsidy between the mean minus the standard deviation and the mean plus the standard deviation.

The mean plus or minus the standard deviation is:

$$\bar{x} - \sigma = 9.377 \; - \; 7.66$$
$$= \$\; 1.717 \text{ million}$$

$$\bar{x} + \sigma = 9.377 + 7.66$$
$$= \$\; 17.037 \text{ million}$$

Let's calculate the percentage of companies that received state subsidies less than $\bar{x} - \sigma$, i.e., less than $ 1.717 million.

Let $P_{<\bar{x}-\sigma}$ be the percentage to find; we use the linear interpolation formula using the data pair $(a_i, f_i \uparrow)$:

$$\bar{x} - \sigma = a_i + (a_{i+1} - a_i)\frac{(P_{<\bar{x}-\sigma} - F_i)}{(F_{i+1} - F_i)}$$
$$1.717 \; = 1 + (2 - 1)\frac{(P_{<\bar{x}-\sigma} - 8.91)}{(17.82 - 8.91)}$$

$$P_{<\bar{x}-\sigma} = 15.3\%$$

The percentage of companies that received state subsidies of less than 1.717 million is equal to 15.3%

Let's calculate the percentage of companies that received state subsidies less than $\bar{x} + \sigma$, i.e., less than \$ 17.037 million.

Let $P_{<\bar{x}+\sigma}$ be the percentage to find; we use the linear interpolation formula using the data pair $(a_i, f_i \uparrow)$:

$$\bar{x} + \sigma = a_i + (a_{i+1} - a_i)\frac{(P_{<\bar{x}+\sigma} - F_i)}{(F_{i+1} - F_i)}$$

$$17.037 = 10 + (20 - 10)\frac{(P_{<\bar{x}+\sigma} - 66.34)}{(87.13 - 66.34)}$$

$$P_{<\bar{x}+\sigma} = 80.97\%$$

The percentage of companies that received state subsidies of less than 17.037 million is equal to 80.97%

The percentage of companies that received state subsidies between $\bar{x} - \sigma$ = \$ 1.717 million and $\bar{x} + \sigma$ = \$ 17.037 million is:

80.97% −15.3% = 65.67%

Which represents 66 companies out of 101.

INTERPRETATION OF THE RESULTS

It has been observed that 65.67% of companies have received a grant amounting between 1.717 million dollars (corresponding to $\bar{x} - \sigma$) and 17.037 million dollars (corresponding to $\bar{x} + \sigma$). This range of intervals represents the proportion of beneficiary companies in our sample.

However, when comparing this distribution with that of a normal distribution, it appears that this range of intervals is less populated. According to the normal distribution, it would be expected that 68.27% of the population falls within this range of intervals. However, our sample shows that only 65.67% of companies have received a grant within this range of intervals, indicating a deviation from the normal distribution in terms of distribution.

Example 58 - Effects of outliers

Let's reconsider **example 56** from the same section, which discusses an insurance company's decision to settle on a remote island with a population of barely 2,000 inhabitants.

Determine the percentage of people with a salary between the mean minus the standard deviation and the mean plus the standard deviation.

Let's calculate the distribution with outliers:

The mean plus or minus the standard deviation is:

$$\bar{x} - \sigma = 8{,}573.69 \; - \; 124{,}023.72$$
$$= - 115{,}450.03 \, \$$$
$$\bar{x} + \sigma = 8{,}573.69 + 124{,}023.72$$
$$= 132{,}597.41\$$$

Since the value of $\bar{x} - \sigma,$ is negative; it is not possible to calculate the percentage of individuals with salaries between the mean minus the standard deviation and the mean plus the standard deviation. This would involve including negative values, which is not applicable in this context.

INTERPRETATION OF THE RESULTS

The presence of outliers has had a significant impact on the statistical results, particularly on the value of the mean minus the standard deviation. Indeed, this value has become negative due to the atypical values that are far from the mean. As a result, it is not possible to calculate the percentage of individuals with salaries between the mean minus the standard deviation and the mean plus the standard deviation, as it would require including negative values, which is not consistent in the context of this study.

Now let's do the calculations on the distribution after eliminating outliers:

The mean plus or minus the standard deviation is:

$$\bar{x} - \sigma = 2{,}388.55 - 1{,}731.53$$
$$= 657.02 \, \$$$
$$\bar{x} + \sigma = 8{,}573.69 + 124{,}023.72$$
$$= 4{,}120.08\$$$

Let’s calculate the percentage of persons with a salary below $\bar{x} - \sigma,$ i.e., less than 657.02 \$.

Let $P_{<\bar{x}-\sigma}$ be this percentage; we use the linear interpolation formula using the data pair $(a_i, f_i \uparrow)$:

$$\bar{x} - \sigma = a_i + (a_{i+1} - a_i)\frac{(P_{<\bar{x}-\sigma} - F_i)}{(F_{i+1} - F_i)}$$
$$657.02 \; = 0 + (1{,}000 - 0)\frac{(P_{<\bar{x}-\sigma} - 0)}{(18.91 - 0)}$$

$$P_{<\bar{x}-\sigma} = 12.42\%$$

The percentage of people earning less than \$ 657.02 equals 12.42%.

Let’s calculate the percentage of persons with a salary below $\bar{x} + \sigma$, i.e., less than 4,120.08 \$.

Let $P_{<\bar{x}+\sigma}$ be this percentage; we use the linear interpolation formula using the data pair $(a_i, f_i \uparrow)$:

$$\bar{x} + \sigma = a_i + (a_{i+1} - a_i)\frac{(P_{<\bar{x}+\sigma} - F_i)}{(F_{i+1} - F_i)}$$

$$4{,}120.08 = 4{,}000 + (5{,}000 - 4{,}000)\frac{(P_{<\bar{x}+\sigma} - 84.13)}{(90.25 - 84.13)}$$

$$P_{<\bar{x}+\sigma} = 84.86\%$$

The percentage of people earning less than $4,120.08 equals 84.86%.

The percentage of people with a salary between the average minus the standard deviation, i.e., $657.02 and the average plus the standard deviation, i.e., 4,120.08 $ is equal to:

$$84.86\% - 12.42\% = 72.44\%$$

INTERPRETATION OF THE RESULTS

The analysis reveals that 72.44% of individuals, totaling 1456 persons, have a salary ranging between 657.02 $ and 4,120.08 $, which corresponds to the mean minus the standard deviation and the mean plus the standard deviation, respectively. This significant proportion of individuals whose salary falls within this range indicates that if the insurance company's project targets this population segment, it has a good chance of becoming viable. Specifically, the project can potentially attract 72.44% of the population, making it a promising investment opportunity.

5.9 Moments

The ***moment of order r*** ($r \in N^*$) or r^{th} order moments with respect to the value q are a statistical tool used to quantify the dispersion and shape of a data distribution. They are defined as the sum of the relative frequencies f_i, where i varies from 1 to k (k being the total number of observations), multiplied by the difference between each observation x_i and the reference value q, raised to the power of r. This sum is then divided by n, which represents the total sample size, to obtain the weighted average of the moments.

Mathematically, the rth order moments with respect to the value q are represented by the following formula:

$$m_{r,q} = \sum_{i=1}^{k} f_i(x_i - q)^r = \frac{1}{n}\sum_{i=1}^{k} n_i(x_i - q)^r$$

Moments of order r with respect to the value q are useful for characterizing how data deviates from the reference value q and how this dispersion varies with power r. They are commonly used in various fields of statistics, such as data analysis, parameter estimation, and modeling of data distributions.

The ***non-central moments*** are statistics used to quantify the dispersion and shape of a data distribution with respect to the reference value $q = 0$. They are defined as the sum of the relative frequencies $f_i$multiplied by the rth power of each observation x_i. In other words, a non-central moment is defined as any moment taken with respect to $q = 0$.

The sum of these products for all observations constitutes the non-central moment. It is called "non-central" because it is not adjusted with respect to a central value, as is the case for centered moments that are calculated with respect to the mean or another reference value.

Mathematically, the non-central moments are represented by the following formula:

$$m_r = \sum_{i=1}^{k} f_i x_i^r = \frac{1}{n} \sum_{i=1}^{k} n_i x_i^r$$

Non-central moments are useful for characterizing how data deviates from the reference value $q = 0$ and how this dispersion varies with the power r. They are commonly used in various fields of statistics, such as data analysis, parameter estimation, and modeling of data distributions.

Example 59

Let's calculate the non-central moment of order 0 to 4:

$$m_0 = \frac{1}{n} \sum_{i=1}^{k} n_i {x_i}^0 = 1$$

$$m_1 = \frac{1}{n} \sum_{i=1}^{k} n_i {x_i}^1$$

$$m_2 = \frac{1}{n} \sum_{i=1}^{k} n_i {x_i}^2$$

$$m_3 = \frac{1}{n} \sum_{i=1}^{k} n_i {x_i}^3$$

$$m_4 = \frac{1}{n} \sum_{i=1}^{k} n_i {x_i}^4$$

A ***Central moment*** is defined as any moment taken with respect to the mean $q = \overline{x}$:

$$\mu_r = \sum_{i=1}^{k} f_i (x_i - \bar{x})^r$$

Hence:

$$\boxed{\mu_r = \frac{1}{n}\sum_{i=1}^{k} n_i (x_i - \bar{x})^r}$$

Example 60

Let's calculate the central moment of order 0 to 4:

$$\mu_0 = \frac{1}{n}\sum_{i=1}^{k} n_i (x_i - \bar{x})^0 = \frac{1}{n}\sum_{i=1}^{k} n_i = m_0 = 1$$

$$\mu_1 = \frac{1}{n}\sum_{i=1}^{k} n_i (x_i - \bar{x})^1 = \frac{1}{n}\sum_{i=1}^{k} (x_i - \bar{x}) = 0$$

$$\begin{aligned}
\mu_2 = \frac{1}{n}\sum_{i=1}^{k} n_i (x_i - \bar{x})^2 &= \sum_{i=1}^{k} f_i (x_i - \bar{x})^2 \\
&= \sum_{i=1}^{k} f_i (x_i^2 - 2x_i\bar{x} + \bar{x}^2) \\
&= \underbrace{\sum_{i=1}^{k} f_i (x_i^2)}_{m_2} - 2\underbrace{\sum_{i=1}^{k} f_i (x_i\bar{x})}_{m_1 m_1 = m_1^2} + \underbrace{\sum_{i=1}^{k} f_i (\bar{x}^2)}_{m_1^2} \\
&= m_2 - 2m_1^2 + m_1^2 \\
&= m_2 - m_1^2
\end{aligned}$$

$$\mu_3 = \frac{1}{n}\sum_{i=1}^{k} n_i(x_i - \bar{x})^3 = \sum_{i=1}^{k} f_i(x_i - \bar{x})^3$$

$$= \sum_{i=1}^{k} f_i(x_i^3 - 3x_i^2\bar{x} + 3x_i\bar{x}^2 - \bar{x}^3)$$

$$= \underbrace{\sum_{i=1}^{k} f_i(x_i^3)}_{m_3} - 3\underbrace{\sum_{i=1}^{k} f_i(x_i^2\bar{x})}_{m_2m_1} + 3\underbrace{\sum_{i=1}^{k} f_i(x_i\bar{x}^2)}_{m_1m_1^2}$$

$$-\underbrace{\sum_{i=1}^{k} f_i(\bar{x}^3)}_{m_1^3}$$

$$= m_3 - 3m_2m_1 + 3m_1^2m_1 - m_1^3$$

$$= m_3 - 3m_2m_1 + 2m_1^3$$

$$\mu_4 = \frac{1}{n}\sum_{i=1}^{k} n_i(x_i - \bar{x})^4 = \sum_{i=1}^{k} f_i(x_i - \bar{x})^4$$

$$= \sum_{i=1}^{k} f_i(x_i^4 - 4x_i^3\bar{x} + 6x_i^2\bar{x}^2 - 4x_i\bar{x}^3 + \bar{x}^4)$$

$$= \underbrace{\sum_{i=1}^{k} f_i(x_i^4)}_{m_4} - 4\underbrace{\sum_{i=1}^{k} f_i(x_i^3\bar{x})}_{m_3m_1} + 6\underbrace{\sum_{i=1}^{k} f_i(x_i^2\bar{x})}_{m_2m_1^2}$$

$$-4\underbrace{\sum_{i=1}^{k} f_i(x_i\bar{x}^3)}_{m_1m_1^3} + \underbrace{\sum_{i=1}^{k} f_i(\bar{x}^4)}_{m_1^4}$$

$$= m_4 - 4m_3m_1 + 6m_1^2m_2 - 4m_1^3m_1 + m_1^4$$

$$= m_4 - 4m_3m_1 + 6m_1^2m_2 - 4m_1^4 + m_1^4$$

$$= m_4 - 4m_1m_3 + 6m_1^2m_2 - 3m_1^4$$

Example 61

Let's look back at **example 58** from **section 5.8** on the statistical distribution of the workers of a company according to their age and determine all the centered and non-central moments of order 0 to 4.

Solution 1

Calculation of center moments of order 0 to 4 from the basic formula:

Classes	x_i	n_i	$n_i(x_i - \bar{x})$	$n_i(x_i - \bar{x})^2$	$n_i(x_i - \bar{x})^3$	$n_i(x_i - \bar{x})^4$
[20;25)	22.5	36	-326.40	2959.32	-26830.94	243265.39
[25;30)	27.5	45	-183.00	744.18	-3026.26	12306.61
[30;35)	32.5	27	25.20	23.52	21.96	20.49
[35;40)	37.5	18	106.80	633.69	3759.96	22309.35
[40;45)	42.5	10	109.33	1195.39	13069.70	142896.29
[45;50)	47.5	8	127.47	2030.99	32360.51	515612.96
[50;55)	52.5	3	62.80	1314.62	27519.50	576076.74
[55;60)	57.5	3	77.80	2017.62	52323.84	1356935.14
Total		150	0	10919.33	99198.27	2869422.98

Finding central moments of order 0 to 4:

$$\mu_0 = \frac{\sum_{i=1}^{i=8} n_i(x_i - \bar{x})^0}{\sum_{i=1}^{i=8} n_i} = \frac{[36(22.5 - 31.566)^0] + [45(27.5 - 31.566)^0]}{150}$$

$$+ \frac{[27(32.5 - 31.566)^0] + [18(37.5 - 31.566)^0]}{150}$$

$$+ \frac{[10(42.5 - 31.566)^0] + [8(47.5 - 31.566)^0]}{150}$$

$$+ \frac{[3(52.5 - 31.566)^0] + [3(57.5 - 31.566)^0]}{150}$$

$$= \frac{150}{150}$$

$$= 1$$

$$\mu_1 = \frac{\sum_{i=1}^{i=8} n_i(x_i - \bar{x})}{\sum_{i=1}^{i=8} n_i} = \frac{[36(22.5 - 31.566)] + [45(27.5 - 31.566)]}{150}$$

$$+\frac{[27(32.5-31.566)]+[18(37.5-31.566)]}{150}$$

$$+\frac{[10(42.5-31.566)]+[8(47.5-31.566)]}{150}$$

$$+\frac{[3(52.5-31.566)]+[3(57.5-31.566)]}{150}$$

$$=\frac{0}{150}$$

$$=0$$

$$\mu_2=\frac{\sum_{i=1}^{i=8} n_i(x_i-\bar{x})^2}{\sum_{i=1}^{i=8} n_i}=\frac{[36(22.5-31.566)^2]+[45(27.5-31.566)^2]}{150}$$

$$+\frac{[27(32.5-31.566)^2]+[18(37.5-31.566)^2]}{150}$$

$$+\frac{[10(42.5-31.566)^2]+[8(47.5-31.566)^2]}{150}$$

$$+\frac{[3(52.5-31.566)^2]+[3(57.5-31.566)^2]}{150}$$

$$=\frac{10{,}919.33}{150}$$

$$=72.80$$

$$\mu_3=\frac{\sum_{i=1}^{i=8} n_i(x_i-\bar{x})^3}{\sum_{i=1}^{i=8} n_i}=\frac{[36(22.5-31.566)^3]+[45(27.5-31.566)^3]}{150}$$

$$+\frac{[27(32.5-31.566)^3]+[18(37.5-31.566)^3]}{150}$$

$$+\frac{[10(42.5-31.566)^3]+[8(47.5-31.566)^3]}{150}$$

$$+\frac{[3(52.5-31.566)^3]+[3(57.5-31.566)^3]}{150}$$

$$= \frac{99,198,27}{150}$$

$$= 661.31$$

$$\mu_4 = \frac{\sum_{i=1}^{i=8} n_i (x_i - \bar{x})^4}{\sum_{i=1}^{i=8} n_i} = \frac{[36(22.5 - 31.566)^4] + [45(27.5 - 31.566)^4]}{150}$$

$$+ \frac{[27(32.5 - 31.566)^4] + [18(37.5 - 31.566)^4]}{150}$$

$$+ \frac{[10(42.5 - 31.566)^4] + [8(47.5 - 31.566)^4]}{150}$$

$$+ \frac{[3(52.5 - 31.566)^4] + [3(57.5 - 31.566)^4]}{150}$$

$$= \frac{2,869,422.98}{150}$$

$$= 19,129.48$$

Solution 2

Calculation of central moments of order 0 to 4 from non-central moments of order 0 to 4 (use of the expended formula):

Classes	x_i	n_i	$n_i x_i$	$n_i x_i^2$	$n_i x_i^3$	$n_i x_i^4$
[20;25)	22.5	36	810.00	18225.00	410062.50	9226406.25
[25;30)	27.5	45	1237.50	34031.25	935859.38	25736132.81
[30;35)	32.5	27	877.50	28518.75	926859.38	30122929.69
[35;40)	37.5	18	675.00	25312.50	949218.75	35595703.13
[40;45)	42.5	10	425.00	18062.50	767656.25	32625390.63
[45;50)	47.5	8	380.00	18050.00	857375.00	40725312.50
[50;55)	52.5	3	157.50	8268.75	434109.38	22790742.19
[55;60)	57.5	3	172.50	9918.75	570328.13	32793867.19
Total		150	4735.00	160387.50	5851468.75	229616484.38

The mean = 31,566 had already been calculated in section **5.5**.

Finding a non-central moment of order 0 to 4:

$$m_0 = \frac{1}{n}\sum_{i=1}^{i=8} n_i x_i^0 = \frac{[36(22.5)^0] + [45(27.5)^0] + [27(32.5)^0]}{150}$$

$$+\frac{[18(37.5)^0] + [10(42.5)^0] + [8(47.5)^0]}{150}$$

$$+\frac{[3(52.5)^0] + [3(57.5)^0]}{150}$$

$$= \frac{150}{150}$$

$$= 1$$

$$m_1 = \frac{1}{n}\sum_{i=1}^{i=8} n_i x_i^1 = \frac{[36(22.5)^1] + [45(27.5)^1] + [27(32.5)^1]}{150}$$

$$+\frac{[18(37.5)^1] + [10(42.5)^1] + [8(47.5)^{12}]}{150}$$

$$+\frac{[3(52.5)^1] + [3(57.5)^1]}{150}$$

$$= \frac{4{,}735}{150}$$

$$= 31.57$$

$$m_2 = \frac{1}{n}\sum_{i=1}^{i=8} n_i x_i^2 = \frac{[36(22.5)^2] + [45(27.5)^2] + [27(32.5)^2]}{150}$$

$$+\frac{[18(37.5)^2] + [10(42.5)^2] + [8(47.5)^2]}{150}$$

$$+\frac{[3(52.5)^2] + [3(57.5)^2]}{150}$$

$$= \frac{160,387.50}{150}$$

$$= 1,069.25$$

$$m_3 = \frac{1}{n}\sum_{i=1}^{i=8} n_i x_i^3 = \frac{[36(22.5)^3] + [45(27.5)^3] + [27(32.5)^3]}{150}$$

$$+ \frac{[18(37.5)^3] + [10(42.5)^3] + [8(47.5)^3]}{150}$$

$$+ \frac{[3(52.5)^3] + [3(57.5)^3]}{150}$$

$$= \frac{5,851,468.75}{150}$$

$$= 39,009.79$$

$$m_4 = \frac{1}{n}\sum_{i=1}^{i=8} n_i x_i^4 = \frac{[36(22.5)^4] + [45(27.5)^4] + [27(32.5)^4]}{150}$$

$$+ \frac{[18(37.5)^4] + [10(42.5)^4] + [8(47.5)^4]}{150}$$

$$+ \frac{[3(52.5)^4] + [3(57.5)^4]}{150}$$

$$= \frac{229,616,484.38}{150}$$

$$= 1,530,776.56$$

Finding central moments of order 0 to 4:

$$\mu_0 = 1$$

$$\mu_1 = 0$$

$$\mu_2 = 1,069.25 - m_1^2$$

$$= 1.315,25 - (31.57)^2$$

$$= 72.8$$

$$\mu_3 = m_3 - 3m_2m_1 + 2m_1^3$$
$$= 39,009.79 - 3(1,069.25)(31.57) + 2(31.57)^3$$
$$= 661.31$$

$$\mu_4 = m_4 - 4m_1m_3 + 6m_1^2m_2 - 3m_1^4$$

$$= 1,530,776.56 - 4(31.57)(39,009.79) + 6(31.57)^2(1,069.25) - 3(31.57)^4$$

$$= 19,129.48$$

5.10 Exercises

5.10.1 Chapter exercises

• **Exercise 18:** Let's consider a newspaper vendor who wants to determine the key factors affecting his sales. To do so, he records the sales figures for 21 consecutive days and notes the following results:

$$47-52-63-55-41-45-61-47-69-58-46-51-64-59-62-48-40-66-59-49-53$$

a. Construct the statistical table distributing the data into classes.

b. Plot the histogram of the distribution.

c. Plot the ascending and descending cumulative frequency curves representing the number of days versus the number of newspapers sold per day.

d. Determine the measures of central tendency for this distribution.

e. Determine the measures of dispersion for this distribution.

f. Calculate the percentage of days where sales fall within one standard deviation below and above the mean.

g. Calculate the centered moments of order 0 to 4 using the developed formula.

Solution

a: The statistical table distributing the data into classes:

Number of newspapers sold per day	x_i	Number of days n_i
[40;45)	42.5	2
[45;50)	47.5	5
[50;55)	52.5	3
[55;60)	57.5	4
[60;65)	62.5	4
[65;70)	67.5	2
Total		20

b: Let's plot the histogram of the distribution.

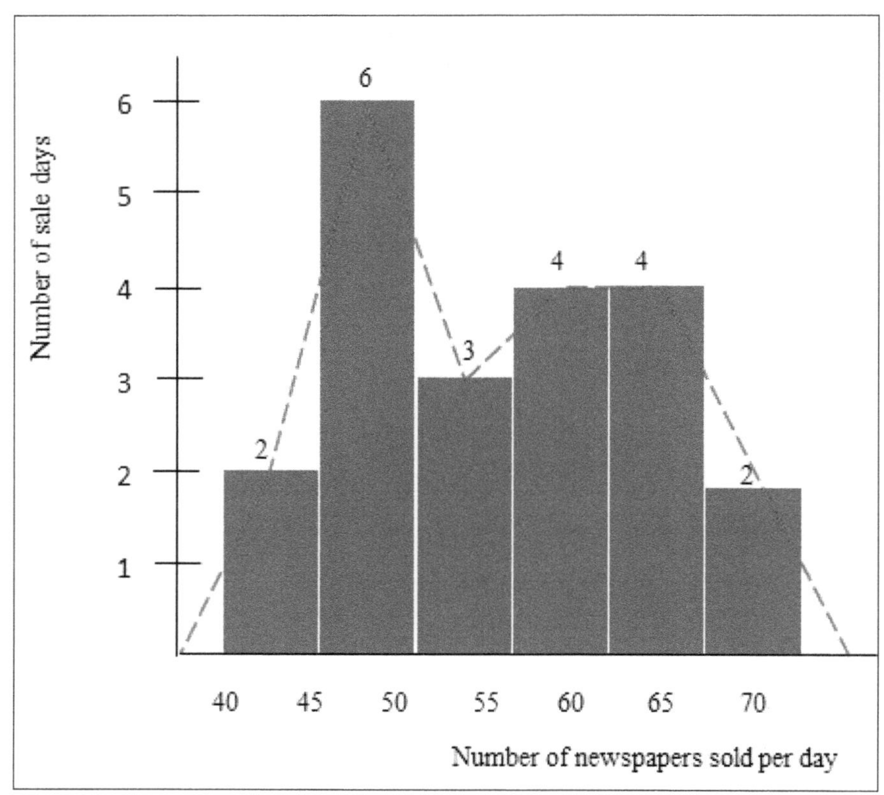

To draw the curves of the cumulative ascending and descending frequencies, we will need the columns of the following table:

Number of newspapers sold per day	x_i	n_i	$n_i x_i$	$n_i x_i^2$	f_i	$f_i \uparrow$	$f_i \downarrow$
[40;45)	42.5	2	85	3612.5	9.52	9.52	100.00
[45;50)	47.5	6	285	13537.5	28.57	38.10	90.48
[50;55)	52.5	3	157.5	8268.75	14.29	52.38	61.90
[55;60)	57.5	4	230	13225	19.05	71.43	47.62
[60;65)	62.5	4	250	15625	19.05	90.48	28.57
[65;70)	67.5	2	135	9112.5	9.52	100.00	9.52
Total		21	1142.5	63381.25	100		

c: The curves of the ascending and descending cumulative frequencies are as follows:

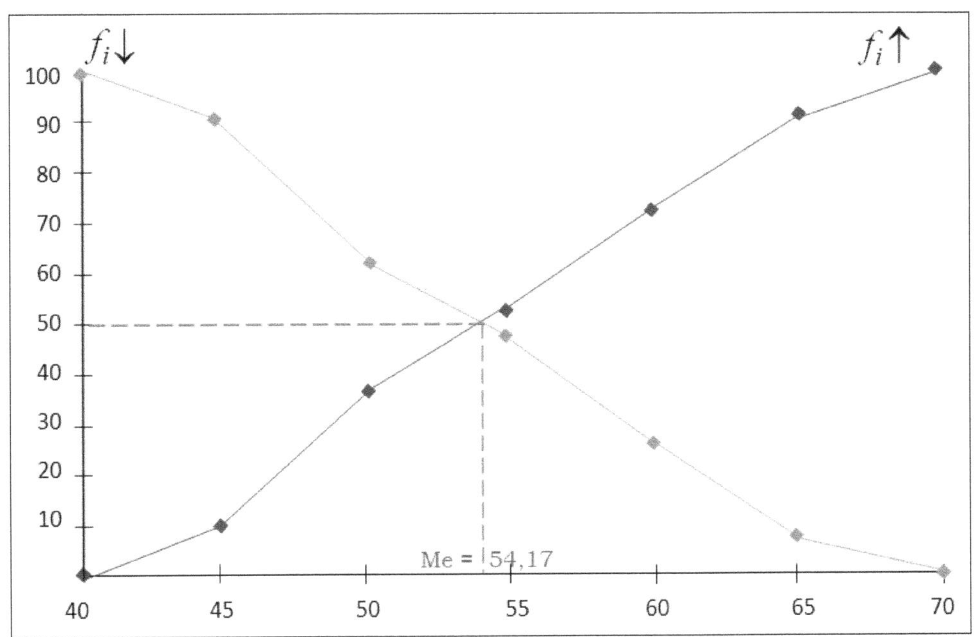

The Mode of this distribution is 47.5, and the modal class is [45,50).

d: Let's calculate the central tendency parameters.

The mean of the distribution is:

$$\bar{x} = \frac{1}{n}\sum_{i=1}^{i=6} n_i x_i = \frac{1,142.5}{21}$$
$$= 54.4047 \text{ newspapers}$$

The median is:

$$M_e = a_i + (a_{i+1} - a_i)\frac{(50 - F_i)}{(F_{i+1} - F_i)}$$
$$= 50 + (55 - 50)\frac{(50 - 38.10)}{(52.38 - 38.10)}$$
$$= 54.17 \text{ newspapers}$$

INTERPRETATION OF THE RESULTS

Based on the newsagent's sales data, we have made the following observations:

d-1. The newspaper vendor typically sells between 45 and 50 newspapers per day, which represents the most frequent range of sales.

d-2. The average daily sales amount to 54.4047 newspapers, providing a measure of central tendency for the distribution.

d-3. It is interesting to note that the number of days where the vendor sold less than 54.17 newspapers is equivalent to the number of days where he sold more, indicating a symmetry in the distribution of sales.

These findings suggest that the newsagent's sales are somewhat volatile, with a relatively large range of sales on any given day. However, the average sales figure indicates that the newsagent is generally successful in attracting customers and generating revenue.

e: Let's calculate the characteristics of dispersion:

Finding the interquartile range:

$$Q_1 = a_i + (a_{i+1} - a_i)\frac{(25 - F_i)}{(F_{i+1} - F_i)}$$

$$= 45 + (50 - 45)\frac{(25 - 9.52)}{(38.1 - 9.52)}$$

$$= 47.71$$

$$Q_3 = a_i + (a_{i+1} - a_i)\frac{(75 - F_i)}{(F_{i+1} - F_i)}$$

$$= 60 + (65 - 60)\frac{(75 - 71.43)}{(90.48 - 71.43)}$$

$$= 60.94$$

$$\text{Interquartile Range} = Q_3 - Q_1$$

$$= 60.94 - 47.71$$

$$= 13.23 \text{ newspapers}$$

The variance of the distribution is:

$$V(x) = \frac{1}{n}\sum_{i=1}^{i=6} n_i x_i^2 - \bar{x}^2$$

$$= \frac{63{,}381.3}{21} - (54.4047)^2$$

$$= 58.28 \text{ newspapers}$$

The standard deviation of the distribution is:

$$\sigma = \sqrt{V(x)}$$
$$= \sqrt{58.28}$$
$$= 7.63 \text{ newspapers}$$

The coefficient of variation is:

$$\text{CV} = \frac{\sigma}{|\bar{x}|}\text{x}100 = \frac{7.63}{54.4047}\text{x}100$$
$$= 14\%$$

f: Let's calculate the percentage of days where sale is between the mean minus the standard deviation and the mean plus the standard deviation:

f-1: The mean plus or minus the standard deviation is:

$$\bar{x} - \sigma = 54.4047 \ - 7.63$$
$$= 46.774 \text{ newspapers}$$

$$\bar{x} + \sigma = 54.4047 \ + 7.63$$
$$= 62.034 \text{ newspapers}$$

f-2: Finding the percentage of days whose sale is less than $\bar{x} - \sigma$, i.e., less than 46.774 newspapers.
Let $P_{<\bar{x}-\sigma}$ be the percentage to find; we use the linear interpolation formula using the data pair $(a_i, f_i \uparrow)$:

$$\bar{x} - \sigma = a_i + (a_{i+1} - a_i)\frac{(P_{<\bar{x}-\sigma} - F_i)}{(F_{i+1} - F_i)}$$
$$46.774 = 45 + (50 - 45)\frac{(P_{<\bar{x}-\sigma} \ - 9.52)}{(38.10 - 9.52)}$$
$$P_{<\bar{x}-\sigma} \ = 19.66\%$$

The percentage of days whose sale is less than 46.774 newspapers is equal to 19.66%

f-3: Finding the percentage of days whose sale is less than $\bar{x} + \sigma$, i.e., less than 62.034 newspapers.

Let $P_{<\bar{x}+\sigma}$ be the percentage to find; we use the linear interpolation formula using the data pair $(a_i, f_i \uparrow)$:

$$\bar{x} + \sigma = a_i + (a_{i+1} - a_i)\frac{(P_{<\bar{x}+\sigma} - F_i)}{(F_{i+1} - F_i)}$$

$$62.034 = 60 + (65 - 60)\frac{(P_{<\bar{x}+\sigma} - 71.43)}{(90.48 - 71.43)}$$

$$P_{<\bar{x}+\sigma} = 79.18\%$$

The percentage of days whose sale is less than 62.034 newspapers is equal to 79.18%

f-4: The percentage of days whose sale is between $\bar{x} - \sigma$ (46.774 newspapers) and $\bar{x} + \sigma$ (62.034 newspapers) is:

$$79.18 - 19.66 = 59.52\%$$

g. Calculate the central moments of order 0 to 4 from the expended formula:

g-1: Let's calculate the non-central moments of order 0 to 4:

$$m_0 = 1$$

$$m_1 = \frac{1}{n}\sum_{i=1}^{6} n_i x_i = \frac{1{,}142.5}{21}$$

$$= 54.404$$

$$m_2 = \frac{1}{n}\sum_{i=1}^{6} n_i {x_i}^2 = \frac{63{,}381.25}{21}$$

$$= 3{,}018.15$$

$$m_3 = \frac{1}{n}\sum_{i=1}^{6} n_i {x_i}^3 = \frac{3{,}582{,}765.63}{21}$$

$$= 170{,}607.89$$

$$m_4 = \frac{1}{n}\sum_{i=1}^{6} n_i {x_i}^4 = \frac{206{,}138{,}945.31}{21}$$

$$= 9{,}816{,}140.25$$

g-2: Let's calculate the central moments of order 0 to 4:

$$\mu_0 = 1$$

$$\mu_1 = 0$$

$$\mu_2 = m_2 - m_1^2$$

$$= 3{,}018.15 - (54.404)^2$$

$$= 58.28$$

$$\mu_3 = m_3 - 3m_1m_2 + 2m_1^3$$

$$= 170{,}607.89 - 3(54.404)(3{,}018.15) + 2(54.404)^3$$

$$= 64.84$$

$$\mu_4 = m_4 - 4m_1m_3 + 6m_1^2m_2 - 3m_1^4$$
$$= 9{,}816{,}140.25 - 4(54.404)(170{,}607.89) + 6(54.404)^2(3{,}018.15) - 3(54.404)^4$$
$$= 6{,}200.39$$

INTERPRETATION OF THE RESULTS

Upon analyzing the newsagent's sales data, we made the following observations:

e. When sorting the sales data by ascending order of the number of newspapers sold, the largest difference in sales within the central half of the series is approximately 13 newspapers. The coefficient of variation is 14%, indicating that the distribution is not highly dispersed around the mean, making it relatively homogeneous in terms of the number of newspapers sold per day. This indicates that there is relatively little variation in sales within this range.

f. We found that 59.52% of the days had sales between 46.774 and 62.034 newspapers sold. Within this interval, the distribution is less populated than that of the normal distribution, with only 59.52% falling within this range compared to the expected 68.27% for a normal distribution. However, given the relatively small coefficient of variation, the sales data still suggest a stable and consistent sales trend.

By assuming that the data provided reflects a real-life situation, several factors could be responsible for the low percentage observed in this interval. Here are some possible hypotheses:

- Seasonal factors: Newspaper sales may be influenced by seasonal factors such as holidays, public holidays, or weather variations. If the observation period includes periods of low newspaper sales activity, this could explain why only 59.52% of days recorded sales were within the given range.
- Market-specific effects: Local economic conditions, competition from other sources of information, or other market-specific factors can also influence newspaper sales. If the market in which the newspaper vendor operates is facing specific challenges during the observation period, this could explain the deviation from a normal distribution.
- Natural variability: Newspaper sales can naturally vary from day to day due to various random factors, such as customer buying behavior, unforeseen events, or demand fluctuations. This natural variability may explain why only 59.52% of days recorded sales were within the given range.
- Measurement errors: Sales figures collected may be subject to measurement errors, such as counting errors, data entry errors, or other human errors. If these errors are present in the collected data, they can influence the observed distribution.

• **Exercise 19:** The table beside shows the distribution of 200 employees in a company based on their length of service in years (years of seniority):

Years of Seniority (x_i)	Numbers (n_i)
[0;10)	30
[10;20)	100
[20;30)	40
[30;40)	20
[40;50)	10
Total	200

a. Specify the observed population and the statistical character used.

b. Plot the histogram of the distribution to visually represent the distribution of length of service values of the employees and observe the shape of the distribution.

c. Calculate the median Q_1 (first quartile) and Q_3. (third quartile) to obtain measures of central tendency and spread of the distribution of employees' length of service.

d. Plot the ascending and descending cumulative frequency curves to analyze the progression of employees with respect to the length of service and graphically find the values of median Q_1 and Q_3. on these curves.

e. Calculate the coefficient of variation to assess the relative variability of employees' length of service compared to the mean and express this variability as a percentage.

f. Calculate the percentage of employees whose length of service falls between the mean minus one standard deviation and the mean plus one standard deviation to evaluate the proportion of employees whose length of service falls within this interval around the mean.

g. Calculate the centered moments of order 0 to 4 using the developed formula to obtain detailed measures of trend and dispersion of the distribution of employees' length of service.

Solution

a: The observed population consists of 200 employees of a company, with their length of service in years used as the statistical character.

b: Let's Plot the histogram of the distribution:

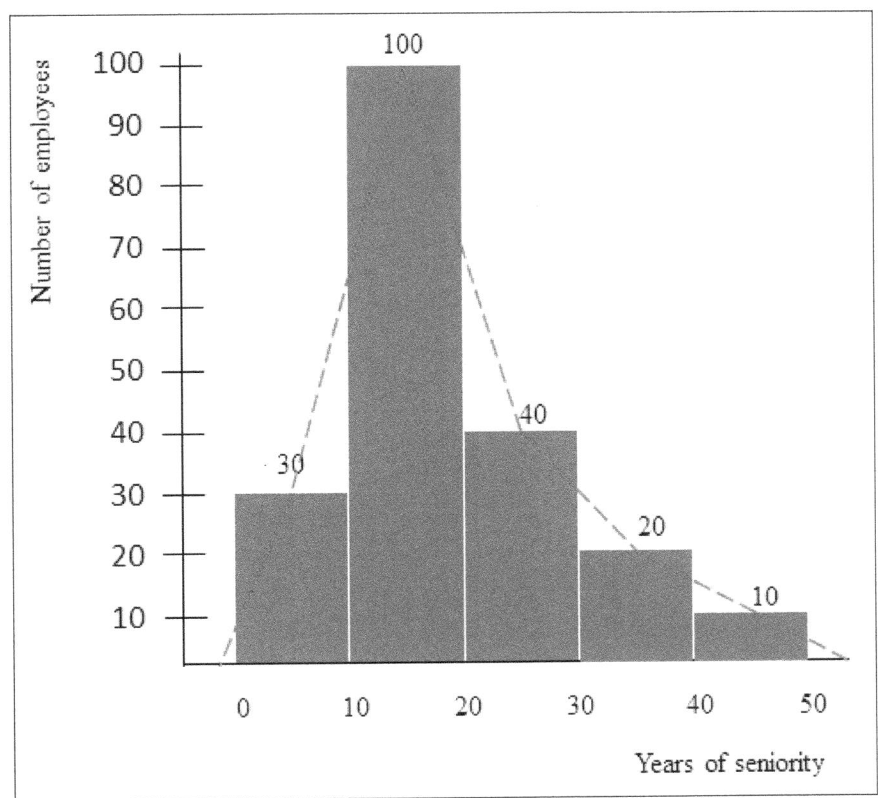

The following table shows the data required to solve this exercise:

Years of seniority	x_i	n_i	$n_i x_i$	$n_i x_i^2$	f_i	$f_i \uparrow$	$f_i \downarrow$
[0;10)	5	30	150	750	15.00	15.00	100.00
[10;20)	15	100	1500	22500	50.00	65.00	85.00
[20;30)	25	40	1000	25000	20.00	85.00	35.00
[30;40)	35	20	700	24500	10.00	95.00	15.00
[40;50)	45	10	450	20250	5.00	100.00	5.00
Total		200	3800	93000			

c: Finding the median Q_1 and Q_3:

$$Me = a_i + (a_{i+1} - a_i)\frac{(50 - F_i)}{(F_{i+1} - F_i)}$$

$$= 10 + (20 - 10)\frac{(50 - 15)}{(65 - 15)}$$

$$= 17 \text{ years}$$

$$Q_1 = a_i + (a_{i+1} - a_i)\frac{(25 - F_i)}{(F_{i+1} - F_i)}$$

$$= 10 + (20 - 10)\frac{(25 - 15)}{(65 - 15)}$$

$$= 12 \text{ years}$$

$$Q_3 = a_i + (a_{i+1} - a_i)\frac{(75 - F_i)}{(F_{i+1} - F_i)}$$

$$= 20 + (30 - 20)\frac{(75 - 65)}{(85 - 65)}$$

$$= 25 \text{ years}$$

d: The curves of the ascending and descending cumulative frequencies are as follows:

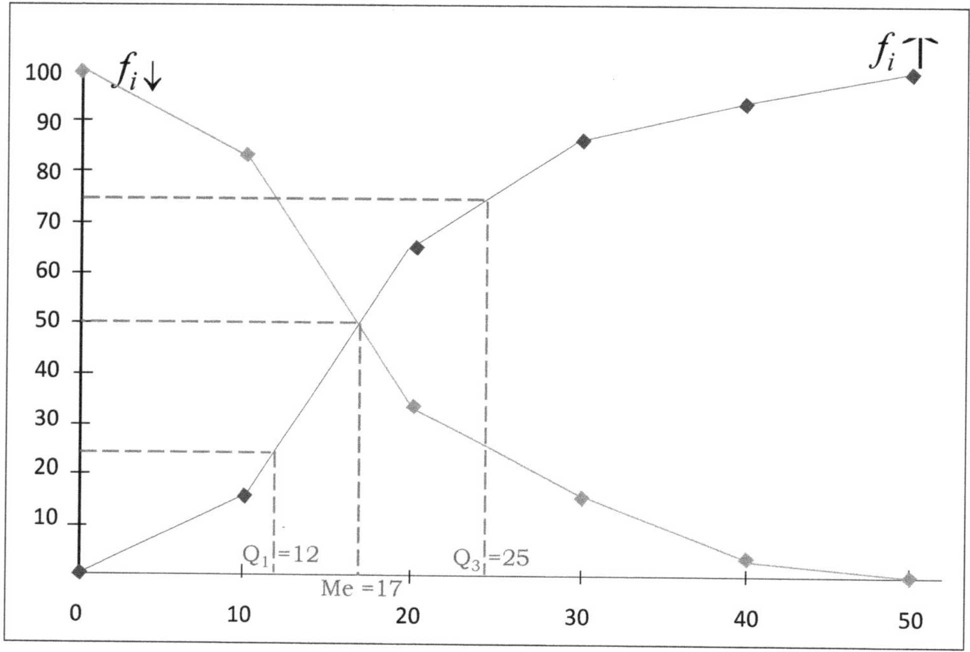

e: Determining the coefficient of variation:

The mean of the distribution is:

$$\bar{x} = \frac{1}{n}\sum_{i=1}^{i=5} n_i x_i = \frac{3{,}800}{200}$$

$$= 19 \text{ years}$$

The variance of the distribution is:

$$V(x) = \frac{1}{n}\sum_{i=1}^{i=5} n_i x_i^2 - \bar{x}^2$$
$$= \frac{93{,}000}{200} - (19)^2$$
$$= 104$$

The standard deviation of the distribution is:

$$\sigma = \sqrt{V(x)}$$
$$= \sqrt{104.04}$$
$$= 10.2 \text{ yars}$$

The coefficient of variation is:

$$\text{CV} = \frac{\sigma}{|\bar{x}|}\text{x}100 = \frac{10.2}{19}$$
$$= 53.68\%$$

f: Let's calculate the percentage of workers with seniority between the mean minus the standard deviation and the mean plus the standard deviation.

f-1: The mean plus or minus the standard deviation is:

$$\bar{x} - \sigma = 10 - 10.2$$
$$= 8.8 \text{ years}$$

$$\bar{x} + \sigma = 10 + 10.2$$
$$= 29.2 \text{ years}$$

f-2: Finding the percentage of workers with seniority less than $\bar{x} - \sigma$, i.e., less than 8.8 years.

Let $P_{<\bar{x}-\sigma}$ be the percentage to find; we use the linear interpolation formula on $(a_i, f_i \uparrow)$:

$$\bar{x} - \sigma = a_i + (a_{i+1} - a_i)\frac{(P_{<\bar{x}-\sigma} - F_i)}{(F_{i+1} - F_i)}$$
$$8.8 = 0 + (10 - 0)\frac{(P_{<\bar{x}-\sigma} - 0)}{(10 - 0)}$$

$$P_{<\bar{x}-\sigma} = 13.2\%$$

The percentage of workers with seniority of less than 8.8 years is equal to 13.2%

f-3: Finding the percentage of workers with seniority less than $\bar{x}+\sigma$, i.e., less than 29.2 years.

Let $P_{<\bar{x}+\sigma}$ be the percentage to find; we use linear interpolation formula on $(a_i, f_i \uparrow)$:

$$\bar{x}+\sigma = a_i + (a_{i+1} - a_i)\frac{(P_{<\bar{x}+\sigma} - F_i)}{(F_{i+1} - F_i)}$$

$$29.2 = 20 + (30 - 20)\frac{(P_{<\bar{x}+\sigma} - 65)}{(85 - 65)}$$

$$P_{<\bar{x}+\sigma} = 83.4\%$$

The percentage of workers with seniority of less than 29.2 years is equal to 83.4%

f-4: The percentage of workers with seniority between 8.8 years and 29.2 years is:

$$83.4 - 13.2 = 80.2\%$$

This corresponds to 160 people out of 200.

g: Let's calculate the central moments of order 0 to 4 from the expended formula:

g-1: Let's calculate the non-central moments of order 0 to 4:

$$m_0 = 1$$

$$m_1 = \frac{1}{n}\sum_{i=1}^{5} n_i x_i = \frac{3{,}800}{200} = 19$$

$$m_2 = \frac{1}{n}\sum_{i=1}^{5} n_i {x_i}^2 = \frac{93{,}000}{200} = 465$$

$$m_3 = \frac{1}{n}\sum_{i=1}^{5} n_i {x_i}^3 = \frac{2{,}735{,}000}{200} = 13{,}675$$

$$m_4 = \frac{1}{n}\sum_{i=1}^{5} n_i {x_i}^4 = \frac{91{,}725{,}000}{200} = 458{,}625$$

g-2: Let's calculate the central moments of order 0 to 4:

$$\mu_0 = 1$$

$$\mu_1 = 0$$

$$\mu_2 = m_2 - m_1^2 = 465 - (19)^2$$

$$= 104$$

$$\mu_3 = m_3 - 3m_1m_2 + 2m_1^3$$
$$= 1{,}3675 - 3(19)(465) + 2(19)^3$$
$$= 888$$
$$\mu_4 = m_4 - 4m_1m_3 + 6m_1^2m_2 - 3m_1^4$$
$$= 458{,}625 - 4(19)(13{,}675) + 6(19)^2(465) - 3(19)^4$$
$$= 35{,}552$$

INTERPRETATION OF THE RESULTS

In this company:

d.

- On average, employees have 19 years of seniority, indicating a certain level of stability in employees' length of service.
- There are as many employees with less than 17 years of seniority as those with more, indicating a balanced distribution of seniority values.
- 25% of employees have less than 12 years of seniority, meaning that 75% of employees have more than 12 years of seniority, highlighting a concentration of employees with a long length of service.
- 75% of employees have less than 25 years of seniority, indicating that 25% of employees have more than 25 years of seniority, showing an uneven distribution of employees' seniority, with some nearing retirement age and others still having a long way to go.

e. The coefficient of variation is 53.68%, indicating that the distribution of seniority is dispersed and shows a significant difference among employees' seniorities.

f. Based on these findings; we can conclude that the seniority of employees in this company varies significantly. However, despite the dispersed distribution, we found that 80.2% of workers have a seniority between 8.8 years and 29.2 years. This range is more populated than the normal distribution, with 80.2% of the population falling within this range compared to the expected 68.27%. This indicates that most employees fall within a standard range of seniority and that the company maintains a relatively stable workforce.

In a real-life situation, such a high concentration within the interval between the mean minus the standard deviation and the mean plus the standard deviation can be justified by the fact that this company may have:

- Employee retention policy and stable tenure: If the company has a policy of employee retention and encourages employees to stay with the company for longer periods of time, this can result in a concentration of tenure values around the mean. Employees who stay longer will accumulate higher tenure, leading to a concentration of tenure values within the interval between the mean minus the standard deviation and the mean plus the standard deviation.
- A system of Benefits or incentives to motivate the employees to stay longer: If the company offers benefits, incentives, or rewards to employees who stay with the company for longer periods of time, this can encourage employees to accumulate higher tenure. For example, benefits such as salary increases, promotions, social benefits, or tenure-related perks can incentivize employees to stay longer, resulting in a concentration of tenure values within the interval between the mean minus the standard deviation and the mean plus the standard deviation.

• **Exercise 20:** In a supply center, for each delivery made, the number of kilometers traveled was recorded. The table below summarizes the observations collected for one week, providing an overview of the distribution of the data:

Trips in Km	Number of deliveries
[0;6)	6
[6;12)	36
[12;18)	52
[18;24)	62
[24;30)	71
[30;36)	78
[36;42)	65
[42;48)	60
[48;54)	40
[54;60)	10
Total	480

a. Plot the histogram of the distribution to visualize the distribution of kilometers traveled.

b. Plot the ascending and descending cumulative frequency curves to better understand the distribution's trend.

c. Determine the modal mileage of the distribution, which is the most frequently observed value.

d. Calculate the average number of kilometers traveled per delivery to get an idea of the typical distance traveled.

e. Find the median number of kilometers traveled per delivery, which represents the value at the center of the distribution and provides an indication of the central tendency of data.

f. Calculate the interquartile range, which measures the difference between the 25th and 75th percentiles of the data, to assess the dispersion of kilometers traveled.

g. Calculate the coefficient of variation, which quantifies the relative variability of the data compared to their mean, to assess the dispersion of data relative to their mean level.

h. Determine the percentage of deliveries that require a distance traveled in kilometers less than the mode in order to understand the proportion of deliveries with a low distance traveled.

i. Calculate the percentage of deliveries that require a distance traveled in kilometers less than the mean to evaluate the proportion of deliveries with a distance traveled below the mean.

j. Calculate the total distance traveled represented by the top 10% of deliveries in terms of mileage to understand the impact of these deliveries on the overall kilometers traveled.

k. Calculate the percentage of deliveries that require a distance traveled in kilometers between $\bar{x} - \sigma$, and $\bar{x} + \sigma$,, to evaluate the proportion of deliveries with a distance traveled within an interval centered on the mean and expanded by one standard deviation.

l. Calculate the centered moments of order 0 to 4 using the developed formula, to obtain more detailed measures of dispersion and skewness on the distribution of kilometers traveled.

Solution

a: Plotting the histogram of the distribution:

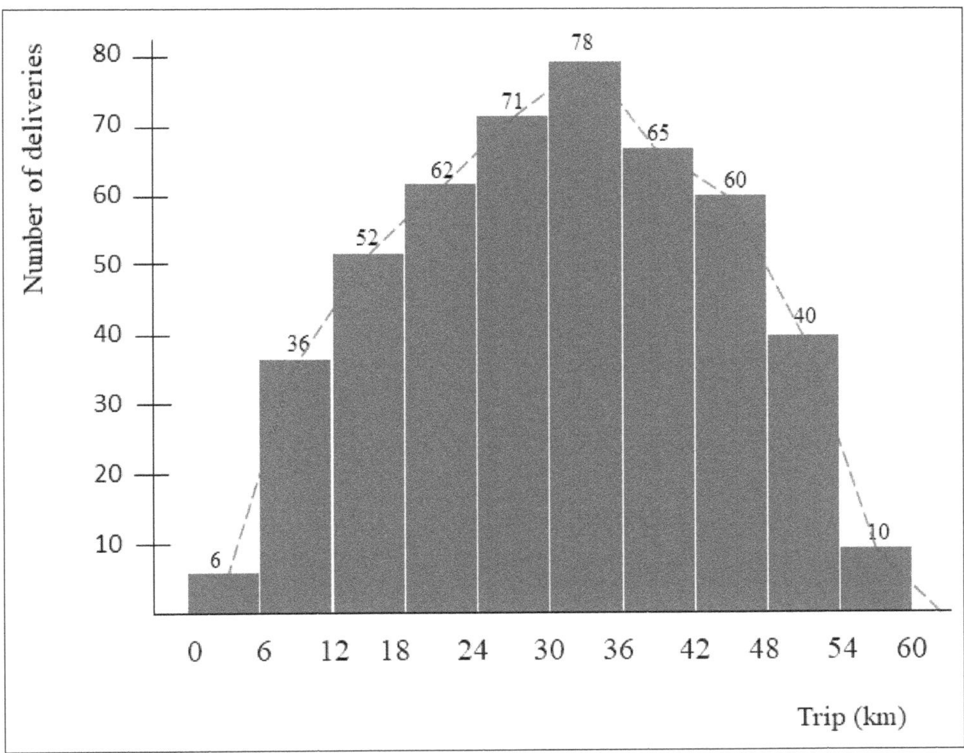

The following table shows the data required to solve this exercise:

Distance	x_i	n_i	$n_i x_i$	$n_i x_i^2$	$n_i x_i^3$	$n_i x_i^4$	f_i	$f_i\uparrow$
[0;6)	3	6	18	54	162	486	1.25	1.25
[6;12)	9	36	324	2916	26244	236196	7.50	8.75
[12;18)	15	52	780	11700	175500	2632500	10.83	19.58
[18;24)	21	62	1302	27342	574182	12057822	12.92	32.50
[24;30)	27	71	1917	51759	1397493	37732311	14.79	47.29
[30;36)	33	78	2574	84942	2803086	92501838	16.25	63.54
[36;42)	39	65	2535	98865	3855735	150373665	13.54	77.08
[42;48)	45	60	2700	121500	5467500	246037500	12.50	89.58
[48;54)	51	40	2040	104040	5306040	270608040	8.33	97.92
[54;60)	57	10	570	32490	1851930	105560010	2.08	100.00
Total		480	14760	535608	21457872	917740368		

b: The curves of the ascending and descending cumulative frequencies are as follows:

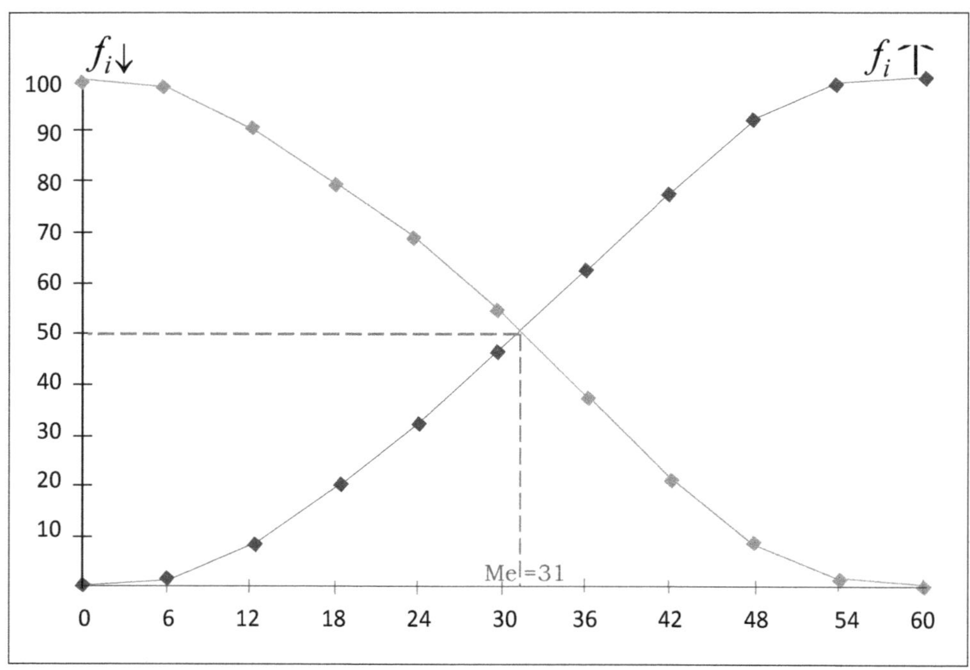

c: The modal class is [30 ; 36) and the modal mileage is 33 Km:

d: The average number of kilometers per delivery is:

$$\bar{x} = \frac{1}{n}\sum_{i=1}^{i=10} n_i x_i = \frac{14{,}760}{480}$$
$$= 30.75 \text{ Km}$$

e: The median number of kilometers per delivery is:

$$Me = a_i + (a_{i+1} - a_i)\frac{(50 - F_i)}{(F_{i+1} - F_i)}$$
$$= 30 + (36 - 30)\frac{(50 - 47.29)}{(63.54 - 47.29)}$$
$$= 31 \text{ Km}$$

f: Finding the Interquartile Range:

$$Q_1 = a_i + (a_{i+1} - a_i)\frac{(25 - F_i)}{(F_{i+1} - F_i)}$$

$$= 18 + (24 - 18)\frac{(25 - 19.58)}{(32.5 - 19.58)}$$

$$= 20.52$$

$$Q_3 = a_i + (a_{i+1} - a_i)\frac{(75 - F_i)}{(F_{i+1} - F_i)}$$

$$= 36 + (42 - 36)\frac{(75 - 63.54)}{(77.08 - 63.54)}$$

$$= 41.08$$

$$\text{The Interquartile Range } = Q_3 - Q_1 = 41.08 - 20.52 = 20.56 \text{ Km}$$

g: Finding the du coefficient of variation of the distribution:

Finding the standard deviation:

$$\sigma = \sqrt{\frac{1}{n}\sum_{i=1}^{i=10} n_i x_i^2 - \bar{x}^2}$$

$$= \sqrt{\frac{535,608}{480} - (30.75)^2}$$

$$= \sqrt{170.28749283}$$

$$= 13.049425 \text{ Km}$$

The coefficient of variation is:

$$\text{CV} = \frac{\sigma}{|\bar{x}|}$$

$$= \frac{13.049425}{30.75} \text{x}100$$

$$= 42.437\%$$

INTERPRETATION OF THE RESULTS

c. The vast majority of deliveries are made within a range of distances ranging from 30 to 36 km, with a modal distance of 33 km for deliveries, which represents the most frequent mileage.

d. On average, each delivery is made at a distance of 30.75 km, which means that if the delivery locations were evenly spaced, they would all be approximately 30.75 km apart.

e. There are as many deliveries made within 31 km as there are deliveries beyond this distance, showing a balanced distribution of deliveries on both sides of this distance.

f. When arranging the delivery distances in ascending order, the largest gap between two delivery distances in the central half of the series is around 20.56 km, highlighting the variability of delivery distances.

g. The coefficient of variation, which measures the relative dispersion of data around their mean, is 42.4%, indicating that the data series has a moderate dispersion but remains overall centered around the mean.

h: Finding the percentage of deliveries requiring a trip with a distance in kilometers less than the mode, i.e., 33 Km.

Let P_{mode} be the percentage to find; we use linear interpolation formula on $(a_i, f_i \uparrow)$:

$$Mode = a_i + (a_{i+1} - a_i)\frac{(P_{<mode} - F_i)}{(F_{i+1} - F_i)}$$

$$33 = 30 + (36 - 30)\frac{(P_{<mode} - 47.29)}{(63.54 - 47.29)}$$

$$P_{<mode} = 55.415\%$$

The percentage of deliveries requiring a trip with a distance in kilometers less than 33 Km is equal to 55.41%, which represents 266 deliveries.

i: Finding the percentage of deliveries requiring a trip with a distance in kilometers less than the average, i.e., 30.75 Km.

Let $P_{<\bar{x}}$ be the percentage to find; we use the linear interpolation formula using the data pair $(a_i, f_i \uparrow)$:

$$\bar{x} = a_i + (a_{i+1} - a_i)\frac{(P_{<\bar{x}} - F_i)}{(F_{i+1} - F_i)}$$

$$30.75 = 30 + (36 - 30)\frac{(P_{<\bar{x}} - 47.29)}{(63.54 - 47.29)}$$

$$P_{<\bar{x}} = 49.32\%$$

The percentage of deliveries requiring a trip with a distance in kilometers less than 30.75 Km is equal to 49.32%, which represents 237 deliveries.

j: Let's calculate the trip represented by the 10% of the most profitable deliveries (in terms of mileage).

Let D_{10} be this distance; we use the linear interpolation formula using the data pair $(a_i, f_i \uparrow)$:

$$D_{10} = 12 + (18 - 12)\frac{(10 - 8.75)}{(19{,}58 - 8.75)}$$

$$D_{10} = 12.69 \text{ Km}$$

The 10% most profitable deliveries are around 12.69 Km.

k: Let's calculate the percentage of deliveries requiring a trip with a distance in kilometers between $\bar{x} - \sigma$ and $\bar{x} + \sigma$.

k-1: The mean plus or minus the standard deviation is:

$$\bar{x} - \sigma = 30.75 - 13.049425$$
$$= 17.701 \text{ Km}$$

$$\bar{x} + \sigma = 30.75 + 13.049425$$
$$= 43.799 \text{ Km}$$

k-2: Finding the percentage of deliveries requiring a trip with a distance in kilometers less than $\bar{x} - \sigma$, i.e., less than 17.701 Km.

Let $P_{<\bar{x}-\sigma}$ be the percentage to find; we use the linear interpolation formula using the data pair $(a_i, f_i \uparrow)$:

$$\bar{x} - \sigma = a_i + (a_{i+1} - a_i)\frac{(P_{<\bar{x}-\sigma} - F_i)}{(F_{i+1} - F_i)}$$

$$17.701 = 12 + (18 - 12)\frac{(P_{<\bar{x}-\sigma} - 8.75)}{(19.58 - 8.75)}$$

$$P_{<\bar{x}-\sigma} = 19.04\%$$

The percentage of deliveries requiring a trip with a distance in kilometers less than 17.701 Km is equal to 19.04. This represents 91 deliveries.

k-3: Finding the percentage of deliveries requiring a trip with a distance in kilometers less than $\bar{x} + \sigma$, i.e., less than 43.799 Km.

Let $P_{<\bar{x}+\sigma}$ be the percentage to find; we use the linear interpolation formula using the data pair $(a_i, f_i \uparrow)$:

$$\bar{x} + \sigma = a_i + (a_{i+1} - a_i)\frac{(P_{<\bar{x}+\sigma} - F_i)}{(F_{i+1} - F_i)}$$

$$43.799 = 24 + (48 - 42)\frac{(P_{<\bar{x}+\sigma} - 32.5)}{(47.29 - 32.5)}$$

$$P_{<\bar{x}+\sigma} = 81.30\%$$

The percentage of deliveries requiring a trip with a distance in kilometers less than 43.799 Km is equal to 81.30%. This represents 390 deliveries.

k-4: The percentage of deliveries requiring a trip with a distance in kilometers between $\bar{x} - \sigma$ = 17.701 Km and $\bar{x} + \sigma$ = 43.799 Km is:

$$81.30 - 19.04 = 62.26\%$$

This is 390-91=299 deliveries.

i: Let's calculate the central moments of order 0 to 4 from the expended formula:

l-1: Determining non-central moments of order 0 to 4:

$$m_0 = 1$$

$$m_1 = \frac{1}{n}\sum_{i=1}^{10} n_i x_i = \frac{14{,}760}{480}$$

$$= 30.75$$

$$m_2 = \frac{1}{n}\sum_{i=1}^{10} n_i {x_i}^2 = \frac{535{,}608}{480}$$

$$= 1{,}115.85$$

$$m_3 = \frac{1}{n}\sum_{i=1}^{10} n_i {x_i}^3 = \frac{21{,}457{,}872}{480}$$

$$= 44{,}703.9$$

$$m_4 = \frac{1}{n}\sum_{i=1}^{10} n_i {x_i}^4 = \frac{917{,}740.368}{480}$$

$$= 1{,}911{,}959.10$$

l-2: Let's calculate central moments of order 0 to 4:

$$\mu_0 = 1$$

$$\mu_1 = 0$$

$$\mu_2 = m_2 - m_1^2$$
$$= 1{,}115.85 - (30.75)^2$$
$$= 170.29$$

$$\mu_3 = m_3 - 3m_1m_2 + 2m_1^3$$
$$= 44{,}703.90 - 3(30.75)(1{,}115.85) + 2(30.75)^3$$
$$= -81.17$$

$$\mu_4 = m_4 - 4m_1m_3 + 6m_1^2m_2 - 3m_1^4$$
$$= 1{,}911{,}959.10 - 4(30.75)(44{,}703.90) + 6(30.75)^2(1{,}115.85) - 3(30.75)^4$$
$$= 61{,}749.57$$

INTERPRETATION OF THE RESULTS

h. Over half of the deliveries, specifically 55.415% of the total (equivalent to 266 deliveries), require a distance traveled of less than 33 kilometers. This indicates that the typical distance traveled for the majority of deliveries is relatively short. This information can be useful in optimizing delivery routes, reducing transportation costs, and improving the overall efficiency of the supply process.

i. The percentage of deliveries requiring a distance of less than 30.75 km is 49.32%, which corresponds to a total of 237 deliveries. This information highlights that nearly half of the deliveries are made over short distances, with less than 30.75 km to travel. This indicates that the majority of deliveries are likely local and involve relatively short trips.

j. It is noteworthy that the average distance for the top 10% most profitable deliveries is 12.69 kilometers. This information highlights the importance of effective route management to maximize the profitability of logistics operations. By optimizing route planning for short distances, it is possible to reduce transportation costs, improve customer satisfaction with shorter delivery times, and maximize overall company profits.

k. It is noteworthy that among all the deliveries made, a proportion of 62.26% of them, totaling 299 deliveries, require a distance ranging from 17.701 kilometers to 43.799 kilometers. This range of distances exhibits an interesting characteristic, as it shows a slight deviation from a normal distribution, where one would generally expect a proportion of 68.27% of the population for this distance range. Thus, this distribution of deliveries demonstrates a slightly lower concentration within this distance interval, with 62.26% of the deliveries falling within it.

In a real-life situation this difference could be significant and may reflect variations in delivery patterns, customer preferences, or specific geographical conditions in the delivery area.

5.10.2 Cross-Chapter Exercises

• **Exercise 21:** Let's continue **13** from section **4.8.1.** It is a table that provides the number of people recovered from a viral pandemic in a country, as well as the length of hospital stay (in days) before total recovery is confirmed and pronounced by the medical body for a given period of time. The data is divided into intervals of hospital stay, ranging from less than 5 days to over 70 days.

Data from Exercise 13

(x_i) = the length of stay in the hospital		(n_i) = number of patients
Mode = 65 days	Mean = 66.75 days	Median = 69.123 days

Time spent in the hospital	x_i	n_i	$n_i x_i$	f_i	$f_i\uparrow$
[0;5)	2.5	12711	31777.5	4.48	4.48
[5;20)	12.5	3512	43900	1.24	5.72
[20;40)	30	13024	390720	4.59	10.31
[40;60)	50	52646	2632300	18.56	28.88
[60;70)	65	65665	4268225	23.15	52.03
[70;100)	85	136051	11564335	47.97	100.00
Total		283609	18931257.5	100.00	

a. Determine the first and ninth decile of the distribution.

b. Determine the interquartile Range.

c. Determine the Variance.

d. Determine the standard deviation.

e. Determine the coefficient of variation of the distribution.

f. Determine the percentage of people cured whose hospital stay time is between $\bar{x} + \sigma$ and $\bar{x} - \sigma$.

g. Determine the central moments of order 0 to 4 from the expended formula.

Solution

The following table shows the data required to solve this exercise:

Classes	x_i	n_i	$n_i x_i$	$n_i x_i^2$	$n_i x_i^3$	$n_i x_i^4$	f_i	$f_i\uparrow$
[0;5)	2.5	12711	31777.5	79443.75	198609	496523	4.48	4.48
[5;20)	12.5	3512	43900	548750	6859375	85742188	1.24	5.72
[20;40)	30	13024	390720	11721600	351648000	10549440000	4.59	10.31
[40;60)	50	52646	2632300	131615000	6580750000	329037500000	18.56	28.88
[60;70)	65	65665	4268225	277434625	18033250625	1172161290625	23.15	52.03
[70;100)	85	136051	11564335	982968475	83552320375	7101947231875	47.97	100.00
Total		283609	18931257.5	1404367894	108525026984	8613781701211	100.00	

a: Determine the first and ninth decile of the distribution:

$$D_1 = a_i + (a_{i+1} - a_i)\frac{(10 - F_i)}{(F_{i+1} - F_i)}$$
$$= 20 + (40 - 20)\frac{(10 - 5.72)}{(10.31 - 5.72)}$$
$$= 38.65 \text{ days}$$
$$= \text{ 38 days and 15 hours}$$

$$D_9 = a_i + (a_{i+1} - a_i)\frac{(90 - F_i)}{(F_{i+1} - F_i)}$$
$$= 70 + (100 - 70)\frac{(90 - 52.03)}{(100 - 52.03)}$$
$$= 93.75 \text{ days}$$
$$= \text{ 93 days and 18 hours}$$

b: Finding the Interquartile Range:

$$Q_1 = a_i + (a_{i+1} - a_i)\frac{(25 - F_i)}{(F_{i+1} - F_i)}$$
$$= 40 + (60 - 40)\frac{(25 - 10.31)}{(28.88 - 10.31)}$$
$$= 55.82 \text{ days}$$
$$= \text{ 55 days and 19 hours}$$

$$Q_3 = a_i + (a_{i+1} - a_i)\frac{(75 - F_i)}{(F_{i+1} - F_i)}$$
$$= 70 + (100 - 70)\frac{(75 - 52.03)}{(100 - 52.03)}$$
$$= 84.37 \text{ days}$$
$$= \text{ 84 days and 8 hours}$$

Interquartile Range $= Q_3 - Q_1$:

$$= 84.37 \text{ days} - 55.82 \text{ days}$$

$$= 28.55 \text{ days}$$
$$= 28 \text{ days and } 13 \text{ hours}$$

c: The variance of the distribution is:

$$V(x) = \frac{1}{n}\sum_{i=1}^{i=6} n_i x_i^2 - \bar{x}^2$$
$$= \frac{1{,}404{,}367{,}894}{283{,}609} - (66.75\,)^2$$
$$= 496.04$$

d: The standard deviation of the distribution is:

$$\sigma = \sqrt{V(x)}$$
$$= \sqrt{496.04}$$
$$= 22.27$$
$$= 22 \text{ days and } 6 \text{ hours}$$

e: The coefficient of variation of the distribution is:

$$\text{CV} = \frac{\sigma}{|\bar{x}|}$$
$$= \frac{22.27}{66.75}\text{x}100$$
$$= 33\%$$

f: Let's calculate the percentage of people healed and whose hospital stay is between $\bar{x} + \sigma$ and $\bar{x} - \sigma$.

f-1: The mean plus or minus the standard deviation is:

$$\bar{x} - \sigma = 66.75 - 22.27$$
$$= 44.48 \text{ days}$$
$$\bar{x} + \sigma = 66.75 + 22.27$$
$$= 88.99 \text{ days}$$

f-2: Finding the percentage of people healed and whose hospital stay is less than $\bar{x} - \sigma$, i.e., less than 44.48 days.

Let $P_{<\bar{x}-\sigma}$ be the percentage to find; we use the linear interpolation formula using the data pair $(a_i, f_i \uparrow)$:

$$\bar{x} - \sigma = a_i + (a_{i+1} - a_i)\frac{(P_{<\bar{x}-\sigma} - F_i)}{(F_{i+1} - F_i)}$$

$$44.48 = 40 + (60 - 40)\frac{(P_{<\bar{x}-\sigma} - 10.31)}{(28.88 - 10.31)}$$

$$P_{<\bar{x}-\sigma} = 14.47\%$$

The percentage of people healed and whose hospital stay is less than 44.48 days is equal to 14.47%. This represents 41,038 people.

f-3: Finding the percentage of people healed and whose hospital stay is less than $\bar{x} + \sigma$, i.e., less than 88.99 days.

Let $P_{<\bar{x}+\sigma}$ be the percentage to find; we use the linear interpolation formula using the data pair $(a_i, f_i \uparrow)$:

$$\bar{x} + \sigma = a_i + (a_{i+1} - a_i)\frac{(P_{<\bar{x}+\sigma} - F_i)}{(F_{i+1} - F_i)}$$

$$88.99 = 70 + (100 - 70)\frac{(P_{<\bar{x}+\sigma} - 52.03)}{(100 - 52.03)}$$

$$P_{<\bar{x}+\sigma} = 82.39\%$$

The percentage of people who are treated and whose hospital stay is less than 88.99 days is equal to 82.39%. This represents 233,665 people.

f-4: The percentage of people healed and whose hospital stay is between $\bar{x} - \sigma = 44.48$ days and $\bar{x} + \sigma = 88.99$ days is:

$$82.39 - 14.47 = 67.92\%$$

This is 233,665 – 41,038= 192,627 people.

g: Let's calculate the central moments of order 0 to 4 from the expended formula.

g-1: Let's calculate non-central moments of order 0 to 4:

$$m_0 = 1$$

$$m_1 = \frac{1}{n}\sum_{i=1}^{i=6} n_i x_i = \frac{18,931,257.5}{283,609}$$

$$= 66.75$$

$$m_2 = \frac{1}{n}\sum_{i=1}^{i=6} n_i {x_i}^2 = \frac{1,404,367,893.75}{283,609}$$

$$= 4,951,77$$

$$m_3 = \frac{1}{n}\sum_{i=1}^{i=6} n_i {x_i}^3 = \frac{108,525,026,984.38}{283,609}$$

$$= 382,657.20$$

$$m_4 = \frac{1}{n}\sum_{i=1}^{i=6} n_i {x_i}^4 = \frac{8,613,781,701,210.94}{283,609}$$

$$= 30,372,032.27$$

g-2: Let's calculate central moments of order 0 to 4:

$$\mu_0 = 1$$

$$\mu_1 = 0$$
$$\mu_2 = m_2 - m_1^2$$
$$= 4{,}951.77 - (66.75)^2$$
$$= 1.38$$

$$\mu_3 = m_3 - 3m_1m_2 + 2m_1^3$$
$$= 382{,}657.20 - 3(66.75)(4{,}951.77) + 2(66.75)^3$$
$$= -14{,}103.16$$
$$\mu_4 = m_4 - 4m_1m_3 + 6m_1^2m_2 - 3m_1^4$$
$$= 30{,}372{,}032.27 - 4(66.75)(382{,}657.20) + 6(66.75)^2(4{,}951.77) - 3(66.75)^4$$
$$= 1{,}022{,}673.86$$

INTERPRETATION OF THE RESULTS

The analysis of the data reveals that:

a. 10% of the patients were discharged from the hospital before spending a total of 38 days and 15 hours in the recovery period. This indicates a rapid improvement in their health condition and an early recovery. On the other hand, the majority of the patients, accounting for 90%, were discharged from the hospital before spending more than 93 days and 18 hours in the recovery period, suggesting a longer length of stay for their recuperation.

b. Upon closer examination of the data, it is observed that the maximum difference between two hospital stays in the median half of the series is approximately 28 days and 13 hours. This finding highlights the fact that hospitalization durations are relatively similar for the majority of patients in this part of the distribution. This low disparity suggests a certain consistency in recovery times within this population.

e. The coefficient of variation is 33%. This value indicates a moderate dispersion of hospitalization times before recovery. In other words, the recovery durations vary relatively moderately around the mean, indicating some level of homogeneity in the distribution of data.

f. The findings reveal that 67.87% of the recovered individuals required a hospital stay duration ranging from 44.48 days to 88.99 days. This time interval represents the period during which these individuals were hospitalized before being considered as recovered. Compared to a normal distribution, this range shows a slight disparity, with a slightly lower data density. Specifically, the proportion of recovered individuals within this range, which is 67.87%, is slightly lower than the expected proportion of 68.27% in a normal distribution based on the studied population.

There could be several reasons for this slightly lower data density in this time range. It is possible that some patients experienced medical complications or underwent more complex treatments, which extended their hospital stay. Other factors such as age, initial health condition, response to treatment, and available medical resources can also influence the duration of hospital stay. It is important to consider these various factors to understand the variations in this specific distribution.

• **Exercise 22:** Let's continue with exercise **14** from section **4.8.1**, which focuses on the company's fleet of vehicles, consisting of a large number of cars. For 100 of them, the odometer readings at the time of maintenance are given in the following table:

Mileage (in km)	x_i	n_i	$n_i x_i$	$f_i \uparrow$
[80000;85000)	82500	5	412500	5.00
[85000;90000)	87500	9	787500	14.00
[90000;95000)	92500	14	1295000	28.00
[95000;100000)	97500	18	1755000	46.00
[100000;105000)	102500	25	2562500	71.00
[105000;110000)	107500	16	1720000	87.00
[110000;115000)	112500	7	787500	94.00
[115000;120000)	117500	6	705000	100.00
Total		100	10025000	

a. Determine the interquartile Range.

b. Determine the Variance.

c. Determine the standard deviation.

d. Determine the coefficient of variation of the distribution.

e. Determine the percentage of vehicles with odometer readings. between $\bar{x} + \sigma$ and $\bar{x} - \sigma$ at the time of routine service.

f. Determine the central moments of order 0 to 4 from the expended formula.

Data from Exercise 14

(x_i) = mileage (in Km)		(n_i) = number of cars
Mode = 102,500 Km	Mean = 100,250 Km	Median = 100,800 Km

Solution

The following table shows the data required to solve this exercise:

Classes	x_i	n_i	$n_i x_i$	$n_i x_i^2$	$f_i\uparrow$
[80000;85000)	82500	5	412500	34031250000	5.00
[85000;90000)	87500	9	787500	68906250000	14.00
[90000;95000)	92500	14	1295000	119787500000	28.00
[95000;100000)	97500	18	1755000	171112500000	46.00
[100000;105000)	102500	25	2562500	262656250000	71.00
[105000;110000)	107500	16	1720000	184900000000	87.00
[110000;115000)	112500	7	787500	88593750000	94.00
[115000;120000)	117500	6	705000	82837500000	100.00
Total		100	10025000	1012825000000	445

a: The Interquartile Range of the distribution is:

$$Q_1 = a_i + (a_{i+1} - a_i)\frac{(25 - F_i)}{(F_{i+1} - F_i)}$$

$$= 90{,}000 + (95{,}000 - 90{,}000)\frac{(25 - 14)}{(28 - 14)}$$

$$= 93{,}928.57 \text{ Km}$$

$$Q_3 = a_i + (a_{i+1} - a_i)\frac{(75 - F_i)}{(F_{i+1} - F_i)}$$

$$= 105{,}000 + (110{,}000 - 105{,}000)\frac{(75 - 71)}{(87 - 71)}$$

$$= 106{,}250 \text{ Km}$$

Interquartile Range $= Q_3 - Q_1$:

$$= 106{,}250 \text{ Km} - 93{,}928.57 \text{ Km}$$

$$= 12{,}321.43 \text{ Km}$$

b: The variance of the distribution is:

$$V(x) = \frac{1}{n}\sum_{i=1}^{i=8} n_i x_i^2 - \bar{x}^2$$

$$= \frac{1{,}012{,}825{,}000{,}000}{100} - (100{,}250\ )^2$$

$$= 78{,}187{,}500$$

c: The standard deviation of the distribution is:

$$\sigma = \sqrt{V(x)}$$

$$= \sqrt{78{,}187{,}500}$$

$= 8{,}842.37$ Km

d: The coefficient of variation of the distribution is:

$$\text{CV} = \frac{\sigma}{|\bar{x}|}\text{x}100$$

$$= \frac{8{,}842.369}{100{,}250}\text{x}100$$

$$= 8.82\%$$

e: Let's calculate the percentage of vehicles with odometer readings between $\bar{x} + \sigma$ and $\bar{x} - \sigma$ at the time of routine service:

e-1: The mean plus or minus the standard deviation is:

$$\bar{x} - \sigma = 100{,}250 - 8{,}842.369$$

$$= 91{,}407.631 \text{ Km}$$

$$\bar{x} + \sigma = 100{,}250 + 8{,}842.369$$

$$= 109{,}092.369 \text{ Km}$$

e-2: Finding the percentage of vehicles with odometer readings less than $\bar{x} - \sigma$, i.e., less than 91,407.631 Km.

Let $P_{<\bar{x}-\sigma}$ be the percentage to find; we use the linear interpolation formula using the data pair $(a_i, f_i \uparrow)$:

$$\bar{x} - \sigma = a_i + (a_{i+1} - a_i)\frac{(P_{<\bar{x}-\sigma} - F_i)}{(F_{i+1} - F_i)}$$

$$91{,}407.631 = 90{,}000 + (95{,}000 - 90{,}000)\frac{(P_{<\bar{x}-\sigma} - 14)}{(28 - 14)}$$

$$P_{<\bar{x}-\sigma} = 17.9\%$$

The percentage of vehicles with odometer readings is less than 91,407.631 Km. This represents 18 vehicles.

e-3: Finding the percentage of vehicles with odometer readings less than $\bar{x} + \sigma$, i.e., less than 109,092.369 Km.

Let $P_{<\bar{x}+\sigma}$ be the percentage to find; we use the linear interpolation formula using the data pair $(a_i, f_i \uparrow)$:

$$\bar{x} + \sigma = a_i + (a_{i+1} - a_i)\frac{(P_{<\bar{x}+\sigma} - F_i)}{(F_{i+1} - F_i)}$$

$$109{,}092.369 = 105{,}000 + (110{,}000 - 105{,}000)\frac{(P_{<\bar{x}+\sigma} - 71)}{(87 - 71)}$$

$$P_{<\bar{x}+\sigma} = 84.09\%$$

The percentage of vehicles with odometer reading less than 109,092.369 Km is 84.09%. This represents 84 vehicles.

e-4: The percentage of vehicles that display odometer readings between $\bar{x} - \sigma = 91.407,631$ Km and $\bar{x} + \sigma = 109.092,369$ Km is:

$$84.09 - 17.9 = 66.19\%$$

This is 84−18 = 66 vehicles.

f: Let's calculate the central moments of order 0 to 4 from the expended formula.

f-1: Let's calculate non-central moments of order 0 to 4:

$$m_0 = 1$$

$$m_1 = \frac{1}{n}\sum_{i=1}^{8} n_i x_i = \frac{10,025,000}{100}$$

$$= 100,250$$

$$m_2 = \frac{1}{n}\sum_{i=1}^{8} n_i {x_i}^2 = \frac{1,012,825,000,000}{100}$$

$$= 10,128,250,000$$

$$m_3 = \frac{1}{n}\sum_{i=1}^{8} n_i {x_i}^3 = \frac{103,099,906,250,000,000}{100}$$

$$= 1,030,999,062,500,000$$

$$m_4 = \frac{1}{n}\sum_{i=1}^{8} n_i {x_i}^4 = \frac{10,571,981,406,250,000,000,000}{100}$$

$$= 105,719,814,062,500,000,000$$

f-2: Let's calculate central moments of order 0 to 4:

$$\mu_0 = 1$$

$$\mu_1 = 0$$

$$\mu_2 = m_2 - m_1^2$$

$$= 10,128,250,000 - (100,250)^2$$

$$= 78,187,500$$

$$\mu_3 = m_3 - 3m_1m_2 + 2m_1^3$$

$$= 1,030,999,062,500,000 - 3(100,250)(10,128,250,000) + 2(100,250)^3$$

$$= -34,593,750,000$$

$$\mu_4 = m_4 - 4m_1m_3 + 6m_1^2m_2 - 3m_1^4$$

$$= 105,719,814,062,500,000,000 - 4(100,250)(1,030,999,062,500,000) + 6(100,250)^2(10,128,250,000) - 3(100,250)^4$$

$$= 15,194,332,031,287,300$$

INTERPRETATION OF THE RESULTS

a. Upon examining the odometer readings of cars at the time of maintenance, it is evident that when arranged in ascending order, the largest difference between two cars in the median half of the series is approximately 12,321.43 km. This measure highlights a limited dispersion of data and a certain consistency in the displayed odometer readings within this range of the series, indicating coherence in the observed data.

d. The coefficient of variation, which is 8.8%, demonstrates that the displayed mileages at the time of maintenance are very minimally dispersed. This low variation indicates a high level of consistency in the mileage covered by the cars before maintenance, which may suggest a coherent and regular usage pattern for these vehicles.

e. The results show that nearly two-thirds of the vehicles, specifically 66.19% of them, display a mileage between 91,407.631 Km and 109,092.369 Km at the time of maintenance. Compared to a normal distribution, this mileage range is slightly less populated, with a difference of 2.08% compared to the expected population of 68.27%. This suggests that the distribution of mileage for the company's vehicles exhibits a slight deviation from normal distribution.

There are several possible reasons that can explain why 66.19% of the vehicles display a mileage between 91,407.631 Km and 109,092.369 Km at the time of maintenance. Here are some possible hypotheses:

- Nature of vehicle usage: It is possible that the company's vehicles are mainly used for short-distance trips, resulting in relatively low mileage accumulation.
- Fleet management policy of the company: The company may have a policy of regular vehicle replacement when they reach a certain mileage, which could explain why the majority of vehicles have relatively similar mileage at the time of maintenance.
- Observation period: The data may have been collected over a specific period, which can influence the distribution of displayed mileage at the time of maintenance. For example, if the data were collected during a period when the company conducted massive vehicle replacements, it may explain the concentration of mileage in a specific range.
- Preventive maintenance: The company may have a policy of regular preventive maintenance for its vehicles, which helps to maintain relatively similar mileage among the vehicles.

• **Exercise 23:** Continuing exercise **16** from section **4.8.2** on the number of production days tires in a factory. The following table gives the number of days compared to the number of tires produced:

Data from Exercise 16

(x_i) = number of tires	(n_i) = number of days

	Mean = 14,216,67 tires	Median = 14,250 tires

Number of tires	x_i	n_i	$n_i x_i$	f_i	$f_i\uparrow$
[12500;13000)	12750	1	12750	6.67	6.67
[13000;13500)	13250	2	26500	13.33	20.00
[13500;14000)	13750	3	41250	20.00	40.00
[14000;14500)	14250	3	42750	20.00	60.00
[14500;15000)	14750	3	44250	20.00	80.00
[15000;15500)	15250	3	45750	20.00	100.00
Total		15	213250	100	

a. Determine the interquartile Range.

b. Determine the Variance.

c. Determine the standard deviation.

d. Determine the coefficient of variation.

e. Determine the percentage of time required (in days) to manufacture between $\bar{x} + \sigma$ and $\bar{x} - \sigma$ tires.

f. Determine the central moments of order 0 to 4 from the expended formula.

Solution

The following table shows the data required to solve this exercise:

Nomber of tires	x_i	n_i	$n_i x_i$	$n_i x_i^2$	$n_i x_i^3$	$n_i x_i^4$	f_i	$f_i\uparrow$
[12500;13000)	12750	1	12750	162562500	2072671875000	26426566406250000	6.67	6.67
[13000;13500)	13250	2	26500	351125000	4652406250000	61644382812500000	13.33	20.00
[13500;14000)	13750	3	41250	567187500	7798828125000	107233886718750000	20.00	40.00
[14000;14500)	14250	3	42750	609187500	8680921875000	123703136718750000	20.00	60.00
[14500;15000)	14750	3	44250	652687500	9627140625000	142000324218750000	20.00	80.00
[15000;15500)	15250	3	45750	697687500	10639734375000	162255949218750000	20.00	100.00
Total		15	213250	3040437500	43471703125000	623264246093750000	100.00	

a: Finding the Interquartile Range of the distribution:

$$Q_1 = a_i + (a_{i+1} - a_i)\frac{(25 - F_i)}{(F_{i+1} - F_i)}$$
$$= 13{,}500 + (14{,}000 - 13{,}500)\frac{(25 - 20)}{(40 - 20)}$$
$$= 13{,}625 \text{ tires}$$

$$Q_3 = a_i + (a_{i+1} - a_i)\frac{(75 - F_i)}{(F_{i+1} - F_i)}$$
$$= 14{,}500 + (15{,}000 - 14{,}500)\frac{(75 - 60)}{(80 - 60)}$$
$$= 14{,}875 \text{ tires}$$

Interquartile Range $= Q_3 - Q_1$:

$$= 14.875 \text{ tires} - 13.625 \text{ tires}$$
$$= 1{,}250 \text{ tires}$$

b: The variance of the distribution is:

$$V(X) = \frac{1}{n}\sum_{i=1}^{i=6} n_i x_i^2 - \bar{x}^2$$
$$= \frac{3{,}040{,}437{,}500}{15} - (14{,}216.67\,)^2$$
$$= 582{,}127.44$$

c: The standard deviation of the distribution is:

$$\sigma = \sqrt{V(x)}$$
$$= \sqrt{582{,}127.44}$$
$$= 763 \text{ tires}$$

d: The coefficient of variation of the distribution is:

$$\text{CV} = \frac{\sigma}{|\bar{x}|}\text{x}100$$
$$= \frac{763}{14{,}217}\text{x}100$$
$$= 5.37\%$$

e: Let's calculate the percentage of time required (in days) to manufacture between $\bar{x} + \sigma$ and $\bar{x} - \sigma$ tires.

e-1: The mean plus or minus the standard deviation is:

$$\bar{x} - \sigma = 14{,}217 - 763$$
$$= 13{,}454 \text{ tires}$$

$$\bar{x} + \sigma = 14{,}217 + 763$$
$$= 14{,}980 \text{ tires}$$

e-2: Let's calculate the percentage of time required (in days) to manufacture $\bar{x} + \sigma$ tires (that is 13,453.97 tires).

Let $P_{<\bar{x}-\sigma}$ be the percentage to find; we use the linear interpolation formula using the data pair $(a_i, f_i \uparrow)$:

$$\bar{x} - \sigma = a_i + (a_{i+1} - a_i)\frac{(P_{<\bar{x}-\sigma} - F_i)}{(F_{i+1} - F_i)}$$

$$13{,}454 = 13{,}000 + (13{,}500 - 13{,}000)\frac{(P_{<\bar{x}-\sigma} - 6.67)}{(20 - 6.67)}$$

$$P_{<\bar{x}-\sigma} = 18.77\%$$

The percentage of time required (in days) to manufacture $\bar{x} + \sigma$ tires (that is 13,453.97 tires) is equal to 18.77%, which represents 2.8 days.

e-3: Let's calculate the percentage of time required (in days) to manufacture $\bar{x} + \sigma$ tires (that is 14,980.03 tires).

Let $P_{<\bar{x}+\sigma}$ be the percentage to find; we use the linear interpolation formula using the data pair $(a_i, f_i \uparrow)$:

$$\bar{x} + \sigma = a_i + (a_{i+1} - a_i)\frac{(P_{<\bar{x}+\sigma} - F_i)}{(F_{i+1} - F_i)}$$

$$14{,}980 = 14{,}500 + (15{,}000 - 14{,}500)\frac{(P_{<\bar{x}+\sigma} - 60)}{(80 - 60)}$$

$$P_{<\bar{x}+\sigma} = 79.20\%$$

The percentage of time required (in days) to manufacture $\bar{x} + \sigma$ tires (that is 14,980.03 tires) is equal to 79.20%, which represents 11.87 days.

e-4: The percentage of time required (in days) to manufacture a quantity of tires included between $\bar{x} - \sigma$ = 13,453.97 tires and $\bar{x} + \sigma$ = 14,980.03 tires is:

$$79.20 - 18.77 = 60.43\%$$

This represents 9 days.

f: Let's calculate the central moments of order 0 to 4 from the expended formula:

f-1: Let's calculate non-central moments of order 0 to 4:

$$m_0 = 1$$

$$m_1 = \frac{1}{n}\sum_{i=1}^{i=6} n_i x_i = \frac{213,250}{15}$$
$$= 14,216.7$$
$$m_2 = \frac{1}{n}\sum_{i=1}^{i=6} n_i {x_i}^2 = \frac{3,040,437,500}{15}$$
$$= 202,695,833$$
$$m_3 = \frac{1}{n}\sum_{i=1}^{i=6} n_i {x_i}^3 = \frac{43,471,703,125,000}{15}$$
$$= 2,898,113,541,666.67$$
$$m_4 = \frac{1}{n}\sum_{i=1}^{i=6} n_i {x_i}^4 = \frac{623,264,246,093,750,000}{15}$$
$$= 41,550,949,739,583,300$$

f-2: Let's calculate central moments of order 0 to 4:

$$\mu_0 = 1$$
$$\mu_1 = 0$$
$$\mu_2 = m_2 - m_1^2$$
$$= 202,695,833 - (14,216.7)^2$$
$$= 582,222$$
$$\mu_3 = m_3 - 3m_1m_2 + 2m_1^3$$
$$= 2,898,113,541,666.67 - 3(14,216.7)(202,695,833) + 2(14,216.7)^3$$
$$= -100,074,074$$
$$\mu_4 = m_4 - 4m_1m_3 + 6m_1^2m_2 - 3m_1^4$$
$$= 33.83 - 4(14,216.7)(10.40) + 6(14,216.7)^2(202,695,833) - 3(14,216.7)^4$$
$$= 678,607,407,376$$

INTERPRETATION OF THE RESULTS

a. The difference of 1250 tires between the production days in the median half of the series can have several interpretations. Firstly, it may indicate that tire production in the factory is relatively stable and consistent, with few significant variations in production days. This could also suggest that production processes are well controlled and optimized, minimizing production gaps. Furthermore, this relatively small difference may reflect effective production planning and balanced distribution of production days across the different ranges of tire quantities produced.

d. The coefficient of variation of 5.37% reflects a very low dispersion in the time required for tire production. This suggests that production times are relatively consistent, with little variation from one tire to another. This low dispersion may indicate effective control of production processes, with consistency in the time required to produce each tire.

e. Approximately 60.43% of the total manufacturing time (equivalent to 9 days) is required to produce a quantity of tires ranging from 13,454 to 14,980 units. This distribution, located within the interval between the mean minus the standard deviation and the mean plus the standard deviation [13,453.97; 14,980.03], has slightly fewer tires than would be expected according to a normal distribution. Specifically, only 60.43% of the total population falls within this interval, whereas in a normal distribution, we would expect to find 68.27%.

There are several possible reasons that could explain the low rate within this interval, including:

- Production variability: The tire manufacturing process may be subject to variations, leading to fluctuations in the quantity produced within this specific interval. Factors such as the quality of raw materials, precision of production machines, or skills of the workers can influence production variability.
- Capacity constraints: It is possible that the production capacity of the factory is not optimized within this range of tire quantities, which could limit production and explain why fewer tires are produced within this interval compared to other intervals.
- Market demand: Tire production can be influenced by market demand, with fluctuations in demand for different quantities of tires. If the demand for specific quantities of tires is lower, it can explain why fewer tires are produced within this interval.
- Seasonal factors: Seasonal variations can also play a role in tire production. For example, if the demand for tires varies depending on the seasons (e.g., winter tires for winter season), it can influence production within this specific interval.

• **Exercise 24:** Continuing exercise **17** from section **4.8.2.** The table below represents the balances recorded in the savings accounts of 300 individuals.

Data from Exercise 17

(x_i) = saving balance		(n_i) = number of savers
Mode = 2 500 euros	Mean = 31,483 euros	Median = 19,670 euros

Saving Balance	x_i	n_i	$n_i xi$	f_i	$f_i\uparrow$
[0;5000)	2500	48	120000	16.00	16.00
[5000;10000)	7500	41	307500	13.67	29.67
[10000;15000)	12500	47	587500	15.67	45.33
[15000;20000)	17500	15	262500	5.00	50.33
[20000;25000)	22500	21	472500	7.00	57.33
[25000;30000)	27500	12	330000	4.00	61.33
[30000;35000)	32500	13	422500	4.33	65.67
[35000;40000)	37500	8	300000	2.67	68.33
[40000;45000)	42500	9	382500	3.00	71.33
[45000; 50000)	47500	9	427500	3.00	74.33
[50000;55000)	52500	10	525000	3.33	77.67
[55000;60000)	57500	6	345000	2.00	79.67
[60000;65000)	62500	9	562500	3.00	82.67
[65000;70000)	67500	5	337500	1.67	84.33
[70000;75000)	72500	2	145000	0.67	85.00
[75000;80000)	77500	13	1007500	4.33	89.33
[80000;85000)	82500	8	660000	2.67	92.00
[85000;90000)	87500	7	612500	2.33	94.33
[90000;95000)	92500	4	370000	1.33	95.67
[95000;100000)	97500	13	1267500	4.33	100.00
Total		300	9445000	100.00	

a. Determine the interquartile Range.

b. Determine the Variance.

c. Determine the standard deviation.

d. Determine the coefficient of variation of the distribution.

e. Determine the percentage of people with savings between $\bar{x} + \sigma$ and $\bar{x} - \sigma$.

f. Determine the central moments of order 0 to 4 from the expended formula.

Data from Exercise 17

(x_i) = saving balance		(n_i) = number of savers
Mode = 2,500 euros	Mean = 31,483 euros	Median = 19,670 euros

Solution

The following table shows the data required to solve this exercise:

Classes	x_i	n_i	$n_i x_i$	$n_i x_i^2$	f_i	$f_i \uparrow$
[0;5000)	2500	48	120000	300000000	16.00	16.00
[5000;10000)	7500	41	307500	2306250000	13.67	29.67
[10000;15000)	12500	47	587500	7343750000	15.67	45.33
[15000;20000)	17500	15	262500	4593750000	5.00	50.33
[20000;25000)	22500	21	472500	10631250000	7.00	57.33
[25000;30000)	27500	12	330000	9075000000	4.00	61.33
[30000;35000)	32500	13	422500	13731250000	4.33	65.67
[35000;40000)	37500	8	300000	11250000000	2.67	68.33
[40000;45000)	42500	9	382500	16256250000	3.00	71.33
[45000;50000)	47500	9	427500	20306250000	3.00	74.33
[50000;55000)	52500	10	525000	27562500000	3.33	77.67
[55000;60000)	57500	6	345000	19837500000	2.00	79.67
[60000;65000)	62500	9	562500	35156250000	3.00	82.67
[65000;70000)	67500	5	337500	22781250000	1.67	84.33
[70000;75000)	72500	2	145000	10512500000	0.67	85.00
[75000;80000)	77500	13	1007500	78081250000	4.33	89.33
[80000;85000)	82500	8	660000	54450000000	2.67	92.00
[85000;90000)	87500	7	612500	53593750000	2.33	94.33
[90000;95000)	92500	4	370000	34225000000	1.33	95.67
[95000;100000)	97500	13	1267500	123581250000	4.33	100.00
Total		300	9445000	555575000000	100.00	

a: Finding the Interquartile Range of the distribution:

$$Q_1 = a_i + (a_{i+1} - a_i)\frac{(25 - F_i)}{(F_{i+1} - F_i)}$$

$$= 5{,}000 + (10{,}000 - 5{,}000)\frac{(25 - 16)}{(29.67 - 16)}$$

$$= 8{,}291.88 \text{ euros}$$

$$Q_3 = a_i + (a_{i+1} - a_i)\frac{(75 - F_i)}{(F_{i+1} - F_i)}$$

$$= 50{,}000 + (55{,}000 - 50{,}000)\frac{(75 - 74.33)}{(77.67 - 74.33)}$$

$$= 51{,}002.99 \text{ euros}$$

$$\text{The Interquartile Range} = Q_3 - Q_1$$
$$= 51{,}002.99 \text{ euros } - 8{,}291.88 \text{ euros}$$
$$= 42{,}711.11 \text{ euros}$$

b: The variance of the distribution is:

$$V(x) = \frac{1}{n}\sum_{i=1}^{i=20} n_i x_i^2 - \bar{x}^2$$
$$= \frac{555{,}575{,}000{,}000}{300} - (31{,}483\,)^2$$
$$= 860{,}716{,}388.9$$

c: The standard deviation of the distribution is:

$$\sigma = \sqrt{V(x)}$$
$$= \sqrt{860{,}716{,}388.9}$$
$$= 29{,}337.96 \text{ euros}$$

d: The coefficient of variation of the distribution is:

$$\text{CV} = \frac{\sigma}{|\bar{x}|} \text{x}100$$
$$= \frac{29{,}337.96}{31.483} \text{x}100$$
$$= 93.186\%$$

e: Let's calculate the percentage of people with savings between $\bar{x} - \sigma$ and $\bar{x} + \sigma$.

e-1: The mean plus or minus the standard deviation is:

$$\bar{x} - \sigma = 31{,}483 - 29{,}337.96$$
$$= 2{,}145.04 \text{ euros}$$

$$\bar{x} + \sigma = 31{,}483 + 29{,}337.96$$
$$= 60{,}820.96 \text{ euros}$$

e-2: Finding the percentage of people with savings less than $\bar{x} - \sigma$, i.e., less than 2,145.04 euros.

Let $P_{<\bar{x}-\sigma}$ be the percentage to find; we use the linear interpolation formula on $(a_i\,, f_i\uparrow)$:

$$\bar{x} - \sigma = a_i + (a_{i+1} - a_i)\frac{(P_{<\bar{x}-\sigma} - F_i)}{(F_{i+1} - F_i)}$$

$$2{,}145.04 = 0 + (5{,}000 - 0)\frac{(P_{<\bar{x}-\sigma} - 0)}{(16 - 0)}$$

$$P_{<\bar{x}-\sigma} = 6.86\%$$

The percentage of people with savings less than $\bar{x} - \sigma$, i.e., less than 2,145.04 euros is 6.86%. This represents 21 people.

e-3: Let's calculate the percentage of people with savings less than $\bar{x} + \sigma$, i.e., less than 60,820.96 euros.

Let $P_{<\bar{x}+\sigma}$ be the percentage to find; we use the linear interpolation formula using the data pair $(a_i, f_i \uparrow)$:

$$\bar{x} + \sigma = a_i + (a_{i+1} - a_i)\frac{(P_{<\bar{x}+\sigma} - F_i)}{(F_{i+1} - F_i)}$$

$$60{,}820.96 = 60{,}000 + (65{,}000 - 60{,}000)\frac{(P_{<\bar{x}+\sigma} - 79.67)}{(82.67 - 79.67)}$$

$$P_{<\bar{x}+\sigma} = 79.72\%$$

The percentage of people with savings less than 60,820.96 euros is equal to 79.72%. This represents 239 people.

e-4: The percentage of people with savings between $\bar{x} - \sigma = 2{,}145.04$ euros and $\bar{x} + \sigma = 60{,}820.96$ euros is:

$$79.72 - 6.86 = 72.86\%$$

This is 239 – 21 = 218 people

f: Let's calculate the central moments of order 0 to 4 from the expended formula:

f-1: Let's calculate non-central moments of order 0 to 4:

$$m_0 = 1$$

$$m_1 = \frac{1}{n}\sum_{i=1}^{i=20} n_i x_i = \frac{9{,}445{,}000}{300}$$

$$= 31{,}483.33$$

$$m_2 = \frac{1}{n}\sum_{i=1}^{i=20} n_i {x_i}^2 = \frac{555{,}575{,}000{,}000}{300}$$

$$= 1{,}851{,}916{,}666.67$$

$$m_3 = \frac{1}{n}\sum_{i=1}^{i=20} n_i {x_i}^3 = \frac{40,734,531,250,120,000}{300}$$
$$= 135,781,770,833,733$$
$$m_4 = \frac{1}{n}\sum_{i=1}^{i=20} n_i {x_i}^4 = \frac{3,275,906,093,750,300,000,000}{300}$$
$$= 10,919,686,979,167,700,000$$

f-2: Let's calculate central moments of order 0 to 4:

$$\mu_0 = 1$$
$$\mu_1 = 0$$
$$\mu_2 = m_2 - m_1^2$$
$$= 1,851,916,666.67 - (31,483.33)^2$$
$$= 860,716,388.89$$

$$\mu_3 = m_3 - 3m_1m_2 + 2m_1^3$$
$$= 135,781,770,833,733 - 3(31,483.33)(1,851,916,666.67) + 2(31,483.33)^3$$
$$= 23,280,819,157,807.40$$
$$\mu_4 = m_4 - 4m_1m_3 + 6m_1^2m_2 - 3m_1^4$$
$$= 10,919,686,979,167,700,000 - 4(31,483.33)(135,781,770,833,733)$$
$$+ 6(31,483.33)^2(1,851,916,666.67) - 3(31,483.33)^4$$
$$= 1,886,523,886,700,400,000$$

INTERPRETATION OF THE RESULTS

a. Upon further analysis of the savings accounts, when arranged in ascending order, it is observed that the largest difference between the savings amounts in the median half of the series is approximately 42,711.11 euros. This means that the savings amounts of individuals in the middle of the distribution vary greatly, ranging from over 42,000 euros below the median to over 42,000 euros above the median. This significant disparity highlights the diversity in savings levels among the individuals in the sample.

d. The coefficient of variation is a statistical indicator that measures the dispersion or relative variability of data within a sample. In our case, a coefficient of variation of 93% indicates that the savings amounts recorded in the savings accounts of the 300 individuals exhibit a high level of diversity, with significant differences between the observed values. This could be attributed to various factors such as differences in income levels, savings behaviors, or personal economic and financial situations.

e. It was observed that 72.86% of individuals have savings between 2,145 euros and 60,821 euros. This distribution exhibits a higher concentration within the range of [2,145; 60,821], with a larger proportion of the population compared to a normal distribution where 68.27% of the population would fall within this range. It is interesting to note that this difference may indicate a specific trend in the studied population, with a higher percentage of individuals having savings levels within this specific interval.

Furthermore, there may be several probable reasons that explain why 72.86% of individuals in the studied sample have savings levels between 2,145 euros and 60,821 euros:

- Specific savings behaviors: Individuals in this population may exhibit specific savings behaviors, such as a preference for short-term savings or a focus on financial security, which leads them to maintain savings levels within this specific interval.
- Income level: This population may have an average or median income level that corresponds to this savings range, which influences the observed savings amounts in the sample.
- Demographic profile: The demographic profile of this population, such as age, education, occupation, can influence savings levels. For example, younger individuals or those with higher income may tend to save more, which can be reflected in the concentration observed in this interval.
- Economic context: The overall economic context and financial conditions can also influence savings levels in this population. For example, a period of economic stability or uncertainty may encourage individuals to save more or maintain higher savings levels.

CHAPTER 6

Characteristics of Shape (Skewness and Kurtosis)

This chapter discusses the shape characteristics of a frequency distribution curve, which can be measured by the coefficients of skewness and kurtosis. Various coefficients are presented, including the YULE's coefficient of skewness, the PEARSON's coefficient of skewness, the FISHER's coefficients of skewness and the FISHER's coefficients of kurtosis. The coefficient of skewness measures the deviation of the curve from a symmetrical shape and helps identify where values are concentrated relative to the central values. In addition, the coefficient of kurtosis describes the degree of flatness or peakedness of the distribution curve compared to the normal curve. Overall, these coefficients provide valuable insights into the shape of a frequency distribution curve and can inform statistical analyses and decision-making processes.

6.1 Symmetry of a Distribution Curve

The symmetry of a distribution curve is an important aspect in the analysis of statistical data. A distribution is considered symmetric when the values of the variables are evenly distributed around the three central measures: the *mean*, *mode*, and *median*. This means that most of the values are located at the center of the distribution, with a balanced distribution of values on the left and right sides of these central measures.

A symmetric distribution can be represented by a distribution curve that is perfectly balanced and exhibits perfect symmetry with respect to a vertical axis.

However, in some cases, the distribution may be asymmetric or skewed, which means that there is a complete lack of symmetry in the distribution of values. For example, a positively skewed distribution is characterized by a concentration of higher values on the right side of the distribution, while a negatively skewed distribution is characterized by a concentration of higher values on the left side of the distribution.

In other cases, the symmetry of the distribution may be altered, resulting in a skewed distribution. Skewness can be caused by extreme values or missing values in the distribution, which alters the distribution of values with respect to the central measures.

Detecting the symmetry or asymmetry of a distribution is important in the interpretation of statistical data. It can have implications for the use of certain statistical methods, understanding relationships between variables, and interpreting the results of statistical analysis.

6.1.1 Analysis of Symmetry with Central Tendency Characteristics

The symmetry of a distribution curve allows for describing the distribution of observed data. In an ideal situation, the distribution is perfectly symmetrical. This would imply that the three measures almost perfectly coincide, indicating a balanced distribution of data.

However, in practice, it is rare to find a perfectly symmetrical distribution. In reality, most distributions exhibit some form of skewness. Generally, three types of skewness are distinguished: ***right-skewed***, ***bell-shaped***, and ***left-skewed***.

6.1.1.1 Right Skewed Distribution

Right skewness, also known as positive skewness, occurs when the curve is stretched towards the right, with a longer right tail containing lower values. This distribution typically appears as a curve leaning towards the left, hence the term "right-skewed" distribution, as this leftward inclination creates a rightward slant, indicating a concentration of higher values on the left side of the distribution. This means that the majority of data points are situated on the left side of the distribution.

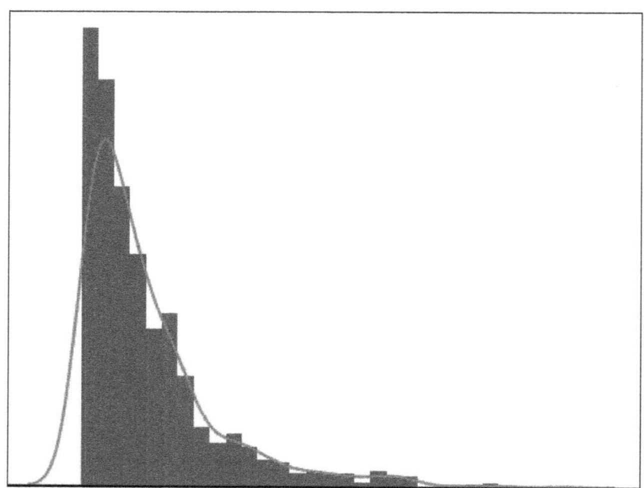

6.1.1.2 Symmetric Distribution

The bell-shaped curve, also known as the Gaussian curve or simply the normal distribution, is a perfectly symmetric distribution in which the data is balanced around the central measure.

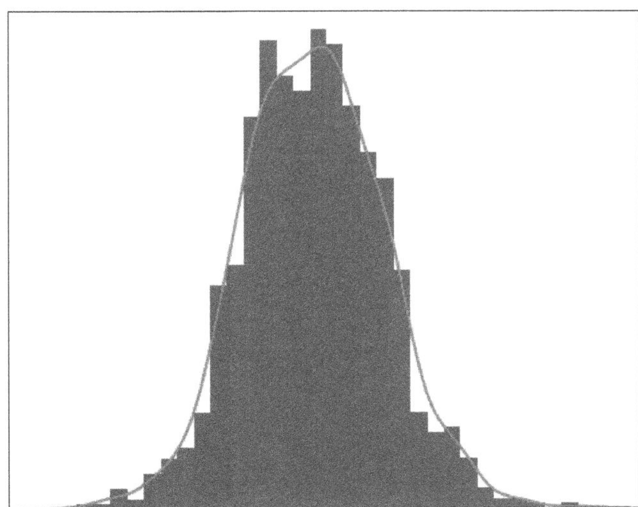

The normal distribution curve forms a bell shape with equal data points on the left and right of the central measure, hence the term "bell-shaped" curve.

6.1.1.3 Left Skewed Distribution

Left skewness, also known as negative skewness occurs when the curve is stretched towards the left, with a longer left tail containing lower values.

This distribution typically appears as a curve leaning towards the right, hence the term "left-skewed" distribution, as this rightward inclination creates a leftward slant, indicating a concentration of higher values on the right side of the distribution. This means that the majority of data points are situated on the right side of the distribution.

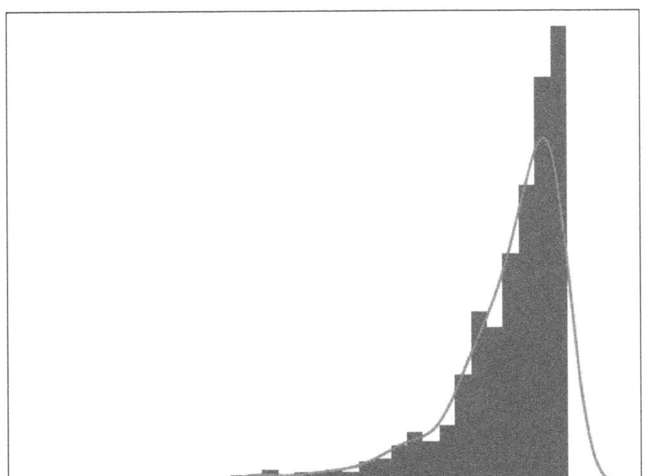

Understanding the symmetry of a distribution curve is important in interpreting data and selecting the appropriate statistical method for data analysis. It can also help identify outliers or potential biases in the data.

RULE OF THUMB

When there is left-skewness, the distribution curve is shifted negatively, with a tail spread towards the left. Lower values are less represented, while higher values are strongly represented. In this case we have:

Mode > Median > Mean

When the distribution curve is perfectly symmetrical, we have:
Mode = Median = Mean

When there is right-skewness, the distribution curve is shifted positively, with a tail spread towards the right. Higher values are less represented, while lower values are strongly represented. In this case we have:

Mode < Median < Mean

Example 62

We will now consider **example 61** from **section 5.9**, which deals with the statistical distribution of workers in a company based on their age. In this example, the following values were calculated for the measures of central tendency: mode is 27.5 years, median is 29.33 years, and mean is 31.566 years. Then, we will interpret the position of these values to gain a better understanding of the age distribution in this company.

INTERPRETATION OF THE RESULTS

Mode < Median < Moyenne

In this company, the mode (27.5 years) and the mean (31.56 years) are both close to the median age of 29.33 years (29 years and 4 months). This means that the majority of workers have an age close to the median, with a slight trend towards the mode and the mean.
On the graph, the effect of the median can be clearly observed, where the number of employees older than the median age is the same as those younger than the median age. This indicates a balanced distribution of data around the median.

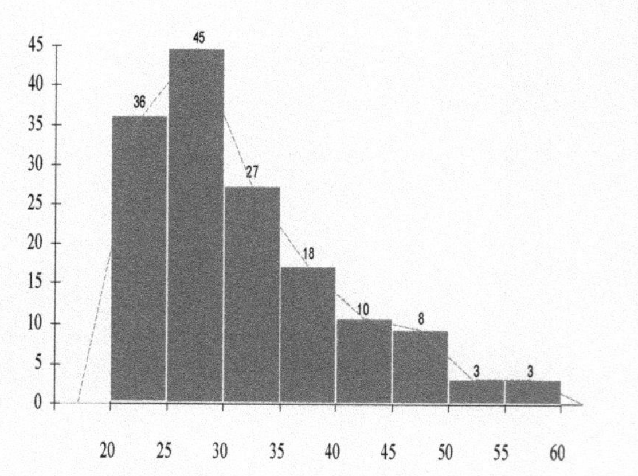

Thus, since these two quantities are identical on both sides of the median, but the age range for older employees (29.33 to 60 years) is three times larger than that for younger employees (20 to 29.33 years), a concentration of younger employees on the left side of the distribution can be easily noticed. This suggests that the majority of employees are younger, with a gradual decrease in the number of employees as age increases.
As for the right-hand side of the distribution, it is widely spread out, meaning that there are few older employees, who are scattered over a wide age range from 29.33 years to 60 years. This indicates a less significant concentration of older employees compared to younger employees in this company.

6.1.2 Analysis of Symmetry with the Coefficients of Asymmetry

6.1.2.1 The YULE-KENDALL's Coefficient of skewness

In the study of statistical distributions, analyzing symmetry is an important aspect to understand the distribution of data. The YULE-KENDALL coefficient of skewness, also known as YULE's coefficient of skewness, is a commonly used indicator to quantify the skewness of a distribution. This coefficient is calculated based on the relative positions of the first quartile (Q_1) and third quartile (Q_3) with respect to the median, which is equivalent to Q_2.

In a symmetrical distribution, the values of Q_1, Q_2, and Q_3 are equidistant from each other, indicating that the median divides the distribution into two equal parts. The YULE's coefficient of skewness in this case is zero, indicating a perfect symmetry of the distribution. However, when the distribution exhibits asymmetry, the values of the quartiles are no longer equidistant, and the YULE's coefficient of skewness takes a non-zero value.

The formula for the YULE-KENDALL's coefficient of skewness is as follows:

$$C_Y = \frac{\text{Q1} - 2\text{Q2} + \text{Q3}}{\text{Q3} - \text{Q1}}$$

The YULE's coefficient of skewness can take different values to indicate the direction and degree of asymmetry of the distribution. If the coefficient is equal to 0, it means that the distribution is perfectly symmetrical. If the coefficient is positive, it indicates the right skewness, where the tail of the distribution is stretched towards the right and the higher values are farther from the median than the lower values. On the other hand, if the coefficient is negative, it indicates left skewness, with a tail stretched towards the left and lower values farther from the median than the higher values.

RULE OF THUMB

If we obtain:

- $C_Y = 0$: Centered curve, there is symmetry.
- $C_Y < 0$: Left skewness, there is long tail on the left side.
- $C_Y > 0$: Right skewness, there is long tail on the right side

Example 63

Let's continue with **example 62**, by moving on to the next step where we will calculate the YULE's coefficient of skewness using the numerical data from this example previously studied on the statistical distribution of ages of workers in a company. Then, we will interpret the obtained value to gain a better understanding of the age distribution in this company.

Solution

Q_1, Q_2 and Q_3 have already been calculated in section **5.3.1**, which are 25.17 , 29.33 and 36.25, respectively. These values represent the quartiles of the studied distribution.

Lat's calculate the YULE's coefficient of skewness:

$$C_Y = \frac{Q1 - 2Q2 + Q3}{Q3 - Q1}$$

$$= \frac{25.17 - 2(29.33) + 36.25}{36.25 - 25.17}$$

$$= \frac{2.76}{11.08}$$

$$= 0.249$$

INTERPRETATION OF THE RESULTS

$C_Y > 0$, this indicates a right-skewed asymmetry in the data distribution. This means that the distribution curve is stretched towards the right, which can be visualized on a graph. In the case of this company, there is a significant representation of young individuals, with a concentration of data on the left-hand side of the graph.

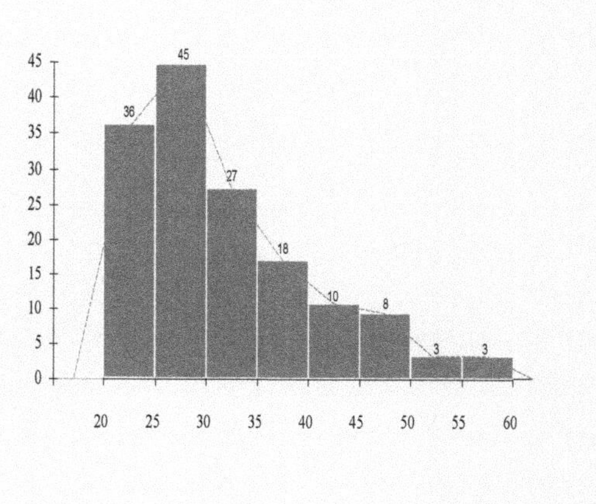

On the other hand, the presence of elderly individuals is less prominent, with a dispersion of data on the right-hand side of the graph. This right-skewed asymmetry can be interpreted as a predominance of young individuals in this company, with less representation of elderly individuals.

6.1.2.2 PEARSON'S coefficient of skewness

The PEARSON's coefficient of skewness, often abbreviated as simply PEARSON skewness, is a statistical measure used to evaluate the asymmetry of a data distribution. It is based on the comparison between the values of the mean (representing the central tendency) and the mode (representing the most frequently observed value). In other words, the PEARSON's coefficient of skewness quantifies the difference between the mean and the mode, providing an estimate of the asymmetry of the distribution.

The formula for PEARSON's coefficient of skewness is as follows:

$$\beta_1 = \frac{(\bar{x} - \text{Mode})}{\sigma}$$

Here, the standard deviation (σ) allows to normalize the difference between the mean and the mode to obtain a relative measure of asymmetry that is independent of the unit of measurement.

The PEARSON's coefficient of skewness can take different values to indicate the direction and degree of asymmetry of the distribution. If the coefficient is equal to 0, it means that the distribution is perfectly symmetrical. If the coefficient is positive, it indicates the right skewness, where the tail of the distribution is stretched towards the right and the higher values are farther from the median than the lower values. On the other hand, if the coefficient is negative, it indicates left skewness, with a tail stretched towards the left and lower values farther from the median than the higher values.

RULE OF THUMB

If we get:

$\beta_1 = 0$: The curve is centered, there is symmetry.
$\beta_1 < 0$: Left skewness, there is long tail on the left side.
$\beta_1 > 0$: Right skewness, there is long tail on the right side

Example 64

Let's continue with **example 63**, by moving on to the next step where we will calculate the PEARSON's coefficient of skewness using the numerical data from this example previously studied on the statistical distribution of ages of workers in a company. Then, we will interpret the obtained value to gain a better understanding of the age distribution in this company.

Solution

The statistics of mode, mean, and standard deviation, with respective values of 27.5, 31.566, and 8.53, have already been calculated in **sections 4.3.2**, **4.4.4** and **5.5**.

Let's calculate the PEARSON's coefficient of skewness:

$$\beta_1 = \frac{(\bar{x} - \text{Mode})}{\sigma} = \frac{(31.566 - 27.5)}{8.53} = \frac{4.06}{8.53} = 0.47667$$

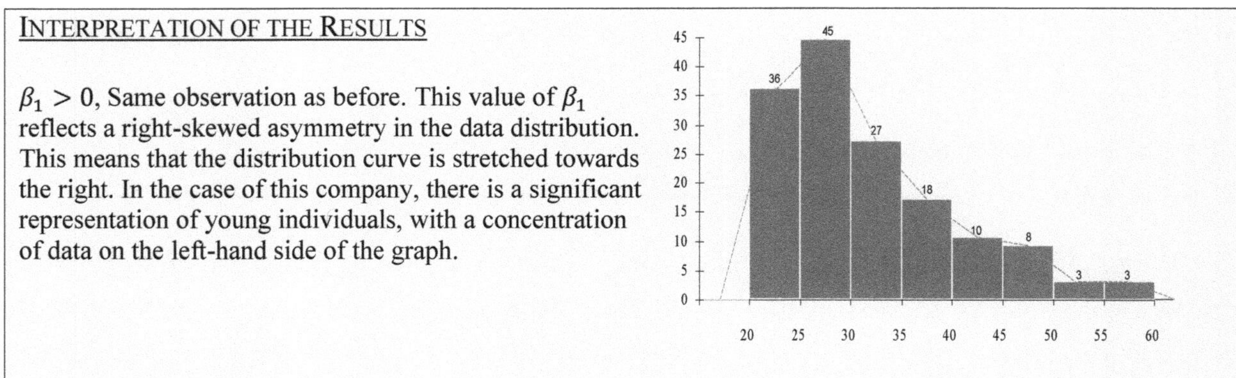

INTERPRETATION OF THE RESULTS

$\beta_1 > 0$, Same observation as before. This value of β_1 reflects a right-skewed asymmetry in the data distribution. This means that the distribution curve is stretched towards the right. In the case of this company, there is a significant representation of young individuals, with a concentration of data on the left-hand side of the graph.

6.1.3 Analysis of Symmetry through the use of moments

6.1.3.1 FISHER's coefficient of Skewness

Odd-order centered moments, such as first-order moment (μ_1), the third-order moment (μ_3) and so on, are often used to quantify the symmetry of a distribution. In a unimodal symmetric distribution, such as a normal distribution, all odd-order centered moments are equal to zero, except for the first-order moment (μ_1), which is null by definition.

To measure the symmetry of a distribution, one can use the FISHER's coefficient of Skewness, also known as the FISHER's skewness. This coefficient is defined as the ratio of the third-order centered moment (μ_3), divided by the standard deviation (σ) raised to the power of 3.

Mathematically, the FISHER's coefficient of Skewness is calculated as follows:

$$\boxed{\gamma_1 = \frac{\mu_3}{\sigma^3}}$$

The FISHER's coefficient of Skewness can take different values to indicate the direction and degree of asymmetry of the distribution. If the coefficient is equal to 0, it means that the distribution is perfectly symmetrical. If the coefficient is positive, it indicates the right skewness, where the tail of the distribution is stretched towards the right and the higher values are farther from the median than the lower values. On the other hand, if the coefficient is negative, it indicates left skewness, with a tail stretched towards the left and lower values farther from the median than the higher values.

Depending on the value of γ_1, three cases will be distinguished which will determine three different curve shapes.

RULE OF THUMB

If we get:

$\gamma_1 = 0$: The curve is centered, there is symmetry.

$\gamma_1 < 0$: Asymmetry to the right (oblique to the right), there is spread to the left.

$\gamma_1 > 0$: Asymmetry on the left (oblique on the left), there is spread to the right.

Example 65

Let's continue with **example 64**, by moving on to the next step where we will calculate the FISHER's coefficient of Skewness using the numerical data from this example previously studied on the statistical distribution of ages of workers in a company. Then, we will interpret the obtained value to gain a better understanding of the age distribution in this company.

Solution

The standard deviation (σ) and the third-order central moment (μ_3) have been previously calculated in **section 5.5** and **5.9**, respectively, with values of 8.53 and 661.31.

Let's calculate the FISHER's coefficient of Skewness:

$$\gamma_1 = \frac{\mu_3}{\sigma^3} = \frac{661.31}{(8.53)^3}$$
$$= 1.0655$$

INTERPRETATION OF THE RESULTS

$\gamma_1 > 0$, the observation is consistent with the previous finding. In this company, there are more young employees who are less than 29.33 years old (29 years, 3 months, and 29 days) compared to those who are older.

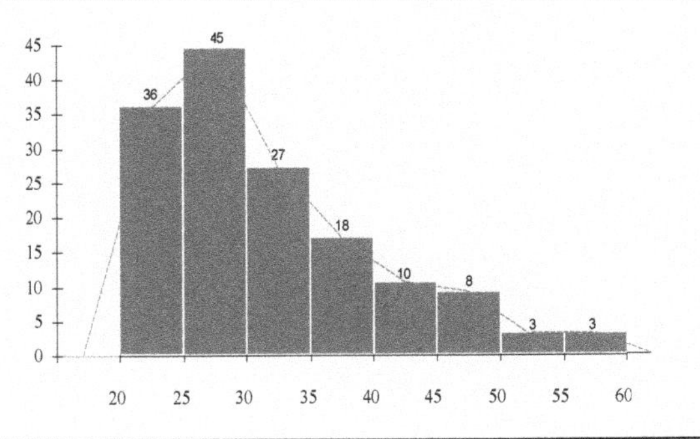

6.2 Kurtosis Distribution

The shape of a distribution is also characterized by its degree of kurtosis in relation to the distribution of the normal distribution. In other words, kurtosis measures how much a distribution deviates from the shape of a normal distribution curve. A distribution is considered ***mesokurtic*** when its curve has a similar shape to that of the normal distribution, with a kurtosis close to zero. This means that the values are relatively concentrated around the mean, with fewer extreme values.

On the other hand, a distribution is referred to as ***platykurtic*** or ***hyponeurmal*** when its curve is flatter than that of the normal distribution. This means that the values are less concentrated around the mean and the distribution has longer tails. As a result, the distribution has fewer extreme values compared to a normal distribution.

Conversely, a distribution is considered ***leptokurtic*** or ***hypernormal*** when its curve is sharper than that of the normal distribution. This means that the values are more concentrated around the mean and the distribution has shorter tails. As a result, the distribution has more extreme values compared to a normal distribution.

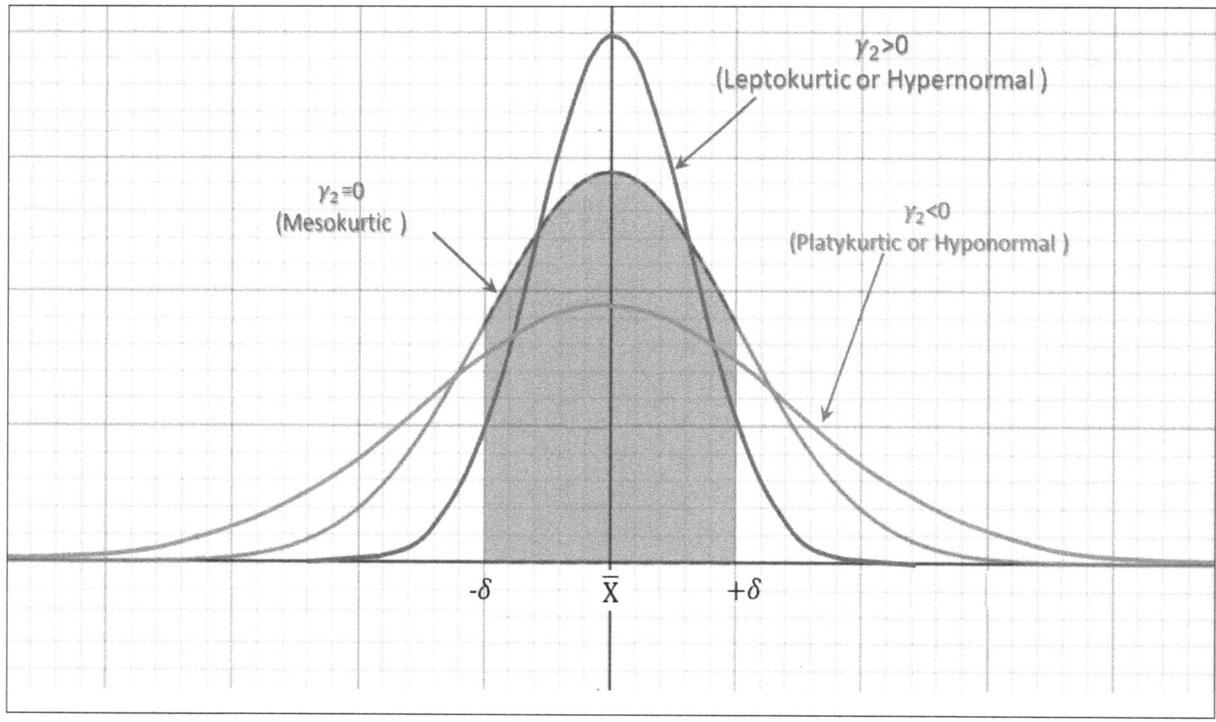

6.2.1 FISHER's coefficient of kurtosis

FISHER's coefficient of kurtosis is a statistical indicator that measures the shape of the distribution of a random variable relative to a normal distribution. It is defined based on the fourth-order moment of the distribution.

Initially, FISHER proposed the quantity $\beta_2 = \frac{\mu_4}{\sigma^4}$ as the kurtosis coefficient. In the case of a perfectly normal distribution, this coefficient has a value of 3. Thus, if β_2 is greater than 3, it indicates that the distribution has thicker tails and is more pointed than the normal distribution. On the other hand, if β_2 is less than 3, it suggests that the distribution has thinner tails and is more flattened than the normal distribution.

However, to maintain consistency with FISHER's coefficient of skewness, which measures the skewness of a distribution relative to a normal distribution, FISHER proposed shifting the reference frame from zero to 3. Thus, he introduced the following formula for the kurtosis coefficient:

$$\boxed{\gamma_2 = \frac{\mu_4}{\sigma^4} - 3}$$

This makes it possible to have a kurtosis coefficient of zero for a perfectly normal distribution, and positive or negative values to indicate either a flattening or an excessive peak of the distribution compared to the normal distribution.

More specifically, FISHER's coefficient of kurtosis is used to evaluate whether a distribution has thicker tails (positive kurtosis) or thinner tails (negative kurtosis) compared to a normal distribution.

A positive kurtosis ($\gamma_2 > 0$) indicates that the distribution has a relatively higher concentration of data around the mean and thicker tails, meaning that extreme values are more frequent than in a normal distribution. On the other hand, a negative kurtosis ($\gamma_2 < 0$) indicates that the distribution has a relatively lower concentration of data around the mean and thinner tails, meaning that extreme values are less frequent than in a normal distribution.

It is important to note that in terminology, the terms "thick tail" and "long tail" are often used to describe the characteristics of a distribution. These terms refer to a distribution having values outside the central range (i.e., far from the mean or the center of the distribution). In other words, these values are far from the mean, indicating that the distribution has extreme values.

FISHER's coefficient of kurtosis is widely used in the fields of statistics, finance, economics, and other disciplines to characterize the shape of data distributions and understand their behavior.

As like the coefficient of skewness, we have 3 possible cases:

RULE OF THUMB

If we get:

$\gamma_2 = 0$: The kurtosis is identical to that of the normal distribution.

$\gamma_2 < 0$: The distribution has less kurtosis than and has thinner tails than the normal distribution.

$\gamma_2 > 0$: The distribution is sharper than and has thicker tails than the normal distribution.

Example 66

Let's continue with **example 65**, by moving on to the next step where we will calculate the FISHER's coefficient of kurtosis using the numerical data from this example previously studied on the statistical distribution of ages of workers in a company. Then, we will interpret the obtained value to gain a better understanding of the age distribution in this company.

Solution

The standard deviation (σ) and the central moment of order 4 (μ_4) which are 8.53 and 19,129.48, respectively had already been calculated in section **5.5** and **5.9**, respectively.

Let's calculate the FISHER's coefficient of kurtosis:

$$\gamma_2 = \frac{\mu_4}{\sigma^4} - 3 = \frac{19{,}129.48}{(8.53)^4} - 3$$

$$= 3.613 - 3$$

$$= 0.613$$

INTERPRETATION OF THE RESULTS

$\gamma_2 > 0$, this means that the distribution curve is sharper than that of the normal distribution. This indicates that the ages of the workers in this company are more concentrated around the mean and have thicker tails, which means that the distribution has more frequent extreme values compared to a normal distribution.

This can be interpreted as a distribution having values that are further from the mean compared to a normal distribution, with a higher concentration around the mean and thicker tails. This may indicate a higher dispersion or variation of the data compared to the normal distribution.

Example 67 - Effect of outliers

Let's continue with **example 58** from **section 5.8** by moving to the next step, where we will calculate the shape coefficients from the numerical data of this example, which concerns an insurance company that wants to establish itself on an island with only 2,000 inhabitants Then, we will interpret the obtained value to gain a better understanding of the age distribution in this company.

Consider the distribution with outliers:

x_i	n_i	$n_i x_i$	$n_i x_i^2$
500	380	190000	95000000
1500	633	949500	1424250000
2500	527	1317500	3293750000
3500	151	528500	1849750000
4500	123	553500	2490750000
5500	85	467500	2571250000
6500	58	377000	2450500000
7500	39	292500	2193750000
8500	8	68000	578000000
9500	6	57000	541500000
2495000	5	12475000	31125125000000
Total	2015	17276000	31142613500000

Let's calculate the YULE's coefficient of skewness:

Q_1, Q_2 and Q_3 have already been calculated in sections **4.7.2** and **5.3.1**, which are 1,195.4, 1,991.4 and 2,945.3, respectively.

$$C_Y = \frac{Q1 - 2Q2 + Q3}{Q3 - Q1}$$

$$= \frac{1{,}195.4 - 2(1{,}991.4) + 2{,}938.2}{2{,}945.3 - 1{,}195.4}$$

$$= \frac{156.9}{1{,}750.9}$$

$$= 0.0864$$

Let's calculate the PEARSON's coefficient of skewness:

The mode is $1,500, and the average is $8,573.69. These values have already been determined in section **4.7**. As for the standard deviation, which is $124,023.72, it has already been calculated in section **5.7**.

$$\beta_1 = \frac{(\bar{x} - \text{Mode})}{\sigma} = \frac{(8{,}573.69 - 1{,}500)}{124{,}023.72}$$

$$= \frac{7{,}073.69}{124{,}023.72}$$

$$= 0.057$$

Resolving this exercise requires the columns in the following table:

x_i	n_i	$n_i x_i$	$n_i(x_i - \bar{x})^3$	$n_i(x_i - \bar{x})^4$
500	380	190000	-199986638300203.00	16146315757751100000.00
1500	633	949500	-224049056193295.00	15848551972462200000.00
2500	527	1317500	-118078285983896.00	717171763282337000.00
3500	151	528500	-19721983895605.70	100063375859417000.00
4500	123	553500	-8315174498814.19	33873503659313300.00
5500	85	467500	-2468324193844.87	7586881337259650.00
6500	58	377000	-517206604081.47	1072529923153570.00
7500	39	292500	-48273651875.12	51831288254010.90
8500	8	68000	-3202168.61	235991086.46
9500	6	57000	4768810683.20	4417362352454.76
2495000	5	12475000	76859362573606200000.00	1911051407140400000000000000.00
	2015	17276000	76858789393428500000	1911051447733510000000000000

Let's calculate the FISHER's coefficient of skewness:

$$\mu_3 = \frac{\sum_{i=1}^{i=11} n_i (x_i - \bar{x})^3}{\sum_{i=1}^{i=11} n_i} = \frac{76{,}858{,}789{,}393{,}428{,}500{,}000}{2015}$$

$$= 38{,}143{,}319{,}798{,}227{,}543.42$$

$$\gamma_1 = \frac{\mu_3}{\sigma^3} = \frac{38{,}143{,}319{,}798{,}227{,}543.42}{(124{,}023.72)^3}$$

$$= 19.99$$

Let's calculate the FISHER's coefficient of kurtosis:

$$\mu_4 = \frac{\sum_{i=1}^{i=11} n_i (x_i - \bar{x})^4}{\sum_{i=1}^{i=11} n_i} = \frac{191{,}105{,}144{,}773{,}351{,}000{,}000{,}000{,}000}{2015}$$

$$= 94{,}841{,}262{,}914{,}814{,}400{,}000{,}000$$

$$\gamma_2 = \frac{\mu_4}{\sigma^4} - 3 = \frac{94{,}841{,}262{,}914{,}814{,}400{,}000{,}000}{(124{,}023.72)^4} - 3$$

$$= 397.84$$

INTERPRETATION OF RESULTS

$$\left.\begin{array}{c} \underbrace{M_0}_{1{,}500} < \underbrace{M_e}_{1{,}991} < \underbrace{\bar{x}}_{8{,}573} \\ \\ C_Y > 0 \\ \beta_1 > 0 \\ \gamma_1 > 0 \end{array}\right\}$$

The interpretation of the results shows that the distribution of salaries on this island is right-skewed, indicating that the majority of the inhabitants earn relatively low salaries. The long tail towards the right suggests that there are a few individuals who earn very high salaries, notably the five billionaire entrepreneurs who each receive a monthly salary of around four million dollars. This is highlighted by the high value of γ_1, which measures the degree of asymmetry in the distribution. The presence of these high salaries accentuates the spreading towards the right of the distribution.

$\gamma_2 > 0$, indicates that the curve of this distribution is sharper than that of a normal distribution. This means that low salaries are very common on this island, causing the distribution to rise quickly when considering the lowest salaries. However, the distribution drops rapidly when moving to higher salaries, indicating that higher salaries are less frequent.

Conclusion

These results suggest that the population of the island is predominantly composed of individuals earning relatively low salaries, with a concentration of low values. The high salaries of the five billionaire entrepreneurs have a significant impact on the distribution of salaries on the island, accentuating the spreading towards the right of the distribution.

Consider the distribution after removal of outliers:

x_i	n_i	$n_i x_i$	$n_i x_i^2$
500	380	190000	95000000
1500	633	949500	1424250000
2500	527	1317500	3293750000
3500	151	528500	1849750000
4500	123	553500	2490750000
5500	85	467500	2571250000
6500	58	377000	2450500000
7500	39	292500	2193750000
8500	8	68000	578000000
9500	6	57000	541500000
	2010	4801000	17488500000

Let's calculate the YULE's coefficient of skewness:

Q_1, Q_2 and Q_3 have already been calculated in **sections 4.7.2** and **5. 7**, which are respectively $ 1,193.39, $ 1,987.30 and $ 2,938.21.

$$C_Y = \frac{\text{Q1} - 2\text{Q2} + \text{Q3}}{\text{Q3} - \text{Q1}}$$

$$= \frac{1{,}193.39 - 2(1{,}987.3\,) + 2{,}938.21}{2{,}938.21 - 1{,}193.39}$$

$$= \frac{157}{1{,}744.82}$$

$$= 0.0899$$

Let's calculate the PEARSON's coefficient of skewness:

The Mode is $1,500, and the average has already been calculated in section 4.7, which is $2,388.55.

$$\beta_1 = \frac{(\bar{x} - \text{Mode})}{\sigma} = \frac{(2{,}388.55 - 1{,}500)}{1{,}731.53}$$

$$= \frac{888.55}{1{,}731.53}$$

$$= 0.513$$

The following table shows the data required to solve this exercise:

x_i	n_i	$n_i x_i$	$n_i(x_i - \bar{x})^3$	$n_i(x_i - \bar{x})^4$
500	380	190000	-2559611399166.15	4833972572753570.00
1500	633	949500	-444078657509.22	394589294682318.00
2500	527	1317500	729401241.12	81286506472.52
3500	151	528500	207318606652.88	230422769782357.00
4500	123	553500	1157825378987.59	2444682043991710.00
5500	85	467500	2560379752735.05	7966475111246270.00
6500	58	377000	4030980960294.94	16573147589988700.00
7500	39	292500	5208289532035.38	26621873956284300.00
8500	8	68000	1826086046886.43	11160020397986500.00
9500	6	57000	2157865698007.33	15345538451401400.00
Total	2010	4801000	14145785320165	85570803474623600

Let's calculate the FISHER's coefficient of skewness:

$$\mu_3 = \frac{\sum_{i=1}^{i=10} n_i(x_i - \bar{x})^3}{\sum_{i=1}^{i=10} n_i} = \frac{14{,}145{,}785{,}320{,}165}{2{,}010}$$

$$= 7{,}037{,}704{,}139.38$$

$$\gamma_1 = \frac{\mu_3}{\sigma^3} = \frac{7{,}037{,}704{,}139{,}38}{(1{,}731.53)^3}$$

$$= 1.35$$

Let's calculate the FISHER's coefficient of kurtosis:

$$\mu_4 = \frac{\sum_{i=1}^{i=10} n_i(x_i - \bar{x})^4}{\sum_{i=1}^{i=10} n_i} = \frac{85{,}570{,}803{,}474{,}623{,}600}{2{,}010}$$

$$= 42{,}572{,}539{,}042{,}101.29$$

$$\gamma_2 = \frac{\mu_4}{\sigma^4} - 3 = \frac{42{,}572{,}539{,}042{,}101{,}29}{(1{,}731.53)^4} - 3$$

$$= 1.744$$

INTERPRETATION OF RESULTS

$$\left.\begin{array}{c} \underbrace{M_0}_{1{,}500} < \underbrace{M_e}_{1{,}991} < \underbrace{\bar{x}}_{8{,}573} \\ C_Y > 0 \\ \beta_1 > 0 \\ \gamma_1 > 0 \end{array}\right\}$$

The distribution of salaries among the inhabitants of the island is positively skewed, with a tail of the distribution spread towards the right. This asymmetry suggests that there is a strong concentration of low salary values, indicating that the majority of the population on the island has relatively low salaries.

$\gamma_2 > 0$, reveals that the distribution of salaries has a sharper peak than that of a normal distribution. This means that low salaries quickly drive up the distribution, but it drops rapidly as the counts move towards higher salaries. Despite some remaining disparities, the distribution of salaries on the island does not exhibit significant outliers, but rather reduced proportions compared to the previous case.

Conclusion

The analysis of the results shows that the population of the island is predominantly composed of individuals with relatively low salaries, with a significant concentration of low salary values. The distribution of salaries exhibits left-skewness and a sharper peak than that of a normal distribution.

6.3 Exercises

6.3.1 Chapter Exercises

- **Exercise 25:** The following table gives the distribution of remuneration by monthly salary brackets in a company.

Salary bracket	Frequency (n_i)
[0;2000)	183
[2000;4000)	319
[4000;6000)	69
[6000;8000)	41
[8000;10000)	24
[10000;12000)	19
[12000;14000)	12
[14000;16000)	7
[16000;18000)	3
[18000;20000)	1
Total	678

a. Draw the histogram of the distribution.

b. Plot the ascending and descending cumulative relative frequency curves of this distribution.

c. Determine the following central tendency characteristics: (mode, mean, median).

d. Determine the following dispersion characteristics: (standard deviation and coefficient of variation).

e. Calculate the percentage of salaries below the mode.

f. Calculate the percentage of salaries below the mean.

g. Calculate the minimum salary among the top 10% of salaries.

h. Determine the percentage of salaries between $\bar{x} - \sigma$ and $\bar{x} + \sigma$.

i. Determine characteristics of shape.

Solution

a: The histogram of the distribution is:

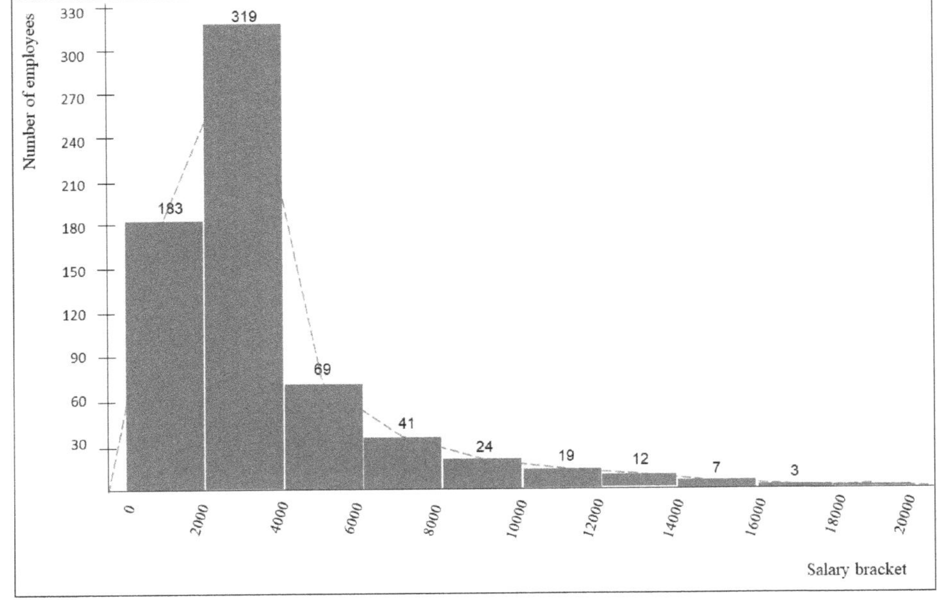

The following table shows the data required to solve this exercise:

Classes	x_i	n_i	$n_i x_i$	$n_i x_i^2$	$n_i(x_i - \bar{x})^3$	$n_i(x_i - \bar{x})^4$	f_i	$f_i\uparrow$	$f_i\downarrow$
[0;2000)	1000	183	183000	183000000	-2071802410213	4652082580101740	26.99	26.99	100.00
[2000;4000)	3000	319	957000	2871000000	-4715867206	1157404576848	47.05	74.04	73.01
[4000;6000)	500	69	34500	17250000	-1427838663866	3920027859764830	10.18	84.22	25.96
[6000;8000)	7000	41	287000	2009000000	2170027622321	8147525538909170	6.05	90.27	15.78
[8000;10000)	9000	24	216000	1944000000	4573517949275	26318639573586300	3.54	93.81	9.73
[10000;12000)	11000	19	209000	2299000000	8859865818752	68704469806301300	2.80	96.61	6.19
[12000;14000)	13000	12	156000	2028000000	11137967294788	108646106933347000	1.77	98.38	3.39
[14000;16000)	15000	7	105000	1575000000	11368902227726	133636582882133000	1.03	99.41	1.62
[16000;18000)	17000	3	51000	867000000	7806610717894	107376591313851000	0.44	99.85	0.59
[18000;20000)	1900	1	1900	3610000	-2435460679	3276736328875	0.15	100.00	0.15
Total		678	2,200,400	13,796,860,000	42,410,099,228,792	461,406,460,628,900,000	100		

b: The curves of the ascending and descending cumulative frequencies are as follows:

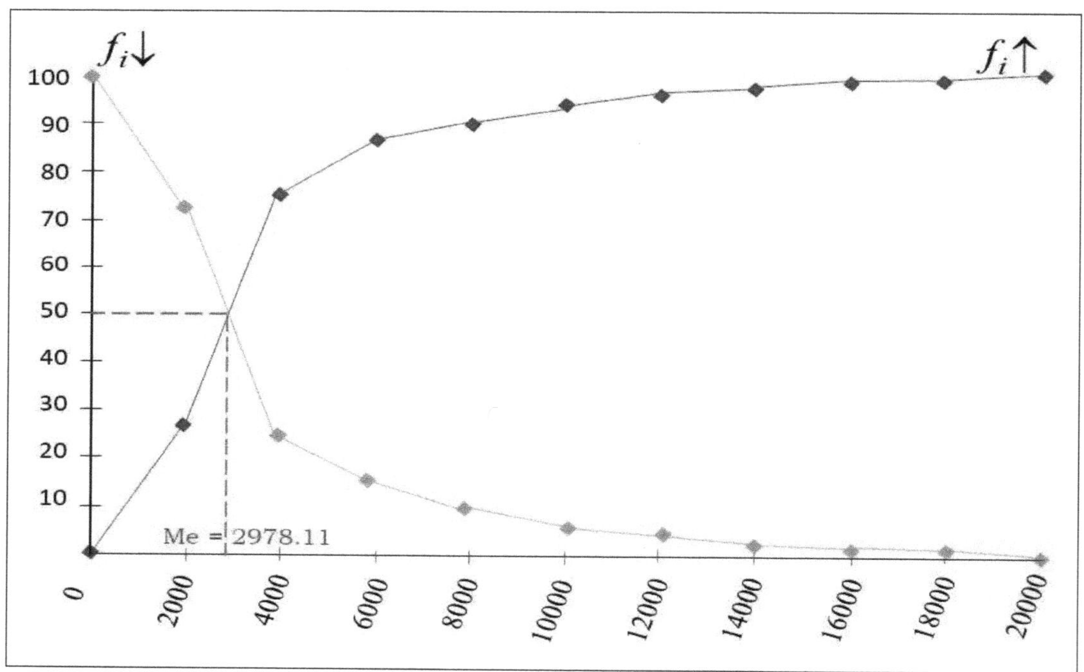

c: Let's calculate the central tendency characteristics.

The modal class is [2000 ; 4000) and the modal wage is 3000 euros:

The average wage is:

$$\bar{x} = \frac{1}{n}\sum_{i=1}^{i=10} n_i x_i = \frac{2{,}200{,}400}{678}$$

$$= 3{,}245.427 \text{ euros}$$

The median wage is:

$$Me = a_i + (a_{i+1} - a_i)\frac{(50 - F_i)}{(F_{i+1} - F_i)}$$

$$= 2{,}000 + (4{,}000 - 2{,}000)\frac{(50 - 26.99)}{(74.04 - 26.99)}$$

$$= 2{,}978.11 \text{ euros}$$

d: Let's calculate the dispersion characteristics:

d-1: Finding the interquartile Range:

$$Q_1 = a_i + (a_{i+1} - a_i)\frac{(25 - F_i)}{(F_{i+1} - F_i)}$$

$$= 0 + (2{,}000 - 0)\frac{(25 - 0)}{(26.99 - 0)}$$

$$= 1{,}852.5 \text{ euros}$$

$$Q_3 = a_i + (a_{i+1} - a_i)\frac{(75 - F_i)}{(F_{i+1} - F_i)}$$

$$= 4{,}000 + (6{,}000 - 4{,}000)\frac{(75 - 74.04)}{(84.22 - 74.04)}$$

$$= 4{,}188.6 \text{ euros}$$

$$\text{Interquartile Range} = Q_3 - Q_1$$

$$= 4{,}188.6 - 1{,}852.5$$

$$= 2{,}336.1 \text{ euros}$$

d-2: The standard deviation is:

$$\sigma = \sqrt{\frac{1}{n}\sum_{i=1}^{i=10} n_i x_i^2 - \bar{x}^2}$$

$$= \sqrt{\frac{13{,}796{,}860{,}000}{678} - (3{,}245.427)^2}$$

$$= \sqrt{9{,}816{,}549.89}$$

$$= 3{,}133.13 \text{ euros}$$

d-3: The coefficient of variation is:

$$CV = \frac{\sigma}{|\bar{x}|} \text{x}100 = \frac{3{,}133.13}{3{,}245.427} \text{x}100$$

$$= 96.54\%$$

e: Let's calculate the percentage of people with a salary less than the mode, i.e., 3,000 euros.

Let P_{mode} be the percentage to find, we use the linear interpolation formula using the data pair $(a_i, f_i \uparrow)$:

$$Mode = a_i + (a_{i+1} - a_i)\frac{(P_{<mode} - F_i)}{(F_{i+1} - F_i)}$$

$$3{,}000 = 2{,}000 + (4{,}000 - 2{,}000)\frac{(P_{<mode} - 26.99)}{(74.04 - 26.99)}$$

$$P_{mode} = 50.515\%$$

Which represents 343 employees.

f: Let's calculate the percentage of people with a salary less than the average, i.e., 3,245.427 euros.

Let $P_{<\bar{x}}$ be the percentage to find, we use the linear interpolation formula using the data pair $(a_i, f_i \uparrow)$:

$$\bar{x} = a_i + (a_{i+1} - a_i)\frac{(P_{<\bar{x}} - F_i)}{(F_{i+1} - F_i)}$$

$$3{,}245.427 = 2{,}000 + (4{,}000 - 2{,}000)\frac{(P_{<\bar{x}} - 26.99)}{(74.04 - 26.99)}$$

$$P_{<\bar{x}} = 56.28\%$$

Which represents 382 employees.

g: To determine the minimum wage of the 10% of the highest salaries, we will calculate the maximum wage of the 90% of the lowest salaries because the two wages are equivalent.

Let $S_{<90}$ the maximum amount of the 90% of the lowest wages, we use the linear interpolation formula using the data pair $(a_i, f_i \uparrow)$:

$$S_{<90} = a_i + (a_{i+1} - a_i)\frac{(90 - F_i)}{(F_{i+1} - F_i)}$$
$$= 6{,}000 + (8{,}000 - 6{,}000)\frac{(90 - 84.22)}{(90.27 - 84.22)}$$

$$S_{<90} = 7{,}910.74 \text{ euros}$$

The same result can directly be obtained by using the descending cumulative relative frequencies, that is to say, the index of F_i will be shifted by 1.

Let $S_{>10}$ the minimum wage of the 10% of the highest wages, we use the linear interpolation formula using the data pair $(a_i, f_i \uparrow)$:

$$S_{>10} = a_i + (a_{i+1} - a_i)\frac{(10 - F_{i+1})}{(F_{i+2} - F_{i+1})}$$
$$= 6\,000 + (8\,000 - 6\,000)\frac{(10 - 15.78)}{(9.73 - 15.78)}$$

$$S_{>10} = 7\,910.74 \text{ euros}$$

10% of employees (67 people) have a wage greater than 7,910.74 euros.

h: Let's calculate the percentage of wages between $\bar{x} - \sigma$ and $\bar{x} + \sigma$:

h-1: The standard deviation plus or minus the mean is:

$$\bar{x} - \sigma = 3{,}245.427 - 3{,}133.13$$
$$= 112.297 \text{ euros}$$

$$\bar{x} + \sigma = 3{,}245.427 + 3{,}133.13$$
$$= 6{,}378.557 \text{ euros}$$

h-2: Let's calculate the percentage of wages less than $\bar{x} - \sigma$, i.e., less than 112.297 euros.

Let $P_{<\bar{x}-\sigma}$ be the percentage to find, we use the linear interpolation formula using the data pair $(a_i, f_i \uparrow)$:

$$\bar{x} - \sigma = a_i + (a_{i+1} - a_i)\frac{(P_{<\bar{x}-\sigma} - F_i)}{(F_{i+1} - F_i)}$$
$$112.297 = 0 + (2{,}000 - 0)\frac{(P_{<\bar{x}-\sigma} - 0)}{(26.99 - 0)}$$

$$P_{<\bar{x}-\sigma} = 1.5\%$$

h-3: Let's calculate the percentage of wages less than $\bar{x} + \sigma$, i.e., less than 6,378.557 euros.

Let $P_{<\bar{x}+\sigma}$ be the percentage to find, we use the linear interpolation formula using the data pair $(a_i, f_i \uparrow)$

$$\bar{x} + \sigma = a_i + (a_{i+1} - a_i)\frac{(P_{<\bar{x}-\sigma} - F_i)}{(F_{i+1} - F_i)}$$

$$6{,}378.557 = 6{,}000 + (8{,}000 - 6{,}000)\frac{(P_{<\bar{x}+\sigma} - 84.22)}{(90.27 - 84.22)}$$

$$P_{<\bar{x}+\sigma} = 85.36\%$$

h-4: The percentage of wages between $\bar{x} - \sigma$ = 112.297 euros and $\bar{x} + \sigma$ = 6,378.557.91 euros is:

$$85.36\% - 1.5\% = 83.86\%$$

Which represents 569 salaries.

INTERPRETATION OF THE RESULTS

The salary distribution in this company is characterized by the following observations:

c-1. The majority of employees earn salaries between 2000 and 4000 euros, with a modal salary of 3000 euros.

c-2. The average salary across all employees is 3,245.427 euros, indicating that each employee would receive this amount if salaries were distributed equally.

c-3. The number of employees earning salaries above 2,978.11 euros is equal to the number earning salaries below that amount.

d. The coefficient of variation is 96.54%, which is close to 100%. This indicates that the distribution is highly scattered around the mean, reflecting significant wage discrepancies among employees. This could suggest that salaries are unevenly distributed within the company, with significant differences among employees in terms of remuneration.

e. A significant portion of employees, 50.51% or 343 people, earn less than 3000 euros.

f. An even larger proportion of employees, 70.199% or 382 people, earn less than the modal salary of 3,245.43 euros.

g. Only 10% of employees, or 67 people, earn a salary of 7,910.74 euros or more, placing them among the top 10% of earners in the company.

h. The salaries of 83.56% of employees fall within the interval between 112,29 euros and 6,378.557 euros. This interval is more densely populated than that of the normal distribution, indicating a concentration of salaries around the mean. Compared to a normal distribution, this salary range is significantly more populated than that of a normal distribution, which is around 68.27%. There could be several possible reasons for this strong concentration in the range between the mean minus one standard deviation and the mean plus one standard deviation:

- Compensation policies: The company may have implemented specific compensation policies that have resulted in a concentration of salaries around the mean plus or minus one standard deviation. This could include regular salary increases, salary caps, or social and financial benefits granted to a large number of employees.
- Organizational structure: The company's structure, such as the distribution of positions and hierarchical levels, can influence the distribution of salaries. If the company has a relatively homogeneous organizational structure with similar salaries for a large number of employees, this can result in a concentration of salaries in this specific range.
- Level of experience and seniority: It is possible that the majority of employees have similar levels of experience and seniority, which can result in relatively similar salaries and thus a concentration of salaries in this range.

i: Let's calculate the characteristics of shape:

i-1 The PEARSON's coefficient of skewness is:

$$\beta_1 = \frac{(\bar{x} - \text{Mode})}{\sigma} = \frac{(3{,}245.427 - 3{,}000)}{3{,}133.13} = 0.07833$$

i-2 The FISHER's coefficient of skewness is:

$$\mu_3 = \frac{\sum_{i=1}^{i=10} n_i (x_i - \bar{x})^3}{\sum_{i=1}^{i=10} n_i} = \frac{42{,}410{,}099{,}228{,}792}{678} = 62{,}551{,}768{,}774.029$$

$$\gamma_1 = \frac{\mu_3}{\sigma^3} = \frac{62{,}551{,}768{,}774.029}{(3{,}133.13)^3} = 2.033767$$

i-3 The FISHER's coefficient of kurtosis is:

$$\mu_4 = \frac{\sum_{i=1}^{i=10} n_i (x_i - \bar{x})^4}{\sum_{i=1}^{i=10} n_i} = \frac{461{,}406{,}460{,}628{,}900{,}000}{678} = 680{,}540{,}502{,}402{,}508$$

$$\gamma_2 = \frac{\mu_4}{\sigma^4} - 3 = \frac{680{,}540{,}502{,}402{,}508}{(3{,}133{,}13)^4} - 3 = 4.0621$$

INTERPRETATION OF THE RESULTS

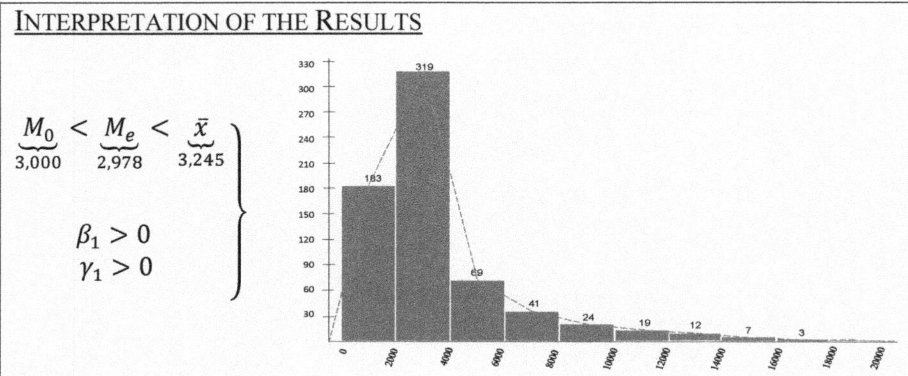

The salary distribution in this company exhibits a right-skewed pattern, with a tail that extends towards the right, indicating that the majority of employees have low salaries.

$\gamma_2 > 0$, the curve of this distribution is sharper than that of a normal distribution. In other words, there is a more pronounced concentration of salaries around the mean, with a rapid drop in salaries as they become higher.

Conclusion

Although the distribution is dispersed with a coefficient of variation of 96.54%, there is a very strong preponderance of the population within the range of the mean plus or minus one standard deviation. This massive preponderance is due to the fact that the majority of employees have low salaries, and the negligible number of individuals with very high salaries significantly extends the tail of the distribution to the right. The high concentration of low salaries within the range of the mean minus one standard deviation and the mean plus one standard deviation, which is around 83.86%, is justified by the value of γ_2, which is significantly above zero. This majority of individuals is located at the top of the distribution curve.

6.3.2 Inter-chapiter Exercises

- **Exercise 26:** Continuing exercise **18** from section **5.10.1** on the newsagent who wanted to know the number of newspapers sold per day and whose distribution table is shown below.

Data from exercise 18

(x_i) = number of newspapers sold per day		(n_i) = number of days
Mean= 54.404 newspapers		Standard deviation = 7.63 newspapers
Q_1 = 47.71 newspapers	Q_2 = 54.17 newspapers	Q_3 = 60.94 newspapers
% Population in $[\bar{x} - \sigma, \bar{x} + \sigma]$ = 59.52%		CV = 14%

Number of newspapers sold per day	x_i	Number of days (n_i)
[40;45)	42.5	2
[45;50)	47.5	6
[50;55)	52.5	3
[55;60)	57.5	4
[60;65)	62.5	4
[65;70)	67.5	2
Total		21

Determine the following shape characteristics:

a. The YULE's coefficient of skewness.

b. The PEARSON's coefficient of skewness.

c. The FISHER's coefficient of skewness.

d. The FISHER's coefficient of kurtosis.

Solution

The following table shows the data required to solve this exercise:

x_i	n_i	$n_i x_i$	$n_i(x_i - \bar{x})^3$	$n_i(x_i - \bar{x})^4$
42.5	2	85	-3374.37	40171.02
47.5	6	285	-1975.14	13637.86
52.5	3	157.5	-20.73	39.49
57.5	4	230	118.62	367.14
62.5	4	250	2122.02	17178.23
67.5	2	135	4491.28	58814.39
	21	1142.5	1361.678005	130208.1309

a: Finding the YULE's coefficient of skewness:

$$C_Y = \frac{Q1 - 2Q2 + Q3}{Q3 - Q1}$$

$$= \frac{47.71 - 2(54.17) + 60.94}{60.94 - 47.71}$$

$$= 0.02343$$

b: Finding the PEARSON's coefficient of skewness:

$$\beta_1 = \frac{(\bar{x} - \text{Mode})}{\sigma}$$

$$= \frac{(54.404 - 47.5)}{7.63}$$

$$= 0.904849$$

c: Finding the FISHER's coefficient of skewness:

$$\mu_3 = \frac{\sum_{i=1}^{i=6} n_i (x_i - \bar{x})^3}{\sum_{i=1}^{i=6} n_i} = \frac{1{,}361.67}{21}$$

$$= 64.84$$

$$\gamma_1 = \frac{\mu_3}{\sigma^3} = \frac{64.84}{(7.63)^3}$$

$$= 0.145$$

d: Finding the FISHER's coefficient of kurtosis:

$$\mu_4 = \frac{\sum_{i=1}^{i=6} n_i (x_i - \bar{x})^4}{\sum_{i=1}^{i=6} n_i} = \frac{130{,}208.131}{21}$$

$$= 6{,}200.39$$

$$\gamma_2 = \frac{\mu_4}{\sigma^4} - 3 = \frac{6{,}200.39}{(7.66)^4} - 3$$

$$= -1.170$$

INTERPRETATION OF THE RESULTS

$$\left.\begin{array}{l} \underbrace{M_0}_{15} < \underbrace{M_e}_{17} < \underbrace{\bar{x}}_{19} \\ C_Y > 0 \\ \beta_1 > 0 \\ \gamma_1 > 0 \end{array}\right\}$$

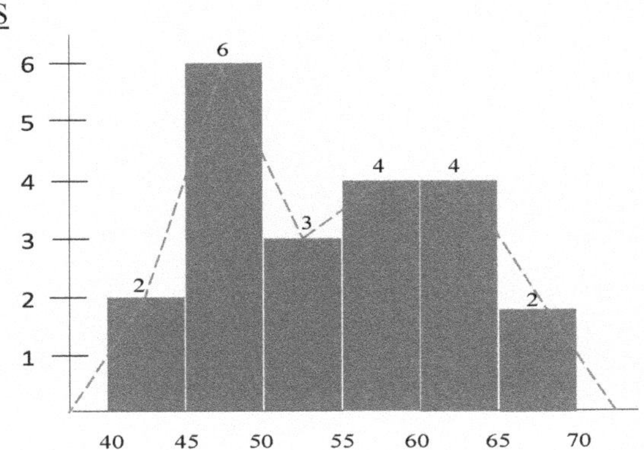

The distribution of the data is right-skewed. This means that the tail of the distribution is slightly skewed to the right, indicating a concentration of low values. In other words, the newspaper vendor sold a large number of newspapers on only a few days, while on other days, the sales were lower.

$\gamma_2 < 0$, the distribution curve is more platykurtic than that of a normal distribution. This suggests that the distribution has fewer extreme values and that the values are more concentrated around the mean. However, despite this flattening, the distribution is still moderately dispersed with a coefficient of variation of 14%.

Conclusion

Although the distribution exhibits a concentration of low values with a tail skewed to the right, it is still moderately dispersed with a relatively high concentration of values around the mean. Only 59.52% of the population falls within the range of the mean plus or minus one standard deviation, suggesting that the distribution may exhibit significant variations.

• **Exercise 27:** Continuing exercise **19** from section **5.10.1** on the distribution of the 200 employees of a company according to their seniority in years, the table of which is given beside. Determine the following shape characteristics:

Years of seniority	x_i	n_i
[0;10)	5	30
[10;20)	15	100
[20;30)	25	40
[30;40)	35	20
[40;50)	45	10
Total		200

a. The YULE's coefficient of skewness.

b. The PEARSON's coefficient of skewness.

c. The FISHER's coefficient of skewness.

d. The FISHER's coefficient of kurtosis.

Data from exercise 19

(x_i) = years of seniority		(n_i) = number of employees
Moyenne = 19 years		Standard deviation = 10.2 years
Q_1 = 12 years	Q_2 = 17 years	Q_3 = 25 years
% Population in $[\bar{x} - \sigma, \bar{x} + \sigma]$ = 80.2%		CV = 53.68%

Solution

The following table shows the data required to solve this exercise:

x_i	n_i	$n_i x_i$	$n_i(x_i - \bar{x})^3$	$n_i(x_i - \bar{x})^4$
5	30	150	-82320.00	1152480.00
15	100	1500	-6400.00	25600.00
25	40	1000	8640.00	51840.00
35	20	700	81920.00	1310720.00
45	10	450	175760.00	4569760.00
	200	3800	177600	7110400

a: Finding the YULE's coefficient of skewness:

$$C_Y = \frac{12 - 2(17) + 25}{25 - 12}$$

$$= \frac{37 - 34}{13}$$

$$= 0.230$$

b: Finding the PEARSON's coefficient of skewness:

$$\beta_1 = \frac{(\bar{x} - \text{ Mode})}{\sigma}$$

$$= \frac{(19 - 15)}{10.20}$$

$$= 0.3921$$

c: Finding the FISHER's coefficient of skewness:

$$\mu_3 = \frac{\sum_{i=1}^{i=5} n_i (x_i - \bar{x})^3}{\sum_{i=1}^{i=5} n_i} = \frac{177{,}600}{200}$$

$$= 888$$

$$\gamma_1 = \frac{\mu_3}{\sigma^3} = \frac{888}{(10.20)^3}$$

$$= 0.837265$$

d: Finding the FISHER's coefficient of skewness:

$$\mu_4 = \frac{\sum_{i=1}^{i=5} n_i (x_i - \bar{x})^4}{\sum_{i=1}^{i=5} n_i} = \frac{7{,}110{,}400}{200}$$

$$= 35{,}552$$

$$\gamma_2 = \frac{\mu_4}{\sigma^4} - 3 = \frac{35{,}552}{(10.20)^4} - 3$$

$$= 0.2869822$$

INTERPRETATION OF THE RESULTS

$$\left.\begin{array}{l} \underbrace{M_0}_{15} < \underbrace{M_e}_{17} < \underbrace{\bar{x}}_{19} \\ C_Y > 0 \\ \beta_1 > 0 \\ \gamma_1 > 0 \end{array}\right\}$$

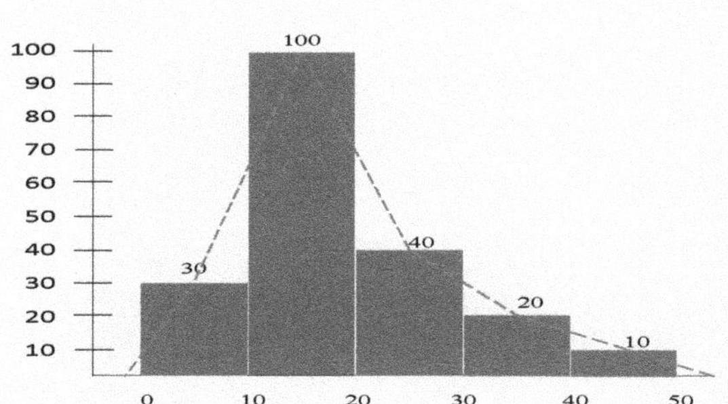

The distribution of years of seniority among the employees of the studied company is positively skewed, indicating that it is skewed towards the right, and there is a concentration of lower values. This suggests that the company is predominantly composed of employees with relatively short seniority, mainly juniors rather than seniors.

$\gamma_2 > 0$, The distribution curve is sharper than that of a normal distribution. In other words, the distribution is narrower and taller than a normal distribution. This may be due to the fact that the younger population, with relatively short seniority, rapidly drives the distribution to a high level, but the distribution drops quickly when reaching older employees.

Conclusion

In terms of dispersion, the distribution of years of seniority is somewhat dispersed but remains concentrated around the mean. This is indicated by the fact that over 80% of the population falls within the mean plus or minus one standard deviation, showing concentration around the mean. However, the coefficient of variation of 53.68% indicates some variability in the data, which can be attributed to the difference in seniority between juniors and seniors.

x_i	n_i
3	6
9	36
15	52
21	62
27	71
33	78
39	65
45	60
51	40
57	10
Total	480

• **Exercise 28:** Continuing exercise **20** from section **5.10.1** on the kilometers driven for each delivery from the supply center. The table beside is a summary of the observations.

Determine the following shape characteristics:

a. The YULE's coefficient of skewness.

b. The PEARSON's coefficient of skewness.

c. The FISHER's coefficient of skewness.

d. The FISHER's coefficient of kurtosis.

e. Provide your interpretation.

Data from exercise 20

(x_i) = mileage		(n_i) = number of deliveries
Mean = 30.75 Km		Standard deviation = 13.049425 Km
Q1 = 20.52 Km	Q2 = 31 Km	Q3 = 41.08 Km
% Population in $[\bar{x} - \sigma, \bar{x} + \sigma]$ = 62.26%		CV = 42.437%

Solution

The following table shows the data required to solve this exercise:

x_i	n_i	$n_i x_i$	$n_i(x_i - \bar{x})^3$	$n_i(x_i - \bar{x})^4$
3	6	18	-128215.41	3557977.52
9	36	324	-370407.94	8056372.64
15	52	780	-203163.19	3199820.20
21	62	1302	-57465.28	560286.49
27	71	1917	-3744.14	14040.53
33	78	2574	888.47	1999.05
39	65	2535	36498.52	301112.75
45	60	2700	173618.44	2474062.73
51	40	2040	332150.63	6726050.16
57	10	570	180878.91	4748071.29
	480	14760	-38961	29639793.38

a: Finding the YULE's coefficient of skewness:

$$C_Y = \frac{Q1 - 2Q2 + Q3}{Q3 - Q1}$$

$$= \frac{20.52 - 2(31) + 41.08}{41.08 - 20.52}$$

$$= -0.019455$$

b: Finding the PEARSON's coefficient of skewness:

$$\beta_1 = \frac{(\bar{x} - \text{Mode})}{\sigma}$$

$$= \frac{(30.75 - 33)}{13.049425}$$

$$= -0.17242$$

c: Finding the FISHER's coefficient of skewness:

$$\mu_3 = \frac{\sum_{i=1}^{i=10} n_i (x_i - \bar{x})^3}{\sum_{i=1}^{i=10} n_i} = \frac{-38{,}961}{480}$$

$$= -81.168$$

$$\gamma_1 = \frac{\mu_3}{\sigma^3} = \frac{-81.168}{(13.049425)^3}$$

$$= -0.036527$$

d: Finding the FISHER's coefficient of kurtosis:

$$\mu_4 = \frac{\sum_{i=1}^{i=10} n_i (x_i - \bar{x})^4}{\sum_{i=1}^{i=10} n_i} = \frac{29{,}639{,}793.38}{480}$$

$$= 61{,}749.569$$

$$\gamma_2 = \frac{\mu_4}{\sigma^4} - 3 = \frac{61{,}749.569}{(13.049425)^4} - 3$$

$$= -0.8705454$$

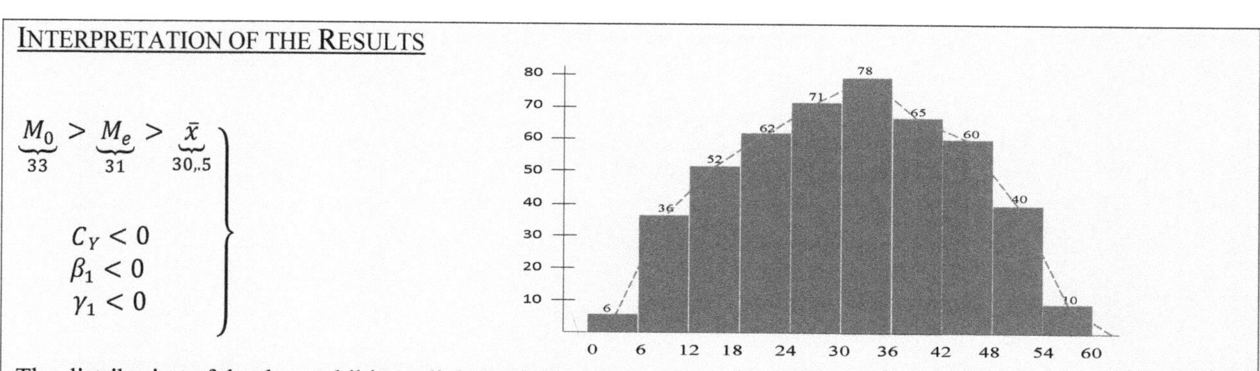

The distribution of the data exhibits a slight left skewness, indicating that the majority of values are concentrated on the left side of the distribution, with a weakly stretched distribution tail to the left. This suggests a small prevalence of long trips in the data under study.

$\gamma_2 < 0$, The curve of this distribution is slightly flatter than that of a normal distribution, indicating that the data may be slightly less stretched and more concentrated around the mean compared to a normal distribution.

Conclusion

Regarding the dispersion of the data, the coefficient of variation of 42.43% indicates that the distribution is moderately dispersed, meaning that the values are relatively close to each other and the variation between values is not too high. The distribution is quasi-symmetric, with negligible spread to the left, indicating that the values are generally balanced on both sides of the mean. This is further supported by the fact that approximately 62.26% of the population studied falls within the range of the mean plus or minus the standard deviation, suggesting an almost equal distribution to the left and right of the mean. This can be observed both by the almost equivalence of the three measures of central tendency (mode, mean, and median) and by the almost absence of dispersion, as the FISHER's coefficient of skewness is -0.03.

- **Exercise 29:** Let's continue with **exercise 21** from **section 5.10.2.** The table beside presents the number of individuals who have recovered from a viral pandemic in a country, along with the length of their hospital stay (in days) before total recovery is confirmed and pronounced by the medical staff, for a given period of time.

x_i	n_i
2.5	12711
12.5	3512
30	13024
50	52646
65	65665
85	136051
Total	283609

Determine the following shape characteristics:

a. The YULE's coefficient of skewness.

b. The PEARSON's coefficient of skewness.

c. The FISHER's coefficient of skewness.

d. The FISHER's coefficient of kurtosis.

e. Provide your interpretation.

Data from exercise 21

(x_i) = time spent in the hospital		(n_i) = number of patients
Mean = 66.75 days		Standard deviation = 22.27 days
Q1 = 55.82 days	Q2 = 69.123 days	Q3 = 84.37 days
% Population in $[\bar{x} - \sigma, \bar{x} + \sigma]$ = 67.92%		CV = 33%

Solution

The following table shows the data required to solve this exercise:

x_i	n_i	n_ix_i	$n_i(x_i-\bar{x})^3$	$n_i(x_i-\bar{x})^4$
2.5	12711	31777.50	-3371511322.35	216623843464.11
12.5	3512	43900.00	-560768931.68	30422419931.62
30	13024	390720.00	-646488810.79	23759277010.92
50	52646	2632300.00	-247461507.35	4145291528.46
65	65665	4268225.00	-352682.79	617638.52
85	136051	11564335.00	826800136.05	15088062456.14
	283609	18931257.50	-3999783118.92	290039512029.76

a: Finding the YULE's coefficient of skewness:

$$C_Y = \frac{Q1 - 2Q2 + Q3}{Q3 - Q1}$$

$$= \frac{55.82 - 2(69.12) + 84.37}{84.37 - 55.82}$$

$$= 0.0683$$

b: Finding the PEARSON's coefficient of skewness:

$$\beta_1 = \frac{(\bar{x} - \text{Mode})}{\sigma}$$

$$= \frac{(66.75 - 85)}{22.27}$$

$$= -0.819$$

c: Finding the FISHER's coefficient of skewness:

$$\mu_3 = \frac{\sum_{i=1}^{i=6} n_i(x_i - \bar{x})^3}{\sum_{i=1}^{i=6} n_i} = \frac{-3{,}999{,}783{,}118.92}{283{,}609}$$

$$= -14{,}103.16$$

$$\gamma_1 = \frac{\mu_3}{\sigma^3} = \frac{-14{,}103.16}{(22.27)^3}$$

$$= -1.276544$$

d: Finding the FISHER's coefficient of kurtosis:

$$\mu_4 = \frac{\sum_{i=1}^{i=6} n_i (x_i - \bar{x})^4}{\sum_{i=1}^{i=6} n_i} = \frac{290{,}039{,}512{,}029.76}{283{,}609}$$

$$= 1{,}022{,}673.86$$

$$\gamma_2 = \frac{\mu_4}{\sigma^4} - 3 = \frac{1{,}022{,}673.86}{(22.27)^4} - 3$$

$$= 1.1561972$$

INTERPRÉTATION DES RÉSULTATS

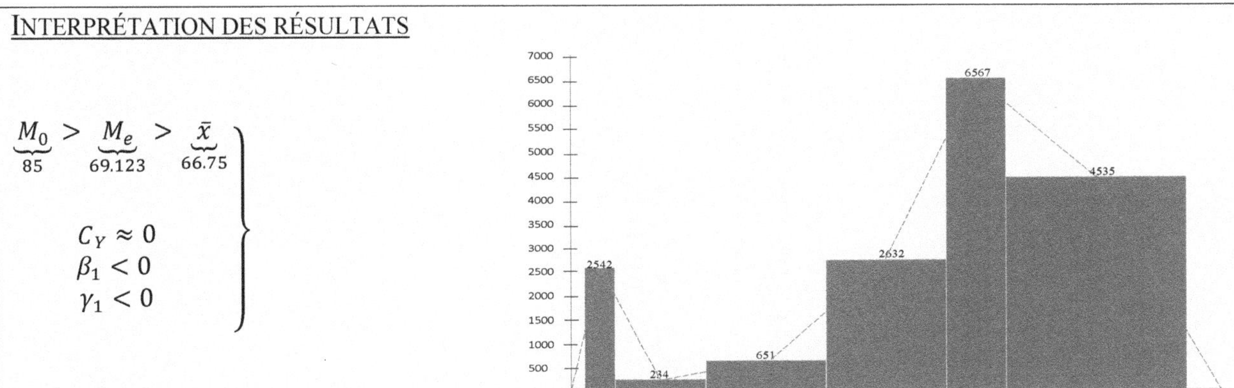

The distribution of hospital stay time shows a left-skewed pattern, indicating that the majority of patients have shorter hospital stays. However, the tail of the distribution is stretched towards the left, suggesting that there is a significant number of patients with longer hospital stays.

$\gamma_2 > 0$, this suggests that the curve of this distribution is sharper than that of a normal distribution. This may be due to the concentration of high values in the tail of the distribution, indicating a predominance of patients with longer hospital stays. Patients with shorter and medium stays contribute to a rapid rise in the distribution, but it drops abruptly when the count shifts to patients with longer stays.

Conclusion

The distribution of hospital length of stay is moderately dispersed, with a coefficient of variation of 33%. However, the high values in the tail of the distribution, corresponding to patients with longer stays, mostly fall within the mean plus or minus one standard deviation interval, accounting for 67.92% of cases. Although lower than the expected percentage according to the normal distribution, this indicates that a rather substantial percentage of recovered patients spent a relatively long time in the hospital before being confirmed as fully cured.

• **Exercise 30 :** Let's continue with exercise **22** from section **5.10.2** which focuses on the company's fleet of vehicles, comprising a significant number of cars. For 100 of these vehicles, the mileage recorded at the time of maintenance is given in the following table. We will calculate the following shape characteristics. Determine the following shape characteristics:

x_i	n_i
82500	5
87500	9
92500	14
97500	18
102500	25
107500	16
112500	7
117500	6
	100

a. The YULE's coefficient of skewness.

b. The PEARSON's coefficient of skewness.

c. The FISHER's coefficient of skewness.

d. The FISHER's coefficient of kurtosis.

e. Provide your interpretation.

Data from exercise 22

(x_i) = mileage (in Km)		(n_i) = number of cars
Mean = 100,250 Km		Standard deviation = 8,842.37 Km
Q1 = 93,928.57 Km	Q2 = 100,800 Km	Q3 = 106,250 Km
% Population in $[\bar{x} - \sigma, \bar{x} + \sigma]$ = 66.19%		CV = 8.82%

Solution

The following table shows the data required to solve this exercise:

x_i	n_i	$n_i x_i$	$n_i(x_i - \bar{x})^3$	$n_i(x_i - \bar{x})^4$
82500	5	412500.00	-27961796875000.00	496321894531250000.00
87500	9	787500.00	-18654046875000.00	237839097656250000.00
92500	14	1295000.00	-6516781250000.00	50505054687500000.00
97500	18	1755000.00	-374343750000.00	1029445312500000.00
102500	25	2562500.00	284765625000.00	640722656250000.00
107500	16	1720000.00	6097250000000.00	44205062500000000.00
112500	7	787500.00	12867859375000.00	157631277343750000.00
117500	6	705000.00	30797718750000.00	531260648437500000.00
	100	10025000.00	-3459375000000.00	1519433203125000000.00

a: Finding the YULE's coefficient of skewness:

$$C_Y = \frac{\text{Q1} - 2\text{Q2} + \text{Q3}}{\text{Q3} - \text{Q1}}$$

$$= \frac{93{,}928.57 - 2(100{,}800) + 106{,}250}{106{,}250 - 93{,}929}$$

$$= -0.115331$$

b: Finding the de PEARSON's coefficient of skewness:

$$\beta_1 = \frac{(\bar{x} - \text{Mode})}{\sigma}$$

$$= \frac{(100{,}250 - 102{,}500)}{8{,}842.37}$$

$$= -0.254456$$

c: Finding the FISHER's coefficient of skewness:

$$\mu_3 = \frac{\sum_{i=1}^{i=8} n_i (x_i - \bar{x})^3}{\sum_{i=1}^{i=8} n_i} = \frac{-3{,}459{,}375{,}000{,}000}{100}$$

$$= -34{,}593{,}750{,}000$$

$$\gamma_1 = \frac{\mu_3}{\sigma^3} = \frac{-34{,}593{,}750{,}000}{(8{,}842.37)^3}$$

$$= -0.0500$$

d: Finding the FISHER's coefficient of kurtosis:

$$\mu_4 = \frac{\sum_{i=1}^{i=8} n_i (x_i - \bar{x})^4}{\sum_{i=1}^{i=60} n_i} = \frac{1{,}519{,}433{,}203{,}125{,}000{,}000}{100}$$

$$= 15{,}194{,}332{,}031{,}250{,}000$$

$$\gamma_2 = \frac{\mu_4}{\sigma^4} - 3 = \frac{15{,}194{,}332{,}031{,}287{,}300}{(8{,}842.37)^4} - 3$$

$$= -0.5145390$$

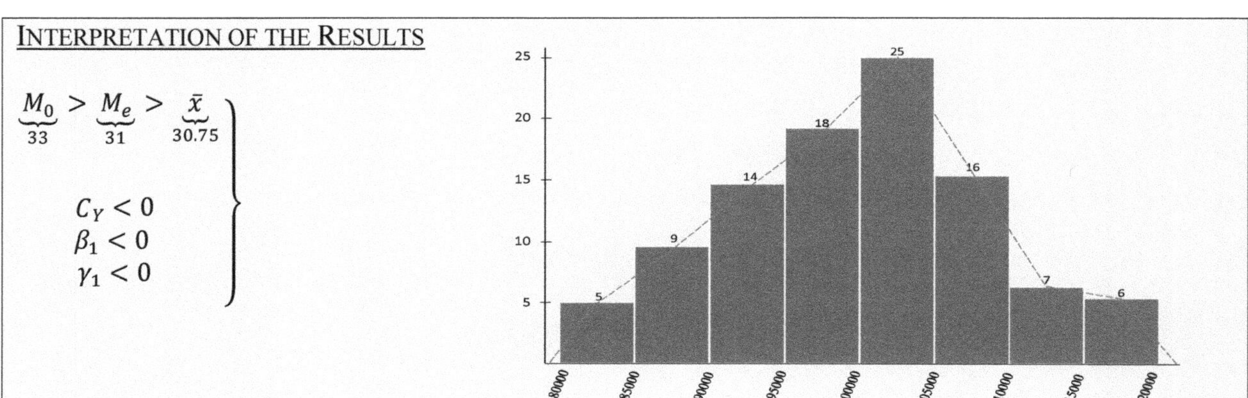

The distribution is skewed to the left, indicating that it is stretched towards the right and the majority of values are concentrated on the left side of the distribution. The distribution tail is slightly spread towards the left, suggesting that there are few vehicles with very high mileage compared to the mean. However, there is a slight concentration of high values, indicating a small prevalence of long trips among the vehicles in the fleet.
$\gamma_2 < 0$, The distribution is skewed to the left. However, the shape of this distribution is slightly flatter than that of a normal distribution, as evidenced by the negative value of the FISHER's coefficient of kurtosis (-0.05). This suggests that the distribution of mileage readings of vehicles before maintenance is slightly less spread out than a normal distribution, with a slight concentration of values around the mean.

Conclusion

The distribution of mileage readings of vehicles before maintenance appears to be almost identical for all cars, with slight differences. This distribution has a coefficient of variation of 8.82%, indicating that it is relatively homogeneous. It is nearly symmetrical, with negligible spreading towards the left, meaning that the values are mostly clustered around the mean. Approximately 66.19% of the population falls within the range of one standard deviation from the mean, and this population is evenly distributed on both sides of the mean. This can be observed through the near-equivalence of the three measures of central tendency (mode, mean, and median), as well as the low value of the FISHER's coefficient of kurtosis (-0.05) indicating a lack of spreading. The shape of this distribution is very similar to a normal distribution, but with a slight leftward asymmetry and a concentration of values around the mean.

• **Exercise 31:** Continuing exercise **23** from section **5.10** on the number of days of tire production. The table below gives the number of days compared to the number of tires produced. Calculate the following shape characteristics.

x_i	n_i
12750	1
13250	2
13750	3
14250	3
14750	3
15250	3
Total	15

a. The YULE's coefficient of skewness.

b. The PEARSON's coefficient of skewness.

c. The FISHER's coefficient of skewness.

d. The FISHER's coefficient of kurtosis.

e. Provide your interpretation.

Data from exercise 23

(x_i) = number of tires		(n_i) = number of days
Mean = 14,216.67 tires		Standard deviation = 736.03 tires
Q1 = 13,625 tires	Q2 = 14,250 tires	Q3 = 14,875 tires
% Population in $[\bar{x} - \sigma, \bar{x} + \sigma]$ = 60.43%		CV = 5.37%

Solution

The following table shows the data required to solve this exercise:

x_i	n_i	$n_i x_i$	$n_i(x_i - \bar{x})^3$	$n_i(x_i - \bar{x})^4$
12750	1	12750.00	-3154962962.96	4627279012345.67
13250	2	26500.00	-1806592592.59	1746372839506.17
13750	3	41250.00	-304888888.89	142281481481.48
14250	3	42750.00	111111.11	3703703.70
14750	3	44250.00	455111111.11	242725925925.93
15250	3	45750.00	3310111111.11	3420448148148.16
	15	213250.00	-1501111111.11	10179111111111.10

a: Finding the YULE's coefficient of skewness:

$$C_Y = \frac{Q1 - 2Q2 + Q3}{Q3 - Q1}$$

$$= \frac{13{,}625 - 2(14{,}250) + 14{,}875}{14{,}875 - 13{,}625}$$

$$= 0$$

b: Finding the PEARSON's coefficient of skewness:

$$\beta_1 = \frac{(\bar{x} - \text{Mode})}{\sigma}$$

$$= \frac{(14{,}216.67 - 15{,}250)}{763.03}$$

$$= -1.353$$

c: Finding the FISHER's coefficient of skewness:

$$\mu_3 = \frac{\sum_{i=1}^{i=6} n_i (x_i - \bar{x})^3}{\sum_{i=1}^{i=6} n_i} = \frac{-1{,}501{,}111{,}111.11}{15}$$

$$= -100{,}074{,}074.07$$

$$\gamma_1 = \frac{\mu_3}{\sigma^3} = \frac{-100{,}074{,}074}{(763.03)^3}$$

$$= -0.225262$$

d: Finding the FISHER's coefficient of kurtosis:

$$\mu_4 = \frac{\sum_{i=1}^{i=6} n_i (x_i - \bar{x})^4}{\sum_{i=1}^{i=6} n_i} = \frac{10{,}179{,}111{,}111{,}111.10}{15}$$

$$= 678.607.407.376$$

$$\gamma_2 = \frac{\mu_4}{\sigma^4} - 3 = \frac{678{,}607{,}407{,}376}{(763.03)^4} - 3$$

$$= -0.9981062$$

INTERPRÉTATION DES RÉSULTATS

$$\left.\begin{array}{c}\underbrace{M_0}_{15{,}250} > \underbrace{M_e}_{14{,}250} > \underbrace{\bar{x}}_{14{,}216} \\ C_Y = 0 \\ \beta_1 < 0 \\ \gamma_1 < 0\end{array}\right\}$$

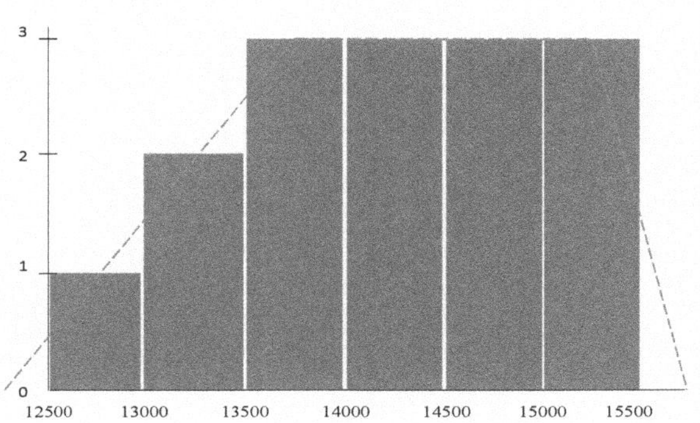

$C_Y = 0$ due to the equidistance between the third quartile (Q_3) and the first quartile (Q_1) from the median. Therefore, it does not provide useful information in interpreting the distribution. However, the coefficients β_1 and γ_1 can be used to further understand the shape of the distribution.
The coefficients by β_1 and γ_1 being < 0, signify that the distribution is left-skewed. This confirms that the tail of the distribution is stretched towards the left, indicating a high concentration of days with high production, implying that tire production is intense during specific days.
$\gamma_2 < 0$, the curve is more flattened than that of the normal distribution.

Conclusion
With an extremely low coefficient of variation of 5.37%, it is homogeneous with a small spread to the right, resulting in a left-skewed distribution tail. There is a high concentration of values around the mean, which is very close to the median and almost coincides with the mode. Furthermore, its multimodal nature amplifies the concentration around these central values. This concentration is observed in the volume of the population within the range of one standard deviation from the mean, which amounts to 60.43%, indicating a certain stability in tire production.

• **Exercise 32:** Continuing exercise **24** from section **5.10.2** on the balances recorded in the savings accounts of 300 individuals.

x_i	n_i
2500	48
7500	41
12500	47
17500	15
22500	21
27500	12
32500	13
37500	8
42500	9
47500	9
52500	10
57500	6
62500	9
67500	5
72500	2
77500	13
82500	8
87500	7
92500	4
97500	13
	300

Determine the following shape characteristics:

a. The YULE's coefficient of skewness.

b. The PEARSON's coefficient of skewness.

c. The FISHER's coefficient of skewness.

d. The FISHER's coefficient of kurtosis.

e. Provide your interpretation.

Data from exercise 24

(x_i) = saving balance		(n_i) = number of savers
Mean = 31,483 euros		Standard deviation = 29,337.96 euros
Q1 = 8,291.88 euros	Q2 = 19,670 euros	Q3 = 51,002.99 euros
% Population in $[\bar{x} - \sigma, \bar{x} + \sigma]$ = 72.86%		CV = 93.18%

Solution

The following table shows the data required to solve this exercise:

x_i	n_i	$n_i x_i$	$n_i(x_i - \bar{x})^3$	$n_i(x_i - \bar{x})^4$
2500	48	120000.00	-1168654759777780.00	33871510454225900000.00
7500	41	307500.00	-565604019810185.00	13565069741780900000.00
12500	47	587500.00	-321525393949074.00	6103623728466590000.00
17500	15	262500.00	-41013174930555.50	573500896112268000.00
22500	21	472500.00	-15224107402777.80	136763231501620000.00
27500	12	330000.00	-758439944444.44	3021119112037030.00
32500	13	422500.00	13660893518.52	13888575077160.60
37500	8	300000.00	1742440037037.04	10483680889506200.00
42500	9	382500.00	12033532541666.70	132569416834028000.00
47500	9	427500.00	36979320041666.70	592285442667361000.00
52500	10	525000.00	92830675046296.30	1950991353889660000.00
57500	6	345000.00	105658930027778.00	2748893162889350000.00
62500	9	562500.00	268551682541667.00	8329578020167360000.00
67500	5	337500.00	233604150023148.00	8413642803333720000.00
72500	2	145000.00	138010168342593.00	5660717071518680000.00
77500	13	1007500.00	1266743898393520.00	58291331724408400000.00
82500	8	660000.00	1062248740037040.00	54192389887556200000.00
87500	7	612500.00	1230409926699070.00	68923462727259800000.00
92500	4	370000.00	908668403351852.00	55443917077852200000.00
97500	13	1267500.00	3740280115060190.00	246920825595890000000.00
	300	9445000.00	6984995747222220.00	565864591024931000000.00

a: Finding the YULE's coefficient of skewness:

$$C_Y = \frac{Q1 - 2Q2 + Q3}{Q3 - Q1}$$

$$= \frac{8{,}291.88 - 2(19{,}670) + 51{,}002.99}{51{,}002.99 - 8{,}291.88}$$

$$= 0.4672$$

b: Finding the PEARSON's coefficient of skewness:

$$\beta_1 = \frac{(\bar{x} - \text{Mode})}{\sigma}$$

$$= \frac{(31{,}483 - 2{,}500)}{29{,}337.97}$$

$$= 0.987910888$$

c: Finding of the FISHER's coefficient of skewness:

$$\mu_3 = \frac{\sum_{i=1}^{i=20} n_i (x_i - \bar{x})^3}{\sum_{i=1}^{i=20} n_i} = \frac{6{,}984{,}995{,}747{,}222{,}220}{300}$$

$$= 23{,}283{,}319{,}157{,}407.40$$

$$\gamma_1 = \frac{\mu_3}{\sigma^3} = \frac{23{,}283{,}319{,}157{,}407.40}{(29{,}337.97)^3}$$

$$= 0.921952$$

d: Finding the FISHER's coefficient of kurtosis:

$$\mu_4 = \frac{\sum_{i=1}^{i=20} n_i (x_i - \bar{x})^4}{\sum_{i=1}^{i=20} n_i} = \frac{565{,}864{,}591{,}024{,}931{,}000{,}000}{300}$$

$$= 1{,}886{,}215{,}303{,}416{,}440{,}000$$

$$\gamma_2 = \frac{\mu_4}{\sigma^4} - 3 = \frac{1{,}886{,}215{,}303{,}416{,}440{,}000}{(29{,}337.97)^4} - 3$$

$$= -0.453$$

INTERPRÉTATION OF THE RESULTS

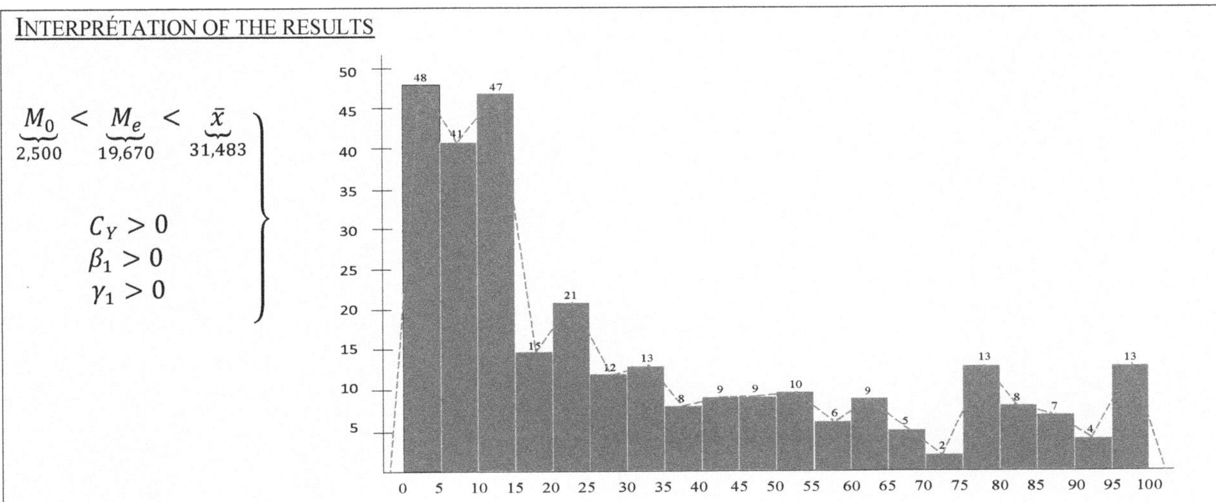

The distribution of savings balances is positively skewed, with a longer tail towards the right. This indicates a concentration of lower values, suggesting that the majority of individuals have very limited savings in their savings.

$\gamma_2 < 0$, the distribution curve exhibits a broader and flatter shape compared to a normal distribution. This means that the values of savings balances are less concentrated around the mean and are more spread out, indicating a larger dispersion of data.

Conclusion

The distribution of savings presents a significant dispersion with a coefficient of variation of 93.18%. We can observe a concentration of the population within the range of the mean plus or minus one standard deviation, aligned with that of the normal distribution which is approximately 68.18%. However, there is also a long tail extending towards the right of the distribution, resulting in a thick spread tail, as there is a nearly constant proportion of savers as the amount of savings increases. The nearly identical concentration of savers within the range of the mean minus one standard deviation and the mean plus one standard deviation is mainly comprised of low savings, as evidenced by the height of the distribution curve. This reveals a form of inequality in the savings of individuals in this population, which is also clearly visible in the comparison of central tendency values.

It is evident that the majority of savers have only 2,500 euros in their savings account, while if these accounts were arranged from the least funded to the most funded, the median account would have 19,670 euros, representing a difference of 17,170 euros between what the majority of people have in savings and what the median account holds. In other words, the median account is 7.8 times more funded than those of the majority of savers. Furthermore, the individual in this majority class is 7.8 times less affluent than the one whose account is in the middle of the population (when ranked from least affluent to most affluent).

Moreover, if all individuals had the same financial capabilities and all accounts were funded identically, each person would have 31,483 euros in their savings account. In other words, if the total sum of money deposited in savings, which is 9,445,000 euros, were evenly distributed among all savers, each of them would have 31,483 euros in savings. This means that even the individual in the middle of the distribution (with 19,670 euros in savings) is not at the average level of savings. They are only at half of the average level of wealth of savers. As for the less affluent individuals, who have on average 2,500 euros in savings, they only have 7.9% of what the average savings represent.

If we consider that the concept of savings is linked to that of wealth, we can conclude that in this case, the majority of individuals in this population are 7.8 times poorer than those in the middle of the population (when ranked from least affluent to most affluent) and 12.5 times poorer than the average level of wealth, which is 31,291 euros. As for the person whose savings are at the median, although they are at the median level of the population, their wealth is only half of the average wealth.

CHAPTER 7

Concentration Characteristics

This chapter provides a comprehensive analysis of concentration characteristics by examining the LORENZ concentration curve and the GINI concentration index. To interpolate the LORENZ curve as a continuous function in the range of zero to one hundred, the chapter explains and utilizes two methods: the Newton method and the Lagrange method. Additionally, the chapter introduces and uses two methods to determine the GINI index by calculating the area under the LORENZ curve: the Trapezoids method, which involves summing the areas of the trapezoids, and the numerical quadrature method, which evaluates the definite integral of the interpolated function within the range of zero to one hundred. By covering these methods and concepts in detail, the chapter offers a comprehensive understanding of concentration analysis and its application in various fields.

7.1 Introduction

In general, the concept of concentration is used in various fields with different definitions, importance, and objectives. It can be found in exact sciences as well as in social sciences and environmental sciences.

In exact sciences, concentration is used in several areas. For example, in chemistry, it refers to the proportion of a solute in a solution. This concentration can be expressed in terms of mass (when comparing mass to volume) or in terms of moles (when comparing the amount of substance to volume). In radio communication or IP communication (Internet Protocol), traffic concentration is used to describe the proportion of communications on the radio interface, or the proportion of IP packets compared to the normal traffic and/or the capacity of the transmission medium. Sometimes, to optimize the use of transmission channels, this traffic concentration can be organized in an orderly manner.

In social sciences, the concept of concentration is also widely used. This is particularly the case in the field of marketing, where a company may define its Strategic Business Area (SBA) by focusing on a less competitive market segment. In economics, market concentration is mentioned when the market departs from its natural oligopolistic or competitive form and evolves towards a more monopolistic form.

In descriptive statistics, the concept of concentration is mainly used to measure the heterogeneity of characteristics within a population and understand potential inequalities within it. Here, the term "population" is used in its statistical sense. These relative data allow us to realize the concentration or lack thereof of the distribution on a small number of statistical individuals. This concept of concentration was introduced by Corrado GINI, an Italian statistician, for the distributions of salaries and incomes. It generally applies to the description of economic units based on their size (e.g., a company based on its revenue, number of employees, etc.), but can also be extrapolated to other areas when the distribution of collective wealth (often monetary) is involved. To measure this heterogeneity, two concentration characteristics are used: the GINI index and the medial.

Today, the importance of these characteristics is observed in various fields of activity. For example, international organizations such as the United Nations, the Organization for Economic Co-operation and Development (OECD), and others publish annual reports on wealth distribution in different countries around the world. These reports present GINI indices for all countries, reflecting how wealth is distributed in these countries. In the same vein, in October 2020, Credit Suisse reported that "*1% of the population owns nearly half of the world's wealth.*"

7.2 Medial of a Distribution

Medial is a statistical concept that determines the median value of a mass in a distribution of variables. Then, it allows for measuring the median position of a data distribution, taking into account the mass of each value, which can be useful for understanding the distribution of values in a given population. Unlike the classical median that is calculated based on the frequency or relative frequency of each value, the medial refers to the cumulative function of the non-centered moment of order zero $(n_i x_i)$. In other words, it takes into account the mass of each value, rather than their frequency of occurrence.

The medial has the particularity of dividing the mass of a variable into two subsets of equal masses. It represents the cutoff point that separates the distribution into two parts of equal weight. It is denoted as $\left(\overset{\approx}{M}_e\right)$ to symbolize this balanced division of the mass.

The media can be used in various fields of statistics, such as economics, demography, biostatistics, etc.

Example 68

Let's consider **example 57** from section **5.8**, which deals with the distribution of the budget of subsidies allocated by the state to companies in an American city, expressed in millions of dollars. In this example, x_i represents the subsidy amounts in millions of dollars, and n_i represents the number of beneficiary businesses.

State Subsidies (in Millions USD)	Class Center (x_i)	Number (n_i)	f_i	Total Budget per class $S_i = n_i x_i$	$q_i = \frac{S_i}{\sum_{Total} S_i}$	$f_i \uparrow$	$q_i \uparrow$
[0.2 ; 0.6)	0.4	4	3.96	1.60	0.17	3.96	0.17
[0;.6 ; 1)	0.8	5	4.95	4.00	0.42	8.91	0.59
[1 ; 2)	1.5	9	8.91	13.50	1.43	17.82	2.02
[2 ; 3)	2.5	7	6.93	17.50	1.85	24.75	3.86
[3 ; 5)	4	13	12.87	52.00	5.49	37.62	9.35
[5 ; 8)	6.5	17	16.83	110.50	11.67	54.46	21.02
[8 ; 10)	9	12	11.88	108.00	11.40	66.34	32.43
[10 ; 20)	15	21	20.79	315.00	33.26	87.13	65.68
[20 ;30)	25	13	12.87	325.00	34.32	100.00	100.00
Total		101		947.10			

In this case, the median of the distribution of subsidies to businesses is an important indicator in the context of fund allocation. It ensures that businesses that have individually received less than the median have collectively received the same amount as those that have received more. In other words, it ensures fairness in the distribution of subsidies by allocating 50% of the total budget mass to subsidies below the median and 50% to subsidies above the median. This helps prevent some businesses from receiving a disproportionate share of funds, while others receive very little.

If the distribution of subsidies is fairly regular, the median subsidy should be very close to the median. The median subsidy, in turn, is the value that divides the distribution of subsidies into two equal parts: 50% of businesses have received a subsidy below the median and 50% have received a subsidy above the median. Thus, the median is a key indicator to ensure that the distribution of subsidies is balanced and that recipient businesses have received a fair and equitable share of the funds allocated by the government.

Let's delve deeper into the concept of budgetary mass corresponding to each class, which represents the total sum of subsidies received by the n_i businesses belonging to that class.

To be more precise, for the ith class, let's denote S_i as the total subsidies received by these businesses, where each subsidy falls between the endpoints a_{i-1} and a_i of that class. For example, if we consider the 13 businesses that received between 3 and 5 million dollars, they have collectively received a total of 52 million dollars in subsidies.

Let's express the column of frequencies of the total budget of a class relative to the total budget, which we will call q_i, in percentage. Consequently, we will introduce a new notion of cumulative increasing frequencies of total budgets, denoted as $q_i \uparrow$, also in percentage. For example, the 13 businesses that each received between 3 and 5 million dollars represent 12.87% of the total number of businesses and share 5.49% of the total budget, amounting to 52 million dollars. Furthermore, 66.34% of the businesses each received less than 10 million dollars, representing 32.43% of the total budget.

The calculation of the median is similar to that of the median. It involves finding the value that divides the distribution into two equal parts, such that 50% of the data points are below and 50% are above this value.

The formula for calculating the median in this specific case is as follows:

$$\overset{\approx}{M}_e = a_i + (a_{i+1} - a_i)\frac{(50 - q_i)}{(q_{i+1} - q_i)}$$

Where a_i represents the value of the subsidy for the $i^{ème}$company, a_{i+1}represents the value of the subsidy for the $(i+1)^{ème}$ company, q_i is the cumulative percentage of companies that have received less than the median up to the $i^{ème}$company, and q_{i+1} is the cumulative percentage of companies that have received less than the median up to the $(i+1)^{ème}$ company.

Thus:

$$\overset{\approx}{M}_e = 10 + (20 - 10)\frac{(50 - 32.43)}{(65.68 - 32.43)}$$
$$= 15.28 \text{ million dollars}$$

The median of the distribution of subsidies to businesses in this example is calculated at $15.28 million. This means that half of the businesses received subsidies lower than $15.28 million, and the other half received subsidies higher than this value. In other words, businesses that have individually received less than $15.28 million have collectively received as much as those that have received more than $15.28 million.

The determination of the median can be done using a graphical method, similar to that used for calculating the median.

7.3 LORENZ Concentration Curve

The LORENZ concentration curve is a widely used graphical tool in economics and statistics to analyze the distribution of wealth within a given population. It allows for a visual representation of the proportion of wealth held by a certain lower percentage of the population, typically on the ordinate, plotted against the cumulative percentage of the population, typically on the abscissa.

The LORENZ concentration curve is constructed from the distribution data of wealth in the form of pairs of points ($f_i \uparrow$, $q_i \uparrow$), where $f_i \uparrow$ represents the cumulative proportion of wealth held by the bottom *i*% of the population, and $q_i \uparrow$ represents the cumulative percentage of the population up to *i*%. Thus, it provides a quick visual snapshot of wealth distribution inequalities within a given population and allows for comparisons of different distributions.

Example 69

Pursuing **example 68,** which deals with the distribution of the budget of subsidies allocated by the state to companies in an American city, expressed in millions of dollars, the concentration curve of the subsidy's distribution is the representative curve of $q_i \uparrow$ as a function of $f_i \uparrow$. It is drawn by connecting the points representing the class ends a_i.

Due to the fact that we are dealing with two cumulative frequencies expressed in percentages ($f_i \uparrow$ and $q_i \uparrow$), the concentration curve will be plotted within a square of side 100, as percentages range between 0 and 100. When both $f_i \uparrow$ and $q_i \uparrow$ reach 100, the curve will pass through the opposite corners of the square, indicating maximum concentration. This means that all the elements under study are present in both cumulative frequencies at their highest level. Therefore, the curve will be located below the diagonal of the square, as a given value of $q_i \uparrow$, which represents the relative cumulative frequency, is always lower than its equivalent $f_i \uparrow$, which represents the absolute cumulative frequency.

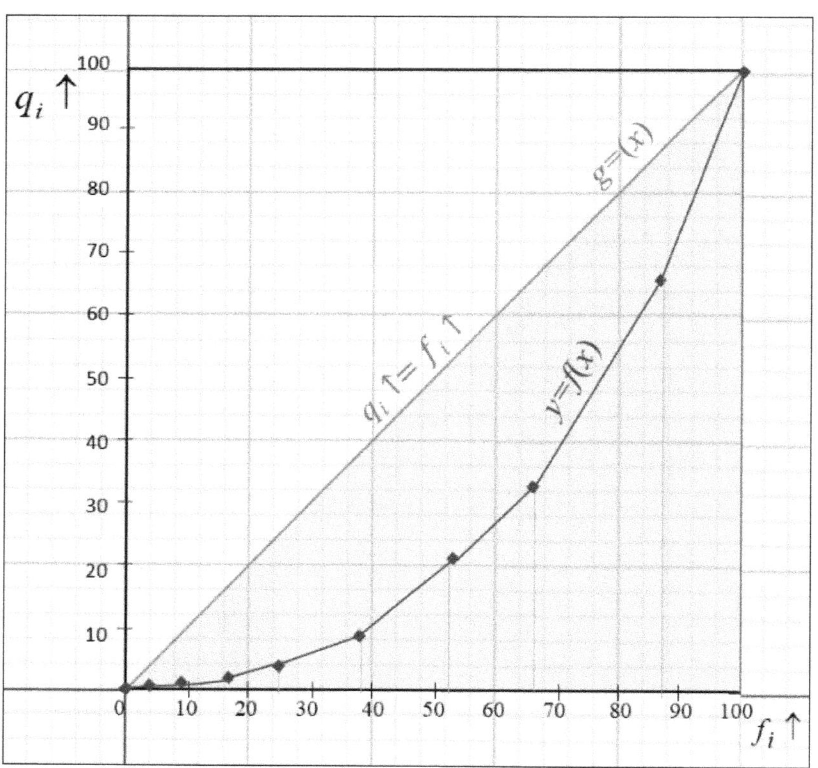

7.4 Calculation of the area under the LORENZ curve

The calculation of the area under the LORENZ curve used to measure the inequality in the distribution of wealth or income in a given population, can be done either graphically using the trapezoidal rule or analytically using the calculation of definite integrals.

The first method is to use a graphical approach, called the trapezoid method. This method involves dividing the area under the LORENZ CURVE into a series of trapezoids, using the points on the curve and the coordinate axes. Then, the area of each trapezoid is calculated by multiplying the average of the sides by the difference between the adjacent abscissa values. Finally, the areas of all the trapezoids are added together to obtain the total area under the curve.

The second method is an analytical approach that uses the calculation of definite integrals. It requires knowing the mathematical equation of the LORENZ curve and integrating this equation over the interval of interest, usually from 0 to 1.

Both methods have their advantages and disadvantages. The trapezoid method is simpler to understand and implement, but it may be less accurate for complex LORENZ curves. On the other hand, the analytical approach based on definite integrals may be more accurate for complex LORENZ curves, but it can be more mathematically complex. It is important to choose the appropriate method based on the specific needs of the analysis and the available data.

7.4.1 Graphical method of calculating the area under the LORENZ curve

The calculation of the area under the LORENZ curve by the graphical method (trapezoidal method) is obtained by carrying out the following three steps:

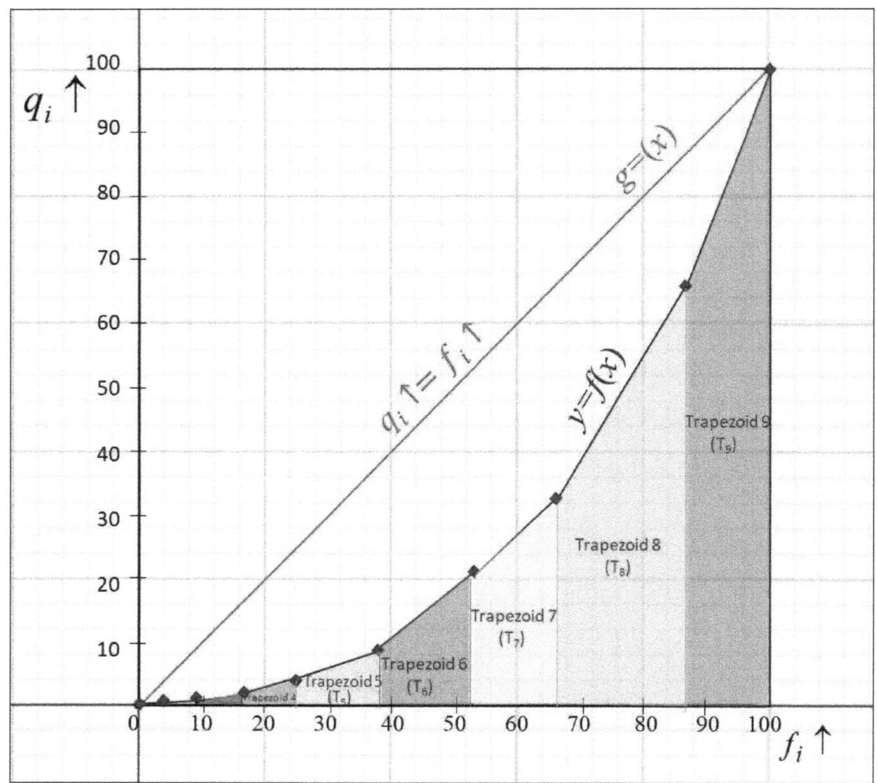

The calculation of the area under the LORENZ curve, also known as the trapezoid method, allows for the estimation of this area by following the detailed steps outlined below:

1. The LORENZ curve graphically represents the cumulative distribution of a variable on the horizontal axis (typically the proportion of the population) against the cumulative proportion of the variable on the vertical axis. To calculate the area under the LORENZ curve, it is divided into small vertical trapezoids, numbered from 0 to *n*, where n is the number of data points on the curve.

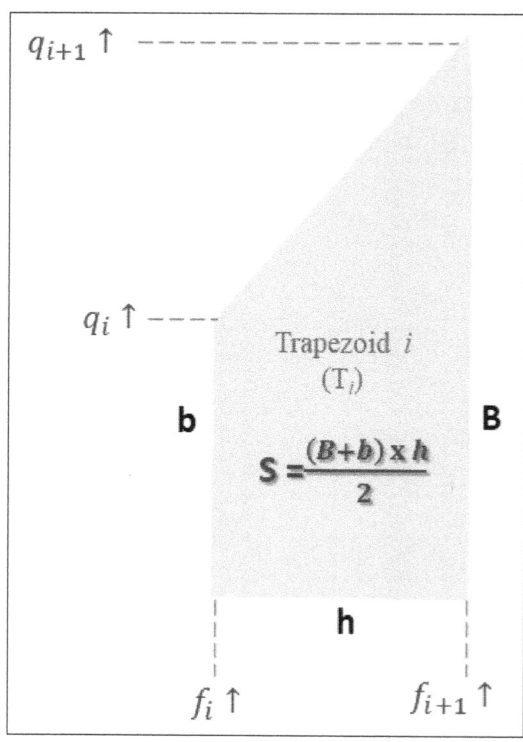

2. Each trapezoid has a height equal to the difference between the values of the LORENZ function $q_i \uparrow$ and $q_{i+1} \uparrow$, which represent the cumulative proportion of the variable at two successive data points. The base of the trapezoid is formed by the two vertical sides $f_i \uparrow$ et $f_{i+1} \uparrow$ of the LORENZ curve that are underpinned by the portion of the curve between the two data points.

   The area of each trapezoid T_i is calculated individually using the following formula:

$$\text{Area}(Ti) = (f_{i+1}\uparrow - f_i\uparrow)\frac{(q_i\uparrow + q_{i+1}\uparrow)}{2}$$

This formula represents the area of a trapezoid, where the height is the difference between the values of the LORENZ function, and the base is the sum of the vertical sides of the LORENZ curve between the two data points.

4. Finally, to obtain the total area under the LORENZ curve, the sum of the areas of all the trapezoids is calculated from $i = 0$ to $i = n$:

$$\text{Total Area} = \sum_{i=0}^{n} \text{Area}(T_i)$$

$$\boxed{\text{Total Area} = \sum_{i=1}^{n} (f_{i+1}\uparrow - f_i\uparrow)\frac{(q_i\uparrow + q_{i+1}\uparrow)}{2}}$$

Example 70

Pursuing **example 69**, which deals with the distribution of the budget of subsidies allocated by the state to companies in an American city, expressed in millions of dollars, let's insert the items listed in the following table (the last three columns):

$f_i\uparrow$	$q_i\uparrow$	$H_i = f_{i+1}\uparrow - f_i\uparrow$	$B_i = \frac{q_i\uparrow + q_{i+1}\uparrow}{2}$	Trapezoid Area (T_i) $T_i = B_i x H_i$
3.96	0.17	4.95	0.38	1.88
8.91	0.59	8.91	1.30	11.62
17.82	2.02	6.93	2.94	20.38
24.75	3.86	12.87	6.61	85.07
37.62	9.35	16.83	15.19	255.65
54.46	21.02	11.88	26.72	317.51
66.34	32.43	20.79	49.06	1019.96
87.13	65.68	12.87	82.84	1066.29
100.00	100.00			
Total				2778.36

Solution

a: Let's decompose the area under the LORENZ curve into T_i.

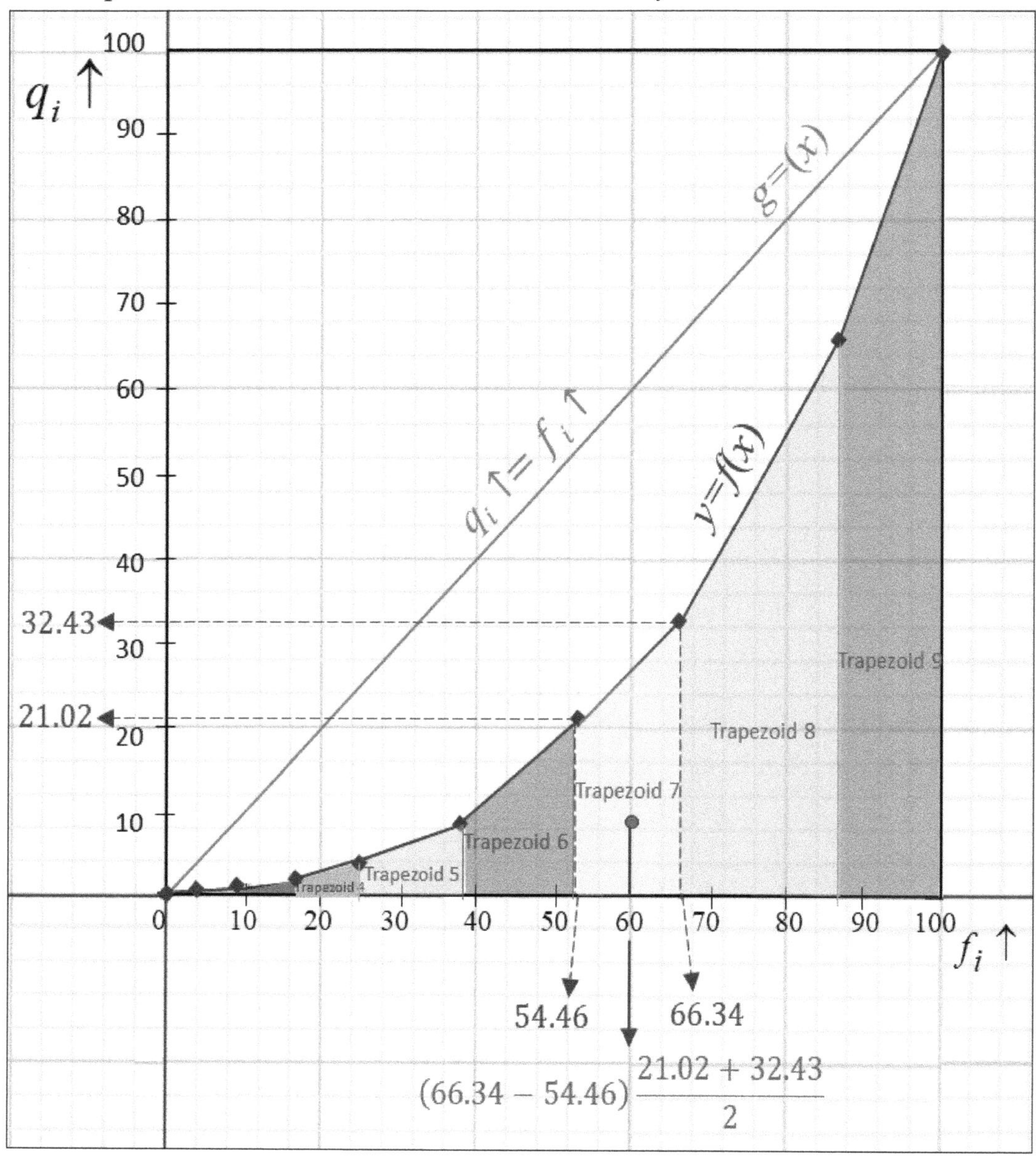

b: Let's calculate the sum of the areas of T_i.

The area under the LORENZ curve is:

$$\sum_{i=1}^{i=8}(f_{i+1}\uparrow - f_i\uparrow)\frac{(q_i\uparrow + q_{i+1}\uparrow)}{2} = (8.91 - 3.6)\frac{(0.17 + 0.59)}{2}$$

$$+(17.82 - 8.91)\frac{(0.59 + 2.02)}{2}$$

$$+(24.75-17.82)\frac{(2.02+3.86)}{2}$$

$$+(37.62-24.75)\frac{(3.86+9.35)}{2}$$

$$+(54.46-37.62)\frac{(9.35+21.02)}{2}$$

$$+(66.34-54.46)\frac{(21.02+32.43)}{2}$$

$$+(87.13-66.34)\frac{(32.43+65.68)}{2}$$

$$+(100-87.13)\frac{(65.68+100)}{2}$$

$$=2{,}778$$

If the data is well organized in the table (as shown in the table above), the result (the area under the LORENZ curve) can be directly obtained from the last column of the table.

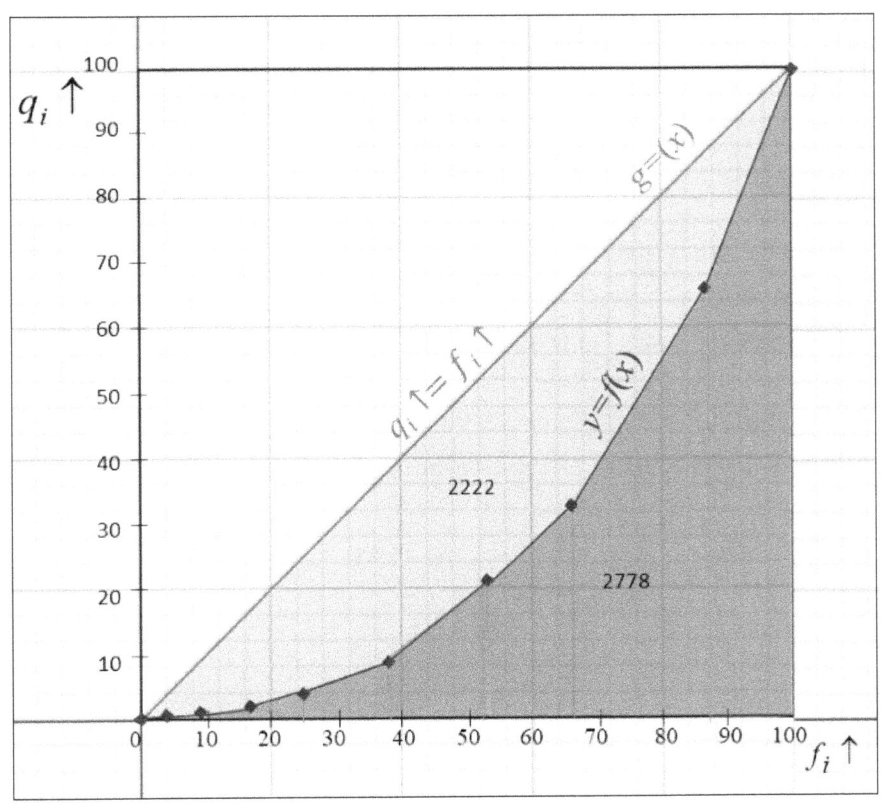

7.4.2 Analytical method for calculating the area under the LORENZ curve

The calculation of the area under the LORENZ curve using the analytical method involves a detailed three-step approach.

1. Firstly, the LORENZ curve is considered as a continuous function in the interval of 0 to 100.

2. Then, the interpolation function of this curve is calculated at the points $(f_i \uparrow, q_i \uparrow)$ using an interpolation polynomial, which can be determined either by NEWTON's interpolation method or LAGRANGE's interpolation method. This step allows finding a smooth and continuous mathematical function that describes the LORENZ curve.

3. Finally, the area under the LORENZ curve is calculated by performing a definite integration of this function between the bounds of 0 and 100. This quantifies the measure of inequality in the distribution of data, where a larger area under the LORENZ curve indicates greater inequality in the distribution of data.

Remark

For the calculation of the interpolation polynomial, we carefully select the points $(f_i \uparrow, q_i \uparrow)$ that are significant to represent the LORENZ curve. These points can be chosen based on the specific characteristics of the data or the objective of the analysis. For example, we may choose the endpoints to capture the maximum and minimum values of the LORENZ function, as well as intermediate points to better represent the variations of the curve.

Furthermore, to improve the readability of the LORENZ curve, we may introduce midpoint points between two existing data points. These intermediate points can be calculated using linear interpolation or other interpolation methods, depending on the needs of the analysis.

The interpolation polynomial obtained in this way allows us to create a continuous approximation of the LORENZ curve, which facilitates the calculation of the area under the curve using integration techniques.

7.4.2.1 LAGRANGE Interpolation Method

The Lagrange polynomial P_n is used in interpolation to obtain a polynomial approximation of the function $f(x)$ at given points $x_0, x_1, \ldots, x_n$.

The polynomial P_n^i associated with the point x_i is obtained using a specific formula that takes into account the differences between the interpolation points.

More precisely, the polynomial P_n^i is the result of dividing the polynomial $(x-x_0)(x-x_1)\ldots\ldots(x-x_{i-1})(x-x_{i+1})\ldots\ldots(x-x_n)$ by the product of the differences between x_i and the other interpolation points, i.e., $(x_i-x_0)(x_i-x_1)\ldots\ldots(x_i-x_{i-1})(x_i-x_{i+1})\ldots\ldots(x_i-x_n)$.

$$P_n^i = \frac{(x-x_0)(x-x_1)\ldots\ldots(x-x_{i-1})(x-x_{i+1})\ldots\ldots(x-x_n)}{(x_i-x_0)(x_i-x_1)\ldots\ldots(x_i-x_{i-1})(x_i-x_{i+1})\ldots\ldots(x_i-x_n)}$$

This division normalizes the polynomial and makes it equal to 1 at the point x_i, and equal to 0 at the other interpolation points.

Thus, to obtain the Lagrange polynomial P_n that interpolates $f(x)$ at points $x_0, x_1, \ldots, x_n$, we multiply the function $f(x_i)$ by the polynomial P_n^i associated with each interpolation point, and sum these terms for all i from 0 to n. This results in a polynomial approximation of $f(x)$ that exactly passes through the interpolation points.

Thus,

$$P_n f(x) = \sum_{i=0}^{n} f(x_i) P_n^i(x)$$

Example 71

Calculate the interpolation polynomial at the following four points:

$$x_0 = 0 \Rightarrow y_0 = 0$$
$$x_1 = 1 \Rightarrow y_1 = 1$$
$$x_2 = 3 \Rightarrow y_2 = 27$$
$$x_3 = 4 \Rightarrow y_3 = 64$$

Let's calculate the P_n^i:

$$P_3^0(x) = \frac{(x-x_1)(x-x_2)(x-x_3)}{(x_0-x_1)(x_0-x_2)(x_0-x_3)}$$

$$P_3^0(x) = \frac{(x-1)(x-3)(x-4)}{(0-1)(0-3)(0-4)}$$

$$P_3^0(x) = \frac{(x-1)(x-3)(x-4)}{-12}$$

$$P_3^1(x) = \frac{(x-x_0)(x-x_2)(x-x_3)}{(x_1-x_0)(x_1-x_2)(x_1-x_3)}$$

$$P_3^1(x) = \frac{x(x-3)(x-4)}{(1-0)(1-3)(1-4)}$$

$$P_3^1(x) = \frac{x(x-3)(x-4)}{6}$$

$$P_3^2(x) = \frac{(x-x_0)(x-x_1)(x-x_3)}{(x_2-x_0)(x_2-x_1)(x_2-x_3)}$$

$$P_3^2(x) = \frac{x(x-1)(x-4)}{(3-0)(3-1)(3-4)}$$

$$P_3^2(x) = \frac{x(x-1)(x-4)}{-6}$$

$$P_3^3(x) = \frac{(x-x_0)(x-x_1)(x-x_2)}{(x_3-x_0)(x_3-x_1)(x_3-x_2)}$$

$$P_3^3(x) = \frac{x(x-1)(x-3)}{(4-0)(4-1)(4-3)}$$

$$P_3^3(x) = \frac{x(x-1)(x-3)}{12}$$

$$P_3f(x) = y_0P_3^0(x) + y_1P_3^1(x) + y_2P_3^2(x) + y_3P_3^3(x)$$

$$P_3f(x) = (0)P_3^0(x) + (1)P_3^1(x) + (27)P_3^2(x) + (64)P_3^3(x)$$

$$= \frac{(0)(x-1)(x-3)(x-4)}{-12} + \frac{(1)x(x-3)(x-4)}{6} + \frac{(27)x(x-1)(x-4)}{-6}$$

$$+ \frac{(64)x(x-1)(x-3)}{12}$$

$$= \frac{x(x^2-7x+12)}{6} + \frac{27x(x^2-5x+4)}{-6} + \frac{64x(x^2-4x+3)}{12}$$

$$= \frac{x^3-7x^2+12x}{6} + \frac{27(x^3-5x^2+4x)}{-6} + \frac{64(x^3-4x^2+3x)}{12}$$

$$= \frac{2x^3-14x^2+24x-54x^3+270x^2-216x+64x^3-256x^2}{12}$$

$$+ \frac{192x}{12}$$

$$= x^3$$

So, we have:

$$P_3f(x) = x^3$$
$$y = x^3$$

7.4.2.2 Newton Interpolation Method (Divided Difference Method)

This method, also known as the divided difference method, is used to interpolate a function f at points $x_0, x_1, \ldots.., x_{k-1}$ and $x_0, x_1, \ldots.., x_k$ using polynomials P_{k-1} and P_k respectively.

Let's consider the polynomial P_{k-1} that interpolates the function f at points $x_0, x_1, \ldots.., x_{k-1}$, and P_k that interpolates the function f at points $x_0, x_1, \ldots.., x_{k-1}, x_k$.

The difference $q_k(x) = p_k(x) - p_{k-1}(x)$ is a polynomial of degree k that vanishes at the k points $x_0, x_1, \ldots.., x_{k-1}$.

Thus, we can write $q_k(x)$ as follows:

$$q_k(x) = A_k(x - x_0)(x - x_1) \ldots\ldots.(x - x_{k-1})$$

$$q_k(x) = A_k \underset{i=0}{\overset{k-1}{\pi}} (x - x_i)$$

where A_k is a real constant.

To determine A_k, we can use the following formula:

$$A_k = \frac{p_k(x) - p_{k-1}(x)}{\underset{i=0}{\overset{k-1}{\pi}} (x - x_i)}$$

In particular, when $x = x_k$, i.e., when evaluating the polynomial A_k at x_k, we have:

$$A_k = \frac{f(x_k) - p_{k-1}(x_k)}{\underset{i=0}{\overset{k-1}{\pi}} (x - x_i)}$$

By setting $A_0 = f(x_0)$, we obtain the following expression for the polynomial $q_k(x)$:

Using the values of A_i, we can construct the polynomials P_k from P_0 to $P_k$to interpolate the function f:

$$q_k(x) = A_0 + A_1(x - x_0) + A_2(x - x_0)(x - x_1) + \ldots. + A_k(x - x_0)(x - x_1)\ldots(x - x_{k-1})$$

$$\begin{aligned}
P_1 f(x) &= P_0 f(x) + A_1(x - x_0) \\
P_2 f(x) &= P_1 f(x) + A_2(x - x_0)(x - x_1) \\
P_3 f(x) &= P_2 f(x) + A_3(x - x_0)(x - x_1)(x - x_2) \\
\ldots\ldots\ldots &= \ldots\ldots\ldots\ldots\ldots\ldots\ldots\ldots\ldots \\
\ldots\ldots\ldots &= \ldots\ldots\ldots\ldots\ldots\ldots\ldots\ldots\ldots \\
P_k f(x) &= P_{k-1} f(x) + A_k(x - x_0)(x - x_1) \ldots\ldots\ldots.(x - x_{k-1})
\end{aligned}$$

By convention, we set:

$$A_i = f[x_0, \ldots., x_i]$$

This expression is called the ***divided difference*** of f at points $x_0, \ldots.., x_i$

$$f[x_0, x_1, \ldots\ldots, x_k] = \frac{f[x_0, x_1, \ldots\ldots x_k]}{x_0 - x_k}$$

Example 72

Let's consider the same **example 71** of **section 7.4.2.1**:

$x_0 = 0 \quad f[x_0] = 0$

$f[x_0, x_1] = 1$

$x_1 = 1 \quad f[x_1] = 1$

$f[x_0, x_1, x_2] = 4$

$f[x_1, x_2] = 13$

$f[x_0, x_1, x_2, x_3] = 1$

$x_2 = 3 \quad f[x_2] = 27$

$f[x_1, x_2, x_3] = 8$

$f[x_2, x_3] = 37$

$x_3 = 4 \quad f[x_3] = 64$

$$P_3(x) = 0 + (x - x_0)(1) + (x - x_0)(x - x_1)(4) + (x - x_0)(x - x_1)(x - x_2)(1)$$

$$P_3(x) = 0 + (1)x + x(x-1)(4) + x(x-1)(x-3)(1)$$

$$P_3(x) = x^3$$

Example 73

Pursuing **example 70** from section **7.4.1**, which deals with the distribution of the budget of subsidies allocated by the state to companies in an American city, expressed in millions of dollars, let's calculate the area under the LORENZ curve:

$f_i \uparrow$	$q_i \uparrow$
3.96	0.17
8.91	0.59
17.82	2.02
24.75	3.86
37.62	9.35
54.46	21.02
66.34	32.43
87.13	65.68
100.00	100.00
Total	

Solution

a: Assuming the LORENZ curve as continuous in the interval$[0\ ;\ 100]$.

b: Let's calculate the interpolation function of this curve at the points $(f_i \uparrow, q_i \uparrow)$:

We will consider the following critical $(f_i \uparrow, q_i \uparrow)$ points:

$$(0,0)-(18,2)-(38,10)-(54,21)-(66,32)-(87,66)-(100,100)$$

Following the computation, the interpolation polynomial is obtained as:

$$y=-\frac{459919879}{421264237693178880}x^6+\frac{9309857477}{28084282512878592}x^5-\frac{1539492468845}{42126423769317888}x^4+\frac{480535464947}{260039652897024}x^3-\frac{18549046730563}{531899290016640}x^2+\frac{10417551663763}{32504956612128}x$$

c: Let's evaluate the area between the curve $q_i \uparrow$ and the $f_i \uparrow$ axes in the interval $[0\ ;100]$:

$$\text{Aire}=\int_0^{100}-\frac{459919879}{421264237693178880}x^6dx+\int_0^{100}\frac{9309857477}{28084282512878592}x^5dx-\int_0^{100}\frac{1539492468845}{42126423769317888}x^4dx+\int_0^{100}\frac{480535464947}{260039652897024}x^3dx-\int_0^{100}\frac{18549046730563}{531899290016640}x^2dx+\int_0^{100}\frac{10417551663763}{32504956612128}xdx$$

After integration,

$$\text{Area}=\left[-\frac{459919879}{421264237693178880}\left(\frac{x^7}{7}\right)+\frac{9309857477}{28084282512878592}\left(\frac{x^6}{6}\right)-\frac{1539492468845}{42126423769317888}\left(\frac{x^5}{5}\right)+\frac{480535464947}{260039652897024}\left(\frac{x^4}{4}\right)-\frac{18549046730563}{531899290016640}\left(\frac{x^3}{3}\right)+\frac{10417551663763}{32504956612128}\left(\frac{x^2}{2}\right)\right]_0^{100}$$

Applying integration bounds, we get:

$$\text{Area}=-\frac{459919879}{421264237693178880}\left(\frac{10^{14}}{7}\right)+\frac{9309857477}{28084282512878592}\left(\frac{10^{12}}{6}\right)-\frac{1539492468845}{42126423769317888}\left(\frac{10^{10}}{5}\right)+\frac{480535464947}{260039652897024}\left(\frac{10^{8}}{4}\right)-\frac{18549046730563}{531899290016640}\left(\frac{10^{6}}{3}\right)+\frac{10417551663763}{32504956612128}\left(\frac{10^{4}}{2}\right)$$

$$\text{Area}=2741$$

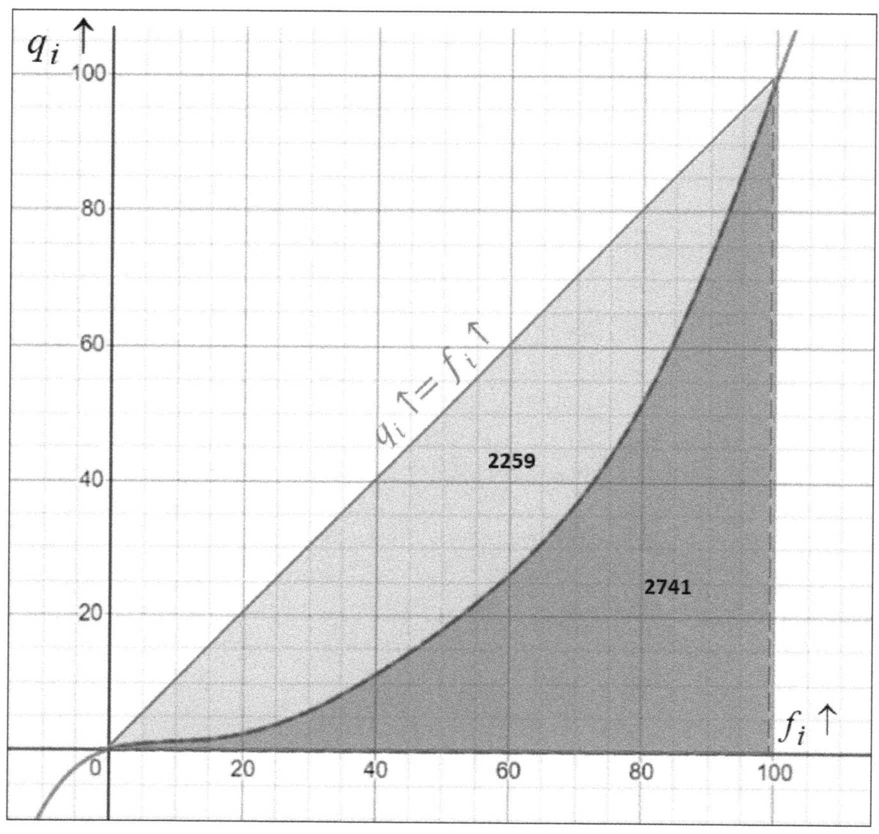

7.4.3 Comparison between Graphical and Analytical Methods

A comparison between the graphical method (trapezoidal rule) and the analytical method (using definite integrals) reveals that both methods yield similar results, accurate up to the hundredth decimal place. To minimize the differences between the two methods, it is advisable to use the maximum number of significant points possible in the interpolation method. Furthermore, as mentioned earlier, the analytical method based on definite integrals may be more accurate for complex LORENZ curves, but it can be more mathematically complex.

7.5 Concentration Index or GINI Index

The ***concentration index***, also known as the ***GINI index***, is a widely used statistical tool to measure inequality or concentration in a distribution. It is widely used in various fields, such as economics, sociology, demography, and other social sciences.

One significant advantage of the GINI index is that it is unit-free, meaning that it is independent of the unit of measurement chosen for the data. This allows for easy comparison of inequality between different distributions, even if they have different units of measurement. For example, the GINI index can be used to compare income inequality between different countries or to assess wealth distribution inequality within a given population.

7.5.1 Interpretation of the GINI Index

The interpretation of the GINI coefficient is based on comparing the concentration curve, which represents the cumulative distribution of the variable being studied, with the 45-degree line, which represents a perfectly equal distribution. If the concentration curve is very close to the 45-degree line, then the GINI coefficient will be close to zero, indicating a nearly equal distribution. On the other hand, the further the concentration curve deviates from the 45-degree line, the higher the GINI coefficient will be, indicating increasing inequality in the distribution of the variable.

Thus, a GINI coefficient close to zero indicates a highly balanced distribution, where each individual or group has an equal share of resources, while a high GINI coefficient indicates significant concentration of resources among certain individuals or groups, reflecting major inequalities in distribution. The GINI coefficient can range from 0 to 1, where 0 represents a perfectly equal distribution and 1 represents a completely unequal distribution, where a single person or group possesses all the resources.

The interpretation of the GINI coefficient can have important implications for public policies and political decisions. A high GINI coefficient may indicate excessive concentration of resources, which may require interventions to reduce inequalities, such as income redistribution policies, social protection programs, or measures to promote access to education and vocational training. On the other hand, a low GINI coefficient can be interpreted as a sign of social and economic equality but may also indicate low social mobility or economic stagnation, requiring policies to stimulate economic growth and promote employment.

7.5.2 Finding the GINI Concentration Index

The GINI concentration index is the GINI index defined as the ratio between, on one hand, twice the area between the 45-degree line and the LORENZ curve, and on the other hand, the area of the square with sides measuring 100 units. This measure thus expresses the proportion of the surface area between the LORENZ curve and the 45-degree line relative to the total surface area of the square, allowing for an assessment of the degree of inequality in the distribution of the variable.

Another way to calculate the GINI index is to define it as the ratio between the area between the 45-degree line and the LORENZ curve, and half the area of the square with sides measuring 100 units. This definition also allows for measuring the deviation from a perfectly equal distribution.

Let A be the area between the bisector and the LORENZ curve, and B be the area under the LORENZ curve. The GINI index (IG) is calculated as the ratio of A to the sum of A and B, i.e.,

$$\text{GINI Index (GI)} = \frac{A}{A+B}$$

Since the values $f_i \uparrow$ and q_i are defined in the range [0 ; 100], the half-area of the square with sides measuring 100 units is 5,000. Therefore, we can write:

$$A + B = 5{,}000 \quad \Rightarrow \quad A = 5000 - B$$

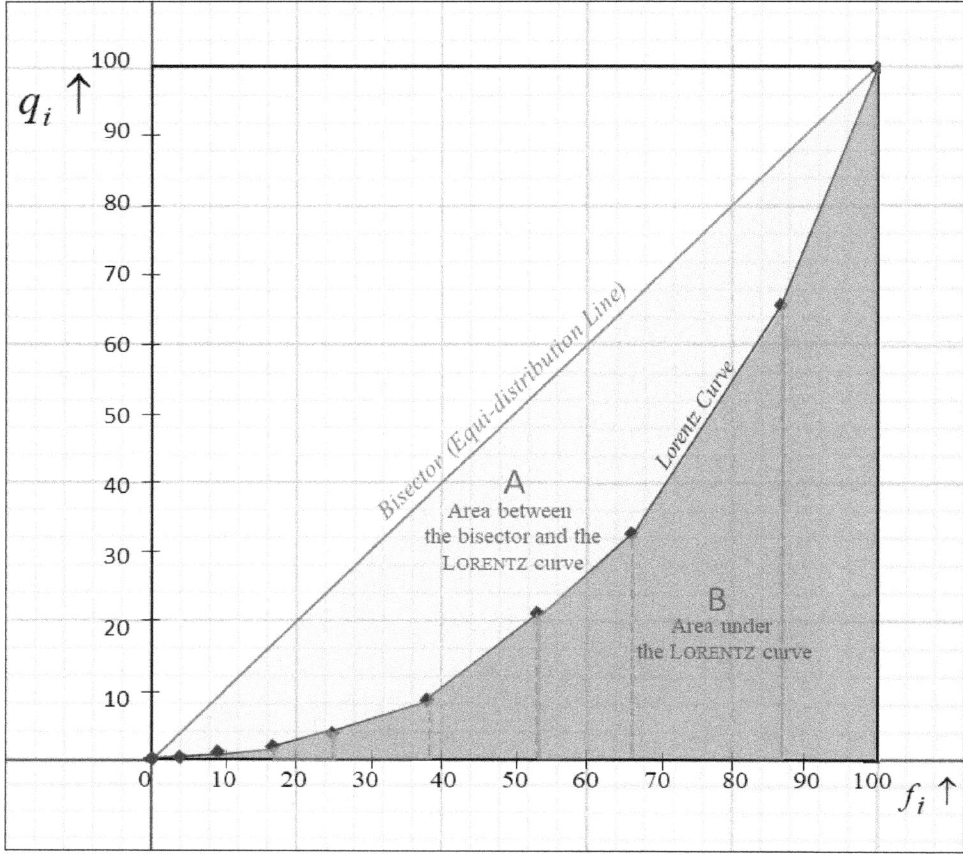

The formula for the GINI index can also be expressed as follows:

$$\text{GINI Index (GI)} = \frac{5{,}000 - B}{5{,}000}$$
$$= 1 - \frac{B}{5{,}000}$$

This ratio is expressed as a percentage.

$$\boxed{\text{GINI Index (GI)} = 100 - \left(\frac{\text{B}}{5{,}000}\right)\text{x}100}$$

Example 74

Pursuing **example 73**, which deals with the distribution of the budget of subsidies allocated by the state to companies in an American city, expressed in millions of dollars, let's see how to calculate the area under the LORENZ curve:

GINI Index by Trapezoids	GINI Index by Integral Calculus
$\text{l'indice de GINI (IG)} = 1 - \frac{B}{5.000} = 1 - \frac{2.741}{5.000} = 45.18\%$	$\text{l'indice de GINI (IG)} = 1 - \frac{B}{5.000} = 1 - \frac{2.778}{5.000} = 44.44\%$

In this example, the difference between the two methods of calculating the GINI index is very small, around 0.7%, which is less than 1%. This means that both methods produce similar results and that the difference between the areas A and B, used to calculate the GINI index, is very minimal. This small difference underscores the robustness and consistency of the GINI index in evaluating the inequality of distribution of a variable.

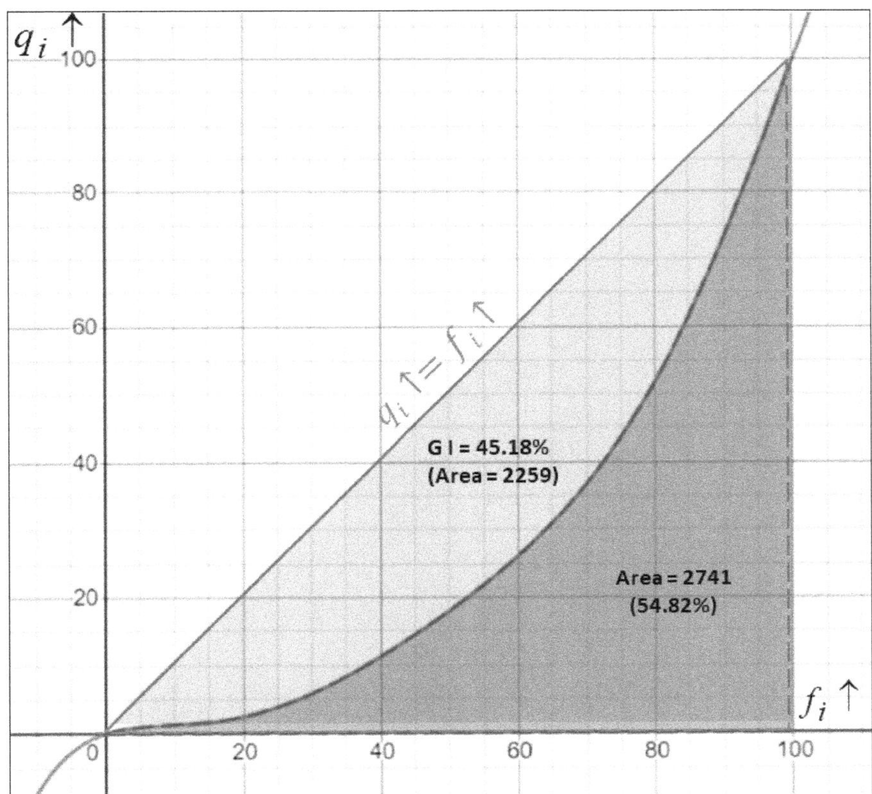

INTERPRETATION OF THE RESULTS

The results reveal a GINI index of 45.18%, indicating a significant inequality in the distribution of budgetary resources. This means that the allocation of funds is not balanced, with a significant concentration of resources among certain companies. Specifically, half of the studied companies receive less than 20% of the total budget allocation, raising concerns about the fairness of resource distribution. In a competitive system, such inequality in the distribution of budgetary resources can have implications for the competitiveness, sustainability, and overall economic growth of the companies and the economy as a whole.

Example 75 showing the GINI index with and without outliers

Let's continue with **example 67** from **section 6.2.1**, by moving to the next step, where we will calculate shape coefficients using numerical data from this example, which involves an insurance company looking to establish itself on an island with only 2,000 inhabitants. We will calculate the GINI coefficient in both cases: with and without outliers.

1. Cases with outliers:

x_i	n_i	$n_i x_i$	$n_i x_i^2$
500	380	190000	95000000
1500	633	949500	1424250000
2500	527	1317500	3293750000
3500	151	528500	1849750000
4500	123	553500	2490750000
5500	85	467500	2571250000
6500	58	377000	2450500000
7500	39	292500	2193750000
8500	8	68000	578000000
9500	6	57000	541500000
2495000	5	12475000	31125125000000
	2015	17276000	31142613500000

a: Calculation of GINI coefficient using the trapezoidal method:

The following table shows the data required to solve this exercise:

x_i	n_i	$n_i x_i$	f_i	q_i	$f_i \uparrow$	$q_i \uparrow$	$H_i = f_{i+1} \uparrow - f_i \uparrow$	$B_i = \frac{q_i \uparrow + q_{i+1} \uparrow}{2}$	Trapezoid Area (T_i) $T_i = B_i x H_i$
500	380	190000	18.86	1.10	18.86	1.10	31.41	3.85	120.88
1500	633	949500	31.41	5.50	50.27	6.60	26.15	10.41	272.23
2500	527	1317500	26.15	7.63	76.43	14.22	7.49	15.75	118.04
3500	151	528500	7.49	3.06	83.92	17.28	6.10	18.88	115.27
4500	123	553500	6.10	3.20	90.02	20.49	4.22	21.84	92.12
5500	85	467500	4.22	2.71	94.24	23.19	2.88	24.28	69.89
6500	58	377000	2.88	2.18	97.12	25.37	1.94	26.22	50.75
7500	39	292500	1.94	1.69	99.06	27.07	0.40	27.26	10.82
8500	8	68000	0.40	0.39	99.45	27.46	0.30	27.63	8.23
9500	6	57000	0.30	0.33	99.75	27.79	0.25	63.89	15.85
2495000	5	12475000	0.25	72.21	100.00	100.00			
Total	2015	17276000	100				81.14143921	240.0150498	874.0855657

The area under the curve is:

$$\sum_{i=1}^{i=11} (f_{i+1} \uparrow - f_i \uparrow) \frac{(q_i \uparrow + q_{i+1} \uparrow)}{2} = (50.27 - 18.86) \frac{(1.1 + 6.6)}{2}$$

$$+ (76.43 - 50.27) \frac{(6.6 + 14.22)}{2}$$

$$+ (83.92 - 76.43) \frac{(14.22 + 17.28)}{2}$$

$$+ (90.02 - 83.92) \frac{(17.28 + 20.49)}{2}$$

$$+ (94.24 - 90.02) \frac{(20.49 + 23.19)}{2}$$

$$+ (97.12 - 94.24) \frac{(23.19 + 25.37)}{2}$$

$$+ (99.06 - 97.12) \frac{(25.37 + 27.07)}{2}$$

$$+ (99.45 - 99.06) \frac{(27.07 + 27.46)}{2}$$

$$+ (99.75 - 99.45) \frac{(27.46 + 27.79)}{2}$$

$$+ (100 - 99.75) \frac{(27.79 + 100)}{2}$$

$$= 874.08$$

It follows that:

$$\text{GINI index} = 100\% - \left(\frac{874.08}{5{,}000}\right)\text{x}100$$

$$= 100\% - 17.48\%$$

$$= 82.51\%$$

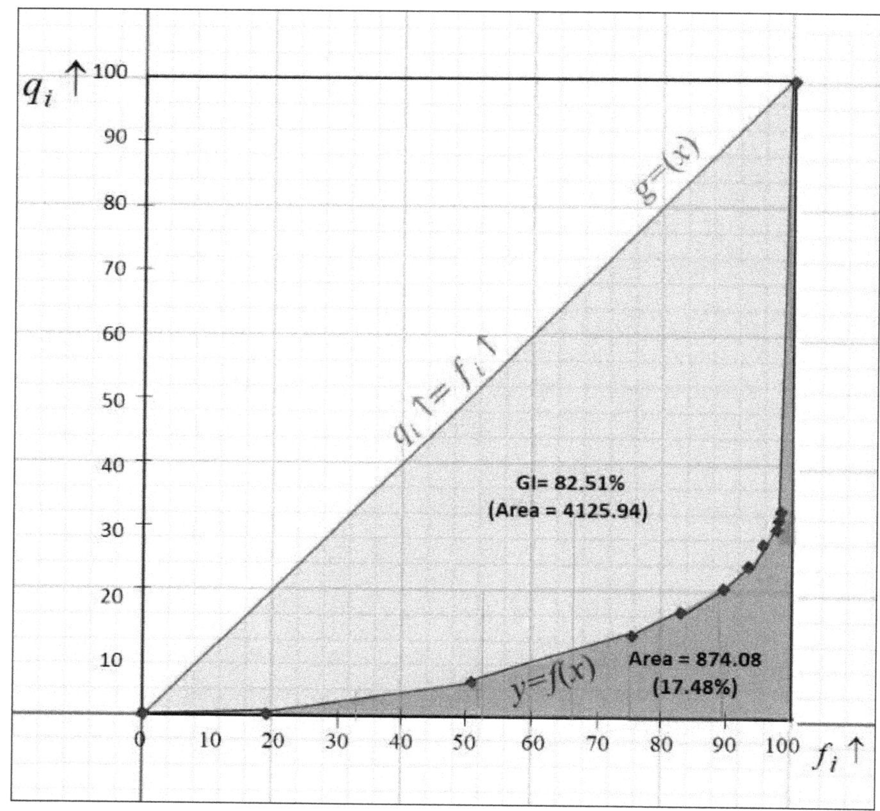

b: Calculation of the GINI index by the defined integrals method

We will consider the following key points: $(f_i \uparrow, q_i \uparrow)$

(5,0) – (10,0) – (15,0) – (19,1) – (30,3) – (40,5) – (50,7) – (76,14) – (84,17) – (90,20) – (97,25) – (100,100)

Following the computation, the interpolation polynomial is:

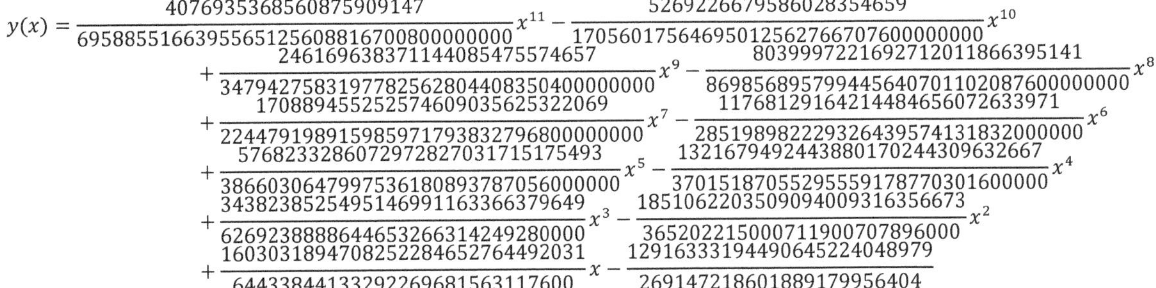

Let's evaluate the area between the $q_i \uparrow$ and the $f_i \uparrow$ in [0 ; 100]:

$$\text{Area} = \int_0^{100} y\,dx = \frac{4076935368560875909147}{69588551663955651256088167008000000}\left(\frac{10^{24}}{8}\right) - \frac{5269226679586028354659}{17056017564695012562766707600000000}\left(\frac{10^{22}}{7}\right)$$
$$+ \frac{246169638371144085475574657}{347942758319778256280440835040000000}\left(\frac{10^{20}}{6}\right) - \frac{8039997221692712011866395141}{8698568957994456407011020876000000000}\left(\frac{10^{18}}{5}\right)$$
$$+ \frac{170889455252574609035625322069}{2244791989159859717938327968000000000}\left(\frac{10^{16}}{4}\right) - \frac{11768129164214484656072633971}{28519898222932643957413183200000}\left(\frac{10^{14}}{3}\right)$$
$$+ \frac{5768233286072972827031715175493}{386603064799753618089378705600000}\left(\frac{10^{12}}{2}\right) - \frac{13216794924438801702443096326 67}{370151870552955591787703016000000}\left(\frac{10^{10}}{6}\right)$$
$$+ \frac{3438238525495146991163366379649}{62692388886446532663142492800000}\left(\frac{10^{8}}{5}\right) - \frac{1851062203509094009316356673}{365202221500071190070789600 0}\left(\frac{10^{6}}{4}\right)$$
$$+ \frac{1603031894708252284652764492031}{64433844133292269681563117600}\left(\frac{10^{4}}{3}\right) - \frac{1291633319449064522404 8979}{269147218601889179956404}(10^{2})$$
$$\text{Area} = 941.8$$

It follows that:

$$\text{GINI Index} = 100\% - \left(\frac{\int_0^{100} f(x)\,dx}{5.000}\right)\text{x}100$$

$$= 100\% - \left(\frac{941.8}{5{,}000}\right)\text{x}100$$
$$= 100\% - 18.84\%$$
$$= 81.16\%$$

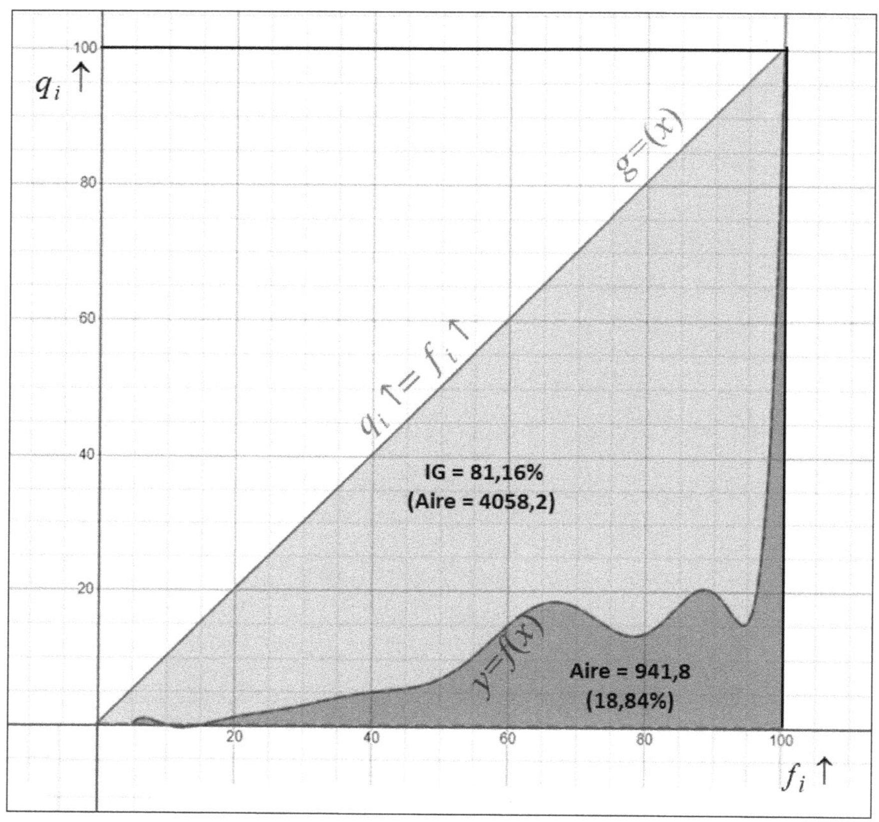

INTERPRETATION OF RESULTS

The results of the study highlight a concerning situation in terms of income inequalities on the island. The GINI coefficient, which measures the concentration of wealth, exceeds 81%, which is extremely high and indicates a very unequal distribution of salaries. The LORENZ curve confirms this trend, clearly showing that the majority of incomes are held by only five billionaires, with a very small portion of the population benefiting from a significant share of the wealth.

Upon examining the LORENZ curve, it can be observed that the distance between the bisector (which represents an equal distribution of incomes) and the curve is very significant from abscissa zero to abscissa 99.95. This indicates that the majority of incomes are concentrated among a handful of individuals, which is further confirmed by the sharp rise of the curve between abscissa 99.95 and 100, illustrating the extremely high incomes of the five billionaires.

These results raise concerns about the viability and fairness of establishing the insurance company on the island. An excessive concentration of wealth in the hands of a few individuals can have significant socio-economic implications, such as creating an economic imbalance, increasing social disparities, and limiting opportunities for the local population. The unusually high standard deviation and coefficient of variation of 1,446% also highlight the significant disparity in salaries, which can have consequences on the stability and sustainability of the island's economic situation.

It is essential for the insurance company to take these results into account in their decision to establish operations on the island and consider measures to mitigate potential inequalities, such as income redistribution policies, local economic development programs, and initiatives to promote more inclusive and sustainable economic growth.

2. Cases without outliers:

x_i	n_i	$n_i x_i$	$n_i x_i^2$
500	380	190000	95000000
1500	633	949500	1424250000
2500	527	1317500	3293750000
3500	151	528500	1849750000
4500	123	553500	2490750000
5500	85	467500	2571250000
6500	58	377000	2450500000
7500	39	292500	2193750000
8500	8	68000	578000000
9500	6	57000	541500000
	2010	4801000	17488500000

a: Calculation of the GINI index by trapezoids

The following table shows the data required to solve this exercise:

x_i	n_i	$n_i x_i$	f_i	q_i	$f_i \uparrow$	$q_i \uparrow$	$H_i = f_{i+1} \uparrow - f_i \uparrow$	$B_i = \frac{q_i \uparrow + q_{i+1} \uparrow}{2}$	Trapezoid Area $T_i = B_i x H_i$
500	380	190000	18.91	3.96	18.91	3.96	31.49	13.85	436.05
1500	633	949500	31.49	19.78	50.40	23.73	26.22	37.46	982.05
2500	527	1317500	26.22	27.44	76.62	51.18	7.51	56.68	425.81
3500	151	528500	7.51	11.01	84.13	62.18	6.12	67.95	415.81
4500	123	553500	6.12	11.53	90.25	73.71	4.23	78.58	332.31
5500	85	467500	4.23	9.74	94.48	83.45	2.89	87.38	252.13
6500	58	377000	2.89	7.85	97.36	91.30	1.94	94.35	183.07
7500	39	292500	1.94	6.09	99.30	97.40	0.40	98.10	39.05
8500	8	68000	0.40	1.42	99.70	98.81	0.30	99.41	29.67
9500	6	57000	0.30	1.19	100.00	100.00			
Total	2010	4801000	100				81.09452736	633.7533847	3095.954305

The area under the curve is:

$$\sum_{i=1}^{i=11} (f_{i+1} \uparrow - f_i \uparrow) \frac{(q_i \uparrow + q_{i+1} \uparrow)}{2} = (50.40 - 18.91) \frac{(3.96 + 23.73)}{2}$$

$$+(76.62-50.40)\frac{(23.73+51.18)}{2}$$

$$+(84.13-76.62)\frac{(51.18+62.18)}{2}$$

$$+(90.25-84.13)\frac{(62.18+73.71)}{2}$$

$$+(94.48-90.25)\frac{(73.71+83.45)}{2}$$

$$+(97.36-94.48)\frac{(83.45+91.30)}{2}$$

$$+(99.30-97.12)\frac{(91.30+97.40)}{2}$$

$$+(99.70-99.30)\frac{(97.40+98.81)}{2}$$

$$+(100-99.70)\frac{(98.81+100)}{2}$$

$$=3095.95$$

It follows that:

$$\begin{aligned}\text{GINI Index} &= 100\% - \left(\frac{3{,}095.95}{5{,}000}\right)\text{x}100\\ &= 100\% - 61.92\%\\ &= 38.08\%\end{aligned}$$

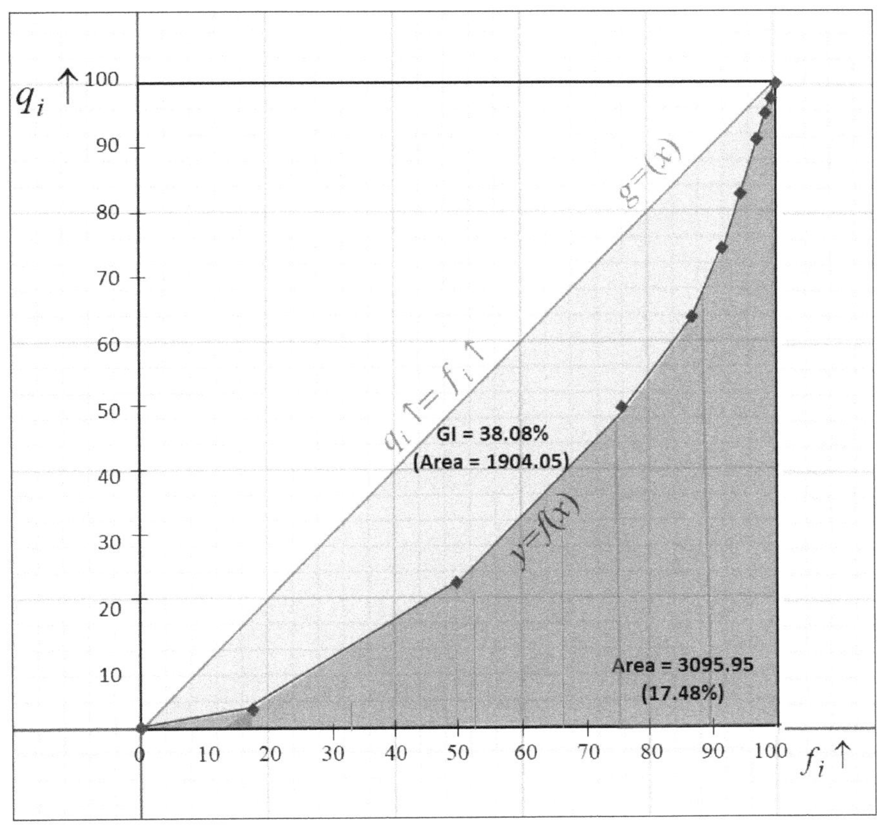

b: Let's calculate the GINI index by the use of the defined integrals.

x_i	n_i	n_ix_i	f_i	q_i	$f_i\uparrow$	$q_i\uparrow$	$H_i = f_{i+1}\uparrow - f_i\uparrow$	$B_i = \frac{q_i\uparrow + q_{i+1}\uparrow}{2}$	Aire Trapèze (T_i) $T_i = B_i x H_i$
500	380	190000	18.91	3.96	18.91	3.96	31.49	13.85	436.05
1500	633	949500	31.49	19.78	50.40	23.73	26.22	37.46	982.05
2500	527	1317500	26.22	27.44	76.62	51.18	7.51	56.68	425.81
3500	151	528500	7.51	11.01	84.13	62.18	6.12	67.95	415.81
4500	123	553500	6.12	11.53	90.25	73.71	4.23	78.58	332.31
5500	85	467500	4.23	9.74	94.48	83.45	2.89	87.38	252.13
6500	58	377000	2.89	7.85	97.36	91.30	1.94	94.35	183.07
7500	39	292500	1.94	6.09	99.30	97.40	0.40	98.10	39.05
8500	8	68000	0.40	1.42	99.70	98.81	0.30	99.41	29.67
9500	6	57000	0.30	1.19	100.00	100.00			
Total	2010	4801000	100				81.09452736	633.7533847	3095.954305

We will consider the following key points: $(f_i\uparrow, q_i\uparrow)$

(5.0) – (10.0) – (15.0) – (19,1) – (30,3) – (40,5) – (50,24)– (76,51)– (84,62) – (90,74)– (97,91)– (100,100)

Following the computation, the interpolation polynomial is obtained as:

$$y = \frac{477408180948827347343}{117487182546731938641648948000000000}x^{11} - \frac{2283287985901577766649 9}{979059854556099488680407900000000000}x^{10}$$

$$+ \frac{848663010452677890000817}{1468589781834149233020611850000000000}x^{9} - \frac{475187080854195944963686296 1}{5874359127336596932082447400000000000}x^{8}$$

$$+ \frac{907614037395183584266930303 1}{1291067940073977347710428000000000000}x^{7} - \frac{907614037395183584266930303 1}{1291067940073977347710428000000000000}x^{6}$$

$$+ \frac{659233163605271472317574668867}{16783883220961705520235564000000000000}x^{5} - \frac{227870457468156309113393144 67}{161162115976312673033812000000000}x^{4}$$

$$+ \frac{3756432219300058966523617672514 3}{117487182546731938641648948000000000}x^{3} - \frac{696415931450982253834781514517}{163176642426016581446734650000 0}x^{2}$$

$$+ \frac{958861588514461496551577833 33}{326353284852033162893469300000}x - \frac{359203814943547577918849}{4957139589155208671580 0}$$

Let's evaluate the area between the curve $q_i \uparrow$ and $f_i \uparrow$ in [0 ; 100]:

$$\text{Area} = \int_0^{100} y\,dx = \frac{477408180948827347343}{117487182546731938641648948000000000}\left(\frac{10^{24}}{8}\right) - \frac{2283287985901577766649 9}{979059854556099488680407900000000000}\left(\frac{10^{22}}{7}\right)$$

$$+ \frac{848663010452677890000817}{1468589781834149233020611850000000000}\left(\frac{10^{20}}{6}\right) - \frac{475187080854195944963686296 1}{5874359127336596932082447400000000000}\left(\frac{10^{18}}{5}\right)$$

$$+ \frac{907614037395183584266930303 1}{1291067940073977347710428000000000000}\left(\frac{10^{16}}{4}\right) - \frac{907614037395183584266930303 1}{1291067940073977347710428000000000000}\left(\frac{10^{14}}{3}\right)$$

$$+ \frac{659233163605271472317574668867}{16783883220961705520235564000000000000}\left(\frac{10^{12}}{2}\right) - \frac{227870457468156309113393144 67}{161162115976312673033812000000000}\left(\frac{10^{10}}{6}\right)$$

$$+ \frac{3756432219300058966523617672514 3}{117487182546731938641648948000000000}\left(\frac{10^{8}}{5}\right) - \frac{696415931450982253834781514517}{163176642426016581446734650000 0}\left(\frac{10^{6}}{4}\right)$$

$$+ \frac{958861588514461496551577833 33}{326353284852033162893469300000}\left(\frac{10^{4}}{3}\right) - \frac{359203814943547577918849}{4957139589155208671580 0}(10^{2})$$

$$\text{Area} = 3102$$

It follows that:

$$\text{GINI Index} = 100\% - \left(\frac{\int_0^{100} f(x)\,dx}{5{,}000}\right) \text{x}100$$

$$= 100\% - \left(\frac{3{,}102}{5{,}000}\right) \text{x}100$$

$$= 100\% - 62.04\%$$

$$= 100\% - 62.04\%$$

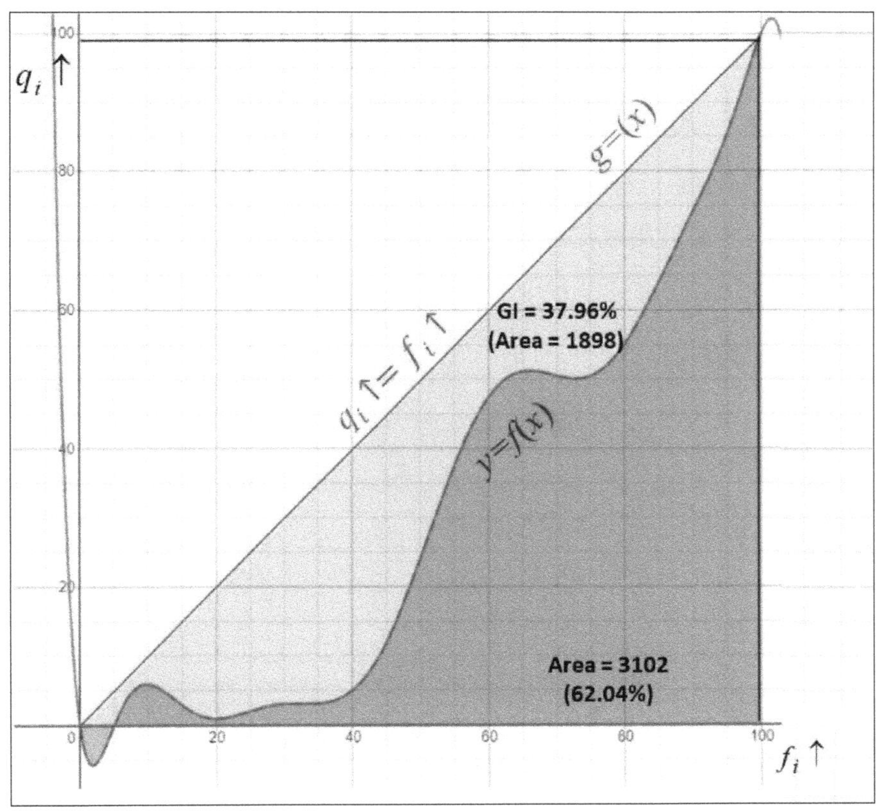

INTERPRETATION OF RESULTS

After excluding the five billionaires from the list of island workers, outliers in the statistical distribution were removed to obtain a more accurate picture of the population's wage situation.

Thus, the GINI index calculated using two different methods showed an index of around 37%, indicating a wage distribution that is considered statistically acceptable.

Although this GINI index reveals some inequalities in wage distribution, they are not so glaring as to raise major concerns about the insurance company's decision to establish itself on the island. In other words, the wage situation of the population appears relatively stable and balanced, despite some disparities. This suggests that the company may consider proceeding with its plan to establish itself on the island without facing major socio-economic challenges related to wage distribution.

These results are consistent with the interpretations of section 5.7 of the study, which highlights a return to a wage distribution with a coefficient of variation of 72.5%. This means that despite the inequalities revealed by the GINI index, wage variation remains relatively stable, which can be considered a positive element in evaluating the socio-economic situation of the island. This indicates that although some inequalities persist, they no longer seem to significantly compromise the economic and social viability of the island.

7.6 Exercises

7.6.1 Chapter Exercise

- **Exercise 33:** The following table shows the distribution of monthly wages (in dollars) in a Canadian company with 212 employees.

Salary Brackets	Frequencies
[0,1000)	29
[1000,2000)	34
[2000,3000)	44
[3000,4000)	36
[4000,6000)	29
[6000,9000)	18
[9000,15000)	13
[15000,25000)	7
[25000,50000)	2
Total	212

a. Draw the histogram of the distribution.

b. Draw the ascendant and descendant cumulative relative frequency curves.

c. Calculate the characteristics of central tendency of the distribution.

d. Calculate the characteristics of dispersion of the distribution.

e. Calculate the characteristics of shape of the distribution.

f. Calculate the medial of the distribution.

g. Calculate the GINI index by the graphical method (trapezoid method).

h. Calculate the GINI index by the analytical method (integral method).

Solution

The following table shows the data required to solve this exercise:

Classes	x_i	n_i	$n_i x_i$	$n_i x_i^2$	$n_i(x_i-\bar{x})^3$	$n_i(x_i-\bar{x})^4$	f_i	$f_i\uparrow$	$f_i\downarrow$
[0;1000)	500	29	14500	7250000	-1846168354804	7371610907272920	13.68	13.68	100.00
[1000;2000)	1500	34	51000	76500000	-911520024002	2728110637872610	16.04	29.72	86.32
[2000;3000)	2500	44	110000	275000000	-348277351805	694090477064405	20.75	50.47	70.28
[3000;4000)	3500	36	126000	441000000	-35241243033	34991894615571	16.98	67.45	49.53
[4000;6000)	5000	29	145000	725000000	3781079492	1917292666777	13.68	81.13	32.55
[6000;9000)	7500	18	135000	1012500000	489446795714	1471803454091910	8.49	89.62	18.87
[9000;15000)	12000	13	156000	1872000000	5499911463865	41288250446893700	6.13	95.75	10.38
[15000;25000)	20000	7	140000	2800000000	26102838821343	404778691628137000	3.30	99.06	4.25
[25000;50000)	37500	2	75000	2812500000	71920241045119	2373876824118970000	0.94	100.00	0.94
Total		212	952,500	10,021,750,000	100,875,012,231,889	2,832,246,290,857,590,000	100		

a. Plotting the histogram of the distribution:

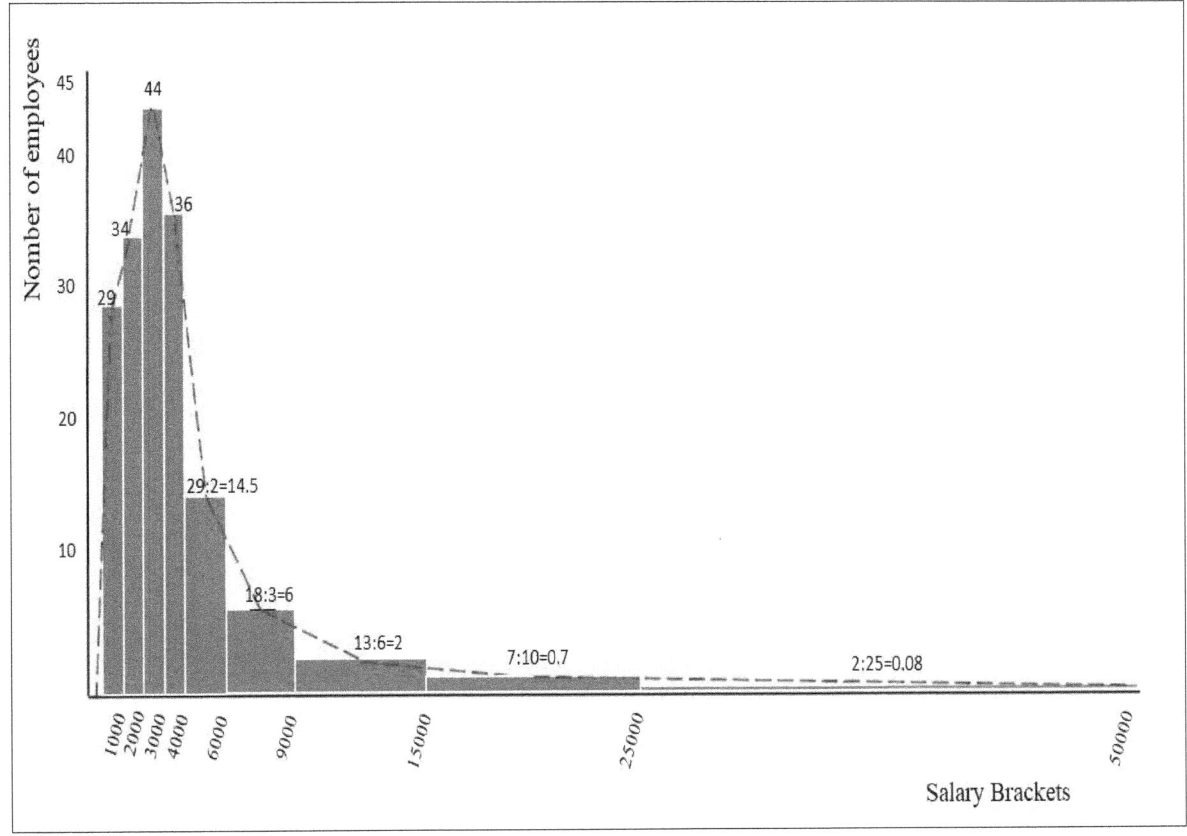

b: The ascending and descending cumulative frequency curves are as follows:

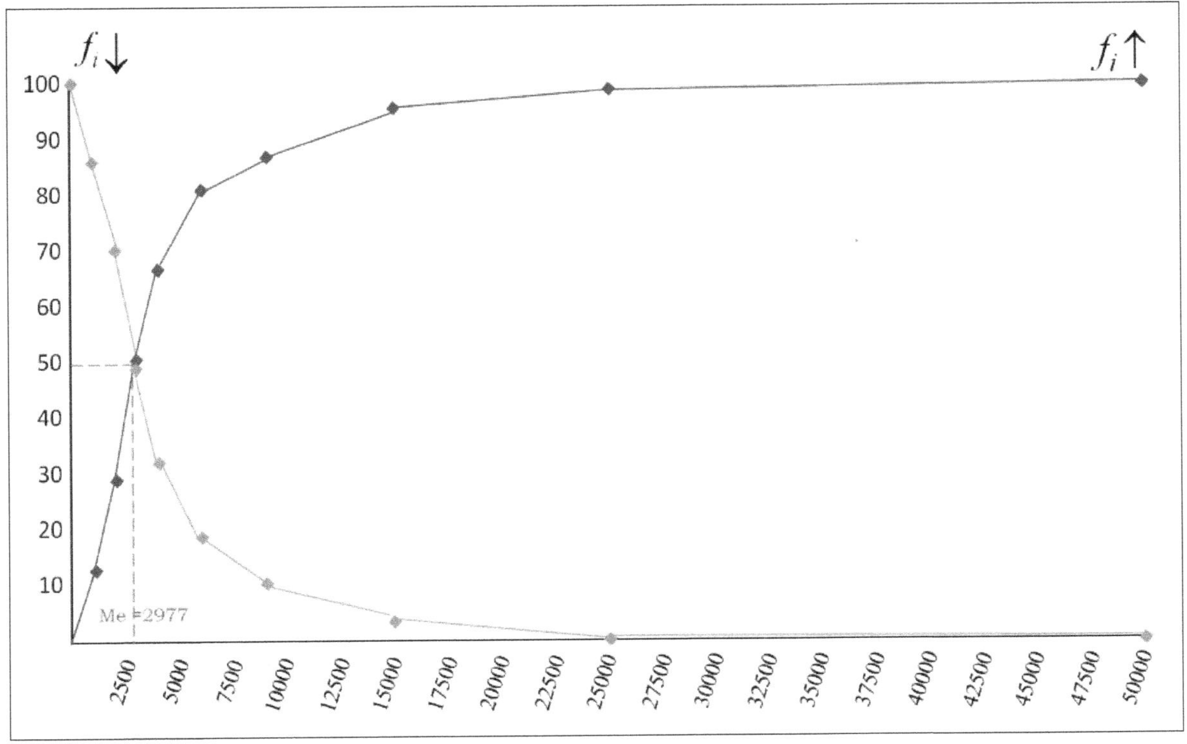

c: Let's calculate the characteristics of central tendency.

Classes	x_i	n_i	Amplitude	Density
[0;1000)	500	29	1000	0.029
[1000;2000)	1500	34	1000	0.034
[2000;3000)	2500	44	1000	0.044
[3000;4000)	3500	36	1000	0.036
[4000;6000)	5000	29	2000	0.015
[6000;9000)	7500	18	3000	0.006
[9000;15000)	12000	13	6000	0.002
[15000;25000)	20000	7	10000	0.001
[25000;50000)	37500	2	25000	0.000
Total		212		

The distribution Mode is 2,500 euros, and the modal class is [2000 ; 3000)

The average salary is:

$$\bar{x} = \frac{1}{n}\sum_{i=1}^{i=9} n_i x_i = \frac{952.500}{212}$$
$$= 4{,}492.92 \text{ euros}$$

The median wage is:

$$M_e = a_i + (a_{i+1} - a_i)\frac{(50 - F_i)}{(F_{i+1} - F_i)}$$
$$= 2{,}000 + (3{,}000 - 2{,}000)\frac{(50 - 29.72)}{(50.47 - 29.72)}$$
$$= 2{,}977.35 \text{ euros}$$

INTERPRETATION OF THE RESULTS

c-1. The mode class is [2,000 ; 3,000[, indicating that the majority of employees in this Canadian company have salaries between $2,000 and $3,000. This can be interpreted as a significant clustering of salaries around this range, with a peak frequency at this mode class.

c-2. The mean salary is $4,492.92, which means that if the salaries in this company were evenly distributed among employees, each employee would have an average salary of $4,492.92.

c-3. The median salary is $2,977.35, which means that there are as many people earning a salary below $2,977.35 as there are earnings above this amount in this company. This indicates that the distribution of salaries is relatively symmetrical around the median, with an equitable distribution of lower and higher salaries around this value.

d: Let's calculate the characteristics of dispersion:

The interquartile distance of the distribution is:

$$Q_1 = a_i + (a_{i+1} - a_i)\frac{(25 - F_i)}{(F_{i+1} - F_i)}$$
$$= 1{,}000 + (2{,}000 - 1{,}000)\frac{(25 - 13.68)}{(29.72 - 13.68)}$$
$$= 1{,}705.74 \text{ euros}$$

$$Q_3 = a_i + (a_{i+1} - a_i)\frac{(75 - F_i)}{(F_{i+1} - F_i)}$$
$$= 4{,}000 + (6{,}000 - 4{,}000)\frac{(75 - 67.45)}{(81.13 - 67.45)}$$
$$= 5{,}103.80 \text{ euros}$$

$$\text{Interquartile Distance} = Q_3 - Q_1$$
$$= 5{,}103.80 - 1{,}705.74$$
$$= 3{,}398.06 \text{ euros}$$

The variance of the distribution is:

$$V(x) = \frac{1}{n}\sum_{i=1}^{i=9} n_i x_i^2 - \bar{x}^2$$
$$= \frac{10{,}021{,}750{,}000}{212} - (4{,}492.92\,)^2$$
$$= 27{,}086{,}034.84$$

The standard deviation of the distribution is:

$$\sigma = \sqrt{V(x)}$$
$$= \sqrt{27{,}086{,}034.84}$$
$$= 5{,}204.42 \text{ euros}$$

The coefficient of variation of the distribution is:

$$\text{CV} = \frac{\sigma}{|\bar{x}|}\text{x}100 = \frac{5{,}204.42}{4{,}492.92}\text{x}100$$
$$= 115.8\%$$

INTERPRETATION OF THE RESULTS

d-1. The interquartile range of the distribution, which is $3,398.06, indicates that there is a large variation in salaries in the central half of the distribution, meaning that the salaries between Q1 and Q3 are spread over a wide range of values. This suggests that there is significant variability in salaries among employees, with some employees receiving much higher salaries than others.

d-2. The standard deviation of the distribution is $5,204.42. This high standard deviation indicates a large variability of values from the mean, with some employees receiving salaries that are significantly different from the mean. This may be a sign of significant wage disparities among employees, with substantial differences between the lowest and highest salaries.

d-3. The coefficient of variation is 115.8%. This coefficient of variation greater than 100% indicates excessive data dispersion. In this case, the very high coefficient of variation confirms the presence of a large variability in salaries in this distribution, with salaries deviating significantly from the mean.

e: Let's calculate the characteristics of the shape:

e-1: Calculation of the YULE's coefficient of skewness:

$$C_Y = \frac{Q_1 - 2Q_2 + Q_3}{Q_3 - Q_1}$$

$$= \frac{1{,}705.74 - 2(2{,}977.35\,) + 5{,}103.80}{5{,}103.80 \; - 1{,}705.74}$$

$$= 0.251567071$$

e-2: Calculation of the PEARSON's coefficient of skewness:

$$\beta_1 = \frac{(\bar{x} - \text{ Mode})}{\sigma}$$

$$= \frac{(4{,}492.92 - \; 2{,}500)}{5{,}204.42}$$

$$= 0.38$$

e-3: Calculation of the FISHER's coefficient of skewness:

$$\mu_3 = \frac{\sum_{i=1}^{i=9} n_i (x_i - \bar{x})^3}{\sum_{i=1}^{i=9} n_i} = \frac{100{,}875{,}012{,}231{,}889}{212}$$

$$= 475{,}825{,}529{,}395.70$$

$$\gamma_1 = \frac{\mu_3}{\sigma^3} = \frac{475{,}825{,}529{,}395{,}70}{(5{,}204.42)^3}$$

$$= 3.375$$

e-4: Calculation of the FISHER's coefficient of kurtosis:

$$\mu_4 = \frac{\sum_{i=1}^{i=9} n_i (x_i - \bar{x})^4}{\sum_{i=1}^{i=9} n_i} = \frac{2{,}832{,}246{,}290{,}857{,}590{,}000}{212}$$

$$= 13{,}359{,}652{,}315{,}366{,}000$$

$$\gamma_2 = \frac{\mu_4}{\sigma^4} - 3 = \frac{13{,}359{,}652{,}315{,}366{,}000}{(5{,}204.42)^4} - 3$$

$$= 15.209$$

INTERPRÉTATION DES RÉSULTATS

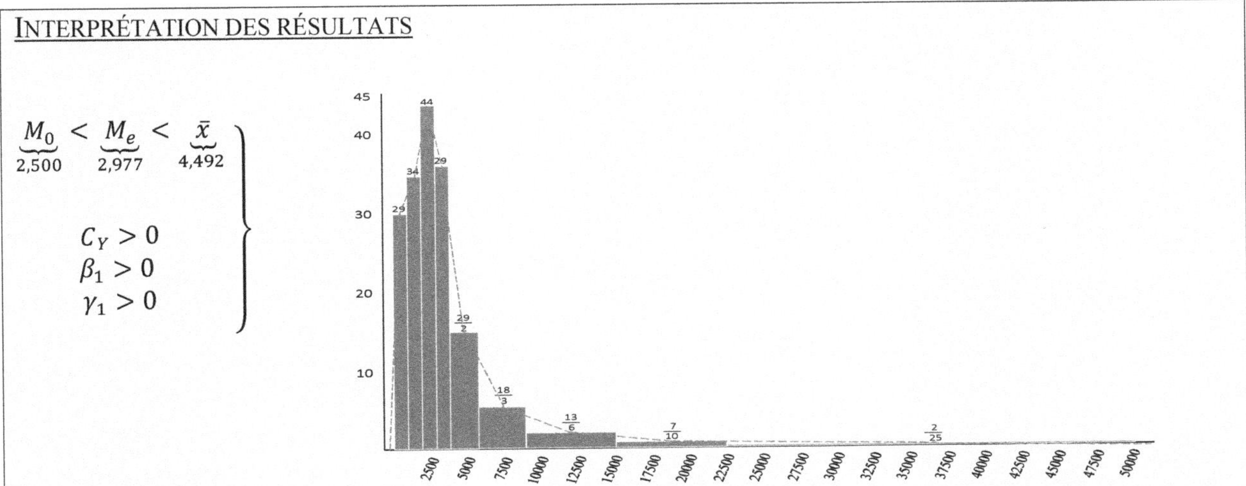

The distribution of salaries is skewed to the right, with a tail stretched towards the right. This means that there is a strong concentration of low salary values, indicating that the majority of employees have relatively low salaries.

$\gamma_2 > 0$, the distribution curve is sharper than that of a normal distribution.

It is notable that salaries increase rapidly for employees with lower salaries but decrease rapidly for those with higher salaries. In other words, there is a sudden increase in the distribution of salaries for low salaries, followed by a rapid decrease as we approach employees with the highest salaries.

f: Let's calculate the medial of concentration:

The following table shows the data required to solve this exercise:

Classes	x_i	n_i	f_i	$S_i = n_i x_i$	$q_i = \frac{S_i}{\sum_{Total} S_i}$	$f_i \uparrow$	$q_i \uparrow$
[0;1000)	500	29	13.68	14,500	1.52	13.68	1.52
[1000;2000)	1500	34	16.04	51,000	5.35	29.72	6.88
[2000;3000)	2500	44	20.75	110,000	11.55	50.47	18.43
[3000;4000)	3500	36	16.98	126,000	13.23	67.45	31.65
[4000;6000)	5000	29	13.68	145,000	15.22	81.13	46.88
[6000;9000)	7500	18	8.49	135,000	14.17	89.62	61.05
[9000;15000)	12000	13	6.13	156,000	16.38	95.75	77.43
[15000;25000)	20000	7	3.30	140,000	14.70	99.06	92.13
[25000;50000)	37500	2	0.94	75,000	7.87	100.00	100.00
Total		212		952,500			

The medial a of this distribution is:

$$\overset{\approx}{M_e} = a_i + (a_{i+1} - a_i)\frac{(50 - FQ_i)}{(Q_{i+1} - Q_i)}$$

$$= 6{,}000 + (9{,}000 - 6{,}000)\frac{(50 - 46.88)}{(61.05 - 46.88)}$$

$$= 6{,}660.55 \text{ euros}$$

INTERPRETATION OF THE RESULTS

f. These data highlight a balanced distribution of the payroll between workers earning salaries below and above 6,660.55 euros. It means that half of the total payroll of the company is allocated to workers whose salaries are below this threshold, while the other half is allocated to workers whose salaries are above it. This equal distribution indicates that the company gives equal importance to workers earning lower salaries and those earning higher salaries in its salary structure.

g: Calculation of the GINI index using the trapezoid method.

The following table shows the data required to solve this exercise:

$f_i \uparrow$	$q_i \uparrow$	$H_i = f_{i+1} \uparrow - f_i \uparrow$	$B_i = \frac{q_i \uparrow + q_{i+1} \uparrow}{2}$	Trapezoid Area (T_i) $T_i = B_i x H_i$
13.68	1.52	16.04	4.20	67.35
29.72	6.88	20.75	12.65	262.57
50.47	18.43	16.98	25.04	425.20
67.45	31.65	13.68	39.27	537.12
81.13	46.88	8.49	53.96	458.18
89.62	61.05	6.13	69.24	424.58
95.75	77.43	3.30	84.78	279.92
99.06	92.13	0.94	96.06	90.63
100.00	100.00			
Total				2545.54

The value of the surface area under the curve is:

$$\sum_{i=1}^{i=9} (f_{i+1} \uparrow - f_i \uparrow) \frac{(q_i \uparrow + q_{i+1} \uparrow)}{2} = 2{,}545.54$$

It follows that:

$$\text{GINI Index} = 100\% - \left(\frac{2{,}545{,}54}{5{,}000}\right) x100$$

$$= 100\% - 50.91\%$$

$$= 49.08\%$$

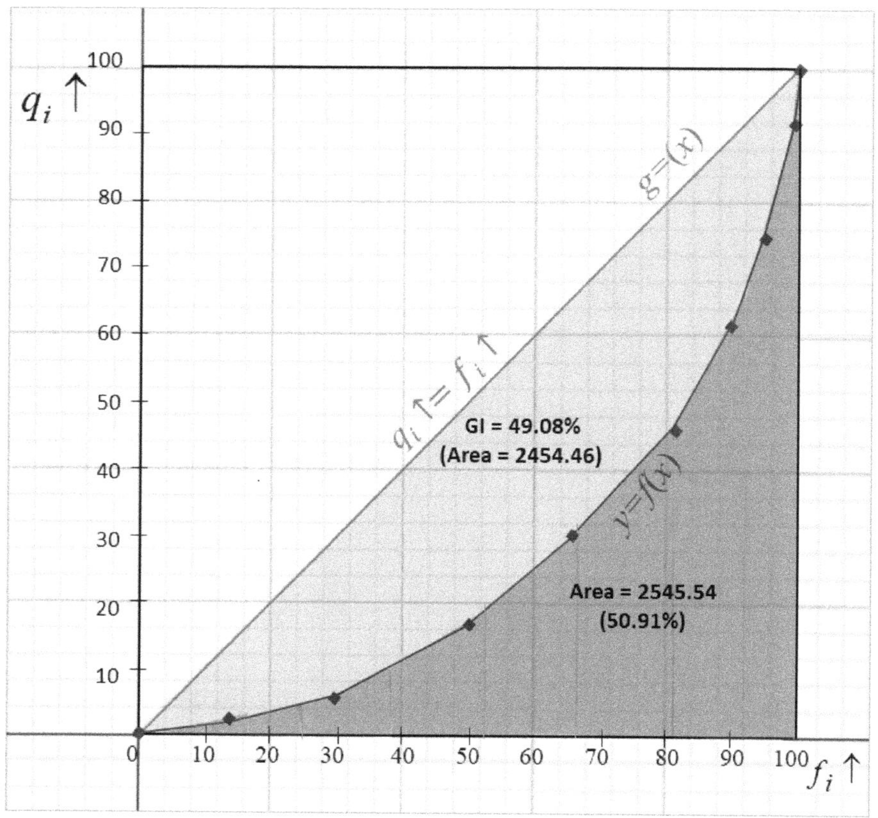

h: Calculation of the GINI index using the method of definite integrals.

We will consider the following key points: $(f_i \uparrow, q_i \uparrow)$:

(0,0) – (14,1) – (30,7) – (50,18) – (67,32) – (81,47) – (90,61) – (96,77) – (99,92) – (100,100)

Following the computation, the interpolation polynomial is obtained as:

$$y = \frac{30681740552358143}{277883647354307623305600000}x^7 - \frac{2852427599569133}{794407225140959472000000}x^6 + \frac{1281850856464148366513}{277883647354307623305600000}x^5 - \frac{5946339813555530129257}{201364961850947553120000}x^4 + \frac{6755778188196601926258 41}{694709118385769058264000000}x^3 - \frac{1113019121219432410511 47}{7718990204286322869600000}x^2 + \frac{5864151317583513188887}{70172638220784753360 0}x$$

Let's evaluate the area between the curve q_i ↑ and the f_i ↑ axes in the interval [0 ; 100]:

$$\begin{aligned}\text{Aire} = \int_0^{100} y\,dx = {} & \frac{30681740552358143}{277883647354307623305600000}\left(\frac{10^{16}}{8}\right) - \frac{30681740552358143}{794407225140959472000000}\left(\frac{10^{14}}{7}\right) \\ & + \frac{128185085646414836651 3}{277883647354307623305600000}\left(\frac{10^{12}}{6}\right) - \frac{5946339813555301292 57}{201364961850947553120000 0}\left(\frac{10^{10}}{5}\right) \\ & + \frac{6755778188196601926258 41}{694709118385769058264000 00}\left(\frac{10^{8}}{4}\right) - \frac{1113019121219432410511 47}{771899020428632286960000}\left(\frac{10^{6}}{3}\right) \\ & + \frac{586415131758351318887}{701726382207847533600}\left(\frac{10^{4}}{2}\right)\end{aligned}$$

$$\text{Aire} = 2.511$$

It follows that:

$$\begin{aligned}\text{GINI Index} &= 100\% - \left(\frac{\int_0^{100} f(x)\,dx}{5{,}000}\right) \text{x}100 \\ &= 100\% - \left(\frac{2{,}511}{5{,}000}\right) \text{x}100 \\ &= 100\% - 50.22\% \\ &= 49.78\%\end{aligned}$$

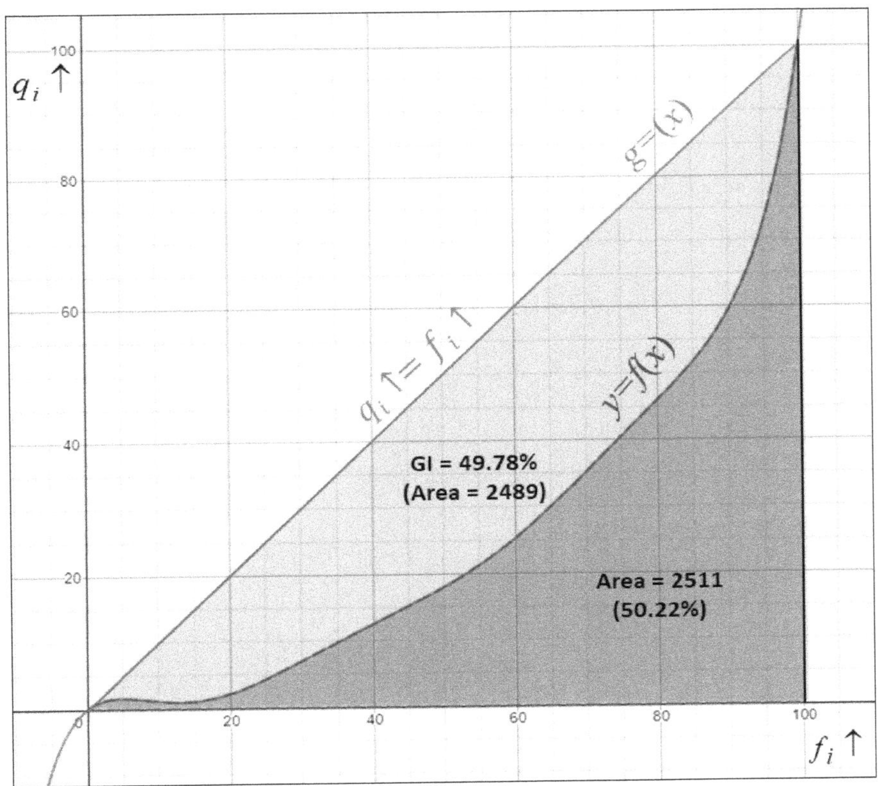

INTERPRETATION OF THE RESULTS

With the GINI index reaching a level close to 50%, it is evident that there is a significant disparity in wage distribution within this company. This indicates that there is a notable difference in salaries among employees, with a concentration of incomes among a minority and a more spread-out distribution among the majority of employees.

These findings corroborate the observations made in the study, particularly in response to the question about the exceptionally high coefficient of variation at 115.8%. This indicates that there are significant gaps in wages, contributing to pronounced inequality in employee compensation.

Furthermore, the results from question **e** also highlight this wage disparity, with a large majority of employees receiving relatively low salaries, while a small group of individuals enjoy extremely high salaries. This underscores the concentration of incomes among a limited minority of individuals, while the majority of employees fall within a lower salary range.

7.6.2 Cross-Chapter Exercise

• **Exercise 34:** Let's continue with exercise **24** of section **6.4.2**, which focuses on the savings account balances of 300 individuals.

Classes	x_i	n_i	n_ix_i	f_i	q_i	$f_i\uparrow$	$q_i\uparrow$
[0;5000)	2500	48	120000	16.00	1.27	16.00	1.27
[5000;10000)	7500	41	307500	13.67	3.26	29.67	4.53
[10000;15000)	12500	47	587500	15.67	6.22	45.33	10.75
[15000;20000)	17500	15	262500	5.00	2.78	50.33	13.53
[20000;25000)	22500	21	472500	7.00	5.00	57.33	18.53
[25000;30000)	27500	12	330000	4.00	3.49	61.33	22.02
[30000;35000)	32500	13	422500	4.33	4.47	65.67	26.50
[35000;40000)	37500	8	300000	2.67	3.18	68.33	29.67
[40000;45000)	42500	9	382500	3.00	4.05	71.33	33.72
[45000;50000)	47500	9	427500	3.00	4.53	74.33	38.25
[50000;55000)	52500	10	525000	3.33	5.56	77.67	43.81
[55000;60000)	57500	6	345000	2.00	3.65	79.67	47.46
[60000;65000)	62500	9	562500	3.00	5.96	82.67	53.41
[65000;70000)	67500	5	337500	1.67	3.57	84.33	56.99
[70000;75000)	72500	2	145000	0.67	1.54	85.00	58.52
[75000;80000)	77500	13	1007500	4.33	10.67	89.33	69.19
[80000;85000)	82500	8	660000	2.67	6.99	92.00	76.18
[85000;90000)	87500	7	612500	2.33	6.48	94.33	82.66
[90000;95000)	92500	4	370000	1.33	3.92	95.67	86.58
[95000;100000)	97500	13	1267500	4.33	13.42	100.00	100.00
Total		300	9445000	100	100		

a: Calculate the medial and the percentage of people with savings less than the medial.

b: Calculate the GINI index by the polynomial interpolation method.

Solution

a-1: Let's calculate the medial:

$$\overset{\approx}{M_e} = a_i + (a_{i+1} - a_i)\frac{(50 - Q_i)}{(Q_{i+1} - Q_i)}$$

$$= 60{,}000 + (65{,}000 - 60{,}000)\frac{(50 - 47.46)}{(53.41 - 47.46)}$$

$$= 62{,}134.45 \text{ euros}$$

a-2: The percentage of people who have savings below the medial is:

Let $P_{<medial}$ be the percentage of people who have a saving less than the medial, and we will use the linear interpolation formula using the data pair $(f_i \uparrow, q_i \uparrow)$:

$$P_{<medial} = f_i + (f_{i+1} - f_i)\frac{(50 - Q_i)}{(F_{i+1} - Q_i)}$$

$$= 79.67 + (82.67 - 79.67)\frac{(50 - 47.46)}{(53.41 - 47.46)}$$

$$= 80.95\%$$

b: Let's calculate the GINI index

b-1: Calculation of the GINI index using the trapezoid method.

The following table shows the data required to solve this exercise:

$f_i \uparrow$	$q_i \uparrow$	$H_i = f_{i+1}\uparrow - f_i\uparrow$	$B_i = \frac{q_i\uparrow + q_{i+1}\uparrow}{2}$	Trapezoid Area *(Ti)* *Ti=BixHi*
16.00	1.27	13.67	24.36	332.98
29.67	4.53	15.67	28.97	453.87
45.33	10.75	5.00	33.87	169.34
50.33	13.53	7.00	36.02	252.17
57.33	18.53	4.00	43.86	175.44
61.33	22.02	4.33	49.10	212.77
65.67	26.50	2.67	54.58	145.54
68.33	29.67	3.00	58.13	174.38
71.33	33.72	3.00	66.86	200.58
74.33	38.25	3.33	19.12	63.75
77.67	43.81	2.00	21.90	43.81
79.67	47.46	3.00	23.73	71.19
82.67	53.41	1.67	26.71	44.51
84.33	56.99	0.67	28.49	19.00
85.00	58.52	4.33	29.26	126.80
89.33	69.19	2.67	34.60	92.25
92.00	76.18	2.33	38.09	88.87
94.33	82.66	1.33	41.33	55.11
95.67	86.58	4.33	43.29	187.59
100.00	100.00			
Total				2909.94

The area under the LORENZ curve is:

$$\sum_{i=1}^{i=20}(f_{i+1}\uparrow - f_i\uparrow)\frac{(q_i\uparrow + q_{i+1}\uparrow)}{2} = 2{,}910$$

It follows that:

$$\begin{aligned}\text{GINI Index} &= 100\% - \left(\frac{2{,}545}{5{,}000}\right)\text{x}100\\ &= 100\% - 50.9\%\\ &= 49.1\%\end{aligned}$$

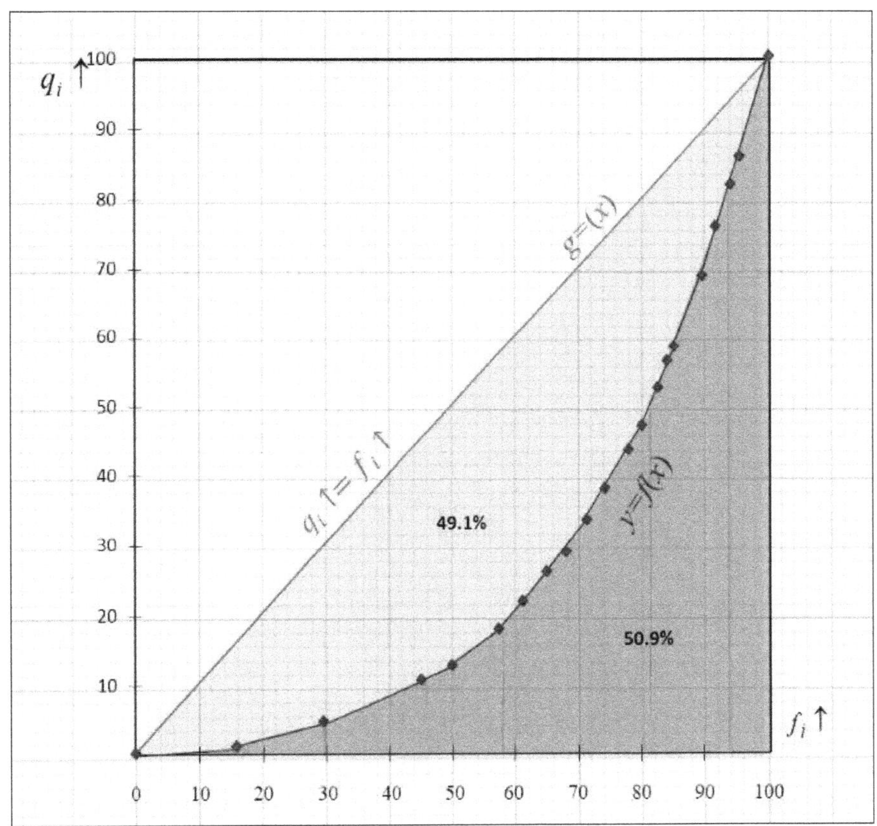

b-2: Calculation of the GINI index using the integration method.

We will consider the following key points: $(f_i\uparrow, q_i\uparrow)$

(0,0) – (8,0) – (16,1) – (30,5) – (45,11) – (61,22) – (74,38) – (83,53) – (92,76) – (100,100)

Following the computation, the interpolation polynomial is obtained as:

$$y = -\frac{2457899533248453868247}{46803928659276767954855896912896000}x^9 + \frac{11928061198454055471469}{537976191485939861550067780608000}x^8 - \frac{97192046809957498132787}{25122881728006853437925870592000}x^7 + \frac{402634866797649188827463}{1117172184252936339774576844800}x^6 - \frac{17128536504742306940381363}{885834065017729728875310336000}x^5 + \frac{111969588869025791546762826221}{185729875632050666487523400448000}x^4 - \frac{15417049365135937413145066222697}{1462622770602398998589246778528000}x^3 + \frac{4910559790161791726878878786407}{48754092353413299952974892617600}x^2 - \frac{3353831960041346913213886229}{9028535621002462954254609744}x$$

Let's evaluate the area between the curve $q_i \uparrow$ and the $f_i \uparrow$ axes in the interval [0 ; 100]:

$$\text{Aire} = \int_0^{100} y\,dx = -\frac{2457899533248453868247}{46803928659276767954855896912896000}\left(\frac{10^{20}}{8}\right) + \frac{11928061198454055471469}{537976191485939861550067780608000}\left(\frac{10^{18}}{8}\right) - \frac{97192046809957498132787}{25122881728006853437925870592000}\left(\frac{10^{16}}{8}\right) + \frac{402634866797649188827463}{1117172184252936339774576844800}\left(\frac{10^{14}}{7}\right) - \frac{17128536504742306940381363}{885834065017729728875310336000}\left(\frac{10^{12}}{6}\right) + \frac{111969588869025791546762826221}{185729875632050666487523400448000}\left(\frac{10^{10}}{5}\right) - \frac{15417049365135937413145066222697}{1462622770602398998589246778528000}\left(\frac{10^{8}}{4}\right) + \frac{4910559790161791726878878786407}{48754092353413299952974892617600}\left(\frac{10^{6}}{3}\right) - \frac{3353831960041346913213886229}{9028535621002462954254609744}\left(\frac{10^{4}}{2}\right) = 2.482.13$$

It follows that:

$$\text{GINI Index} = 100\% - \left(\frac{\int_0^{100} f(x)\,dx}{5{,}000}\right)x100 = 100\% - \left(\frac{2{,}482}{5{,}000}\right)x100 = 100\% - 50.22\% = 50.36\%$$

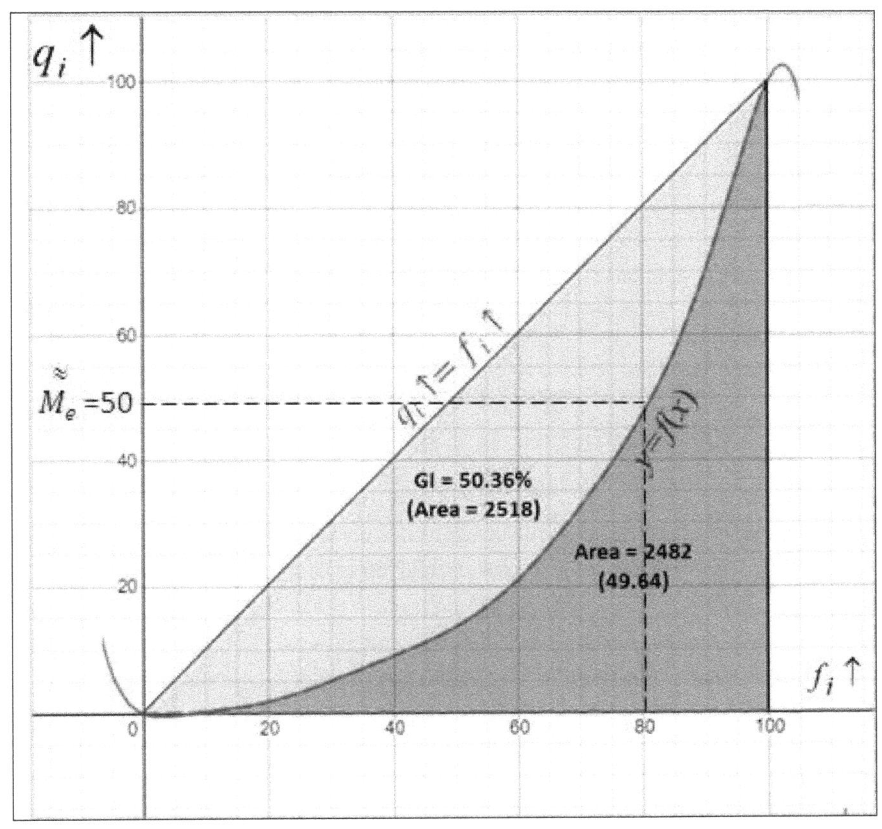

INTERPRETATION OF THE RESULTS

In this analysis, several noteworthy observations can be made. Firstly, it is remarkable that the sum of individual savings below 62,134.45 euros is equivalent to the sum of individual savings above this amount. In other words, those who have saved less than 62,134.45 euros have accumulated as much savings as those who have saved more. This observation highlights a certain disparity in the distribution of savings within the studied sample.

Furthermore, it is also interesting to note that the percentage of people with savings below the medial is 80.95%. This means that nearly 20% of people (or 60 individuals) have accumulated as much savings as 80% of the remaining people (or 240 individuals). This disparity in the distribution of savings is striking and underscores the fact that the majority of savings are concentrated in the hands of a small group of people.

Finally, these observations lead to a compelling conclusion: in this statistical sample, 20% of savers hold as much wealth as the remaining 80%. In other words, wealth is heavily concentrated in the hands of a small group of individuals, highlighting significant inequality in the distribution of wealth within this society.

CHAPTER 8

Two-character statistical series

This chapter provides an in-depth study of statistics with two variables, delving into several key concepts in detail. It explores marginal statistical distributions, which allow for examining the individual properties of each variable, as well as conditional statistical distributions, which investigate the relationships between variables based on certain conditions. Furthermore, marginal and conditional characteristics, such as means and variances, are thoroughly analyzed to better understand their behavior in the context of studying two variables. The complex relationships between these marginal and conditional characteristics are also explored, highlighting potential links and dependencies between the variables. Finally, the concept of independence of variables is discussed in detail to assess how variables may mutually influence their statistical distributions.

8.1 Contingency Table

The contingency table allows for analysis of relationships between two variables within a given population and for a detailed examination of the frequency of combinations of modalities of variables *X* and *Y*. Specifically, for a population of *n* individuals, it gathers data on the number of individuals who exhibit a particular modality of variable *X* in conjunction with a specific modality of variable *Y*, represented respectively by the variables x_i and y_i. By organizing this data into a contingency table, it becomes possible to obtain a clear and structured representation of potential associations between the two variables, facilitating their analysis and interpretation.

The contingency table can be used to calculate various measures of association and dependence between variables *X* and *Y*, such as marginal frequencies, conditional frequencies, odds ratios, and many others. It is also useful for conducting statistical tests and exploring the significance of observed associations.

Consider a population of *n* individuals simultaneously described by two variables, *X* and *Y*, where x_i represents the *p* modalities of variable *X* and y_j represents the *q* modalities of variable *Y*. The number of individuals in the population who exhibit both the y_j modality of variable *X* and the y_j

modality of variable Y can be denoted as n_{ij} . A contingency table is commonly used to represent this information.

X \ Y	y_1	y_2	....	y_j	....	y_q	Horizontal Totals
x_1	n_{11}	n_{12}	....	n_{1j}	....	n_{1q}	n_{1*}
x_2	n_{21}	n_{22}	....	n_{2j}	....	n_{2q}	n_{2*}
....	....	....	....	....	....	....	....
x_i	n_{i1}	n_{i2}	....	n_{ij}	....	n_{iq}	n_{i*}
....	....	....	....	....	....	....	....
x_p	n_{p1}	n_{p2}	....	n_{pj}	....	n_{pq}	n_{p*}
Vertical Totals	n_{*1}	n_{*2}	....	n_{*j}	....	n_{*q}	n_{**}

Points refer to the totalization according to the i or j index. So,

- n_{i*} corresponds to the sum of frequencies n_{ij} for each index j, which represents the number of individuals in the population who exhibit the modality x_i of variable X independently of the modalities of variable Y that they exhibit.

- n_{i*} corresponds to the sum of frequencies n_{ij} for each index j, which represents the number of individuals in the population who exhibit the modality x_i of variable X independently of the modalities of variable Y that they exhibit.

We have:

$$n_{i*} = \sum_{j=1}^{j=q} n_{ij} \text{ and } n_{*j} = \sum_{i=1}^{i=p} n_{ij}$$

$$n_{ij} = \sum_{i=1}^{i=p} \sum_{j=1}^{j=q} n_{ij} = \sum_{i=1}^{i=p} n_{i*} = \sum_{j=1}^{j=q} n_{*j} = n$$

Example 76

The table below presents the values of weight in relation to height for 10,470 individuals in a given city. Height measurements are given in centimeters (cm), and weight measurements in kilograms (kg). This study was conducted to understand the distribution of weights based on heights within this population. In a real-life situation, the results of such a study may also be used to implement health and nutrition programs tailored to this population.

Y(Weight) / X(Height)	[0;30)	[30;60)	[60;90)	[90;120)	[120;150)	[150;180)	[180;210)	Total	
[0;20)	13	1	0	0	0	0	0	14	
[20;40)	27	2	0	0	0	0	0	29	
[40;60)	34	12	3	0	0	0	0	49	
[60;80)	32	19	8	5	2	1	0	67	
[80;100)	2	556	113	9	4	1	0	685	n_{5*}
[100;120)	0	1634	148	9	7	2	0	1800	
[120;140)	0	1073	384	27	5	2	1	1492	
[140;160)	0	467	643	112	49	11	2	1284	
[160;180)	0	215	986	1003	234	53	8	2499	
[180;200)	0	111	843	672	312	21	4	1963	
[200;220)	0	41	347	165	31	3	1	588	
Total	108	4131	3475	2002	644	94	16	10470	
							n_{*7}		

INTERPRETATION OF THE TABLE

A detailed analysis of the table data reveals several interesting findings:

- 556 individuals have a height between 80 and 100 cm and a weight between 30 and 60 kg. This suggests that in this city, there is a significant group of individuals with relatively small height and moderate weight.
- Only 3 individuals have a height between 40 and 60 cm, which is quite rare and unusual, and a weight between 60 and 90 kg. This finding may require further investigation to understand the reasons behind this peculiar characteristic.
- 11 individuals have a height between 140 and 160 cm and a weight between 150 and 180 kg. This group of individuals exhibits a combination of relatively high height and high weight, which may indicate a population with a prevalence of overweight or obesity.
- It is interesting to note that no individuals were recorded with a height between 60 and 80 cm and a weight between 180 and 210 kg. This may suggest that such a combination of height and weight is rare in this given population.

In addition, the additional data extracted from the table is also relevant:

1. $n_{5*} = \sum_{j=1}^{j=7} n_{5j} = 2 + 556 + 113 + 9 + 4 + 1 + 0 = 685$

In the given population of 10,470 individuals, 685 individuals exhibit the x_5 modality of the X character. Specifically, these individuals have a height ranging from 80 to 100 cm, regardless of the modality of character Y they exhibit, i.e., regardless of their weight.

2. $n_{*7} = \sum_{i=1}^{i=11} n_{i7} = 0 + 0 + 0 + 0 + 0 + 0 + 0 + 1 + 2 + 8 + 4 + 1 = 16$

Out of the total population of 10,470 individuals, only 16 exhibit the y_7 modality of the Y character. This indicates that these individuals have a weight ranging from 180 to 210 kg, regardless of the modalities of character X they exhibit, i.e., regardless of their height.

For any couple of modalities x_i and y_j there exists a relative frequency that is defined as the proportion of individuals who simultaneously present the modalities x_i and y_j, this is denoted by:

For each pair of modalities x_i and y_j, the frequency f_{ij} is calculated as the ratio between the number of individuals who simultaneously exhibit the modalities x_i and y_j:

$$\boxed{f_{ij} = \frac{n_{ij}}{n}}$$

This infers:

$$f_{i*} = \sum_{j=1}^{j=q} f_{ij} = \frac{n_{i*}}{n} \quad \text{and} \quad f_{*j} = \sum_{i=1}^{i=p} f_{ij} = \frac{n_{*j}}{n}$$

and

$$\sum_{i=1}^{p} f_{ij} = \sum_{j=1}^{q} f_{ij} = \sum_{i=1}^{p}\sum_{j=1}^{q} f_{ij} = 1$$

Furthermore, it is important to note that the sum of frequencies f_{ij} for each modality x_i over all possible modalities y_j is equal to 1, which means that the entire population is accounted for.

Similarly, the sum of frequencies f_{ij} for each modality y_j over all possible modalities x_i is also equal to 1, ensuring that all possible combinations of characteristics x_i and y_j are covered.

Example 77

Considering the same table as in the previous example, we will have the following frequency table:

Y(Weight) / X(Height)	[0;30)	[30;60)	[60;90)	[90;120)	[120;150)	[150;180)	[180;210)	Total (%)	
[0;20)	0.12	0.01	0.00	0.00	0.00	0.00	0.00	0.13	
[20;40)	0.26	0.02	0.00	0.00	0.00	0.00	0.00	0.28	
[40;60)	0.32	0.11	0.03	0.00	0.00	0.00	0.00	0.47	
[60;80)	0.31	0.18	0.08	0.05	0.02	0.01	0.00	0.64	
[80;100)	0.02	5.31	**1.08**	0.09	0.04	0.01	0.00	6.54	f_{5*}
[100;120)	0.00	15.61	1.41	0.09	0.07	0.02	0.00	17.19	
[120;140)	0.00	10.25	3.67	0.26	0.05	0.02	0.01	14.24	
[140;160)	0.00	4.46	6.14	1.07	0.47	0.11	0.02	12.24	
[160;180)	0.00	**2.05**	9.42	9.58	2.23	0.51	0.08	23.79	
[180;200)	0.00	1.06	8.05	6.42	2.98	0.20	**0.04**	18.71	
[200;220)	0.00	0.39	3.31	1.58	0.30	0.03	0.01	5.61	
Total	1.03	39.46	33.19	19.12	6.15	0.90	0.16	100	
							f_{*7}		

INTERPRETATION OF THE TABLE

Based on the provided data, we can observe that in the studied population:

- 2.05% of the studied population falls within a range of height between 160 and 180 cm and a weight between 30 and 60 kg. These ranges of height and weight fall within the range of body mass index (BMI) from 11.5 to 18.5. A BMI below 18.5 is considered an indicator of thinness or underweight.
- This data can be interpreted as indicating the presence of a small proportion of people with low weight compared to their height, which may be associated with a risk of malnutrition or health problems related to being underweight. This could be due to several factors such as eating disorders, underlying medical conditions, or an unhealthy lifestyle.
- 1.08% of individuals have a height ranging from 80 to 100 cm and a weight ranging from 60 to 90 kg. This group could include individuals of medium to large stature with higher weight, such as average to robust-sized adults.
- 0.04% of individuals have a height ranging from 180 to 200 cm and a weight ranging from 180 to 210 kg. This group represents a small proportion of the population and may include very tall and heavy individuals, such as professional athletes or individuals with severe obesity.

In addition, the additional data extracted from the table is also relevant:

1. $f_{5*} = \sum_{j=1}^{j=7} f_{5j} = 0{,}02 + 5{,}31 + 1{,}08 + 0{,}09 + 0{,}04 + 0{,}01 + 0 = 6{,}54$

Out of the population of 10,470 individuals, 6.54% exhibit the x_5 modality of the X character. This indicates that these individuals have a height ranging from 80 to 100 cm, regardless of the modality of character Y they exhibit, i.e., regardless of their weight.

2. $f_{*7} = \sum_{i=1}^{i=11} n_{i7} = 0 + 0 + 0 + 0 + 0 + 0 + 0{,}01 + 0{,}02 + 0{,}08 + 0{,}04 + 0{,}01 = 0.16$

In the population of 10,470 individuals, only 0.16% exhibit the y_7 modality of the Y character. This indicates that these individuals have a weight ranging from 180 to 210 kg, regardless of the modalities of character X they exhibit, i.e., regardless of their height.

8.2 Marginal Statistical Distributions

The marginal statistical distribution of the X character is the table that associates any modality x_i of X its number n_{ij}. It represents the frequency of occurrence of each value or modality x_i of the character X in a dataset. It is usually presented in the form of a table or a graph, where each modality x_i is associated with its frequency count n_{ij}, which is the number of times it appears in the dataset.

The marginal statistical distribution allows for examining the distribution of values of an individual character without considering other characters in the dataset. It can be used to study the frequency of occurrence of each modality, detect possible outliers, evaluate data variability or dispersion, or construct statistical models.

It is important to note that the marginal statistical distribution can be calculated for any character X, whether it is a qualitative or quantitative variable. In the case of a qualitative variable, the marginal distribution will simply represent the frequency of occurrence of each modality. In the case of a quantitative variable, the marginal distribution can be obtained by grouping the values into classes or using other discretization methods.

X	Frequencies	Relative Frequencies
x_1	n_{1*}	f_{1*}
x_2	n_{2*}	f_{2*}
....	....	....
x_i	n_{i*}	f_{i*}
....	....	....
x_p	n_{p*}	
Total	n_{**}	1

Similarly, the marginal statistical distribution of the character Y can be defined. This distribution represents the frequency of occurrence of each value or modality y_i of the character Y in a given dataset. It can be presented in the form of a table or a graph, where each modality y_i is associated with its frequency count n_{ij}, which is the number of times it appears in the dataset.

Example 78

Considering the same table as in the previous example, we can obtain the following marginal distribution tables:

Marginal Height Distribution (X)		
Height	Frequencies	Proportions
[0;20)	14	0.13
[20;40)	29	0.28
[40;60)	49	0.47
[60;80)	67	0.64
[80;100)	685	6.54
[100;120)	1800	17.19
[120;140)	1492	14.25
[140;160)	1284	12.26
[160;180)	2499	23.87
[180;200)	1963	18.75
[200;220)	588	5.62
Total	10470	100

Marginal Weight Distribution (Y)		
Weight	Frequencies	Proportions
[0;30)	108	1.03
[30;60)	4131	39.46
[60;90)	3475	33.19
[90;120)	2002	19.12
[120;150)	644	6.15
[150;180)	94	0.90
[180;210)	16	0.15
Total	10470	100

INTERPRETATION OF THE TABLES

By examining the given examples, the following information can be noted for each modality:

- For the modality of measurement between 160 cm and 180 cm of character X, it can be observed that there is a total of 2499 individuals in the total population who fall within this range of values. This represents 23.87% of the total population, indicating that this range of values is fairly represented in the overall data.
- For the modality of weight between 60 kg and 90 kg of character Y, it can be observed that there is a total of 3475 individuals in the total population who fall within this range of values. This represents 33.19% of the total population, indicating that this range of values is also well represented in the data.

This information provides a better understanding of the proportion of the total population that falls within each range of values for characters X and Y, and thus evaluate their distribution and representativeness in the overall data.

8.3 Conditional Statistical Distributions

Conditional statistical distributions are used to study the relationships between different variables in a dataset by focusing on a specific column given a particular row, or vice versa. They are used to analyze how the distribution of one variable may vary based on the values of another variable. For example, in a study on academic performance of students, one may be interested in the distribution of Descriptive Statistics scores only for male students.

The notation $n_{i/j}$ is used to represent the frequency or count of variable i conditional on variable j. It indicates how many times variable i appears in the data when variable j takes a specific value. This approach allows for exploring conditional relationships between different variables and identifying specific trends or patterns that may not be apparent in the overall distribution of the data.

Example 79

Let's consider the distribution of eye color based on hair color as an example. This statistical study aims to analyze how eye color varies depending on hair color in a given dataset. We can examine the relative proportions of different possible combinations, such as the frequency of blue eyes among individuals with blonde, brunette, black, red hair, etc. This approach allows for visualizing and exploring potential relationships between these two characteristics and identifying specific trends or patterns that may exist in the population being studied.

		Hair				
		Blonde	Brunette	Black	Red	Horizontal Totals
Eyes	Blue	1768	807	189	47	2811
	Gray green	946	1387	746	53	3132
	Brown	115	438	288	16	857
	Vertical Totals	2829	2638	1223	116	6800

The conditional distribution looking at the color of the hair, knowing that the eyes are blue is:

		Hair				
		Blonde	Brunette	Black	Red	Total
Eyes	Blue	1768	807	189	47	2811

So, we have as many conditional distributions as there are modalities.

8.4. Marginal and conditional characteristics

8.4.1 Marginal Mean and Marginal Variance

8.4.1.1 Marginal Mean of X Character

The marginal mean of the variable *X*, denoted as $\bar{\bar{x}}$, is calculated by considering the frequency of occurrence of each value x_i of *X* in the table. Each value x_i is multiplied by its frequency n_{i*} to obtain the sum of the products. This sum is then divided by the total number of individuals n_{**} to obtain the marginal mean $\bar{\bar{x}}$.

Thus, the marginal mean provides an estimation of the average value of *X* in the population under study, taking into account the frequency of each specific value.

$$\text{Mean }(x) = \bar{\bar{x}} = \frac{1}{n_{**}} \sum_{i=1}^{i=p} n_{i*} x_i = \sum_{i=1}^{i=p} f_{i*} x_i$$

8.4.1.2 Marginal Variance of X Character

The marginal variance of the variable *X*, denoted as *V(X)*, measures the dispersion of the values of *X* around its marginal mean $\bar{\bar{x}}$. To calculate it, we subtract the marginal mean $\bar{x}$ from each value x_i of *X*, square the difference, multiply the result by the frequency n_{i*} of each value, and sum these products. Finally, this sum is divided by the total number of individuals n_{**} to obtain the marginal variance *V(x)*.

Thus, the marginal variance allows for evaluating the variability of the values of *X* in the population under study, taking into account the frequency of each specific value and the distance between each value and the marginal mean.

$$\text{Variance }(X) = V(X) = \frac{1}{n_{**}} \sum_{i=1}^{i=p} n_{i*} (x_i - \bar{\bar{x}})^2 = \sum_{i=1}^{i=p} f_{i*} (x_i - \bar{\bar{x}})^2$$

Example 80

Let's consider the same table that presents the marginal distribution of heights, as in **example 78**:

Height	x_i	n_i	$n_i x_i$	$n_i(x_i - \bar{\bar{x}})^2$
[0;20 [	10	14	140	276328
[20;40 [	30	29	870	26100
[40;60 [	50	49	2450	122500
[60;80 [	70	67	4690	328300
[80;100 [	90	685	61650	5548500
[100;120 [	110	1800	198000	21780000
[120;140 [	130	1492	193960	25214800
[140;160 [	150	1284	192600	28890000
[160;180 [	170	2499	424830	72221100
[180;200 [	190	1963	372970	70864300
[200;220 [	210	588	123480	25930800
Total		10470	1575640	251202728

$$\bar{\bar{x}} = \frac{1}{n}\sum_{i=1}^{i=11} n_i x_i = \frac{1,575,640}{10,470}$$

$$= 150.5 \text{ Cm}$$

$$V(X) = \frac{1}{n}\sum_{i=1}^{i=p} n_i(x_i - \bar{\bar{x}})^2 = \frac{251,202,728}{10\ 470}$$

$$= 23,992.62$$

INTERPRETATION OF THE TABLES

The interpretation of the tables provides an in-depth understanding of the height situation in this particular city. The marginal average height of the population, 150.5 cm, indicates that this value is the most frequently observed in the studied population. It can serve as a reference point to understand the general trend of height distribution in this city.

The marginal variance of 23,992.5 indicates the dispersion of individual heights around the mean for each modality. This higher variance suggests a greater variability of individual heights, with larger deviations from the mean for certain modalities. This may suggest that there are significant differences in individual heights within this population, with some groups showing more pronounced variations from the mean than others.

In a real-life situation, this information can be used to further explore the distribution of heights in this specific population and generate hypotheses about factors that may influence the observed variations. For example, it can be used to study environmental, genetic, socio-economic, or other factors that may contribute to the variability of heights observed in this city. It can also serve as a basis for comparative studies with other populations or to evaluate the effectiveness of health policies or other interventions aimed at improving the health and well-being of the inhabitants of this city.

8.4.1.3 Marginal Mean of Y Character

The marginal mean of the variable *Y*, denoted as $\bar{\bar{y}}$, is calculated by considering the frequency of occurrence of each value y_i of *Y* in the table. Each value y_i is multiplied by its frequency n_{*j} to obtain the sum of the products. This sum is then divided by the total number of individuals n_{**} to obtain the marginal mean $\bar{\bar{y}}$.

Thus, the marginal mean provides an estimation of the average value of *Y* in the population under study, taking into account the frequency of each specific value.

$$\text{Mean } (Y) = \bar{\bar{y}} = \frac{1}{n_{**}} \sum_{j=1}^{j=q} n_{*j} y_j = \sum_{j=1}^{j=q} f_{*j} y_j$$

8.4.1.4 Marginal Variance of Y Character

The marginal variance of the variable *Y*, denoted as *V(Y)*, measures the dispersion of the values of *Y* around its marginal mean $\bar{\bar{y}}$. To calculate it, we subtract the marginal mean $\bar{y}$ from each value y_i of *Y*, square the difference, multiply the result by the frequency n_{*j} of each value, and sum these products. Finally, this sum is divided by the total number of individuals n_{**} to obtain the marginal variance *V(y)*.

Thus, the marginal variance allows for evaluating the variability of the values of *Y* in the population under study, taking into account the frequency of each specific value and the distance between each value and the marginal mean.

$$\text{Variance } (Y) = V(Y) \frac{1}{n_{**}} \sum_{j=1}^{j=q} n_{*j} (y_j - \bar{\bar{y}})^2 = \sum_{j=1}^{j=q} f_{*j} (y_j - \bar{\bar{y}})^2$$

Example 81

Let's consider the same table that presents the marginal distribution of weights, as in **example 78**:

Weight	y_i	n_i	n_iy_i	$n_j(y_j - \bar{y})^2$
[0;30)	15	108	1620	24300
[30;60)	45	4131	185895	8365275
[60;90)	75	3475	260625	19546875
[90;120)	105	2002	210210	22072050
[120;150)	135	644	86940	11736900
[150;180)	165	94	15510	2559150
[180;210)	195	16	3120	608400
Total		10470	763920	64912950

$$\bar{\bar{y}} = \frac{1}{n}\sum_{j=1}^{j=7} n_i y_j = \frac{763{,}920}{1{,}0470}$$

$$= 72{,}96 \text{ Cm}$$

$$V(y) = \frac{1}{n}\sum_{j=1}^{j=7} n_j\left(y_j - \bar{\bar{y}}\right)^2 = \frac{64{,}912{,}950}{10{,}470}$$

$$= 6{,}199{,}90 \text{ Kg}$$

$$= 6\ 199{,}90 \text{ Kg}$$

INTERPRETATION OF THE TABLES

By analyzing the tables, several important conclusions can be drawn. First, the marginal mean weight of the population is 72.96 Kg. This means that, considering only the weight variable, the weighted average of the population's weight is approximately 72.96 Kg. This information can be useful in understanding the overall trend of individuals' weights in this city.

As for the marginal variance, which is a measure of the dispersion of values around the mean, it is 6,199.90. In the present case, a marginal variance of 6,199.90 may indicate that the weights of individuals in this population are relatively scattered, with significant variations around the mean.

In a real-life situation, this can have important implications in various contexts, such as public health, nutrition, or medical resource planning.

8.4.2 Conditional Mean and Conditional Variance

In a statistical contingency table, the jth column represents the n_{*j} individuals who have the value y_j of the variable *Y* based on the variable *X*. This column describes the one-dimensional conditional variable, i.e., the variable *X* considered marginally.

8.4.2.1 Conditional Mean of X Character

The conditional mean of *X*, denoted as $\bar{x}_j$, is calculated by taking the sum of the products of the values of *X* (x_i) with their frequencies n_{ij}, for *i* ranging from 1 to *p*, where *p* is the total number of possible values for *X*. Then, this sum is divided by the total number of individuals n_{*j}, representing the number of observations of the variable *Y* for the value y_j of *Y*.

Mathematically, the conditional mean of X can be expressed as follows:

$$\text{Mean}\ (X) = \bar{x}_j = \frac{1}{n_{*j}} \sum_{i=1}^{i=p} n_{ij} x_i = \sum_{i=1}^{i=p} f_{ij} x_i$$

8.4.2.2 Conditional Variance of X Character

The conditional variance of *X*, denoted as Variance (*X*), is calculated by taking the sum of the products of the squared differences between the values of *X* (x_i) and the conditional mean of *X* $(\bar{x}_j)$, weighted by their frequencies n_{ij}, for *i* ranging from 1 to *p*. Then, this sum is divided by the total number of individuals n_{*j}.

Mathematically, the conditional variance of *X* can be expressed as follows.

$$\text{Variance}\ (X) = V_j\ (X) = \frac{1}{n_{*j}} \sum_{i=1}^{i=p} n_j \left(x_i - \bar{x}_j\right)^2 = \sum_{i=1}^{i=p} f_{ij} \left(x_i - \bar{x}_j\right)^2$$

8.4.3.3 Conditional Mean of Y Character

The conditional mean of *Y*, denoted as $\bar{y}_i$, is calculated by taking the sum of the products of the values of *Y* (y_j) with their frequencies n_{ij}, for *j* ranging from 1 to *q*, where *q* is the total number of possible values for *Y*. Then, this sum is divided by the total number of individuals n_{i*}, representing the number of observations of the variable *X* for the value x_i of *X*. Mathematically, the conditional mean of *Y* can be expressed as follows:

$$\text{Mean }(Y) = \bar{y}_i = \frac{1}{n_{i*}} \sum_{j=1}^{i=q} n_{ij} y_j = \sum_{j=1}^{i=q} f_{ij} y_j$$

8.4.2.4 Conditional Variance of Y Character

The conditional variance of Y, denoted as Variance (Y), is calculated by taking the sum of the products of the squared differences between the values of Y (y_j) and the conditional mean of Y $(\bar{y}_i)$, weighted by their frequencies n_{ij}, for j ranging from 1 to q. Then, this sum is divided by the total number of individuals n_{i*}.

Mathematically, the conditional variance of Y can be expressed as follows.

$$\text{Variance }(Y) = V_i\ (Y) = \frac{1}{n_{i*}} \sum_{j=1}^{i=q} n_i \left(y_j - \bar{y}_i\right)^2 = \sum_{j=1}^{i=q} f_{ij} \left(y_j - \bar{y}_i\right)^2$$

Example 82

The following table presents the values of height in relation to age for 10,470 individuals residing in the same city as in **example 76**. Height measurements are expressed in centimeters (cm) and age measurements in years. This study was conducted to understand the distribution of heights based on age within this population. Analyzing this distribution will help to better understand the factors that influence growth and development in this population. In a real-life situation, the results of such a study can be used to improve health policies and nutrition programs for this population.

y_i	[18 ; 20)		[20 ; 25)		[25 ; 30)		[30 ; 40)		[40 ; 50)		[50 ; 60)		[60 ; 70)	
x_i	n_1	n_1x_i	n_2	n_2x_i	n_3	n_3x_i	n_4	n_4x_i	n_5	n_5x_i	n_6	n_6x_i	n_7	n_7x_i
10	13	130	1	10	0	0	0	0	0	0	0	0	0	0
30	27	810	2	60	0	0	0	0	0	0	0	0	0	0
50	34	1700	12	600	3	150	0	0	0	0	0	0	0	0
70	32	2240	19	1330	8	560	5	350	2	140	1	70	0	0
90	2	180	556	50040	113	10170	9	810	4	360	1	90	0	0
110	0	0	1634	179740	148	16280	9	990	7	770	2	220	0	0
130	0	0	1073	139490	384	49920	27	3510	5	650	2	260	1	130
150	0	0	467	70050	643	96450	112	16800	49	7350	11	1650	2	300
170	0	0	215	36550	986	167620	1003	170510	234	39780	53	9010	8	1360
190	0	0	111	21090	843	160170	672	127680	312	59280	21	3990	4	760
210	0	0	41	8610	347	72870	165	34650	31	6510	3	630	1	210
	108	5060	4131	507570	3475	574190	2002	355300	644	114840	94	15920	16	2760
	$\bar{x}_1 =$ 47		$\bar{x}_2 =$ 123		$\bar{x}_3 =$ 165		$\bar{x}_4 =$ 177		$\bar{x}_5 =$ 178		$\bar{x}_6 =$ 169		$\bar{x}_7 =$ 173	
	$V_1(X) =$ 2625.9259		$V_2(X) =$ 15751.61		$V_3(X) =$ 28184.259		$V_4(X) =$ 31820.679		$V_5(X) =$ 32153.416		$V_6(X) =$ 29155.319		$V_7(X) =$ 30100	

INTERPRETATION OF THE TABLES

The analysis of these tables highlights significant differences in height depending on the age of individuals residing in this city:

- Young people between the ages of 18 and 20 have an average height of 47 cm.
- People between the ages of 20 and 25 have an average height of 123 cm.
- People between the ages of 25 and 30 have an average height of 165 cm.
- People between the ages of 30 and 40 have an average height of 177 cm.
- People between the ages of 40 and 50 have an average height of 178 cm.
- People between the ages of 50 and 60 have an average height of 169 cm.
- People between the ages of 60 and 70 have an average height of 173 cm.

These results suggest that height growth continues after the age of 18, with a significant increase in height until the age of 30. However, after this age, height growth slows down and becomes almost negligible. Additionally, there appears to be a significant variation in average height among the different age groups. For example, young adults aged 18 to 20 have a much lower average height than those aged 30 to 40, with an average difference of 30 cm. This difference is even more pronounced when comparing young adults aged 18 to 20 with those aged 40 to 50, who have an average height that is 31 cm higher.

These results could be influenced by several factors, including genetics, nutrition, lifestyle, and environment. It is possible that dietary habits and exposure to different chemicals and pollutants could affect height growth, especially in young adults. Additionally, factors such as the quality of sleep and level of physical activity may also play a role in height development.

8.4.3 Marginal and Conditional Characteristics Relationships

8.4.3.1 Linking Variable

The linking variable is a variable that allows the study of the relationships between two other variables by eliminating the possible effect of other factors that could influence the relationship. This variable is used in statistical analyses to control for confounding variables, which are variables that can bias the results by influencing the studied relationship. By including a control variable in the analysis, one can eliminate the possible effect of these confounding factors and better understand the relationship between the two other variables.

In medical research, for example, the control variable can be used to control for the effects of age, sex, diet, or other factors that may influence the relationship between two variables such as disease and treatment. Similarly, in economics, the control variable can be used to eliminate the influence of factors such as education level, age, sex, or profession on the relationship between income and consumption.

8.4.3.2 Marginal Mean and Conditional Mean Relationship

When marginal and conditional means are linked by a linking variable, the marginal mean is equal to the weighted average of the conditional means, weighted by the marginal frequencies of the linking variable. In other words, the marginal mean is a weighted combination of the conditional means, depending on the proportion of each subpopulation.

To put it more clearly, the mean of a mixed population is calculated by taking the weighted average of conditional means by the proportions of the mixture. In other words, marginal means are equal to the means of conditional means weighted by the marginal frequencies of the linking variable.

This mathematical formula helps to better understand how these different means are related to each other and understand how a linking variable affects the mean of the mixed population.

$$\bar{\bar{x}} = \frac{\sum_{j=1}^{q} f_{*j}\bar{x}_j}{\sum_{j=1}^{q} f_{*j}} = \sum_{j=1}^{q} f_{*j}\bar{x}_j$$

Example 83

Consider the same table as in the previous example; let's calculate the conditional mean and variance:

	y_i													
	[18;20)		[20;25)		[25;30)		[30;40)		[40;50)		[50;60)		[60;70)	
x_i	n_1	n_1x_i	n_2	n_2x_i	n_3	n_3x_i	n_4	n_4x_i	n_5	n_5x_i	n_6	n_6x_i	n_7	n_7x_i
10	13	130	1	10	0	0	0	0	0	0	0	0	0	0
30	27	810	2	60	0	0	0	0	0	0	0	0	0	0
50	34	1700	12	600	3	150	0	0	0	0	0	0	0	0
70	32	2240	19	1330	8	560	5	350	2	140	1	70	0	0
90	2	180	556	50040	113	10170	9	810	4	360	1	90	0	0
110	0	0	1634	179740	148	16280	9	990	7	770	2	220	0	0
130	0	0	1073	139490	384	49920	27	3510	5	650	2	260	1	130
150	0	0	467	70050	643	96450	112	16800	49	7350	11	1650	2	300
170	0	0	215	36550	986	167620	1003	170510	234	39780	53	9010	8	1360
190	0	0	111	21090	843	160170	672	127680	312	59280	21	3990	4	760
210	0	0	41	8610	347	72870	165	34650	31	6510	3	630	1	210
	108	5060	4131	507570	3475	574190	2002	355300	644	114840	94	15920	16	2760
f_j	1.03%		39.45%		33.19%		19.12%		6.15%		0.89%		0.15%	
	$\bar{x}_1$= 47		$\bar{x}_2$= 123		$\bar{x}_3$= 165		$\bar{x}_4$= 177		$\bar{x}_5$= 178		$\bar{x}_6$= 169		$\bar{x}_7$= 173	
$f_j\bar{x}_j$	0.48		48.47		54.84		33.93		10.97		1.51		0.26	
	$\bar{\bar{x}}$= 150													

Thus, we can observe that the marginal mean is equivalent to the average of the conditional means, which are weighted by the marginal frequencies of the associated variable.

Indeed,

$$\bar{\bar{x}} = \frac{(47 \text{ x } 1.03) + (123\text{x}39.45) + (165\text{x}33.19)}{100}$$

$$+\frac{(177\text{x}19.12) + (178\text{x}6.15) + (169\text{x}0.89) + (1.73\text{x}0.15)}{100}$$

$$= 150.46$$

INTERPRETATION OF THE TABLES

The table above provides various pieces of information about heights and weights in this city:

- Firstly, we can see that the most populous age group is the 20-30 age range, comprising a total of 72.64% of the overall population. We can also observe that the 60-70 age range is the least represented, accounting for only 0.15% of the total population.
- Furthermore, if we look at column f*j, we can see that the heaviest age group is the 20-25 age range, representing 39.45% of the total population and contributing 48.47% to the overall average.
- Conversely, the lightest age group is the 60-70 age range, contributing only 0.26% to the overall average.
- We can also note that the average height is 150.5 cm, but this does not mean that the majority of individuals measure exactly 150.5 cm. Indeed, the distribution of heights varies across different age groups, as highlighted by the conditional averages within each age range.

Similarly, we have:

$$\bar{\bar{x}} = \frac{\sum_{j=1}^{q} f_{*j}\bar{x}_j}{\sum_{j=1}^{q} f_{*j}} = \sum_{j=1}^{q} f_{*j}\bar{x}_j$$

Using these two previous formulas, we can deduce that the marginal mean is always housed in the bounded interval of the lower and upper conditional means. Thus, if we calculate the mean of x for each subset of data, we will obtain a series of conditional means. The range of values between the minimum and maximum conditional means represents the extent of the marginal mean of x. Similarly, if we calculate the mean of y for each subset of data, we will obtain a series of conditional means. The range of values between the minimum and maximum conditional means represents the extent of the marginal mean of y.

In other words, the mean of x is located within the interval delimited by the minimum and maximum values of x, and the mean of y is located within the interval delimited by the minimum and maximum values of y.

$$\bar{\bar{x}} \in [Min(\bar{x}_j)\ ;\ Max(\bar{x}_j)]$$

$$\bar{\bar{y}} \in [Min(\bar{y}_i)\ ;\ Max(\bar{y}_i)]$$

Example 84

Consider the same data table as in the previous example. The value 150.46 is located within the interval bounded by the minimum and maximum mean values, that is between 47 and 178.

8.4.3.3 Relationship between Marginal Variance and Conditional Mean

The variance of a combination is equal to the weighted variance of variances increased by the weighted average of the averages. It is obtained by calculating the weighted variance of each subset of data and adding to it the weighted mean of the deviations from the mean for each subset. The weights associated with each subset are represented by the relative frequencies, that is, the proportion of data in each subset.

Formula to calculate the Variance of *X*:

$$V(X) = \sum_{j=1}^{q} f_{*j} V_j(X) + \sum_{j=1}^{q} f_{*j} (\bar{x}_j - \bar{\bar{x}})^2$$

Formula to calculate the Variance of *Y*:

$$V(Y) = \sum_{i=1}^{p} f_{i*} V_i(Y) + \sum_{i=1}^{p} f_{i*} (\bar{y}_i - \bar{\bar{y}})^2$$

The first part of the formula calculates the weighted variance of the variances of the subsets. The weighted variance measures the variability of each subset relative to its weighted mean, taking into account the size of each subset.

The second part of the formula calculates the weighted mean of the deviations from the mean for each subset. This part measures the contribution of each subset to the overall variability of the mixed data set.

By combining these two parts of the formula, we obtain the total variance of the mixed data set. This variance measures the overall dispersion of the data around the weighted mean of the data set.
This formula is particularly useful in cases where data is grouped into several subsets with different relative frequencies, such as data from different sources or different samples.

More generally, this formula allows for the calculation of the variance of a mixed set of data, meaning a combination of distinct data subsets. Therefore, it helps to better understand the distribution of variance in a mixed data set, taking into account the variance of each subset and the contribution of each subset to the overall mean.

8.5 Independence of variables

The concept of independence is very useful in many fields, especially in statistics and probability. It allows for simpler analyses and a better understanding of the relationships between different variables. For example, if we know that two variables are independent, we can use simpler statistical methods to analyze them, as we do not need to take into account their interactions.

If character *X* is independent of character *Y*, this means that the distribution of variable *X* remains the same, regardless of the different modalities of variable *Y*. Similarly, the distribution of variable *Y* remains the same, regardless of the different modalities of variable *X*. In other words, knowledge of one character's modality does not allow prediction of the modality of the other character.

Moreover, it is important to note that independence is a symmetrical concept. This means that if *X* is independent of *Y*, then *Y* is also independent of *X*. In other words, the order in which we consider the variables does not matter.

Finally, when we talk about the independence between two variables, we can also examine the proportionality between the different modalities of these variables. Thus, if the two variables are independent, all the rows and columns of the contingency table will be proportional, which can be verified by comparing the different frequencies.

It follows that we can determine if two characters are independent by comparing the conditional frequencies and the marginal distribution. If all the conditional frequencies are equal to the marginal distribution, this indicates that the two characters are independent.

Indeed, we say that character *X* is independent of character *Y* if all the conditional frequencies are such that:

$$f_{i/y_1} = f_{i/y_2} = f_{i/y_3} = \dots\dots\dots\dots = f_{i/y_q}$$

$$\frac{n_{i1}}{n_{.1}} = \frac{n_{i2}}{n_{.2}} = \dots\dots\dots = \frac{n_{iq}}{n_{.q}} = \frac{n_{i1} + n_{i2} + \dots\dots\dots n_{iq}}{n_{.1} + n_{.2} + \dots\dots\dots n_{.q}} = \frac{n_{i.}}{n} = f_{i.}$$

That is, all conditional distributions are equal to the marginal distribution.
At independence:

$$\boxed{f_{ij} = f_{i.} f_{.j}}$$

8.7 Exercise

• **Exercise 35:** Let's continue **exercise 33** from **section 7.7.1** on the distribution of monthly salaries (in euros) in a company of 212 employees. In this exercise, we will study the relationship between the monthly salaries of employees in a company and their age. To do this, we will add the component of employee age. We will thus have two statistical variables, X for monthly salary and Y for age.

The objective of this exercise is to analyze the distribution of salaries according to the age of the employees. We will thus determine if there is a correlation between the age and salary of employees. We will thus be able to identify the age groups for which salaries are the highest and those for which they are the lowest. This analysis will allow us to better understand the salary dynamics of the company in question.

Y (Age) / *X (Wage)*	[18;20)	[20;25)	[25;30)	[30;40)	[40;50)	[50;60)	[60;70)
[0;1000)	23	6	0	0	0	0	0
[1000;2000)	2	15	9	3	2	1	1
[2000;3000)	1	6	9	15	7	2	4
[3000;4000)	0	4	7	8	8	5	4
[4000;6000)	0	3	4	7	7	5	3
[6000;9000)	0	0	2	5	7	2	3
[9000;15000)	0	0	1	3	5	2	2
[15000;25000)	0	0	0	0	2	3	2
[25000;50000)	0	0	0	0	1	1	0
Total	26	34	32	41	39	21	19

We now wish to study the distribution of monthly salaries (in euros) in this company of 212 employees according to the age of the employees.

1. To do this, we will develop contingency tables of frequencies and relative frequencies for this series.

2. We will then provide the marginal distribution table for salaries as well as the marginal distribution table for age, to better understand the distribution of salaries and age in our sample of 212 employees.

3. We will also give the conditional distribution of age given that the salary is between 3,000 and 4,000 euros, in order to study the relationship between age and salary in this salary range.

4. Similarly, we will provide the conditional distribution of salary given that the age is between 40 and 50 years old, to better understand the distribution of salaries in this age group.

5. Next, we will calculate the marginal characteristics of salaries, namely the mean and variance, to better understand the distribution of salaries in our sample.

6. We will also calculate the marginal characteristics of the age of employees, namely the mean and variance, to better understand the distribution of age in our sample.

7. Finally, we will verify that the marginal means are equal to the means of conditional means weighted by the marginal frequencies of the linking variable, to ensure the consistency of our analysis.

Solution

1: Let's create contingency tables for the frequencies and relative frequencies regarding this series.

1-a: The frequency contingency table is:

X (Wage) \ *Y (Age)*	[18;20)	[20;25)	[25;30)	[30;40)	[40;50)	[50;60)	[60;70)	Total
[0;1000)	23	6	0	0	0	0	0	29
[1000;2000)	2	15	9	3	2	1	1	33
[2000;3000)	1	6	9	15	7	2	4	44
[3000;4000)	0	4	7	8	8	5	4	36
[4000;6000)	0	3	4	7	7	5	3	29
[6000;9000)	0	0	2	5	7	2	3	19
[9000;15000)	0	0	1	3	5	2	2	13
[15000;25000)	0	0	0	0	2	3	2	7
[25000;50000)	0	0	0	0	1	1	0	2
Total	26	34	32	41	39	21	19	212

INTERPRETATION OF THE RESULTS

The analysis of the data reveals that there are:

- 23 employees aged between 18 and 20 earn less than 1000 euros.
- 15 employees aged between 20 and 25 earn between 1000 and 2000 euros.
- 5 employees aged between 40 and 50 earn between 9,000 and 15,000 euros.
- There are no employees aged between 30 and 40 who earn between 15,000 and 25,000 euros.

Furthermore, the additional data extracted from the table is also relevant:

1. $n_{7*} = \sum_{j=1}^{j=7} n_{7j} = 0 + 0 + 1 + 3 + 5 + 2 + 2 = \mathbf{13}$

This means that out of the 212 employees in this company, 13 belong to the x_7 modality of character *X*, indicating that they have a salary between 9,000 and 15,000 euros, regardless of the modalities of character *Y* they possess, i.e., regardless of their age.

2. $n_{*6} = \sum_{i=1}^{i=9} n_{i6} = 0 + 1 + 2 + 5 + 5 + 2 + 2 + 3 + 1 = \mathbf{21}$

This means that among the 212 employees in this company, 21 belong to the y_6 modality of the character *Y*, indicating that they are between 50 and 60 years old, irrespective of the modalities of the character *X* they belong to, i.e., regardless of their salary.

1-b: Frequency Contingence Table

Y (Age) / *X (Wage)*	[18;20)	[20;25)	[25;30)	[30;40)	[40;50)	[50;60)	[60;70)	Total
[0;1000)	10.85	2.83	0.00	0.00	0.00	0.00	0.00	13.68
[1000;2000)	0.94	7.08	4.25	1.42	0.94	0.47	**0.47**	15.57
[2000;3000)	0.47	2.83	4.25	7.08	3.30	0.94	1.89	20.75
[3000;4000)	0.00	1.89	3.30	**3.77**	3.77	2.36	1.89	16.98
[4000;6000)	0.00	1.42	1.89	3.30	3.30	2.36	1.42	13.68
[6000;9000)	0.00	**0.00**	0.94	2.36	3.30	**0.94**	1.42	8.96
[9000;15000)	0.00	0.00	0.47	1.42	2.36	0.94	0.94	6.13
[15000;25000)	0.00	0.00	0.00	0.00	0.94	1.42	0.94	3.30
[25000;50000)	0.00	0.00	0.00	0.00	0.47	0.47	0.00	0.94
Total	12.26	16.04	15.09	19.34	18.40	9.91	8.96	100.00

INTERPRETATION OF THE RESULTS

As per the above data, we can observe that the analysis of contingency tables and marginal distributions has allowed us to paint a picture of the distribution of salaries based on the age of employees in this company. For example, we can observe that the majority of employees earn between 2,000 and 3,000 euros per month, and that this proportion gradually decreases with increasing salary.
Similarly, we can notice that the distribution of salaries varies considerably depending on the age of employees. Younger employees tend to earn less, while older employees have a greater proportion of higher salaries.

Indeed,

- 3.77% of employees who fall between the age bracket of 30 and 40, have a salary range between 3,000 and 4,000 euros.
- 0.47% of employees who fall between the age bracket of 60 and 70, have a salary range between 1,000 and 2,000 euros.
- 0.94% of employees who fall between the age bracket of 50 and 60, have a salary range between 6,000 and 9,000 euros.
- There are no employees who fall between the age bracket of 20 and 25 and have a salary range between 6,000 and 9,000 euros.

2: Let's create a table for the marginal distribution of salaries as well as the table for the marginal distribution of age.

2-a:

Marginal Distribution of Weights (X)		
Wage	Frequencies	Proportions
[0;1000)	29	13.68
[1000;2000)	33	15.57
[2000;3000)	44	20.75
[3000;4000)	36	16.98
[4000;6000)	29	13.68
[6000;9000)	19	8.96
[9000;15000)	13	6.13
[15000;25000)	7	3.30
[25000;50000)	2	0.94
Total	212	100

2-b:

Marginal Distribution of Age (Y)		
Age	Frequencies	Proportions
[18;20)	26	12.26
[20;25)	34	16.04
[25;30)	32	15.09
[30;40)	41	19.34
[40;50)	39	18.40
[50;60)	21	9.91
[60;70)	19	8.96
Total	212	100

INTERPRETATION OF THE TABLES

Based on the table above, the following information can be observed:

- The population size with a salary ranging from 6,000 to 9,000 euros is 19 individuals, accounting for 8.96% of the total population.
- The population size of individuals aged between 20 and 25 is 34, representing 16.04% of the total population.

By analyzing these contingency tables and marginal distributions, we can observe interesting trends and relationships between salaries and the ages of employees in the company. For example, we can see that there is a significant number of employees earning salaries between 2,000 and 4,000 euros, and there is a strong concentration of employees aged 25 to 30 in this salary range. Additionally, we can observe that there are very few employees earning more than 10,000 euros, and they are all aged 40 or above.

Assuming that the data in question reflects the situation of a real company, this information can help the company better understand the distribution of salaries and ages within the company. They can help decision-makers make informed decisions regarding human resource management and salary policies. For example, by observing that people aged 30 to 40 represent a significant percentage of the company's population, leaders can decide to implement training and professional development programs for this age group. Similarly, by noticing that the average salary of people aged 18 to 25 is lower than that of other age groups, decision-makers may consider salary adjustments for this category of employees.

3: Conditional law of the age knowing that the wage is between 3,000 and 4,000 euros:

Age	[3000,4000)
[18;20)	0
[20;25)	4
[25;30)	7
[30;40)	8
[40;50)	8
[50;60)	5
[60;70)	4
Total	36

INTERPRETATION OF THE RESULTS

We can count:

- No employee whose age is between 18 and 20 years old.
- 4 employees are between 20 and 25 years old.
- 7 employees are between 25 and 30 years old.
- 8 employees are between 30 and 40 years old.
- 8 employees are between 40 and 50 years old.
- 5 employees are between 50 and 60 years old.
- 4 employees are between 60 and 70 years old.

We can observe that by setting the salary between 3,000 and 4,000 euros, there is a majority of employees between 30 and 50 years old, with a peak among the 30-40 age group. However, we also notice the presence of older employees, with 5 employees between 50 and 60 years old and 4 employees between 60 and 70 years old, which can be interpreted as a sign of job stability in this company.

4: Conditional law of the wage knowing that the age is between 40 and 50 years:

Wage	[40,50)
[0;1000)	0
[1000;2000)	2
[2000;3000)	7
[3000;4000)	8
[4000;6000)	7
[6000;9000)	7
[9000;15000)	5
[15000;25000)	2
[25000;50000)	1
Total	39

INTERPRETATION OF THE RESULTS

By setting the age between 40 and 50 years old, we observe that:

- No employee earns between 0 and 1,000 euros.
- 2 employees earn between 1,000 and 2,000 euros.
- 7 employees earn between 2,000 and 3,000 euros.
- 8 employees earn between 3,000 and 4,000 euros.
- 7 employees earn between 4,000 and 6,000 euros.
- 7 employees earn between 6,000 and 9,000 euros.
- 5 employees earn between 9,000 and 15,000 euros.
- 2 employees earn between 15,000 and 25,000 euros.
- 1 employee earns between 25,000 and 50,000 euros.

Therefore, it can be concluded that employees aged between 40 and 50 years old have relatively high salaries on average in this company, with the majority of salaries ranging from 2,000 to 9,000 euros per month. It is also important to note that very high salaries are rare in this age category, with only 2 employees earning between 15,000 and 25,000 euros per month and only one employee earning more than 25,000 euros per month.

The conditional laws of age and salary allow for a better understanding of the relationship between these two variables. For example, if we know that the salary is between 3,000 and 4,000 euros, the conditional law of age indicates that it is more likely that the employee is between 30 and 50 years old. Similarly, if we know that the age is between 40 and 50 years old, the conditional law of salary shows that the employee is more likely to earn between 3,000 and 4,000 euros.

5: Table to help calculate the marginal characteristics of wage:

Wage	x_i	n_i	$n_i x_i$	$n_i x_i^2$
[0;1000)	500	29	14,500	7,250,000.00
[1000;2000)	1500	33	49,500	74,250,000.00
[2000;3000)	2500	44	110,000	275,000,000.00
[3000;4000)	3500	36	126,000	441,000,000.00
[4000;6000)	5000	29	145,000	725,000,000.00
[6000;9000)	7500	19	142,500	1,068,750,000.00
[9000;15000)	12000	13	156,000	1,872,000,000.00
[15000;25000)	20000	7	140,000	2,800,000,000.00
[25000;50000)	37500	2	75,000	2,812,500,000.00
Total	Total	212	958,500	10,075,750,000.00

5-a: Finding the marginal average of wage:

$$\bar{\bar{x}} = \frac{1}{n}\sum_{i=1}^{i=9} n_i x_i = \frac{958,500}{212}$$

$$= 4{,}521.61 \text{ euros}$$

5-b: Finding the marginal variance of wage:

$$V(x) = \frac{1}{n}\sum_{i=1}^{i=9} n_i x_{i^2} - \bar{x}^2$$

$$= \frac{10{,}075{,}750{,}000}{212} - (4{,}474.06\,)^2$$

$$= 27{,}509{,}909.75$$

INTERPRETATION OF THE RESULTS

Upon examining the results, we observe that the marginal mean of x, i.e., the average salary, is equal to 4,521.61 euros. This value represents the average salary of all employees in the company, regardless of age group. Similarly, the marginal variance of x, which measures the average difference between each salary and the average salary, is equivalent to 27,509,909.75.

It is interesting to note that these results perfectly match the values obtained in exercise **33** of section **7.6.1**, which attests to the reliability of the calculations and the relevance of the data collected.

Analyzing these results provides a better understanding of the company's salary situation. Indeed, the information collected provides an overview of the salaries paid to all employees. Moreover, the obtained variance value is an important indicator for establishing compensation policies and for human resource management.

6: In the same way, we can determine the marginal characteristics of the age:

Age	y_i	n_i	$n_i y_i$	$n_i y_i^2$
[18;20)	19	26	494	9,386
[20;25)	23	34	765	17,213
[25;30)	28	32	880	24,200
[30;40)	35	41	1,435	50,225
[40;50)	45	39	1,755	78,975
[50;60)	55	21	1,155	63,525
[60;70)	65	19	1,235	80,275
Total	Total	212	7,719	323,799

6-a: Finding the marginal average of age:

$$\bar{\bar{y}} = \frac{1}{n}\sum_{i=1}^{i=7} n_i y_i = \frac{7{,}719}{212}$$

$$= 36.41 \text{ years}$$

6-b: Finding the marginal variance of age:

$$V(y) = \frac{1}{n}\sum_{i=1}^{i=7} n_i y_{i^2} - y^2$$

$$= \frac{323{,}799}{212} - (36.41)^2$$

$$= 201.64 \text{ years}$$

INTERPRETATION OF THE RESULTS

Similarly, just like in the case of salaries, we observe that the marginal mean of y is 36.41 years and the marginal variance of y is 201.64 years.

7: Let's verify that the marginal means are equal to the averages of conditional means weighted by the marginal proportions of the binding variable

We will check it for the variable X and this will also be valid for the variable Y.

We know that the marginal mean calculated above is 4474.06 and check that the mean of the weighted conditional means equals the same value.

	y_i													
	[18 ; 20)		[20 ; 25)		[25 ; 30)		[30 ; 40)		[40 ; 50)		[50 ; 60)		[60 ; 70)	
x_i	n_1	n_1x_i	n_2	n_2x_i	n_3	n_3x_i	n_4	n_4x_i	n_5	n_5x_i	n_6	n_6x_i	n_7	n_7x_i
500	23	11500	6	3000	0	0	0	0	0	0	0	0	0	0
1,500	2	3000	15	22500	9	13500	3	4500	2	3000	1	1500	1	1500
2,500	1	2500	6	15000	9	22500	15	37500	7	17500	2	5000	4	10000
3,500	0	0	4	14000	7	24500	8	28000	8	28000	5	17500	4	14000
5,000	0	0	3	15000	4	20000	7	35000	7	35000	5	25000	3	15000
7,500	0	0	0	0	2	15000	5	37500	7	52500	2	15000	3	22500
12,000	0	0	0	0	1	12000	3	36000	5	60000	2	24000	2	24000
20,000	0	0	0	0	0	0	0	0	2	40000	3	60000	2	40000
37,500	0	0	0	0	0	0	0	0	1	37500	1	37500	0	0
	26	17000	34	69500	32	107500	41	178500	39	273500	21	185500	19	127000
$f_{\bullet i}$	12.26%		16.04%		15.09%		19.34%		18.40%		9.91%		8.96%	
	$\bar{x}_1$= 654		$\bar{x}_2$= 2044		$\bar{x}_3$= 3359		$\bar{x}_4$= 4354		$\bar{x}_5$= 7013		$\bar{x}_6$= 8833		$\bar{x}_7$= 6684	
	80.16		327.88		506.93		842.00		1290.36		875.38		598.91	
	$\bar{\bar{x}}$= 4521.61													

INTERPRETATION OF THE RESULTS

In this city, the average salary is 4,521 euros. Further, we can see that:

- People aged between 18 and 20 have an average weight of 654 euros, representing 1.03% of the total population.
- People aged between 20 and 25 have an average weight of 2,044 euros, representing 39.45% of the total population.

Furthermore, we can observe that the marginal mean of the variable x is equal to the mean of the conditional means weighted by the marginal frequencies of the associate variable. In this case, the marginal mean is 4,474.06 euros, which was also calculated **in question 5.a.** However, since this value is expressed in thousands of euros, it is equivalent to the calculated value.

Indeed,

$$\bar{\bar{x}} = \frac{(654 \text{ x } 12.26) + (2{,}044\text{x}16.04) + (3{,}359\text{x}15.09)}{100}$$

$$+ \frac{(4{,}354\text{x}19.34) + (7{,}013\text{x}18.4) + (8{,}833\text{x}9.91) + (6{,}684\text{x}8.96)}{100}$$

$$= 4{,}521.61$$

CHAPTER 9

Linear adjustment and Correlation

This chapter introduces the concept of linear regression and presents two different cases: one where one variable is controlled and another where both variables are not controlled. Several methods have been explored in this chapter for linear regression, including empirical method, moving average method, Mayer's method, and least squares method. These methods have been applied to illustrate the predictive determination of future values and provide a better understanding of the concept of linear regression.

9.1 Introduction

In a statistical population, individuals have common traits that are referred to as "characteristics". These characteristics can be singular or multiple. In statistical studies, it is common to have data with several characteristics simultaneously. The characteristics can be related to each other, and their analysis can help to understand the relationships that exist between them.

To study these relationships, several statistical tools and techniques are used, such as correlation analysis and multiple regression. These methods allow to determine if there is a correlation between the characteristics and to identify cause-and-effect relationships between them.

Example 85

To study the statistical population of patients in a hospital, it is often desirable to study a population based on multiple characteristics in order to better understand the relationships between these characteristics and their impact on the population being studied. In the case of studying hospital patients, relevant characteristics such as height, weight, age, and marital status are included to provide a comprehensive view of the situation. However, in this book, we will focus on the study of two statistical characteristics.

It is important to note that two distinct situations may arise when studying two statistical characteristics. In the first case, at least one of the characteristics is controlled by the experimenter, which allows for determining the effect of this characteristic on the other. In the second case, the experimenter has no control over both characteristics, making the study more complex as it is difficult to determine the cause and effect between the two characteristics.

9.2 Cases where at least one of the characters is controlled

In the context of a statistical study, when the experimenter controls at least one of the variables, they have the power to determine its values. This can be done by setting the value of this parameter to a certain level or by varying it according to a pre-defined experimental plan. This variable can take different forms such as speed, size, date, temperature, etc.

To further develop this point, it is important to understand that controlling a variable in a statistical study allows for a better understanding of its impact on other characteristics of the population being studied. For example, by controlling the temperature, one can study its influence on the growth of certain plants. Specifically, if we study the effect of temperature on plant growth, the experimenter can choose to maintain the temperature at a constant level or to vary it according to a series of predefined values. In all cases, controlling at least one of the variables allows the experimenter to better understand the relationships between the different variables studied and to obtain more precise and meaningful results.

Example 86

To study the relationship between fuel consumption and the speed of a car, we seek to determine if these two variables are linked by a mathematical function of the $y = f(x)$. In other words, we want to know if fuel consumption varies with the speed of the car, which in turn is influenced by the car's acceleration. In this study, we are particularly interested in the case where the relationship between these two variables is linear.

To further elaborate on this concept, it is important to understand that the search for a linear relationship between fuel consumption and the speed of a car is common in many studies in mechanical and automotive engineering. Indeed, this relationship can help determine the energy efficiency of a car at different speeds and identify the factors that influence fuel consumption.

9.3 Cases where none of the characters is controlled

When the experimenter has no control over both variables, the statistical study becomes more complex as it is difficult to determine the causal relationship between these variables. In this situation, the variables are not fixed by the experimenter and their evolution is free, it is therefore often necessary to collect data from natural observations or existing data sources. For example, if we study the relationship between age and mental health, we can collect data from questionnaires or interviews to assess the mental state of individuals of different ages.

It is important to note that this method has its limitations as the collected data can be influenced by many external factors that are not controlled by the experimenter. Nevertheless, this method can be useful for identifying correlations between variables and for formulating hypotheses that can be tested later using other experimental methods.

Example 87

When studying two variables simultaneously, such as the population and surface area of municipalities in a country, it is important to determine if there is a correlation between these variables. If one variable depends on the other, then they are correlated. In this case, the data can be represented as a scatter plot, where each point represents a pair of data (x, y).

In cases where it is not possible to fix one of the variables, we cannot study the functional relationship between the variables. However, we can still determine whether there is a linear relationship between the variables. For this, we use the coefficient of linear correlation, which measures the strength and direction of the linear relationship between the variables.

There are three major cases to consider when studying the correlation between two variables. The first case is when the scatter plot does not show any particular shape, which means that there is no linear relationship between the variables. In the second case, the scatter plot is linear but not parallel to the axes, indicating a linear correlation but with a slope different from zero. Finally, in the third case, the scatter plot is not linear, suggesting a non-linear relationship between the variables.

Indeed, studying the correlation between two variables allows us to understand how these variables are related to each other. If the variables are correlated, we can use this information to predict one variable from the other or to understand how changes in one variable affect the other variable.

9.3.1 The has no particular shape

When a scatter plot does not show any particular shape, it means that the points are scattered randomly without following a specific trend or direction. Thus, it is difficult to find a relationship between the variables x and y. Indeed, the absence of a particular shape can be due to several reasons, such as the presence of outliers, the non-linearity of the relationship between the variables, or the random variability. In this case, it can be concluded that the variables x and y are independent of each other. In other words, the variations in variable x have no effect on variable y and vice versa. However, it is

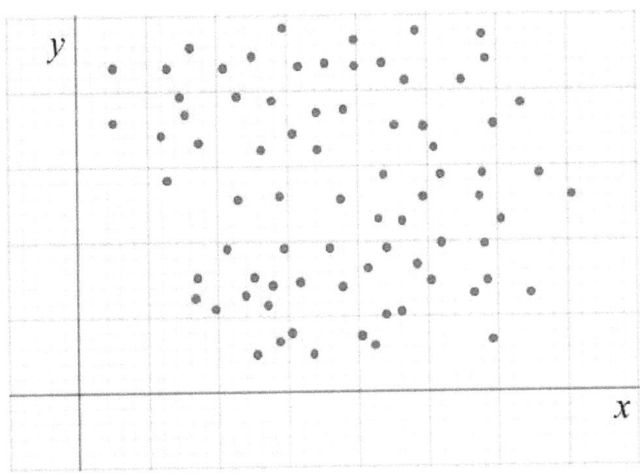

important to note that the absence of a linear relationship does not necessarily mean the absence of a relationship between the variables. It is possible that there is a non-linear relationship between the variables, or that other variables influence their relationship.

We should note that when the scatter plot does not show any particular shape, there is no correlation between the variables x and y. This result can be used to eliminate certain hypotheses and guide subsequent statistical analyses.

9.3.2 The cloud is rectilinear and not parallel to the axes

When the scatter plot is linear but not parallel to the axes, this indicates a linear correlation between the variables x and y. In other words, there is a functional relationship between the two variables. This means that as the variable x increases, the variable y tends to increase (or decrease, depending on the direction of the correlation). This relationship can be either positive or negative, depending on the direction of the slope of the scatter plot.

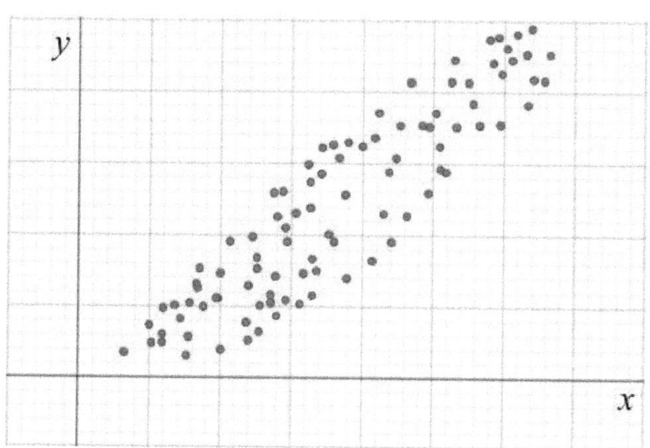

Specifically, the slope of the scatter plot represents the strength of the linear relationship between the variables. If the slope is steep, this indicates a strong correlation, while if the slope is shallow, this indicates a weaker correlation. Similarly, the direction of the slope (up or down) indicates whether the correlation is positive or negative.

It is important to note that even in this case, there may be random variations in the data, which can create noise in the scatter plot. However, if the linear relationship is strong enough, it can be used to predict the values of one variable from the other. For example, if the correlation is positive, one can expect that the variable y will increase as the variable x increases.

We should note that when a scatter plot is linear but not parallel to the axes, this indicates a linear correlation between the variables x and y, with a direction and strength that can be evaluated from the slope of the scatter plot.

9.3.3 The cloud is not rectilinear

When the scatter plot is not straight, it indicates that the relationship between the variables x and y is not linear. In other words, the variations in x are not proportional to the variations in y. In this case, it is important to study the shape of the scatter plot to determine the nature of the relationship between the variables.

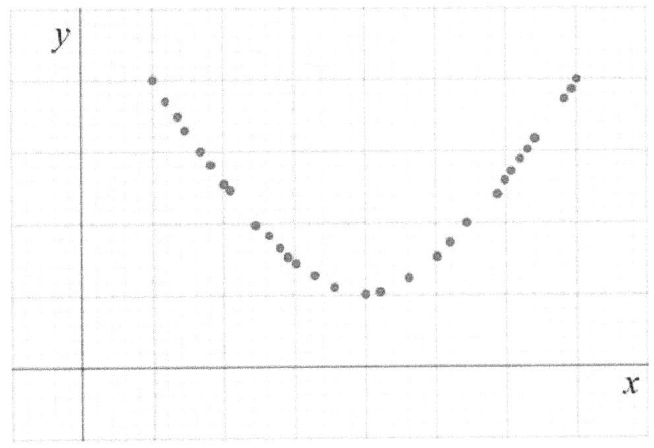

There are several shapes of non-linear scatter plots, such as U-shaped, J-shaped, S-shaped, or scatter plots presenting a bell-shaped curve. In these cases, it is possible that a mathematical transformation of the variables (such as the introduction of logarithms, powers or roots) may help find a functional relationship between x and y.

However, it is important to note that the absence of a linear relationship does not necessarily mean the absence of a relationship between the variables. It is possible that there is a complex non-linear relationship between the variables, which requires more advanced statistical methods to be identified and analyzed.

In summary, when the scatter plot is not straight, it is necessary to explore in more detail the relationship between the variables x and y using appropriate mathematical and statistical tools.

9.4 Linear Adjustment

Linear regression is a very useful method for modeling the relationship between two quantitative variables. This method is particularly suitable when the scatter plot suggests a linear relationship between the variables. Indeed, in this case, it is possible to draw a line that passes as close as possible to all points, which allows for a simplified representation of the relationship between the variables.

There are several methods to obtain a linear regression. The empirical method consists in drawing a line that passes as close as possible to the points, based on visual intuition rather than on a rigorous mathematical method. The moving averages method, on the other hand, consists in calculating averages on subsets of neighboring points to obtain a line that represents the overall trend of the scatter plot. The MAYER method is a more sophisticated method that allows to obtain a line by minimizing a specific cost function. Finally, the least squares method is a widely used method to obtain a linear regression. It consists in minimizing the sum of the squares of the residuals between the points and the regression line.

9.4.1 Methods of Moving Averages

The moving average method is a commonly used technique to attenuate random variations in a scatter plot and to highlight a general trend. This method is particularly useful when significant noise or fluctuations are observed around the general trend in the data.

The principle of the moving average method consists of replacing certain terms in the series with the arithmetic mean of a group of n terms. In other words, we calculate the average of each group of n consecutive terms and replace it with the central value of those terms. The choice of the group size n depends on the frequency of variation in the data and the desired precision for the general trend.

The result of the moving average method is a smoothed series that allows for a better visualization of the general trend in the data by eliminating random fluctuations. This technique is particularly useful for analyzing time series, where random fluctuations can make it difficult to detect longer-term trends.

Example 88

Consider the table opposite providing the monthly gold expertise for the year 2021 declared by the national counters of an African country in $ dollars.

Months	Values in $
January	624,578
February	107,077
March	646,409
April	586,660
May	886,767
June	1,181,417
July	1,060,053
August	1,116,891
September	880,692
October	867,777
November	361,990
December	242,199
Total	8,562,510

Considering the arithmetic average for one quarter, three months (January, February, March), then the one corresponding to the months (February, March, April) and then the one corresponding to the months (March, April, but) etc. We obtain the following table.

Months	Valuers in $	Moving Average
January	624,578	
February	107,077	459,354.67
March	646,409	446,715.33
April	586,660	706,612.00
May	886,767	884,948.00
June	1,181,417	1,042,745.67
July	1,060,053	1,119,453.67
August	1,116,891	1,019,212.00
September	880,692	955,120.00
October	867,777	703,486.33
November	361,990	490,655.33
December	242,199	
Total	8,562,510	

By plotting the graph of the smoothed series, we can observe a general upward trend in the value of gold in the middle of the year 2021, specifically between the months of May, June, and July.

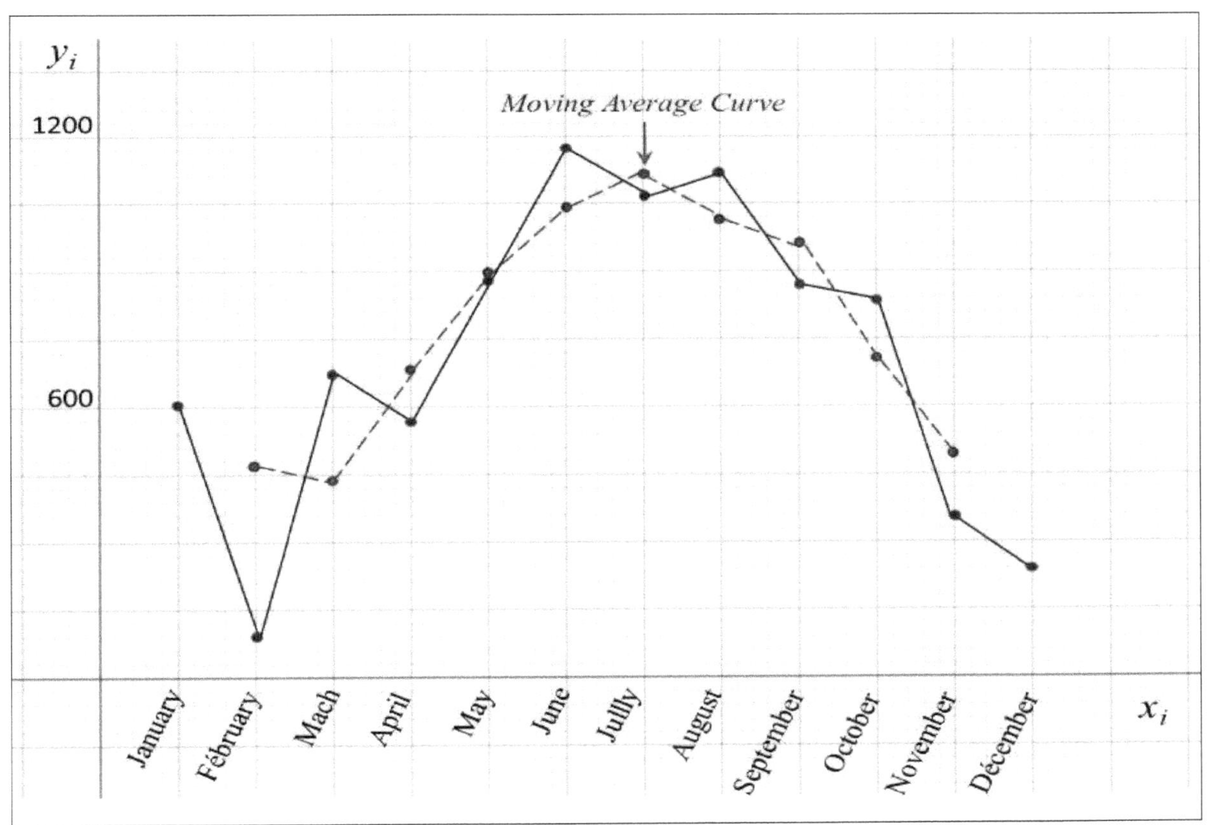

9.4.2 Empirical Method

The empirical method is a commonly used technique in statistics and social sciences for studying complex phenomena. Although considered less rigorous than other methods such as statistical methods, it can still be very useful in giving a general overview of a phenomenon.

The empirical method consists of drawing a line *d* that best represents the phenomenon being studied. To do this, it is important to carefully choose the position of the line so that it best reflects the trend of the data. Then, two points *A* and *B*, with respective coordinates (x_1, y_1) and (x_2, y_2) are taken, where x_1 represents the years and y_1 represents the values of the phenomenon. Using these points, we can write the equation of the corresponding line:

$$y - y_1 = \frac{y_2 - y_1}{x_2 - x_1}(x - x_1)$$

This equation can then be used to predict the values of the phenomenon from the corresponding years. Although the empirical method is not as precise as other more advanced methods, it can provide satisfactory results for relatively simple data. It is also useful for getting a general idea of the trend of the data and can be used to determine if data should be analyzed in more detail using more advanced methods.

Example 89

In this example, we have data on the population of a large city from 2001 to 2020. Using the empirical method, we will draw the regression line by choosing two points on the curve. These points will be chosen arbitrarily, but we will choose points that represent the trend of the curve well.

Year (xi)	Population (yi) in thousands
2001	22.30
2002	26.60
2003	30.00
2004	31.50
2005	36.60
2006	38.50
2007	40.30
2008	43.30
2009	50.00
2010	51.60
2011	54.60
2012	61.50
2013	62.50
2014	63.30
2015	66.20
2016	68.30
2017	70.80
2018	71.00
2019	74.80
2020	78.20
Total	1,041.00

Next, we use the equation of the regression line to predict the value of the population for the missing years.

a. Point cloud corresponding to the evolution of the population of the big city from 2001 to 2020 is:

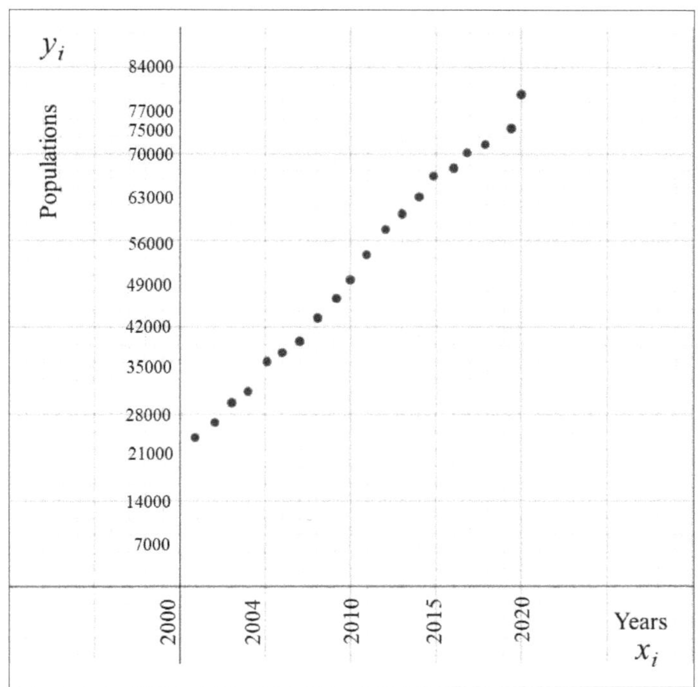

By examining the scatter plot, we can observe that the points are not randomly distributed, but they follow a clear trend by forming a line. This observation indicates a gradual increase or decrease in the population of the city over the years.

b. Determination of the regression line of the scatter plot.

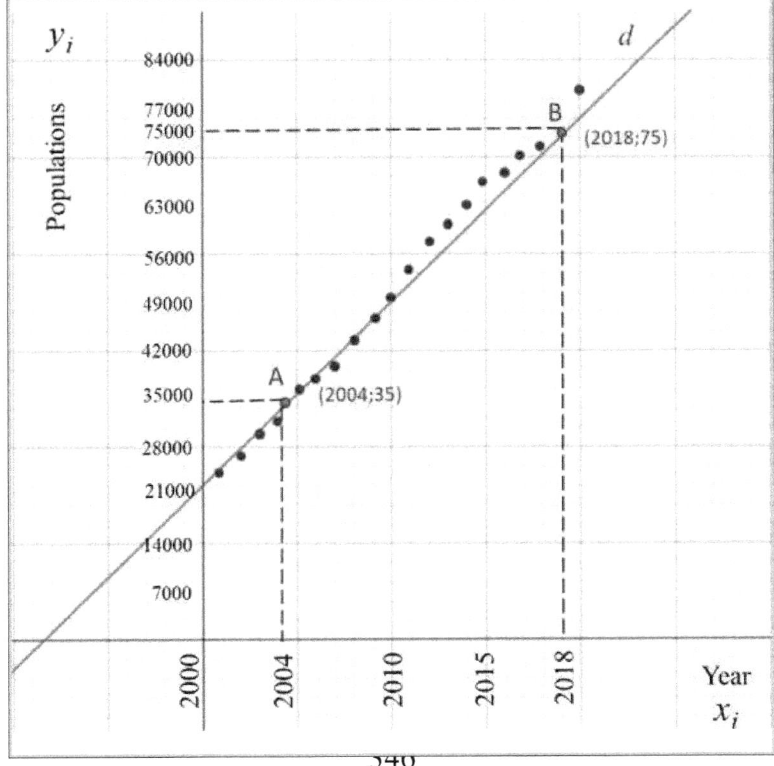

We have plotted these points and drawn the regression line d that best represents the trend of events. We will determine the equation of this line by considering two points A and B with respective coordinates ("2004", "35") and (2004 ; 35) et (2018 ; 75) that best fit the line. This equation allows us to model the relationship between the years and the population of the city, and thus better understand the general trend.

$$y - y_1 = \frac{y_2 - y_1}{x_2 - x_1}(x - x_1)$$

$$y - 35 = \frac{75 - 35}{2018 - 2004}(x - 2004)$$

$$y = \frac{20}{7}(x - 2004) + 35$$

$$= \frac{20x}{7} - \frac{39{,}835}{7}$$

c. Projection of the population

By using the equation of the regression line plotted on the scatter plot, the projection of the population of the city at a specific given year can be obtained without any experimentation. With this equation, we can estimate the future population of the city over time, without the need for conducting new experiments.

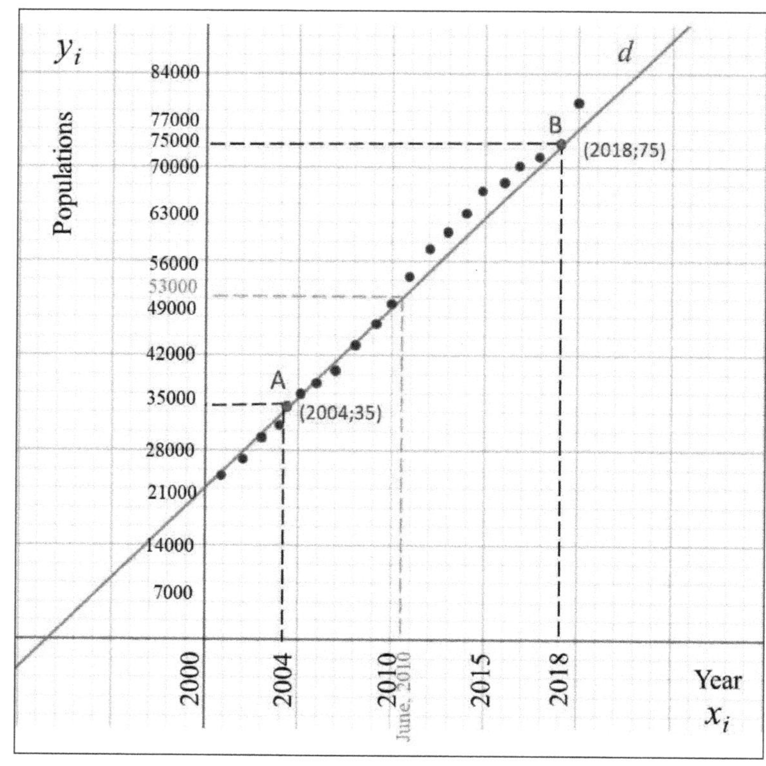

Thus, in the middle of the year (June) 2010, the population was:

$$y = \frac{20(2{,}010.5)}{7} - \frac{39{,}835}{7}$$
$$= 53{,}571$$

So, we had 53,571 people in this city (as there is a ratio of 1000).

Prediction of the population:

$$y = \frac{20(2\,050)}{7} - \frac{39{,}835}{7}$$
$$= 166{,}428$$

It can therefore be predicted that by the year 2,050 the population of this city would be 166,428 people.

The empirical method was used to predict the population of the city in 2050. According to the equation of the regression line, the population of the city will reach approximately 166,428 people by 2050. However, it is important to keep in mind that the empirical method is less precise than other more advanced modeling methods. It should only be used to give a general idea of the trend of the data.

9.4.3 MAYER Method

The MAYER method, also known as the half- MAYER method, is a useful tool for visualizing the correlation between two variables in a scatter plot. It can be used to predict the values of the dependent variable for values of the independent variable that are not present in the scatter plot, that is, to study the evolution of one variable as a function of another.

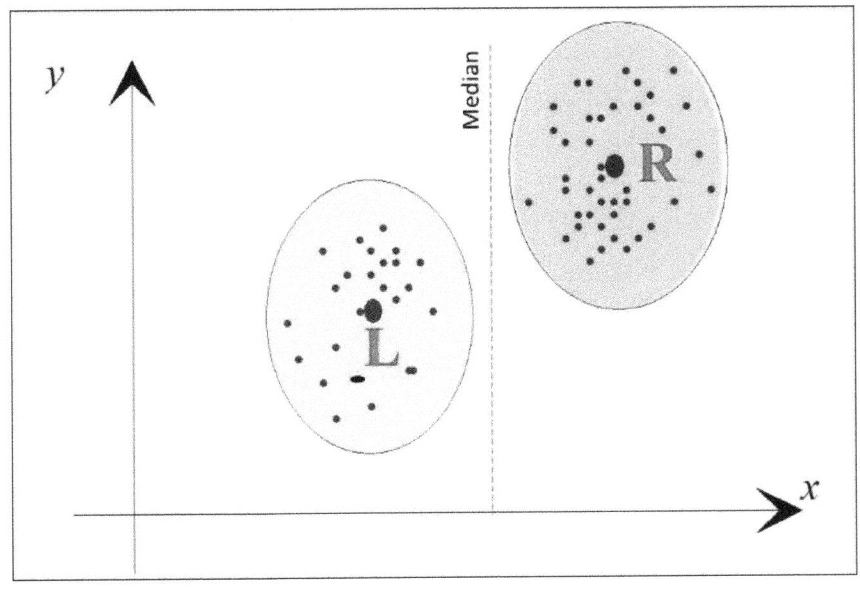

Indeed, it is a graphical representation technique that allows to divide a scatter plot into two subplots of equal importance. The division of the plot into two sub-plots is based on the position of the abscissa values of the points. The points to the right of the median vertical line (R) have an abscissa value greater than those to the left of the line (L). Thus, the plot can be divided into two equal sub-plots, each containing the same proportion of points but with a different distribution of abscissa values.

The MAYER method allows for the study of the correlation between the two variables represented on the graph axes by evaluating the regression line slopes for each sub-plot. The correlation coefficients calculated for each of the two sub-plots can be compared to evaluate the homogeneity of the correlation in the entire plot.

The MAYER adjustment line is a line that passes through points L and R with respective coordinates $(\bar{x}_L, \bar{y}_L)$ et $(\bar{x}_R, \bar{y}_R)$. This line is calculated using the mean values of the coordinates of the points in the left and right sub-clouds. It is calculated using the following formula:

$$y - \bar{y}_L = \frac{\bar{y}_R - \bar{y}_L}{\bar{x}_R - \bar{x}_L}(x - \bar{x}_L)$$

It is important to note that the MAYER method may be less precise than other more advanced methods for analyzing correlation, such as methods based on generalized linear models. However, this method remains useful for obtaining a general idea of the distribution of data in a scatter plot and for studying the overall trends of correlation.

Example 90

Let's consider **example 89** from **section 9.4.2**, which is about the evolution of population in a big city from 2001 to 2020, expressed in thousands of people. To apply the MAYER method, we will first plot a scatterplot representing the population data against time. Then, we will divide the scatterplot into two equal subgroups by drawing a vertical line through the middle of the scatterplot. The points to the right of this line will belong to the subgroup (R) while the points to the left will belong to the subgroup (L). Using the mean values of the coordinates of each subgroup, we will then calculate the MAYER regression line.

a. Let's plot a scatterplot representing the population data against time and divide the cloud into two sub-clouds of equal importance

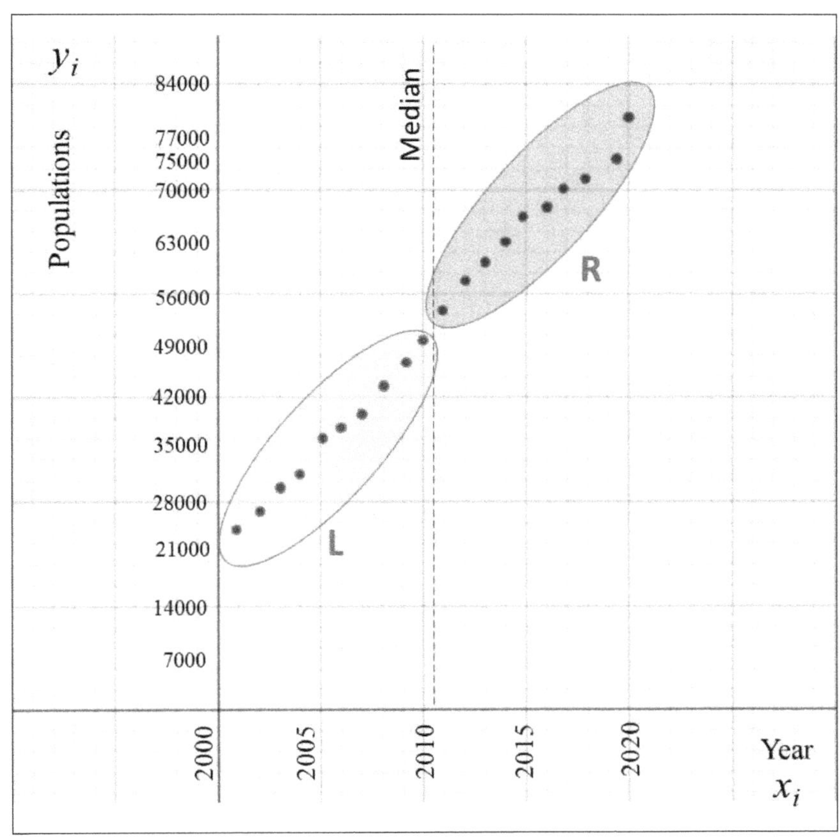

Let's calculate the coordinates of L and R by calculating the mean values of the coordinates of each subgroup:

Point L

	Year	Population
	2001	22.30
	2002	26.60
	2003	30.00
	2004	31.50
	2005	36.60
	2006	38.50
	2007	40.30
	2008	43.30
	2009	50.00
	2010	51.60
Average	2005.5	37.07

Point R

	Year	Population
	2011	54.60
	2012	61.50
	2013	62.50
	2014	63.30
	2015	66.20
	2016	68.30
	2017	70.80
	2018	71.00
	2019	74.80
	2020	78.20
Average	2015.5	67.12

$(\bar{x}_L = 2005,5 \ ; \ \bar{y}_L = 37{,}07)$ $(\bar{x}_R = 2015,5 \ ; \ \bar{y}_R = 67{,}12)$

b. Let us now calculate the MAYER adjustment line that passes through the points $(\bar{x}_L, \bar{y}_L)$ et $(\bar{x}_R, \bar{y}_R)$:

$$y - 37.07 = \frac{67.12 - 37.07}{2015.5 - 2005.5}(x - 2005.5)$$

$$y = 3.005(x - 2005.5) + 37.07$$

$$= 3.005x - 5{,}989.4575$$

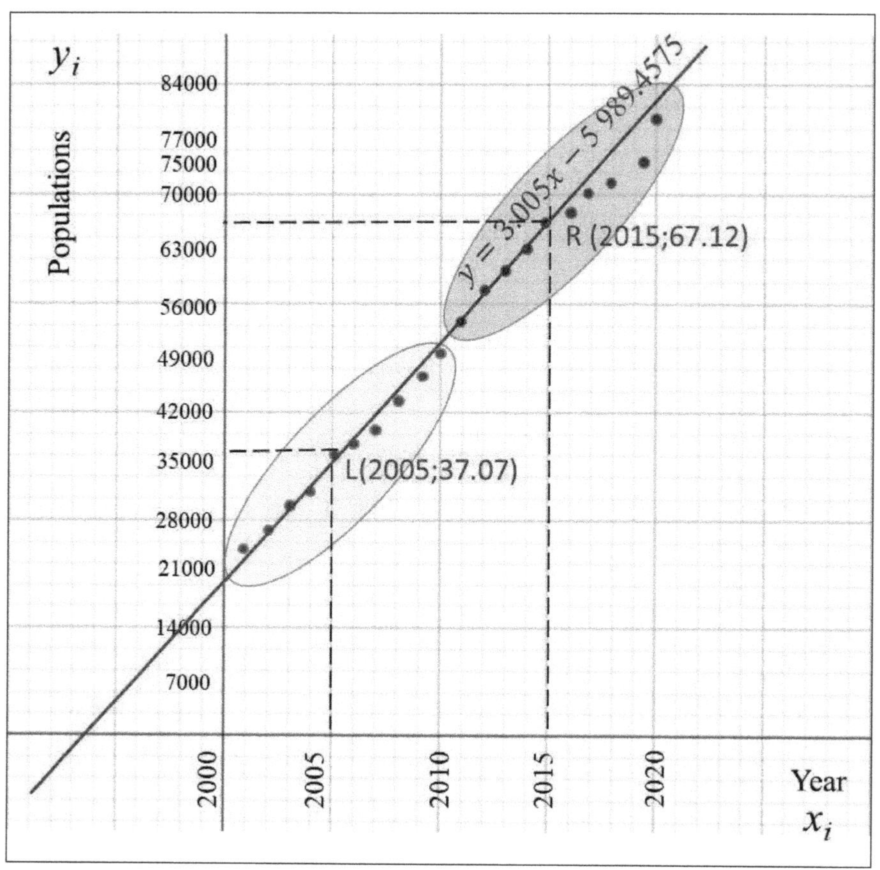

c. Let's make a projection of the future population of the city:

Let's recalculate the population for the middle of the year (June) 2010

$$y = 3.005(2010.5) - 5{,}989.4575 = 52.095$$

According to the MAYER method, the population of the city in June 2010 was 52,095 people. Using this data and the linear trend of the MAYER line of best fit, we can estimate that the population of the city in 2050 will be 170,925 people.

9.4.4 Least Squares Method

9.4.4.1 Least squares regression line

The least squares regression line is a statistical method used to find the best linear approximation of data in a scatter plot. This involves finding the best possible line for a set of data. The line is obtained by minimizing the sum of the squared differences between the observed values and the values predicted by the line, that is, the predicted values of the dependent variable y as a function of the independent variable x.

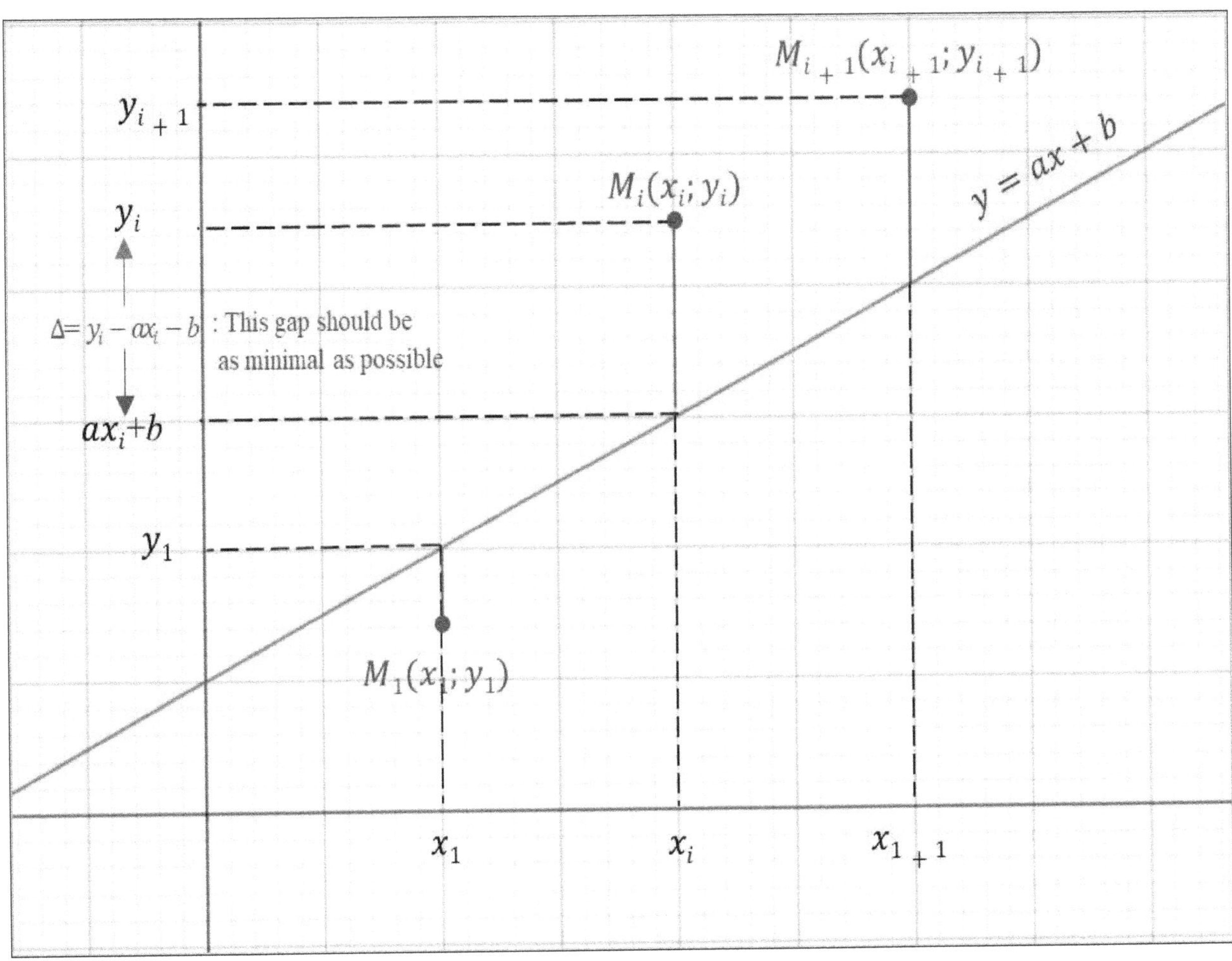

Specifically, given a set of data points, the least squares regression line, considered the best line in the representing plane, is the regression line of the form $y = ax + b$, where a and b are coefficients determined in such a way as to minimize the sum of the squared differences of the observed values y_i at points $ax_i + b$ belonging to it.

More precisely, the coefficients a and b are determined to minimize the following quantity:

$$d(a,b) = \sum_{i}^{p}\sum_{j}^{q} f_{ij}(y_j - ax_i - b)^2$$

To find the optimal values of a and b, we will calculate the partial derivatives of the function $\varphi(a,b)$ with respect to a and b and set them equal to zero.

Using the derivative with respect to b.

$$\frac{d}{db}[\varphi(a,b)] = -2\sum_{i}^{p}\sum_{j}^{q} f_{ij}(y_j - ax_i - b)$$

$$= -2\left(\sum_{i}^{p}\sum_{j}^{q} f_{ij}y_j - a\sum_{i}^{p}\sum_{j}^{q} f_{ij}x_i - b\sum_{i}^{p}\sum_{j}^{q} f_{ij}\right)$$

Knowing that:

$$\sum_{i}^{p}\sum_{j}^{q} f_{ij}x_i = \bar{\bar{x}}, \quad \sum_{i}^{p}\sum_{j}^{q} f_{ij}y_j = \bar{\bar{y}} \quad \text{et} \quad \sum_{i}^{p}\sum_{j}^{q} f_{ij} = 1$$

Thus,

$$\frac{d}{db}[\varphi(a,b)] = -2(\bar{\bar{y}} - a\bar{\bar{x}} - b) = 0$$

To make this expression zero, we must have:

$$\bar{\bar{y}} - a\bar{\bar{x}} - b = 0$$

Then,

$$b = \bar{\bar{y}} - a\bar{\bar{x}}$$

with

$\bar{\bar{x}}$ = marginal average of x

$\bar{\bar{y}}$ = marginal average of y

Using the derivative with respect to a.

$$\frac{d}{da}[\varphi(a,b)] = -2\sum_{i}^{p}\sum_{j}^{q} f_{ij}x_i(y_j - ax_i - b) = 0$$

$$= -2\left(\sum_{i}^{p}\sum_{j}^{q} f_{ij}x_i y_j - a\sum_{i}^{p}\sum_{j}^{q} f_{ij}x_i^2 - b\sum_{i}^{p}\sum_{j}^{q} f_{ij}x_i\right)$$

To make this expression zero, we must have:

$$\sum_{i}^{p}\sum_{j}^{q} f_{ij}x_i y_j - a\sum_{i}^{p}\sum_{j}^{q} f_{ij}x_i^2 - b\bar{\bar{x}} = 0$$

Replacing b with its value $\bar{\bar{y}} - a\bar{\bar{x}}$ obtained by setting the partial derivative of $\varphi(a, b)$ with respect to b to zero:

$$\sum_{i}^{p}\sum_{j}^{q} f_{ij}x_i y_j - a\sum_{i}^{p}\sum_{j}^{q} f_{ij}x_i^2 - (\bar{\bar{y}} - a\bar{\bar{x}})\bar{\bar{x}} = 0$$

$$\sum_{i}^{p}\sum_{j}^{q} f_{ij}x_i y_j - a\sum_{i}^{p}\sum_{j}^{q} f_{ij}x_i^2 - \bar{\bar{x}}\bar{\bar{y}} - a\bar{\bar{x}}^2 = 0$$

$$\sum_{i}^{p}\sum_{j}^{q} f_{ij}x_i y_j - \bar{\bar{x}}\bar{\bar{y}} - a\left(\sum_{i}^{p}\sum_{j}^{q} f_{ij}x_i^2 - \bar{\bar{x}}^2\right) = 0$$

Then,

$$\boxed{a = \frac{\sum_i^p \sum_j^q f_{ij}x_i y_j - \bar{\bar{x}}\bar{\bar{y}}}{\sum_i^p \sum_j^q f_{ij}x_i^2 - \bar{\bar{x}}^2}}$$

Let's calculate the covariance, which is a measure of how two variables vary together:

$$\mathrm{COV}(X, X) = V(X) = \sum_{i}^{p}\sum_{j}^{q} f_{ij}x_i^2 - \bar{\bar{x}}^2$$

$$\mathrm{COV}(X, Y) = \sum_{i}^{p}\sum_{j}^{q} f_{ij}(x_i - \bar{\bar{x}})(y_j - \bar{\bar{y}})$$

$$= \sum_{i}^{p}\sum_{j}^{q} f_{ij}x_i y_j - \bar{\bar{x}}\bar{\bar{y}}$$

It follows that:

$$\begin{cases} a = \dfrac{\text{COV}(X,Y)}{V(X)} \\ b = \bar{\bar{y}} - \left(\dfrac{\text{COV}(X,Y)}{V(X)}\right)\bar{\bar{x}} \end{cases}$$

Thus, the linear regression line using the method of least squares has the equation:

$$\boxed{y = \left(\frac{\text{COV}(X,Y)}{V(X)}\right)x + \left[\bar{\bar{y}} - \left(\frac{\text{COV}(X,Y)}{V(X)}\right)\bar{\bar{x}}\right]}$$

With

$$\begin{cases} \text{COV}(X,Y) = \dfrac{1}{n}\displaystyle\sum_{i=1}^{i=n} x_i y_i - \bar{\bar{x}}\bar{\bar{y}} \\ \text{V}(X) = \text{COV}(X,X) = \dfrac{1}{n}\displaystyle\sum_{i=1}^{i=n} x_i^2 - \bar{\bar{x}}^2 \end{cases} \quad \text{and} \quad \begin{cases} \bar{\bar{x}} = \dfrac{1}{n}\displaystyle\sum_{i=1}^{i=n} x_i \\ \bar{\bar{y}} = \dfrac{1}{n}\displaystyle\sum_{i=1}^{i=n} y_i \end{cases}$$

Example 91

Let's take **example 90** from **section 9.4.3**, which deals with the population trend of a large city from 2001 to 2020. We will calculate the equation of this line using the method of least squares.

a. The following table shows the data required to solve this exercise:

x_i	y_j	$x_i y_j$	x_i^2	y_i^2
2001	22.30	44,622.30	4,004,001	497.29
2002	26.60	53,253.20	4,008,004	707.56
2003	30.00	60,090.00	4,012,009	900.00
2004	31.50	63,126.00	4,016,016	992.25
2005	36.60	73,383.00	4,020,025	1,339.56
2006	38.50	77,231.00	4,024,036	1,482.25
2007	40.30	80,882.10	4,028,049	1,624.09
2008	43.30	86,946.40	4,032,064	1,874.89
2009	50.00	100,450.00	4,036,081	2,500.00
2010	51.60	103,716.00	4,040,100	2,662.56
2011	54.60	109,800.60	4,044,121	2,981.16
2012	61.50	123,738.00	4,048,144	3,782.25
2013	62.50	125,812.50	4,052,169	3,906.25
2014	63.30	127,486.20	4,056,196	4,006.89
2015	66.20	133,393.00	4,060,225	4,382.44
2016	68.30	137,692.80	4,064,256	4,664.89
2017	70.80	142,803.60	4,068,289	5,012.64
2018	71.00	143,278.00	4,072,324	5,041.00
2019	74.80	151,021.20	4,076,361	5,595.04
2020	78.20	157,964.00	4,080,400	6,115.24
40210	1,041.90	2,096,689.90	80,842,870	60,068.25

b. Determination of the linear adjustment line:

$$\bar{\bar{x}} = \frac{1}{n}\sum_{i=1}^{i=20} x_i = \frac{40{,}210}{20} = 2{,}010.5$$

$$\bar{\bar{y}} = \frac{1}{n}\sum_{i=1}^{i=20} y_i = \frac{1{,}041.9}{20} = 52.095$$

$$\bar{\bar{x}}\bar{\bar{y}} = 2{,}010.5 \text{ x } 52.095 = 104{,}736.99$$

$$V(X) = \frac{1}{n}\sum_{i=1}^{i=20} x_i^2 - \bar{\bar{x}}^2 = \frac{80{,}842.870}{20} - (2{,}010.5)^2$$

$$= 33.25$$

$$V(Y) = \frac{1}{n}\sum_{i=1}^{i=20} x_i^2 - \bar{\bar{x}}^2 = \frac{60{,}068.25}{20} - (52.095)^2$$

$$= 290.44$$

$$\text{COV}(X,Y) = \frac{1}{n}\sum_{i=1}^{i=20} x_i y_i - \bar{\bar{x}}\bar{\bar{y}} = \frac{2{,}096{,}689.90}{20} - 104{,}736.99$$

$$= 97.49$$

The coefficient of the adjustment line is:

$$a = \frac{\text{COV}(X,Y)}{V(X)} = \frac{97.49}{33.25}$$

$$= 2.93$$

The adjustment line is, therefore:

$$y = 2.93x + b$$
$$b = y - 2.93x$$

The average point$(\bar{\bar{x}}, \bar{\bar{y}})$ is on this line:

$$b = 52.095 - 2.93(2{,}010.5)$$
$$= -5{,}842.69$$

The adjustment line has as its equation:

$$y = 2.93x - 5{,}842.69$$

Where x is the year and y is the population in thousands of inhabitants.

c. Population projection

The population is recalculated for the middle of the year (June) 2010:

$$y = 2.93(2010.5) - 5{,}842.691 = 52.095$$

The least squares method estimates the population of this period at 52,095 (as there is a ratio of 1000) people.

Similarly, it can be predicted that by the year 2,050 the population of this city would be 167,909 people.

9.4.4.2 Linear Correlation Coefficient (ρ)

The linear correlation coefficient is a useful measure for statisticians, researchers, and professionals who need to analyze data. It detects linear relationships between variables by determining if two quantitative variables are linearly correlated, which can be important for prediction, modeling, and decision-making.

The correlation coefficient is a number between -1 and 1, which expresses the strength and direction of the relationship between the two variables. If the coefficient is close to 1, it means that there is a strong positive correlation between the two variables, that is, when one increases, the other also increases. If the coefficient is close to -1, it means that there is a strong negative correlation between the two variables, that is, when one increases, the other decreases. If the coefficient is close to zero, it means that there is no linear correlation between the two variables.

Symbolized by the letter ρ, the formula for the linear correlation coefficient uses the covariance and standard deviations of the two variables where the covariance is divided by the product of the standard deviations:

$$\rho = \frac{\sum_{i=1}^{p} \sum_{j=1}^{q} f_{ij}(x_i - \bar{\bar{x}})(y_j - \bar{\bar{y}})}{\sqrt{\sum_{i=1}^{p} f_{i.}(x_i - \bar{\bar{x}})^2} \sqrt{\sum_{j=1}^{q} f_{.j}(y_j - \bar{\bar{y}})^2}}$$

Hence the formula:

$$\boxed{\rho = \frac{\mathrm{COV}(X,Y)}{\sigma_x \sigma_y} = \frac{\mathrm{COV}(X,Y)}{\sqrt{V(X)V(Y)}}}$$

$$-1 \leq \rho \leq 1$$

To calculate the linear correlation coefficient in practice, a data table containing the values of the two variables is needed. The data must be paired, meaning that each value of one variable must be associated with a value of the other variable. From this table, the mean and standard deviation of each variable, as well as the covariance, can be calculated.

However, it should be noted that the linear correlation coefficient only measures linear relationships between variables and does not account for non-linear or causal relationships.

RULE OF THUMB

If

$\rho = 0$: There is no correlation (the closer ρ is to 0, the weaker the linear relationship between the variables).
$\rho = 1$: There is a positive functional link (the closer ρ is to 1, the stronger the positive linear relationship between the variables)
$\rho = -1$: There is a negative functional link (the closer ρ is to -1, the stronger the negative linear relationship between the variables)

Example 92

Let's take **example 91** from **section 9.4.4.1**, which deals with the population trend of a large city from 2001 to 2020. We will calculate and interpret the value of the linear correlation coefficient.

$$\rho = \frac{\text{COV}(X, Y)}{\sigma_x \sigma_y} = \frac{\text{COV}(X, Y)}{\sqrt{V(X)V(Y)}}$$

The values of these terms have already been calculated in section **9.4.4.1**:

$$\text{COV(X,Y)} = 97.49$$

$$\text{V(X)} = 33.25$$

$$\text{V(Y)} = 290.03$$

Then,

$$\rho = \frac{\text{COV}(X, Y)}{\sqrt{V(X)V(Y)}} = \frac{97.49}{\sqrt{(33.25)(290.44)}}$$

$$= \frac{97.49}{98.27}$$

$$= 0.9920$$

INTERPRETATION OF THE RESULTS

The interpretation of the results shows that the linear correlation coefficient of 0.992 indicates a very strong linear correlation, meaning that there is a very strong positive linear relationship between the variables, namely that the population almost constantly increases over time.

In other words, this means that if time increases by a certain amount, then the population will also increase by a similar amount, to some extent. This strong linear correlation can be useful for predicting future population based on time, provided that the same conditions remain in the future.

This data can be utilized in several ways. Firstly, they can help to predict future population behavior based on time. By using the established linear relationship between the two variables, projections and forecasts can be made for the coming years. This can be useful for planners and decision-makers who need to make decisions regarding resource allocation or the development of public policies. Similarly, these data can be used to perform sensitivity analyses that examine the impact of various variables on the linear relationship between population and time. This can help identify factors that have the most impact on population growth, which can be useful in guiding policies and programs aimed at regulating demographic growth.

This correlation coefficient can also be determined from the data extracted from the following table:

x_i	y_j	$(x - \bar{\bar{x}})$	$(y - \bar{\bar{y}})$	$(x - \bar{\bar{x}})(y - \bar{\bar{y}})$	$(x - \bar{\bar{x}})^2$	$(y - \bar{\bar{y}})^2$
2001	22.30	-9.50	-29.80	283.05	90.25	887.74
2002	26.60	-8.50	-25.50	216.71	72.25	650.00
2003	30.00	-7.50	-22.10	165.71	56.25	488.19
2004	31.50	-6.50	-20.60	133.87	42.25	424.15
2005	36.60	-5.50	-15.50	85.22	30.25	240.10
2006	38.50	-4.50	-13.60	61.18	20.25	184.82
2007	40.30	-3.50	-11.80	41.28	12.25	139.12
2008	43.30	-2.50	-8.79	21.99	6.25	77.35
2009	50.00	-1.50	-2.09	3.14	2.25	4.39
2010	51.60	-0.50	-0.49	0.25	0.25	0.25
2011	54.60	0.50	2.51	1.25	0.25	6.28
2012	61.50	1.50	9.41	14.11	2.25	88.45
2013	62.50	2.50	10.41	26.01	6.25	108.26
2014	63.30	3.50	11.21	39.22	12.25	125.55
2015	66.20	4.50	14.11	63.47	20.25	198.95
2016	68.30	5.50	16.21	89.13	30.25	262.60
2017	70.80	6.50	18.71	121.58	42.25	349.88
2018	71.00	7.50	18.91	141.79	56.25	357.40
2019	74.80	8.50	22.71	192.99	72.25	515.52
2020	78.20	9.50	26.11	248.00	90.25	681.47
40210	1041.9	0	0.00	1949.95	665.00	5790.47

$$\rho = \frac{\mathrm{COV}(X,Y)}{\sigma_x \sigma_y} = \frac{\sum_{i=1}^{p}[(x_i - \bar{\bar{x}})(y_i - \bar{\bar{y}})]}{\sqrt{\sum_{i=1}^{p}(x_i - \bar{\bar{x}})^2 * \sum_{i=1}^{p}(y_i - \bar{y})^2}}$$

$$= \frac{1949.95}{\sqrt{(665) * (5790.47)}}$$

$$= \frac{1949.95}{\sqrt{3,850,662.55}}$$

$$= \frac{1949.95}{1962.31}$$

$$= 0.993$$

9.5 Cases where one can fall back to a linear adjustment

In some situations, it may be difficult to find an obvious linear relationship between two variables when they are plotted on a graph. However, it is sometimes possible to transform the data to achieve a linear fit by changing the scale of one or both variables. For example, by applying a logarithmic transformation to one or both variables, one can often obtain a scatter plot that exhibits an apparent linear relationship. This can be very useful for performing statistical analyses and for understanding the relationship between the variables in question.

x_i	y_i
2004	42.10
2005	34.90
2006	37.20
2007	76.00
2008	52.20
2009	23.80
2010	46.70
2011	85.00
2012	82.70
2013	104.00
2014	264.00
2015	2,154.00
2016	4,333.00
2017	8,827.80
2018	6,029.70
2019	8,000.00
2020	9,797.00

This technique is particularly useful in scientific fields, where the relationships between variables can be complex and non-linear. For example, in biology, increasing the dose of a drug may not have a linear relationship with its effectiveness. However, by using a logarithmic scale to represent the dose, a linear relationship between dose and effectiveness can often be obtained.

Similarly, in economics, the relationship between two variables such as income and expenditure may not be linear. By using a logarithmic scale to represent the values, a linear relationship between these two variables can often be obtained.

Example 93

The table provided allows us to track the evolution of the inflation rate in a developing country over several years. To analyze the trends more visually, we will plot a scatter plot using the data provided: the years will be represented on the horizontal axis (x_i) and the inflation rate on the vertical axis (y_i). Then, we will determine the linear regression line, which will model the relationship between these two variables.

To simplify the calculations, we will choose the year 2004 as the reference point and designate it as year 1. Using the provided data, we can then calculate the values of the other years relative to this reference year. Then, by plotting the linear regression line, we can visualize the general trend of inflation rate over this period. This analysis can be useful for making decisions in economic and financial policy in this country.

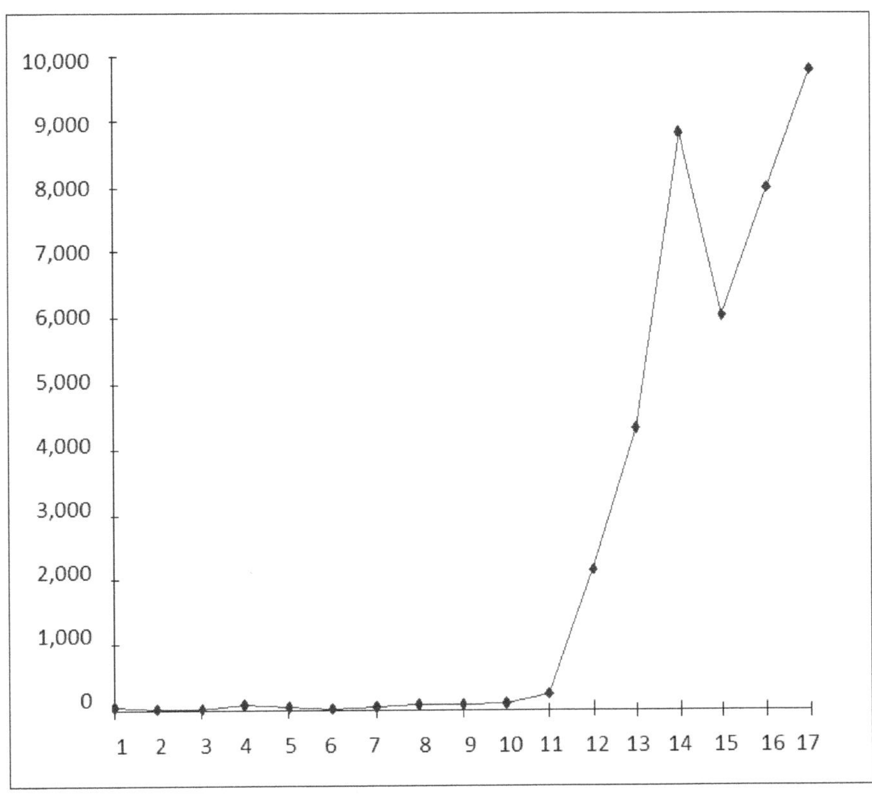

The shape of the curve connecting the points shows that a linear fit for the inflation rate of this country is not possible. For this type of series, it is recommended to make a scale change and represent the scatter plot on a semi-logarithmic graph. The x-axis will have a normal metric scale, while the y-axis will carry the value $Y = \log_{10}(y)$. This will give us the following scatter plot:

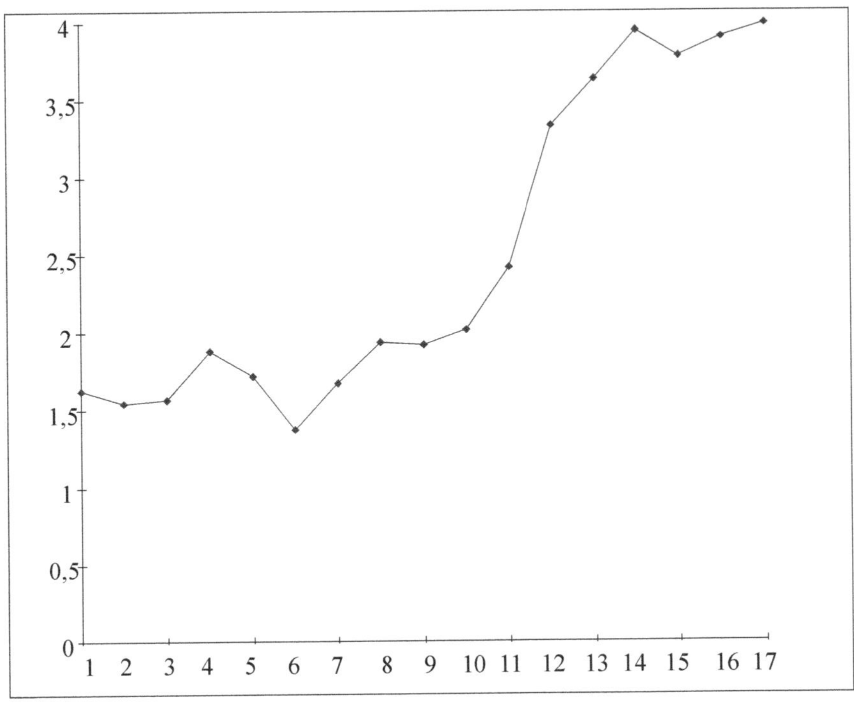

On this graph, we can see that a linear fit is almost possible for the point cloud $(x_i, Y_i = \log(y_i))$. Although the relationship between the two variables is not perfectly linear, it can be approximated by a linear regression line. This line will allow us to model the relationship between years and inflation rates and can be used to make predictions about future inflation rate values.

x_i	y_i	$Y_i = log(y_i)$	x^2	$x_i log(y_i)$
1	42.10	1.624282096	1	1.624282096
2	34.90	1.542825427	4	3.085650854
3	37.20	1.57054294	9	4.71162882
4	76.00	1.880813592	16	7.523254369
5	52.20	1.717670503	25	8.588352515
6	23.80	1.376576957	36	8.259461742
7	46.70	1.669316881	49	11.68521816
8	85.00	1.929418926	64	15.43535141
9	82.70	1.91750551	81	17.25754959
10	104.00	2.017033339	100	20.17033339
11	264.00	2.421603927	121	26.6376432
12	2,154.00	3.333245699	144	39.99894839
13	4,333.00	3.636788689	169	47.27825296
14	8,827.80	3.945852485	196	55.24193479
15	6,029.70	3.780295705	225	56.70443557
16	8,000.00	3.903089987	256	62.44943979
17	9,797.00	3.991093108	289	67.84858284
153		42.25795577	1,785	454.5003205

$$\bar{\bar{x}} = \frac{1}{n}\sum_{i=1}^{i=17} x_i = \frac{153}{17} = 9$$

$$\bar{\bar{y}} = \frac{1}{n}\sum_{i=1}^{i=17} y_i = \frac{42.26}{17} = 2.4859$$

$$\bar{x}\bar{y} = 9 \text{ x } 2.4859 = 22.373$$

$$V(X) = \frac{1}{n}\sum_{i=1}^{i=17} x_i^2 - \bar{\bar{x}}^2 = \frac{1785}{17} - (9)^2$$

$$= 24$$

$$COV(X,Y) = \frac{1}{n}\sum_{i=1}^{i=17} x_i y_i - \bar{\bar{x}}\bar{\bar{y}} = \frac{454.5}{17} - 22.373$$

$$= 4.362$$

The coefficient of the adjustment line is:

$$a = \frac{COV(X,Y)}{V(X)} = \frac{4.362}{24}$$

$$= 0.181810585$$

The coefficient of the adjustment line is:

$$Y = 0.181810585x + b$$

$$b = Y - 0.181810585x$$

The average point$(\bar{\bar{x}}, \bar{\bar{y}})$ is on this line:

$$b = 2.485762104 - 9(0.181810585)$$

$$= 0.849466842$$

The equation of the adjustment line is therefore:

$$Y = 0.181810585x + 0.849466842$$

Substituting back the value of $y = \log_{10}(y)$, we get:

$$\log_{10}(y) = 0.181810585x + 0.849466842$$

Thus

$$y = 10^{0.181810585x+0,849466842} = 10^{0.849466842}[(10^{0.181810585})^x]$$

Finally,

$$y = 7.070772139(1.519884496^{(x)})$$

We find that this country's inflation rate cloud adjustment curve is an exponential curve of base 1.5.

x_i	y_i	x_iy_i	$7{,}07x1{,}5^x$
2004	42.10	84368.40	10.75
2005	34.90	69974.50	16.33
2006	37.20	74623.20	24.83
2007	76.00	152532.00	37.73
2008	52.20	104817.60	57.35
2009	23.80	47814.20	87.16
2010	46.70	93867.00	132.48
2011	85.00	170935.00	201.35
2012	82.70	166392.40	306.03
2013	104.00	209352.00	465.13
2014	264.00	531696.00	706.94
2015	2154.00	4340310.00	1074.47
2016	4333.00	8735328.00	1633.07
2017	8827.80	17805672.60	2482.08
2018	6029.70	12167934.60	3772.47
2019	8000.00	16152000.00	5733.72
2020	9797.00	19789940.00	8714.59
	39990.10	80697557.50	25456.46

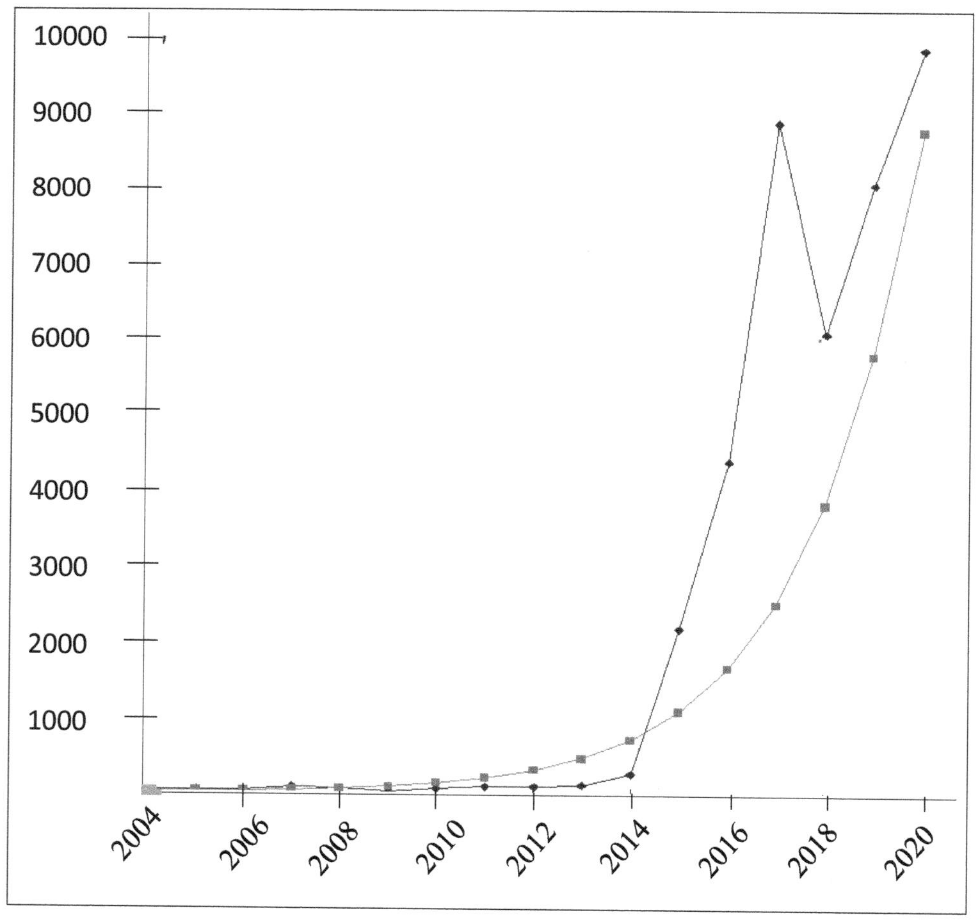

INTERPRETATION OF THE RESULTS

By using the scaling transformation mentioned earlier, we plotted the curve of best fit for the point cloud representing the inflation rate in this developing country. We can observe that the resulting curve is an exponential curve with a base of 1.5. This means that the evolution of the inflation rate follows an exponential trend rather than a linear one. This observation can be useful for better understanding the underlying causes of inflation in this country and for developing effective strategies to control it.

9.6 Exercises

• **Exercise 36:** The ocean-going ship commanded by Captain Gabriel, during the last six years of its transoceanic voyages, carried the following cargo (in tens of thousands of tons):

Years	2015	2016	2017	2018	2019	2020
Goods Transported	12.20	13.10	13.50	14.40	15.00	15.30

a. Adjust this cloud by the least square method.

b. What is the expected production for the year 2030?

Solution

Solving this exercise requires the topics in the following table.

Years (x_i)	Merchamdise tonnage (n_i)	$n_i x_i$	x_i^2
1	12.20	12.2	1
2	13.10	26.2	4
3	13.50	40.5	9
4	14.40	57.6	16
5	15.00	75	25
6	15.30	91.8	36
21	83.50	303.3	91

a: To adjust this scatter plot using the least squares method, we first need to calculate the coefficient value of the adjustment line:

We took the year 2015 as the original year:

$$\bar{\bar{x}} = \frac{1}{n}\sum_{i=1}^{i=6} x_i = \frac{21}{6}$$

$$= 3.5$$

$$\bar{\bar{y}} = \frac{1}{n}\sum_{i=1}^{i=6} y_i = \frac{83.5}{6}$$

$$= 13.92$$

$$V(X) = \frac{1}{n}\sum_{i=1}^{i=6} x_i^2 - \bar{\bar{x}}^2 = \frac{91}{6} - (3.5)^2$$

$$= 2.91$$

$$\text{COV}(X,Y) = \frac{1}{n}\sum_{i=1}^{i=6} x_i y_i - \bar{\bar{x}}\bar{\bar{y}} = \frac{303.3}{6} - (3.50\text{x}13.92)$$

$$= 1.830$$

The coefficient of the adjustment line is:

$$a = \frac{\text{COV}(X,Y)}{V(X)} = \frac{1.830}{2.916}$$
$$= 0.63$$

The adjustment line is, therefore:

$$y = 0.63x + b$$

The average point $(\bar{\bar{x}}, \bar{\bar{y}})$ is on this line:

$$13.92 = 0.63(3.5) + b$$
$$b = 11.72$$

The adjustment line has as its equation:

$$y = 0.63x + 11.72$$

b: Having taken 2015 as the year 0, the abscissa of the year 2030 is 15:

The tonnage transported will be:

$$y = 0.63(15) + 11.72 = 21.17$$

Then,

211,700 tons of merchandise

(the unit being the tens of thousands)

• **Exercise 37:** The following distribution shows the electricity production in a country from 2001 to 2020 (in 10^9 KWH):

x_i	y_i
2001	22.80
2002	25.80
2003	28.90
2004	29.90
2005	33.00
2006	38.10
2007	40.50
2008	41.40
2009	45.60
2010	49.60
2011	53.80
2012	57.40
2013	61.60
2014	64.50
2015	72.10
2016	76.50
2017	83.80
2018	88.20
2019	93.80
2020	102.20
	1,109.50

a. Plot the Point cloud $M_i(x_i, y_i)$.

b. Present a statistical table where the "production" column is replaced by a column $Y_i = \log_{10} y_i$ starting the years with 1 and determine the linear adjustment line between x and y by the least square method.

c. Find the linear adjustment line between x and y by the method of the least square method

d. Deduce the equation to determine the production as a function of the rank of the year.

e. Assuming that the observed growth has continued to date, forecast production by the year 2030.

f. Plot the point cloud and draw the curve on the same graph.

Solution

a: To plot the points $M_i(x_i, y_i)$, we represent the year x_i on the x-axis and the electricity production y_i on the y-axis, for each year from 2001 to 2020:

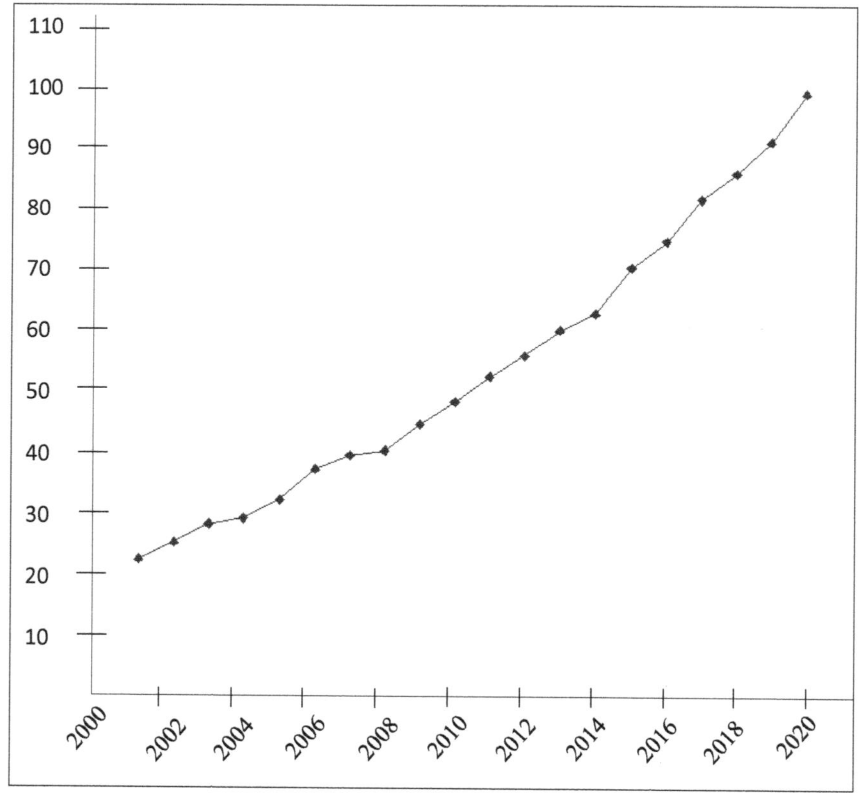

b: The statistical table where the "production" column is replaced by the column $y_i^* = \log_{10} y_i$. (To accommodate the calculation, we will take)

$x_i^* = x_i$	$y_i^* = log(y_i)$	$x_i log(y_i)$	x_i^2
1	1.357934847	1.357934847	1
2	1.411619706	2.823239412	4
3	1.460897843	4.382693528	9
4	1.475671188	5.902684753	16
5	1.51851394	7.592569699	25
6	1.580924976	9.485549854	36
7	1.607455023	11.25218516	49
8	1.617000341	12.93600273	64
9	1.658964843	14.93068358	81
10	1.695481676	16.95481676	100
11	1.730782276	19.03860503	121
12	1.758911892	21.10694271	144
13	1.789580712	23.26454926	169
14	1.809559715	25.333836	196
15	1.857935265	27.86902897	225
16	1.883661435	30.13858296	256
17	1.923244019	32.69514832	289
18	1.945468585	35.01843453	324
19	1.972202838	37.47185393	361
20	2.009450896	40.18901792	400
210	34.06526202	379.74436	2870

c: The linear adjustment line between x_i^* and y_i^* by the method of the least square method is:

$$\overline{\overline{x^*}} = \frac{1}{n}\sum_{i=1}^{i=20} x^*{}_i = \frac{210}{20}$$

$$= 10.5$$

$$\overline{\overline{y^*}} = \frac{1}{n}\sum_{i=1}^{i=20} {y^*}_i = \frac{34.065}{20}$$

$$= 1.7034$$

$$\overline{\overline{x^*}}\,\overline{\overline{y^*}} = 10.5 \text{ x } 1.7034 = 17.8854$$

$$V(X^*) = \frac{1}{n}\sum_{i=1}^{i=20} (x_i^*)^2 - (\overline{\overline{x^*}})^2 = \frac{2{,}870}{20} - (10.5)^2$$

$$= 33.25$$

$$\mathrm{COV}(X^*, Y^*) = \frac{1}{n}\sum_{i=1}^{i=20} x_i^* y_i^* - \overline{\overline{x^*}}\,\overline{\overline{y^*}} = \frac{379.744}{20} - 17.8854$$

$$= 1.1072$$

The coefficient of the adjustment line is:

$$a = \frac{\mathrm{COV}(X^*, Y^*)}{V(X^*)} = \frac{1.1072}{33.25}$$

$$= 0.0332$$

The adjustment line is, therefore:

$$y^* = 0.0332x^* + b$$

The average point $(\overline{\overline{x^*}}, \overline{\overline{y^*}})$ is on this line:`

$$1.7034 = 0.0332(10{,}5) + b$$

$$b = 1.355$$

The equation of the linear adjustment line is therefore:

$$y^* = 0{,}0332x^* + 1{,}355$$

d: The equation allowing production according to the rank of the year is:

As we have $y^* = \log_{10} y$ and $x^* = x$,

$$\log_{10}(y) = 0.0332x + 1.355$$

$$y = 10^{(0.0332x+1.355)}$$
$$= 10^{1.355} \text{ x } (10^{0.0332})^x$$

and

$$y = 22.65(1.08^x)$$

The initial cloud adjustment curve is therefore an exponential base curve of 1.08

e: The 2030 production forecast is:

$$y = 22.65(1.08^{30}) = 227.91$$

$$227.91\text{x } 10^{9K} \text{ WH}$$

f: Representing the curve and the point cloud:

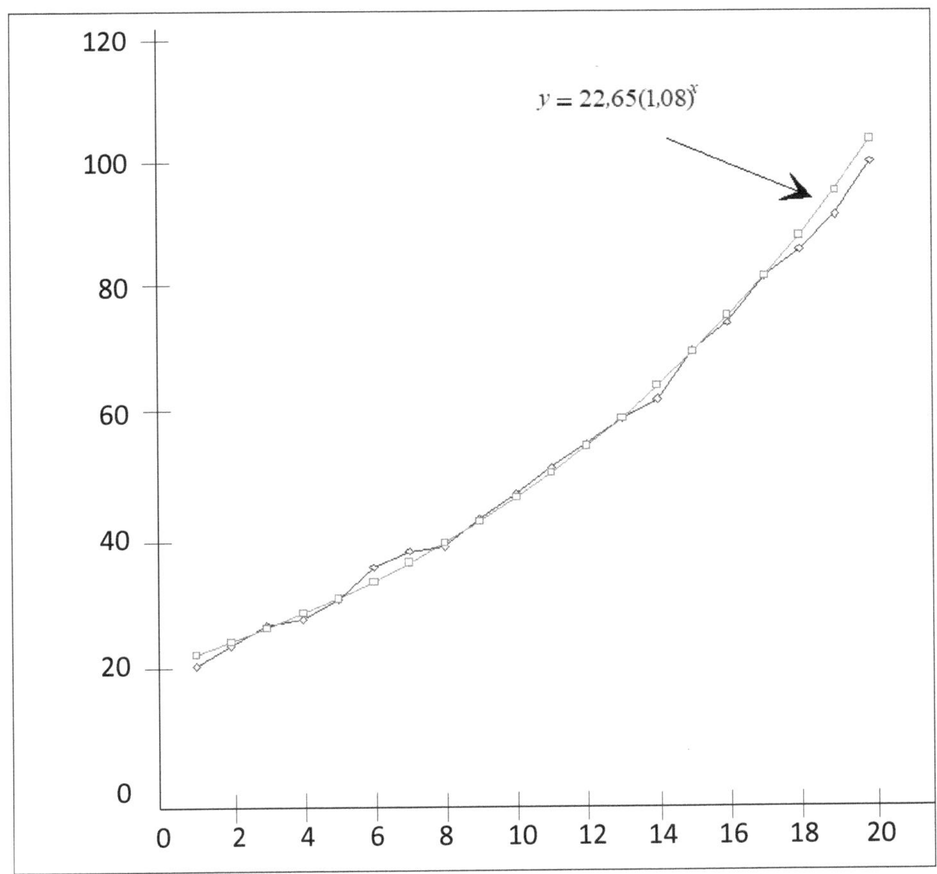

We can see how the two representations are practically overlapped.

• **Exercise 38:** In May 2021, cryptocurrencies suffered a dramatic fall in their values. The table below shows the overall evolution of bitcoin cryptocurrency for two weeks (From May 10 to 23, 2021) at midnight.

x_i	Value in $
10 May 2021, 00:00:00	$59,564.46
11 May 2021, 00:00:00	$55,154.55
12 May 2021, 00:00:00	$57,934.71
13 May 2021, 00:00:00	$50,756.20
14 May 2021, 00:00:00	$49,260.33
15 May 2021, 00:00:00	$49,704.13
16 May 2021, 00:00:00	$48,433.88
17 May 2021, 00:00:00	$42,792.56
18 May 2021, 00:00:00	$44,703.31
19 May 2021, 00:00:00	$40,701.65
20 May 2021, 00:00:00	$38,300.00
21 May 2021, 00:00:00	$40,649.59
22 May 2021, 00:00:00	$36,878.51
23 May 2021, 00:00:00	$37,379.34

a. Plot the point cloud and draw the curve representing this series on the same graph.

b. For each of the following methods, determine the linear adjustment line between x and y, give the predicted value of Bitcoin for May 23, 2021 at midnight and the absolute error rate between the theoretical value provided by the method and the value actual registered $ 35,200.

- The method of moving averages.
- The empirical method of linear interpolation.
- The MAYER'S method.
- The method of the smallest squares.

Solution

A: Plotting the curve and the point cloud.

b: Let's calculate the linear adjustment line by the different methods.

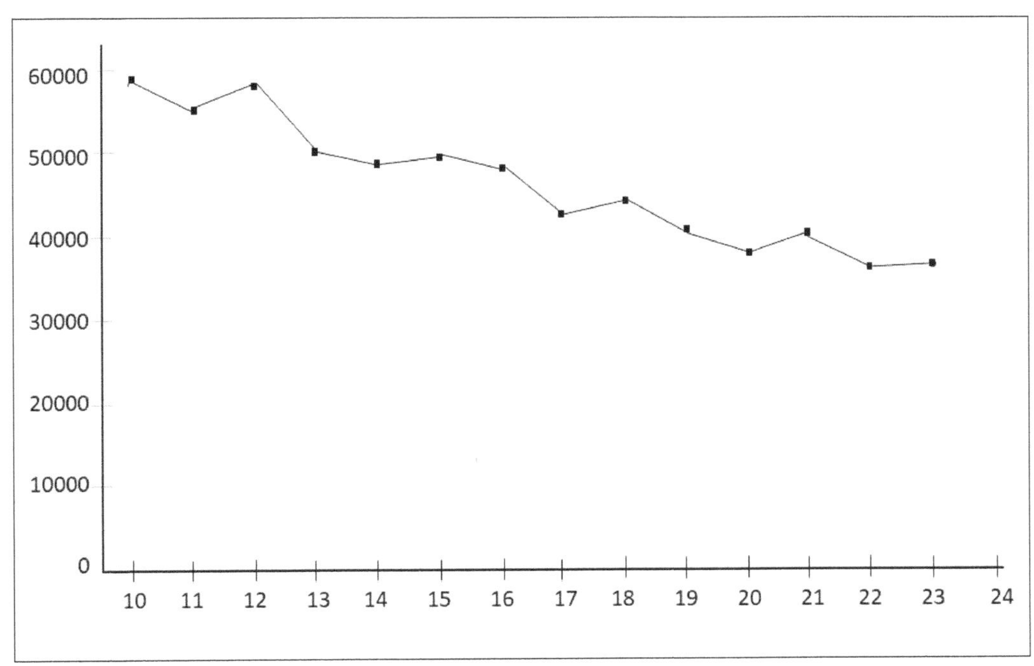

b-1: Let's calculate the linear adjustment line by the moving average method:

b-1-1: Consider the arithmetic average for a group of three consecutive days, and you get the following table:

x_i	Value in USD	Moving Average
10 May 2021, 00:00:00	$59,564.46	
11 May 2021, 00:00:00	$55,154.55	
12 May 2021, 00:00:00	$57,934.71	
13 May 2021, 00:00:00	$50,756.20	$57,551.24
14 May 2021, 00:00:00	$49,260.33	$54,615.15
15 May 2021, 00:00:00	$49,704.13	$52,650.41
16 May 2021, 00:00:00	$48,433.88	$49,906.89
17 May 2021, 00:00:00	$42,792.56	$49,132.78
18 May 2021, 00:00:00	$44,703.31	$46,976.86
19 May 2021, 00:00:00	$40,701.65	$45,309.92
20 May 2021, 00:00:00	$38,300.00	$42,732.51
21 May 2021, 00:00:00	$40,649.59	$41,234.99
22 May 2021, 00:00:00	$36,878.51	$39,883.75
23 May 2021, 00:00:00	$37,379.34	$38,609.37
		$38,302.48

The moving average is shifted for forecast purpose.

Hence the following graph:

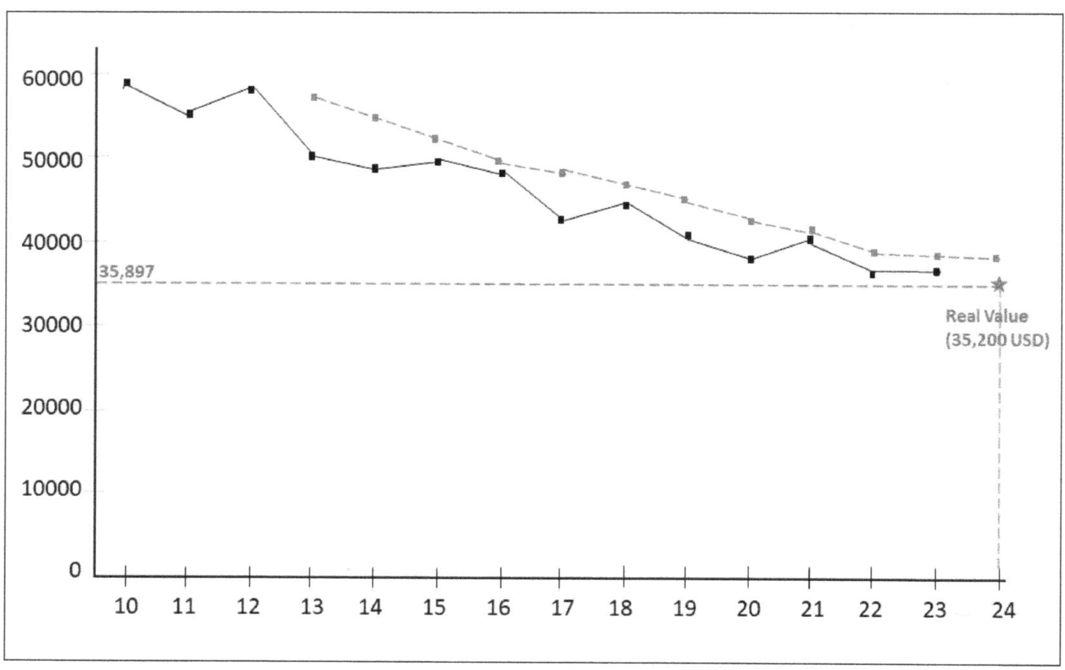

b-1-2: The prediction for May 24, 2021, is:

Based on the linear fitting curve obtained using the method of moving averages shown above, the value of the Bitcoin cryptocurrency on May 24, 2021, is $ 38,302.

b-1-3: When comparing the prediction of the moving averages method with the actual value of the Bitcoin cryptocurrency at midnight on the Market on May 24, this method shows an absolute error of:

$$\text{Absolute Error} = \frac{|38{,}302 - 35{,}200|}{35200} = 8.09\%$$

This means that the moving averages method overestimated the value of the cryptocurrency on this date by more than 8%. This error is relatively significant and may have consequences on investment decisions.

b-2: Let's calculate the linear adjustment line by the empirical method:

b-2-1: We have drawn the line that best reflects the evolution of the Cryptocurrency Bitcoin during the period from 10 to 23 May 2021 from the points provided in the data table. We then determined the equation of this line by considering two points, *A* and *B,* of coordinates $(11\,;55155)$ and $(23\,;37379)$ respectively. The equation of this line is:

$$y - y_1 = \frac{y_2 - y_1}{x_2 - x_1}(x - x_1)$$

$$y - 55{,}155 = \frac{37{,}379 - 55{,}155}{23 - 11}(x - 11)$$

$$y = -\frac{17{,}776}{12}(x - 11) + 55{,}155$$

$$= -\frac{17{,}776}{12}x + \frac{857{,}396}{12}$$

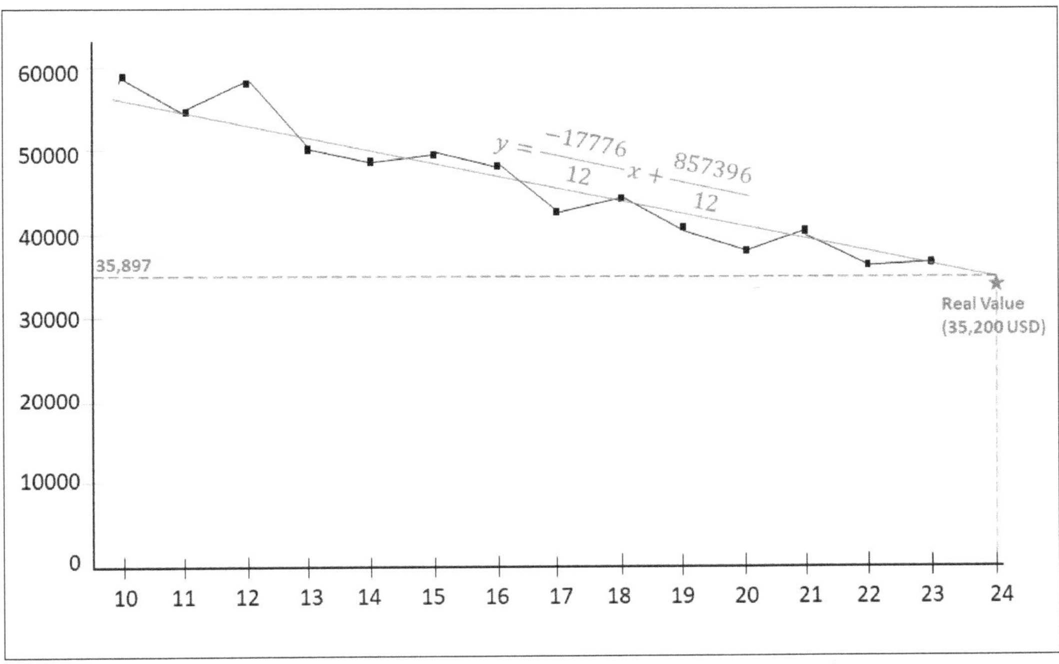

b-2-2: The prediction for May 24, 2021, is:

Based on the linear fitting curve obtained using the empirical method shown above, the value of the Bitcoin cryptocurrency on May 24, 2021, is 35,897 $.

b-2-3: By comparing this value with the real value of the Bitcoin cryptocurrency at midnight on the Market on May 24, this method shows an absolute error of:

$$\text{Absolute Error} = \frac{|35{,}897 - 35{,}200|}{35{,}200} = 1.98\%$$

This error is relatively low, indicating that the empirical method of linear interpolation is accurate enough to estimate the value of the Bitcoin cryptocurrency over this period.

b-3: Let's calculate the linear adjustment line by MAYER's method:

b-3-1: Finding the coordinates of L and R:

$$\bar{x}_L = 13, \qquad \bar{y}_R = 52{,}972.61$$

$$\bar{x}_L = 20, \qquad \bar{y}_R = 40{,}200.71$$

The equation of the line going through these two points is:

$$y - 52{,}972.61 = \frac{40{,}201.71 - 52{,}972.61}{20 - 13}(x - 13)$$

$$= -\frac{12{,}771.9}{7}(x - 13) + 52{,}972.61$$

$$= \frac{-12{,}771.9}{7}x + \frac{536{,}842.98}{7}$$

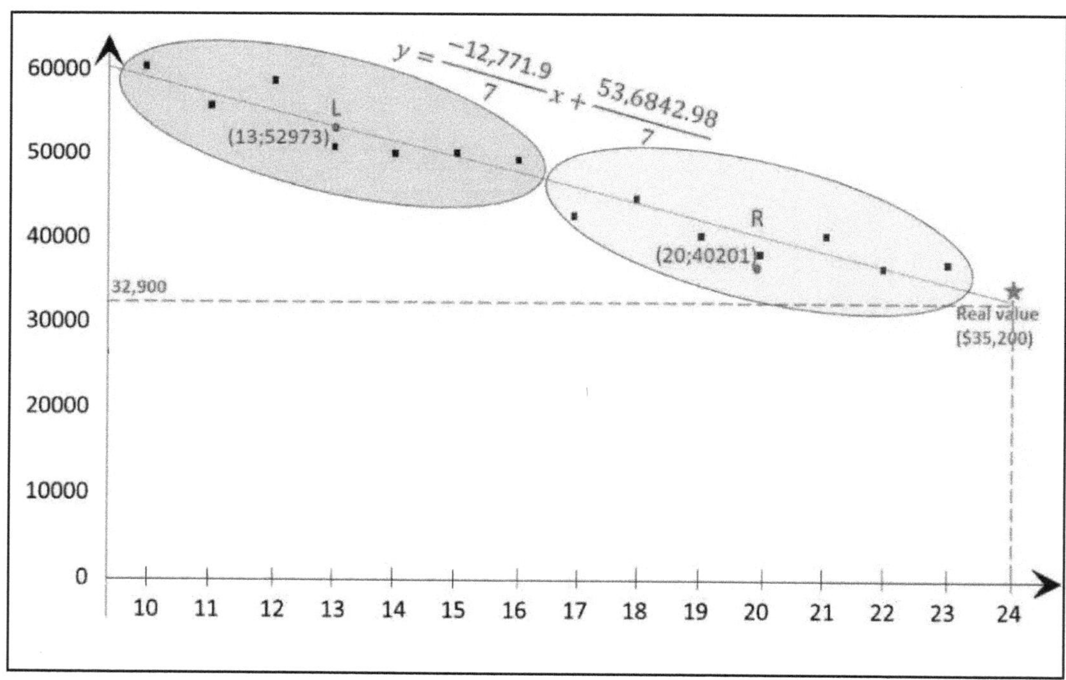

b-3-2: The prediction for May 24, 2021

Based on the linear fitting curve obtained using MAYER'S method shown above, the value of the Bitcoin cryptocurrency on May 24, 2021, is 32,902 USD.

b-3-3: By comparing this value with the real value of the Bitcoin cryptocurrency at midnight on the market on May 24, this method shows an absolute error of:

$$\text{Absolute Error} = \frac{|32{,}902 - 35{,}200|}{35{,}200} = 6.52\%$$

Although this error is not negligible, it remains relatively small compared to the amplitude of price fluctuations in the cryptocurrency market.

b-4: Let's calculate the linear adjustment line by the least square method:

x_i	y_i	$x_i y_i$	x_i^2
10	$59,564.46	595,644.63	100
11	$55,154.55	606,700.00	121
12	$57,934.71	695,216.53	144
13	$50,756.20	659,830.58	169
14	$49,260.33	689,644.63	196
15	$49,704.13	745,561.98	225
16	$48,433.88	774,942.15	256
17	$42,792.56	727,473.55	289
18	$44,703.31	804,659.50	324
19	$40,701.65	773,331.40	361
20	$38,300.00	766,000.00	400
21	$40,649.59	853,641.32	441
22	$36,878.51	811,327.27	484
23	$37,379.34	859,724.79	529
231	652,213.22	10,363,698.35	4,039

b-4-1: Let's calculate the linear adjustment line:

$$\bar{\bar{x}} = \frac{1}{n}\sum_{i=1}^{i=14} x_i = \frac{231}{14}$$

$$= 16.5$$

$$\bar{\bar{y}} = \frac{1}{n}\sum_{j=1}^{j=14} y_i = \frac{652{,}213.22}{14}$$

$$= 46{,}586.65$$

$$\bar{\bar{x}}\bar{\bar{y}} = 16.5 \text{ x } 46{,}586.65 = 768{,}679.866$$

$$V(X) = \frac{1}{n}\sum_{i=1}^{i=14} x_i^2 - \bar{\bar{x}}^2 = \frac{4039}{14} - (16.5)^2$$

$$= 16.25$$

$$\text{COV}(X, Y) = \frac{1}{n}\sum_{i=1}^{i=20} x_i y_i - \bar{\bar{x}}\bar{\bar{y}} = \frac{10{,}363{,}698.35}{14} - 768{,}679{,}866$$

$$= -28{,}415.698$$

The coefficient of the adjustment line is:

$$a = \frac{\text{COV}(X, Y)}{V(X)} = \frac{-28{,}415.698}{16.25}$$

The adjustment line is, therefore:

$$y = \frac{-28{,}415.698}{16.25} x + b$$

As the average point$(\bar{\bar{x}}, \bar{\bar{y}})$ is on this line:

$$b = y - x\left(\frac{-28{,}415.698}{16.25}\right)$$

$$= 46\ 586{,}65 - 16{,}5\left(\frac{-28{,}415.698}{16.25}\right)$$

$$= \frac{1\ 225\ 892.0795}{16.25}$$

Equation of the linear adjustment line by the smallest square:

$$y = \left(\frac{\text{COV}(X, Y)}{V(X)}\right) x + \left[\bar{\bar{y}} - \left(\frac{\text{COV}(X, Y)}{V(X)}\right)\bar{\bar{x}}\right]$$

$$= \left(\frac{-28{,}415.698}{16.25}\right)x + \left[46{,}586.65 - \left(\frac{-28{,}415.698)}{16.25}\right)16.5\right]$$

$$= \frac{-28{,}415{,}698}{16.25}x + \frac{1{,}225{,}892.0795}{16.25}$$

This equation could have been found directly by the formula:

$$y = \left(\frac{\mathrm{COV}(X,Y)}{V(X)}\right)x + \left[\bar{\bar{y}} - \left(\frac{\mathrm{COV}(X,Y)}{V(X)}\right)\bar{\bar{x}}\right]$$

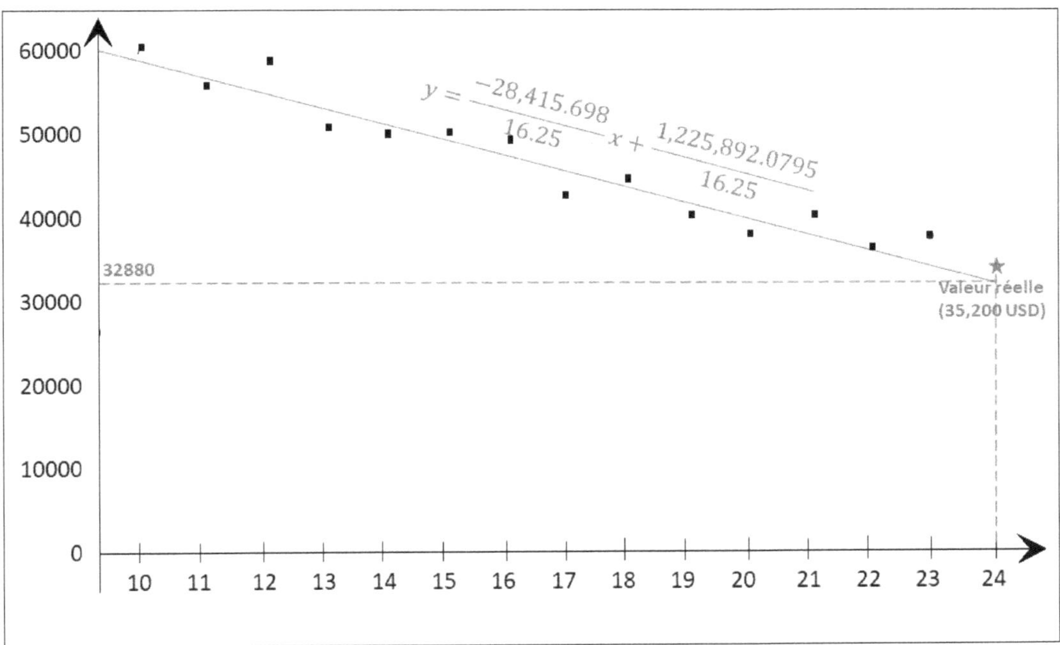

b-4-2: The prediction for May 23, 2021, is:

Based on the linear fitting curve obtained using the least square method shown above, the value of the Bitcoin cryptocurrency on May 24, 2021, is 32,880 USD.

b-4-3: By comparing this value with the real value of the Bitcoin cryptocurrency at midnight on the market on May 24, this method shows an absolute error of:

$$\text{Absolute Error} = \frac{|32{,}880 - 35{,}200|}{35{,}200} = 6.5\%$$

Although this error is not negligible, it remains relatively small compared to the amplitude of price fluctuations in the cryptocurrency market.

• **Exercise 39:** One proposes to study the fossils of a fish found in a dry lake. Beside are the values recorded on the length of their skulls and those of their dorsal fins in centimeters.

	Skull length (cm) x_i	Dorsal fin length (cm) y_i
1	11.2	16.7
2	6.1	9.4
3	8.7	12.9
4	3.5	4.8
5	16.3	24.4
6	5.7	8.3
7	19.4	28.8
8	22.1	33.5
9	40.9	61.2
10	7.7	11.8
11	12.2	18.6
12	17.5	26.3
13	33.3	50.7
14	10.5	15.7
15	25.4	38.2
16	2.8	4.2
17	52.7	79.5
18	19.9	30.1
19	46.6	80.8
20	13.2	19.5
21	35.5	52.7
Total	411.2	628.10
Means	19.58	29.91

a. Determine the variance of the skull.

b. Determine Dorsal Fin Variance

c. Determine the covariance of the skull and the dorsal fin.

d. Determine by the method of least squares, the equation of the regression line of the dorsal fin in relation to the skull.

e. Calculate the linear correlation coefficient and provide its comment.

f. We found a skull with a length of 58.29 Cm without a dorsal fin, determine the predicted length of this missing dorsal fin.

Solution

	x_i	y_i	x_iy_i	x_i^2	y_i^2
1	11.2	16.7	187.04	125.44	278.89
2	6.1	9.4	57.34	37.21	88.36
3	8.7	12.9	112.23	75.69	166.41
4	3.5	4.8	16.80	12.25	23.04
5	16.3	24.4	397.72	265.69	595.36
6	5.7	8.3	47.31	32.49	68.89
7	19.4	28.8	558.72	376.36	829.44
8	22.1	33.5	740.35	488.41	1,122.25
9	40.9	61.2	2,503.08	1,672.81	3,745.44
10	7.7	11.8	90.86	59.29	139.24
11	12.2	18.6	226.92	148.84	345.96
12	17.5	26.3	460.25	306.25	691.69
13	33.3	50.7	1,688.31	1,108.89	2,570.49
14	10.5	15.7	164.85	110.25	246.49
15	25.4	38.2	970.28	645.16	1,459.24
16	2.8	4.2	11.76	7.84	17.64
17	52.7	79.5	4,189.65	2,777.29	6,320.25
18	19.9	30.1	598.99	396.01	906.01
19	46.6	80.8	3,765.28	2,171.56	6,528.64
20	13.2	19.5	257.40	174.24	380.25
21	35.5	52.7	1,870.85	1,260.25	2,777.29
Total	411.2	628.10	18,915.99	12,252.22	29,301.27
Mean	19.58	29.91			

From the data table above, we can read:

$$\bar{\bar{x}} = 19.58$$

$$\bar{\bar{y}} = 29.91$$

a: Let's calculate the variance of the skull:

$$V(x) = 12{,}252.22 - (19.58)^2 = 11{,}868.81$$

b: Let's calculate the dorsal fin variance:

$$V(y) = 29{,}301.27 - (29.91)^2 = 28{,}406.69$$

c: Let's calculate the skull and dorsal fin covariance:

$$COV(x,y) = 18{,}915.99 - (19.58 \text{ x } 29.91) = 18{,}330.33$$

d: Let's calculate the equation of the regression line of the dorsal fin in relation to the skull:

$$a = \frac{COV(x,y)}{V(x)} = \frac{18{,}330.33}{11{,}868.81}$$

$$= 1.54$$

The regression line from the dorsal fin to the skull is of the form:

$$y = 1.54x + b$$

The average point $(\bar{\bar{x}}, \bar{\bar{y}})$ is on the line.

$$29.91 = 1.54(19.58) + b$$
$$b = -0.2432$$

The equation of the linear adjustment line is therefore:

$$y = 1.54x - 0.2432$$

e: Let's calculate the linear correlation coefficient:

$$\rho = \frac{\text{COV}(X,Y)}{\sqrt{V(X)V(Y)}} = \frac{18{,}330.33}{\sqrt{11{,}868.81 \text{ x } 28{,}406.69}}$$

$$= \frac{18{,}330.33}{18{,}361.74}$$

$$= 0.99829$$

INTERPRETATION OF THE RESULTS

The value of the linear correlation coefficient found is very close to 1, which means that there is a strong positive correlation between the length of the skull and that of the dorsal fin of the studied fossils.
This means that the larger the length of the skull, the larger the length of the dorsal fin as well. Thus, it can be affirmed that these fossils indeed belong to the same fish species.

f. Let's calculate the length of the missing dorsal fin, which they found skull measures 58.29 Cm:

Turning back to the equation of the linear adjustment line

$$y = 1.54x - 0.2432$$

We have:

$$y = 1.54 \text{ x } (58.29) - 0.2432$$

$$= 89.52 \text{ Cm}$$

The missing dorsal fin measures 89.52 Cm

9.7 Summing-up Exercise

• **Exercise 40:**

9.7.1 Origin of Data

The data used in this summing-up exercise are real data on the 216 municipalities of an African country. We have coded them numerically to preserve their anonymity. They have been gathered, organized, and analyzed for use in this study, and the results and interpretations presented here are the responsibility of the authors alone. The choice of these data is based on their completeness and their ability to generate different types of statistical series, allowing many aspects to be addressed in a single exercise.

9.7.2 Presentation of the Study

This study is divided into three parts. The first part focuses on analyzing the population distribution across various municipalities in the country. The second part examines the distribution of surface area across all municipalities. The third part is a two-dimensional analysis that studies both population and surface area of the country, revealing the existence or absence of functional dependence. In each case, statistical results were carefully analyzed, and conclusions were drawn based on the findings.

The table below presents the 216 municipalities of the country:

Nbr	Pop.	Area (km^2)
1	252181	81.7
2	49297	3.9
3	17360	192.2
4	69147	4.7
5	74708	2.9
6	49173	2.9
7	52820	358.9
8	126589	23.2
9	97214	6.8
10	74888	5
11	160719	6.6
12	82303	3.4
13	113968	5.3
14	108939	5.6
15	159775	23.7
16	74447	3.2
17	128197	27.1
18	104904	4.9
19	117774	16.6
20	157010	11.4
21	158080	69.7
22	353209	76.9
23	28963	1076.8
24	52676	7948.8
25	20785	88.7
26	337368	12.8
27	55645	8.5
28	32815	28.5
29	50345	11.4
30	11824	25.1
31	102633	4265
32	244900	3090
33	122782	3620
34	183709	3270
35	386121	8507
36	197675	8190
37	140825	6784
38	234291	7968
39	92956	4680
40	54899	3371
41	34405	20
42	12988	112
43	16249	90
44	150788	24184
45	85640	12070
46	85590	41824
47	232528	18773
48	92693	25074
49	88154	5318
50	555124	13404
51	428962	14327
52	214231	16898
53	477458	18926
54	291310	14572
55	22648	36
56	50198	20
57	35093	18
58	41357	18
59	268960	18126
60	114839	19187
61	92597	19264
62	278346	26648
63	99583	6749
64	75632	274.9
65	61659	185.1
66	84484	21239
67	98246	24598
68	90255	17328
69	119993	13842
70	45824	8608
71	23445	7091
72	35760	10736
73	439079	11488
74	221932	13434
75	220854	12848
76	122012	12833
77	12273	753.4
78	6093	291.6
79	92568	10559
80	112436	15397
81	203243	17411
82	119627	13277
83	214404	14733
84	359450	15498
85	149605	27910
86	157760	19718
87	69718	16797
88	111806	17494
89	137746	22567
90	140875	19996
91	47566	36385
92	64664	768
93	31662	59.9
94	67378	18.5
95	62298	809.5
96	62622	23.9
97	28957	230.2
98	91226	24430
99	59646	47087
100	119637	42196
101	110411	26665
102	245548	15770
103	65927	19073
104	116871	22436
105	74972	18098
106	93434	25417
107	116538	38075
108	48862	34704
109	99419	9128
110	112233	22909
111	215679	8605
112	60138	7204
113	122499	32446
114	158258	13108
115	109269	16015
116	227268	10305
117	295107	8730
118	84031	36783
119	551137	8184
120	422919	5221
121	396062	6740
122	115659	405
123	137065	23475
124	619827	18096
125	604210	7484
126	479064	5289
127	478450	4734
128	149164	21181
129	56017	24953
130	60448	19805
131	52907	16055
132	171600	14542
133	246959	16201
134	112580	19513
135	54958	27
136	64274	10
137	48718	23
138	365675	1800
139	320022	3148
140	204843	15786
141	215895	11172
142	173948	25216
143	226811	5707
144	92247	281
145	340597	1960
146	148363	14.3
147	64230	34.2
148	126074	36.1
149	75070	15
150	88732	4
151	26581	2
152	35780	641.4
153	27805	53.3
154	121836	40.6
155	44743	39
156	20078	102.1
157	97281	197.5
158	123425	142.5
159	95809	23081
160	99607	17861
161	161174	24963
162	114081	30363
163	82844	24700
164	171008	40214
165	80850	13400
166	179480	20503
167	242629	14222
168	217054	19865
169	180164	30512
170	213230	24350
171	171453	34198
172	98788	15250
173	184633	13403
174	74227	1727
175	88492	12059
176	84363	21675
177	105887	26676
178	79052	25894
179	131765	22673
180	123366	22466
181	55212	14.2
182	96950	20.4
183	75644	4.8
184	120016	10.3
185	138413	14
186	125085	1747
187	63575	1480
188	86960	2397
189	127120	1836
190	191635	2021
191	58366	10530
192	84808	17682
193	76056	26346
194	169955	25490
195	77438	12229
196	258568	12054
197	265237	11747
198	21811	2100
199	227801	5726
200	188112	14373
201	136463	22480
202	50756	221.8
203	65635	24
204	42612	44
205	41593	153
206	98097	300
207	178573	8601
208	119993	14702
209	266863	12881
210	187593	8961
211	139748	13223
212	111133	8450
213	530257	26217
214	167258	15634
215	231440	20155
216	56695	25175

9.7.3 Questions

9.7.3.1 Study of the Population

1. In processing *"the number of municipalities whose population is between two values,"* it is requested to compile, break down the series, group the data, and represent the series (we will consider the classes of amplitudes 60,000 inhabitants)

2. Represent the frequency histogram as well as the frequency polygon.

3. Define the distribution function and plot the cumulative relative frequency curves.

4. Determine all the characteristics central tendency:

a. The modal population.
b. The average population.
c. The median population.

5. What is the percentage of municipalities with a below-average population

6. Determine all characteristics of dispersion:

a. Interquartile Range.
b. The variance.
c. The standard deviation.
d. The coefficient of variation.

7. Determine the percentage of municipalities:

a. Occupied by 50% of the least populated municipalities.
b. Occupied by 25% of the least populated municipalities.
c. Occupied by 10% of the most populated municipalities.

8. Determine the percentage of municipalities with a population

a. Lower than the mean minus the standard deviation.
b. Lower than the mean plus the standard deviation.
c. Comprises between the mean minus the standard deviation and the mean plus the standard deviation.
d. Lower than that of Guadeloupe (French Caribbean Island with 395,700 inhabitants).

9. Determine all characteristics of shape

a. The YULE's coefficient of skewness.
b. The PEARSON's coefficient of skewness.
c. The FISHER's coefficient of skewness.
d. The FISHER's coefficient of kurtosis.

10. Determine all characteristics of concentration:

a. The medial of the distribution.
b. The percentage of municipalities with a population is below the medial.
c. Plot the LORENZ curve
d. Determine the GINI index by the graphical method (trapezoid method).
e. Determine the GINI index by the definite integral calculus.

9.7.3.2 Study of the Area

1. In processing *"number of municipalities whose area is between two* values" characters, it is requested to split the series, to group the data and to represent the series (we'll consider the classes of amplitudes Km^2 5000)

2. Represent the number histogram as well as the number polygon.

3 Define the distribution function and plot the cumulative relative frequency curves.

4. Determine all the central trend characteristics:

a. The modal area.
b. The average area.
c. The median area.

5. What is the percentage of municipalities with a below-average area.

6. Determine all characteristics of dispersion:

a. Interquartile Range.
b. The variance.
c. The standard deviation.
d. The coefficient of variation.

7. Determine the percentage of municipalities:

a. Occupied by 50% of the smallest municipalities.
b. Occupied by 25% of the smallest municipalities.
c. Occupied by 10% of the biggest municipalities.

8. Calculate the percentage of municipalities with an area:

a. Lower than the mean minus the standard deviation.
b. Lower than the mean plus the standard deviation.
c. Between the mean minus the standard deviation and the mean plus the standard deviation.
d. Greater than that of the Paris region which has 12012 Km^2.

9. Determine all characteristics of shape:

a. The YULE's coefficient of skewness.
b. The PEARSON's coefficient of skewness.
c. The FISHER's coefficient of skewness.
d. The FISHER's coefficient of kurtosis.

10. Determine all characteristics of concentration:

a. The medial of the distribution.
b. The percentage of municipalities with an area below the medial.
c. Plot the LORENZ curve
d. Determine the GINI index by the graphical method (trapezoid method).
e. Determine the GINI index by the definite integral calculus.

9.7.3.3 Study of Area and Population (two-dimensional series)

Considering the same data as in the two series, we wish to study the distribution of the population of these administrative entities according to their area.

Let X be the statistical variable corresponding to the population and Y the statistical variable corresponding to the area.

1. Develop contingency tables (frequencies and proportions) from raw information (population and area).

2. Give the table of the marginal distribution of X as well as that of the marginal distribution of Y.

3. Give the conditional law of Y knowing that the population of a municipality is between 60,000 and 120,000 people.

4. Give the conditional law of X knowing that the area of a municipality is between 10,000 and 15,000 Km^2.

5. Determine the marginal characteristics of X (mean and variance).

6. Determine the marginal characteristics of Y (mean and variance).

7. Verify that the marginal means are equal to the means of the conditional means weighted by the marginal proportions of the link variable.

8. Retake the raw data on the area and population of municipalities:

a. Create a two-entry statistical table representing the surface area and population.
b. Sort the table in ascending order of surface area.
c. Calculate the variance of the surface area of municipalities.
d. Calculate the variance of the population of municipalities.
e. Calculate the covariance of surface area and population of municipalities.
f. Determine, by the method of least squares, the regression line equation of surface area with respect to population of municipalities.
g. Calculate the linear correlation coefficient and provide an interpretation.

9.7.4 Solutions

9.7.4.1 Study of the Population

1-a: To count this series, we will first sort the municipalities in increasing order of population

Nbr	Population
1	6093
2	11824
3	12273
4	12988
5	16249
6	17360
7	20078
8	20785
9	21811
10	22648
11	23445
12	26581
13	27805
14	28957
15	28963
16	31662
17	32815
18	34405
19	35093
20	35760
21	35780
22	41357
23	41593
24	42612
25	44743
26	45824
27	47566
28	48718
29	48862
30	49173
31	49297
32	50198
33	50345
34	50756
35	52676
36	52820
37	52907
38	54899
39	54958
40	55212
41	55645
42	56017
43	56695
44	58366
45	59646
46	60138
47	60448
48	61659
49	62298
50	62622
51	63575
52	64230
53	64274
54	64664
55	65635
56	65927
57	67378
58	69147
59	69718
60	74227
61	74447
62	74708
63	74888
64	74972
65	75070
66	75632
67	75644
68	76056
69	77438
70	79052
71	80850
72	82303
73	82844
74	84031
75	84363
76	84484
77	84808
78	85590
79	85640
80	86960
81	88154
82	88492
83	88732
84	90255
85	91226
86	92247
87	92568
88	92597
89	92693
90	92956
91	93434
92	95809
93	96950
94	97214
95	97281
96	98097
97	98246
98	98788
99	99419
100	99583
101	99607
102	102633
103	104904
104	105887
105	108939
106	109269
107	110411
108	111133
109	111806
110	112233
111	112436
112	112580
113	113968
114	114081
115	114839
116	115659
117	116538
118	116871
119	117774
120	119627
121	119637
122	119993
123	119993
124	120016
125	121836
126	122012
127	122499
128	122782
129	123366
130	123425
131	125085
132	126074
133	126589
134	127120
135	128197
136	131765
137	136463
138	137065
139	137746
140	138413
141	139748
142	140825
143	140875
144	148363
145	149164
146	149605
147	150788
148	157010
149	157760
150	158080
151	158258
152	159775
153	160719
154	161174
155	167258
156	169955
157	171008
158	171453
159	171600
160	173948
161	178573
162	179480
163	180164
164	183709
165	184633
166	187593
167	188112
168	191635
169	197675
170	203243
171	204843
172	213230
173	214231
174	214404
175	215679
176	215895
177	217054
178	220854
179	221932
180	226811
181	227268
182	227801
183	231440
184	232528
185	234291
186	242629
187	244900
188	245548
189	246959
190	252181
191	258568
192	265237
193	266863
194	268960
195	278346
196	291310
197	295107
198	320022
199	337368
200	340597
201	353209
202	359450
203	365675
204	386121
205	396062
206	422919
207	428962
208	439079
209	477458
210	478450
211	479064
212	530257
213	551137
214	555124
215	604210
216	619827

1-b: By grouping the municipalities by classes of magnitude 60,000 inhabitants, we have the statistical table:

Population	Number of Municipalities
[0;60000)	45
[60000;120000)	78
[120000;180000)	39
[180000;240000)	23
[240000;300000)	12
[300000;360000)	5
[360000;420000)	3
[420000;480000)	6
[480000;540000)	1
[540000;600000)	2
[600000;660000)	2
Total	216

2: Plotting the histogram and the polygon of frequencies of the distribution:

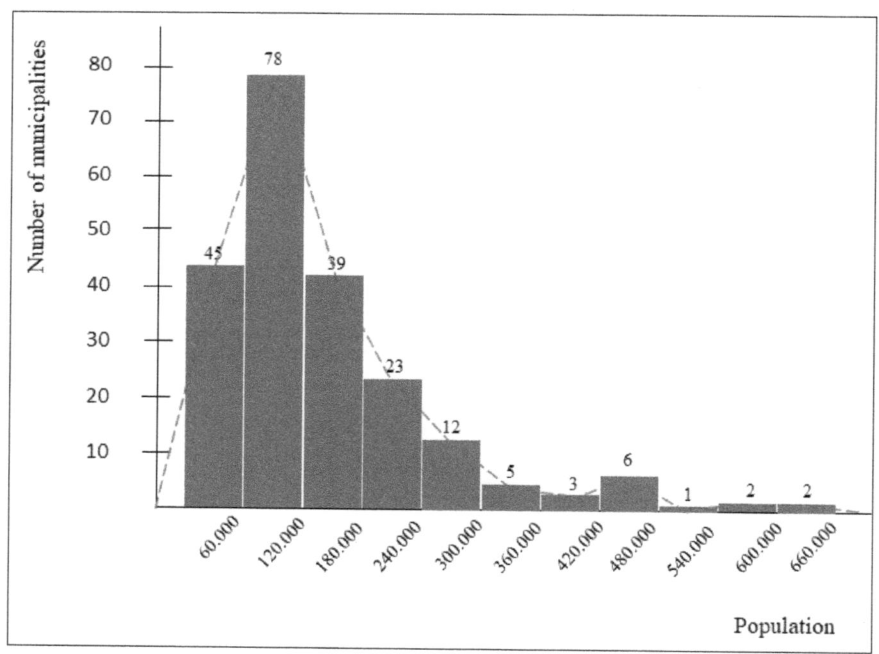

The frequency polygon was obtained by joining the mid-stages of the histogram. The polygon was closed by taking the middle of the fictitious rectangles lying to the right and left of the histogram.

Since the classes have been chosen of equal amplitudes, the rectangles of the histogram each have a height proportional to the size of its class.

3: Let's define the distribution function and then draw the curves of the cumulative relative frequencies:

3-1: The distribution function associated with this statistical series is that which in any number x associates the number of municipalities with a strictly less area than x (continuous character).

Indeed

if

$x < 0$	$F(x) \leq 0$
$x\epsilon[0\,,60000]$	$F(x) = 45$
$x\epsilon[60000\,,120000]$	$F(x) = 45 + 78 = 123$
$x\epsilon[120000\,,180000]$	$F(x) = 45 + 78 + 39 = 162$
$x\epsilon[180000\,,240000]$	$F(x) = 45 + 78 + 39 + 23 = 185$
$x\epsilon[240000\,,300000]$	$F(x) = 45 + 78 + 39 + 23 + 12 = 197$
$x\epsilon[300000\,,360000]$	$F(x) = 45 + 78 + 39 + 23 + 12 + 5 = 202$
$x\epsilon[360000\,,420000]$	$F(x) = 45 + 78 + 39 + 23 + 12 + 5 + 3 = 205$
$x\epsilon[420000\,,480000]$	$F(x) = 45 + 78 + 39 + 23 + 12 + 5 + 3 + 6 = 211$
$x\epsilon[480000\,,540000]$	$F(x) = 45 + 78 + 39 + 23 + 12 + 5 + 3 + 6 + 1 = 212$
$x\epsilon[540000\,,600000]$	$F(x) = 45 + 78 + 39 + 23 + 12 + 5 + 3 + 6 + 1 + 2 = 214$
$x\epsilon[600000\,,660000]$	$F(x) = 45 + 78 + 39 + 23 + 12 + 5 + 3 + 6 + 1 + 2 + 2 = 216$

3-2: Plotting the cumulative relative frequency curves.

Solving this exercise requires the following table:

Population	Nb Municipalities (n_i)	$n_i\uparrow$	f_i	$f_i\uparrow$	$f_i\downarrow$
[0;60000)	45	45	20.83	20.83	100.00
[60000;120000)	78	123	36.11	56.94	79.17
[120000;180000)	39	162	18.06	75.00	43.06
[180000;240000)	23	185	10.65	85.65	25.00
[240000;300000)	12	197	5.56	91.20	14.35
[300000;360000)	5	202	2.31	93.52	8.80
[360000;420000)	3	205	1.39	94.91	6.48
[420000;480000)	6	211	2.78	97.69	5.09
[480000;540000)	1	212	0.46	98.15	2.31
[540000;600000)	2	214	0.93	99.07	1.85
[600000;660000)	2	216	0.93	100.00	0.93
Total	216		100.00		

The increasing and decreasing cumulative relative frequency curves are:

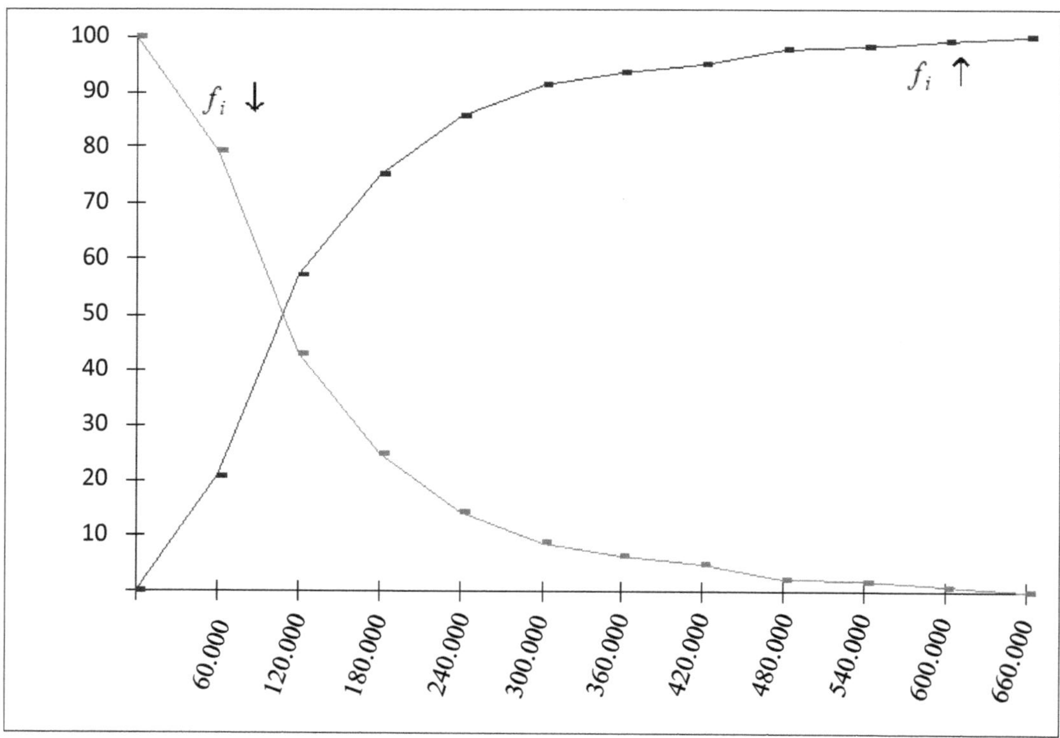

4: Calculation of the central trend characteristics

The following table shows the data required to solve this exercise:

Population	x_i	n_i	$x_i n_i$	$x_i n_i^2$	f_i	q_i	$f_i\uparrow$	$q_i\uparrow$
[0;60000)	30000	45	1350000	40500000000	20.83	4.39	20.83	4.39
[60000;120000)	90000	78	7020000	631800000000	36.11	22.85	56.94	27.25
[120000;180000)	150000	39	5850000	877500000000	18.06	19.04	75.00	46.29
[180000;240000)	210000	23	4830000	1014300000000	10.65	15.72	85.65	62.01
[240000;300000)	270000	12	3240000	874800000000	5.56	10.55	91.20	72.56
[300000;360000)	330000	5	1650000	544500000000	2.31	5.37	93.52	77.93
[360000;420000)	390000	3	1170000	456300000000	1.39	3.81	94.91	81.74
[420000;480000)	450000	6	2700000	1215000000000	2.78	8.79	97.69	90.53
[480000;540000)	510000	1	510000	260100000000	0.46	1.66	98.15	92.19
[540000;600000)	570000	2	1140000	649800000000	0.93	3.71	99.07	95.90
[600000;660000)	630000	2	1260000	793800000000	0.93	4.10	100.00	100.00
Total	Total	216	30720000	7358400000000	100.00	100.00		

4-a: Finding the mode

Mode = 90,000
Modal class = [60,000 ; 120,000)

4-b: Determining the mean

$$\bar{x} = \frac{1}{n}\sum_{i=1}^{i=11} n_i x_i = \frac{30{,}720{,}000}{216}$$

$$= 142{,}222 \text{ people}$$

4-c: Determining the median

$$M_e = a_i + (a_{i+1} - a_i)\frac{(50 - F_i)}{(F_{i+1} - F_i)}$$

$$= 60{,}000 + (120{,}000 - 60{,}000)\frac{(50 - 20.83)}{(56.94 - 20.83)}$$

$$= 108{,}468.57 \text{ people}$$

5: Determining the percentage of municipalities with a below-average population

We know that the average is $\bar{x} = 142{,}222$ people.

Let $P_{<\bar{x}}$ be the percentage to find, and we use the linear interpolation formula using the data pair $(a_i, f_i \uparrow)$:

$$\bar{x} = a_i + (a_{i+1} - a_i)\frac{(P_{<\bar{x}} - F_i)}{(F_{i+1} - F_i)}$$

$$142{,}222 = 120{,}000 + 60{,}000\frac{(P_{<\bar{x}} - 56.94)}{(75 - 56.94)}$$

$$P_{<\bar{x}} = 63.63\%$$

INTERPRETATION OF THE RESULTS

4-a. The modal population, which is the most frequent value of the population of the municipalities in this country, is 90,000 people, and it corresponds to the modal class [60,000; 120,000). This indicates that the majority of the municipalities in the country have a population that falls within this class.

4-b. The average population of all the municipalities in the country is 142,222 people, meaning this is the population that each municipality would have if the entire population of the country were distributed equally among all the municipalities. In this case, this value provides a general idea of the size of the municipalities in the country, but it should not be considered a relevant measure of central tendency if the distribution of populations is highly asymmetrical.

4-c. The median population, which divides the distribution of municipality populations into two equal parts, is 108,468.57 inhabitants. In other words, there are as many municipalities with a population below the median as there are municipalities with a population above the median. In this case, this measure of central tendency is more robust than the mean, as it is not affected by extreme values.

5. There are 63.63% of municipalities with a population below the average. In contrast, only 36.37% of municipalities have a population above the average. In other words, two-thirds of the municipalities have a population below the average, and only one-third of the municipalities have a population above the average. This could predict disparities, but at this stage, we cannot draw any conclusions as these measures are not sufficient to make such a claim.

6: Calculation of the characteristics of dispersion

The following table shows the data required to solve this exercise:

x_i	n_i	$n_i x_i$	$n_i x_i^2$
30000	45	1350000	40500000000
90000	78	7020000	631800000000
150000	39	5850000	877500000000
210000	23	4830000	1014300000000
270000	12	3240000	874800000000
330000	5	1650000	544500000000
390000	3	1170000	456300000000
450000	6	2700000	1215000000000
510000	1	510000	260100000000
570000	2	1140000	649800000000
630000	2	1260000	793800000000
	216	30,720,000	7,358,400,000,000

6-a: Finding the interquartile range:

$$Q_1 = a_i + (a_{i+1} - a_i)\frac{(25 - F_i)}{(F_{i+1} - F_i)}$$

$$= 60{,}000 + (120{,}000 - 60{,}000)\frac{(25 - 20.83)}{(56.94 - 20.83)}$$

$$= 66{,}928.82$$

$$Q_3 = a_i + (a_{i+1} - a_i)\frac{(75 - F_i)}{(F_{i+1} - F_i)}$$

$$= 180{,}000 + (240{,}000 - 180{,}000)\frac{(75 - 75)}{(85.75 - 75)}$$

$$= 180.000$$

$$\text{Distance Interquartile} = Q_3 - Q_1$$

$$= 180{,}000 - 66{,}928.82$$

$$= 113{,}071.18 \text{ people}$$

6-b: Finding the variance:

$$V(x) = \frac{1}{n}\sum_{i=1}^{i=11} n_i x_i^2 - \bar{x}^2$$

$$= \frac{7{,}358{,}400{,}000{,}000}{216} - (142{,}222.2)^2$$

$$= 13{,}839{,}506{,}173$$

6-c: Finding the standard deviation:

$$\sigma = \sqrt{V(x)}$$

$$= \sqrt{13{,}839{,}506{,}173}$$

$$= 117{,}641.43 \text{ people}$$

6-d: Finding the coefficient of variation:

$$\text{CV} = \frac{\sigma}{|\bar{x}|}$$

$$= \frac{117{,}641.43}{|142{,}222|} \text{ x } 100\%$$

$$= 82.7\%$$

7: Let's calculate the percentage of the population occupied by:

7-a: 50% of the least populated municipalities:

Let $P_{<50\%}$ be the percentage to find, and we use the linear interpolation formula using the data pair $(q_i \uparrow, f_i \uparrow)$:

$$P_{<50\%} = Q_i + (Q_{i+1} - Q_i)\frac{(50 - F_i)}{(F_{i+1} - F_i)}$$

$$= 4.39 + (27.25 - 4.39)\frac{(50 - 20.83)}{(56.94 - 20.83)}$$

$$= 22.86\%$$

7-b: 25% of the least populated municipalities

Let $P_{25\%<}$ be the percentage to find and use linear interpolation formula using the data pair $(q_i \uparrow, f_i \uparrow)$:

$$P_{>25\%} = Q_i + (Q_{i+1} - Q_i)\frac{(25 - F_i)}{(F_{i+1} - F_i)}$$

$$= 4.39 + (27.25 - 4.39)\frac{(25 - 20.83)}{(56.94 - 20.83)}$$

$$= 7.03\%$$

7-c: The percentage of the population occupied by 10% of the most populated municipalities.

Let $P_{10\%>}$ be the percentage of the population occupied by 10% of the most populated municipalities. This population can be obtained by subtracting from 100% the percentage of the population occupied by 90% of the least populated municipalities that is denoted by $P_{<90\%}$. We will then use the linear interpolation formula using the data pair $(q_i \uparrow, f_i \uparrow)$:

$$P_{>10\%} = 100 - P_{<90\%}$$

$$= 100 - \left[Q_i + (Q_{i+1} - Q_i)\frac{(90 - F_i)}{(F_{i+1} - F_i)}\right]$$

$$= 100 - \left[62.01 + (72.56 - 62.01)\frac{(90 - 85.65)}{(91.20 - 85.65)}\right]$$

$$= 100 - 70.28$$

$$= 29.72\%$$

The population of the 10% of the most populous municipalities represents about 30% of the total population of the country.

This result can be obtained by using the decreasing cumulative proportions.

8: Calculate the percentage of municipalities with a population:

8-a: Lower than average minus the standard deviation:

$$\bar{x} - \sigma = 142{,}222.2 - 117{,}641.43$$
$$= 24{,}580.77$$

Let $P_{<\bar{x}-\sigma}$ be the percentage to find and use the linear interpolation formula using the data pair $(a_i, f_i \uparrow)$:

$$\bar{x}-\sigma = a_i + (a_{i+1} - a_i)\frac{(P_{<\inf} - F_i)}{(F_{i+1} - F_i)}$$

$$24.580{,}77 = (60.000 - 0)\frac{(P_{<\bar{x}-\sigma} - 0)}{(20.83 - 0)}$$

$$P_{<\bar{x}-\sigma} = 8.53\%$$

8-b: Lower than average plus the standard deviation:

$$\bar{x} + \sigma = 142{,}222.2 + 117{,}641.43$$
$$= 259{,}863.63$$

Let $P_{<\bar{x}+\sigma}$ be the percentage to find and use the linear interpolation formula using the data pair $(a_i, f_i \uparrow)$:

$$\bar{x} + \sigma = a_i + (a_{i+1} - a_i)\frac{(P_{<\bar{x}+\sigma} - F_i)}{(F_{i+1} - F_i)}$$

$$259{,}863.63 = 240{,}000 + (300{,}000 - 240{,}000)\frac{(P_{<\bar{x}+\sigma} - 85.65)}{(91 - 85.65)}$$

$$P_{<\bar{x}+\sigma} = 87.42\%$$

8-c: Between the mean plus the standard deviation and the mean minus the standard deviation:

$$P_{<\bar{x}+\sigma} - P_{<\bar{x}+\sigma} = 87.42\% - 8.53\%$$
$$= 78.89\%$$

8-d: Lower than Guadeloupe (Caribbean Island with 395,700 inhabitants):

Let $P_{<\text{Guad}}$ be the percentage to find and use the linear interpolation formula using the data pair $(a_i, f_i \uparrow)$:

$$\text{Pop\_Guadeloupe} = a_i + (a_{i+1} - a)\frac{(F - F_i)}{(F_{i+1} - F_i)}$$

$$395{,}700 = 390{,}000 + (450{,}000 - 390{,}000)\frac{P_{<\text{Guad}} - 94.91}{(97.69 - 94.91)}$$

$$P_{<\text{Guad}} = 95.17\%$$

95.17% of the municipalities of this country have a lower population than Guadeloupe (that has 395,700 inhabitants).

INTERPRETATION OF THE RESULTS

6-a. The interquartile range of the distribution of populations of municipalities in the country is 113,071 people, which means that the central half of the distribution, i.e., the area between the first quartile (Q1) and the third quartile (Q3), has a range of 113,071 people.

6-d. The coefficient of variation of the distribution is 82.7%, which is a bit high, confirming the existence of variation in the size of populations in different municipalities. This measurement confirms the observation made in question 5 that two-thirds of municipalities have a population below the average, while only one-third have a population above the average.

7-a. When analyzing the least populated municipalities, it can be seen that half of them represent only 22.86% of the country's total population. This indicates a strong concentration of the population in the most populous municipalities of the country.

7-b. By examining the 25% least populated municipalities, it can be seen that they represent only 7.03% of the country's total population. This figure confirms the strong concentration of the population in the most populous municipalities.

7-c. The population of the top 10% most populous municipalities represents approximately 30% of the country's total population. This observation reinforces the idea that the distribution is massively dominated by a few of the most populous municipalities. In a normal situation, such a concentration of the population can have important implications for economic and social development, as it can lead to geographical disparities in access to resources and services.

8-c. Approximately 79% of municipalities in the country have a population between the mean minus the standard deviation and the mean plus the standard deviation, i.e., in the range of [24,581; 259,863.63] people. This observation suggests the existence of a relative concentration of the population of municipalities around the mean, which may reflect similar socio-economic characteristics in these municipalities.

8-d. Finally, it is interesting to note that 95.17% of municipalities in the country have a population lower than that of Guadeloupe, a French island in the Caribbean with a population of 395,700 inhabitants.

9: Calculation of the Characteristics of shape.

The following table shows the data required to solve this exercise:

x_i	n_i	$n_i x_i$	$n_i x_i^2$	$n_i(x_i-\bar{x})^3$	$n_i(x_i-\bar{x})^4$
30,000	45	1350000	40500000000	-63598827160493800	71372017146776400000000
90,000	78	7020000	631800000000	-11108633744856000	5801175400091450000000
150,000	39	5850000	877500000000	18349794238683	142720621856425000
210,000	23	4830000	1014300000000	7161266117969820	485374703551288000000
270,000	12	3240000	874800000000	25034979423868300	3198914037494280000000
330,000	5	1650000	544500000000	33105685871056200	6216512124676120000000
390,000	3	1170000	456300000000	45636078189300400	11307606040237800000000
450,000	6	2700000	1215000000000	174929489711934000	53839409611339700000000
510,000	1	510000	260100000000	49745803840877900	18295401190367300000000
570,000	2	1140000	649800000000	156561385459534000	66973481557689400000000
630,000	2	1260000	793800000000	232111163237311000	113218667401311000000000
	216	30,720,000	7,358,400,000,000	649,596,740,740,741,000	281,252,828,641,975,000,000,000

9-a: Calculation of the YULE's coefficient of skewness:

$$C_Y = \frac{Q_1 - 2Q_2 + Q_3}{Q_3 - Q_1}$$

$$= \frac{66{,}928.82 - 2(108{,}468.57) + 180{,}000}{180{,}000 - 66{,}928.82}$$

$$= 0.265246016$$

9-b: Calculation of the PEARSON's coefficient of skewness:

$$\beta_1 = \frac{(\bar{x} - \text{Mode})}{\sigma}$$

$$= \frac{(142{,}222 - 90{,}000)}{117{,}641.43}$$

$$= 0.4439$$

9-c: Calculation of the FISHER's coefficient of skewness:

$$\mu_3 = \frac{\sum_{i=1}^{i=11} n_i (x_i - \bar{x})^3}{\sum_{i=1}^{i=11} n_i} = \frac{649{,}596{,}740{,}740{,}741{,}000}{216}$$

$$= 3{,}007{,}392{,}318{,}244{,}171{,}29$$

$$\gamma_1 = \frac{\mu_3}{\sigma^3} = \frac{3{,}007{,}392{,}318{,}244{,}171{,}29}{(117{,}641.43)^3}$$

$$= 1.85$$

9-d: Calculation of the FISHER's coefficient of kurtosis:

$$\mu_4 = \frac{\sum_{i=1}^{i=11} n_i (x_i - \bar{x})^4}{\sum_{i=1}^{i=11} n_i} = \frac{281{,}252{,}828{,}641{,}975{,}000{,}000{,}000}{216}$$

$$= 1{,}302{,}096{,}428{,}898{,}032{,}407{,}407{,}4$$

$$\gamma_2 = \frac{\mu_4}{\sigma^4} - 3 = \frac{1{,}302{,}096{,}428{,}898{,}032{,}407{,}407{,}4}{(117{,}641.43)^4} - 3$$

$$= 3.7982$$

INTERPRETATION OF THE RESULTS

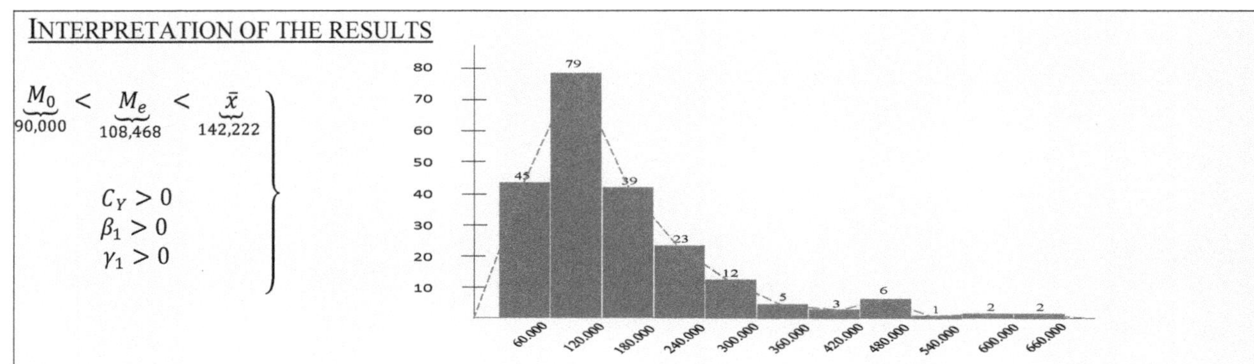

The distribution of the population of municipalities is left-skewed, indicating a higher concentration of low values and a tail spreading to the right. The distribution of populations of municipalities in the country is right-skewed, meaning that the tail of the distribution is spread to the right. In other words, the majority of municipalities have low populations, while the most populous municipalities are much fewer. This observation confirms the results obtained **in question 5,** where it was found that two-thirds of municipalities have a population lower than the average.

$\gamma_2 > 0$, this indicates that the curve of this distribution is sharper than that of the normal distribution. Low populations grow the distribution up to 79 municipalities, and then it drops sharply as the count moves to higher populations. In other words, there is a strong concentration of municipalities with populations below the average.

These results underline the importance of understanding population disparities between municipalities in the country. They may have significant implications for economic and social development as they can lead to geographical disparities in access to resources and services.

10: Let's calculate the characteristics of concentration

10-a: To determine the medial, we use linear interpolation formula using the data pair $(a_i, q_i \uparrow)$ using the table below:

Population	x_i	n_i	$x_i n_i$	f_i	q_i	$f_i \uparrow$	$q_i \uparrow$
[0;60000)	30000	45	1350000	20.83	4.39	20.83	4.39
[60000;120000)	90000	78	7020000	36.11	22.85	56.94	27.25
[120000;180000)	150000	39	5850000	18.06	19.04	75.00	46.29
[180000;240000)	210000	23	4830000	10.65	15.72	85.65	62.01
[240000;300000)	270000	12	3240000	5.56	10.55	91.20	72.56
[300000;360000)	330000	5	1650000	2.31	5.37	93.52	77.93
[360000;420000)	390000	3	1170000	1.39	3.81	94.91	81.74
[420000;480000)	450000	6	2700000	2.78	8.79	97.69	90.53
[480000;540000)	510000	1	510000	0.46	1.66	98.15	92.19
[540000;600000)	570000	2	1140000	0.93	3.71	99.07	95.90
[600000;660000)	630000	2	1260000	0.93	4.10	100.00	100.00
Total	Total	216	30720000	100.00	100.00		

$$\overset{\approx}{M}_e = a_i + (a_{i+1} - a_i)\frac{(50 - Q_i)}{(Q_{i+1} - Q_i)}$$

$$= 180{,}000 + (240{,}000 - 180{,}000)\frac{(50 - 46.29)}{(62.01 - 46.29)}$$

$$= 194{,}160.305 \text{ people}$$

10-b: Finding the percentage of municipalities with a population below the medial

Let $P_{<\text{medial}}$ be the percentage to find and use linear interpolation formula using the data pair $(f_i \uparrow, q_i \uparrow)$:

$$P_{<\text{medial}} = f_i + (f_{i+1} - f_i)\frac{(50 - Q_i)}{(Q_{i+1} - Q_i)}$$

$$= 75 + (85.65 - 75)\frac{(50 - 46.29)}{(62.01 - 46.29)}$$

$$= 77.5\%$$

10-c: Plotting the LORENZ curve.

The plot of the LORENZ curve uses the previous table:

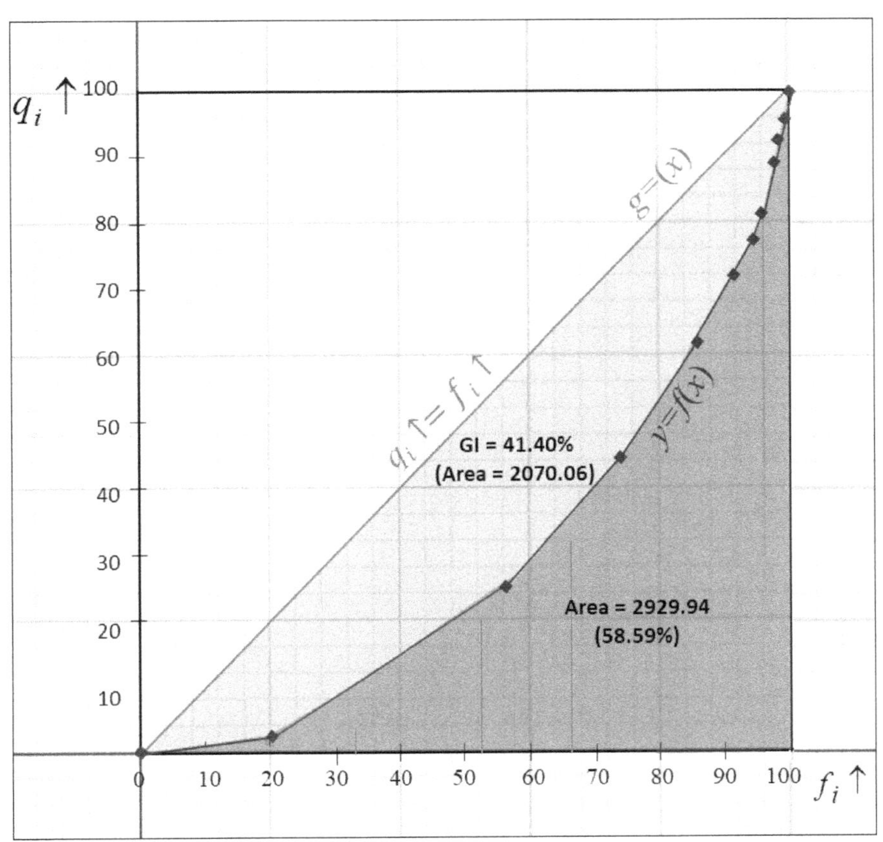

10-d: Let's calculate the GINI index by the Trapezoid method

The following table shows the data required to solve this exercise:

$f_i \uparrow$	$q_i \uparrow$	$H_i = f_{i+1}\uparrow - f_i\uparrow$	$B_i = \frac{q_i \uparrow + q_{i+1} \uparrow}{2}$	Trapezoid Area (T_i) $T_i = B_i x H_i$
20.83	4.39	36.11	15.82	571.29
56.94	27.25	18.06	36.77	663.86
75.00	46.29	10.65	54.15	576.60
85.65	62.01	5.56	67.29	373.81
91.20	72.56	2.31	75.24	174.18
93.52	77.93	1.39	79.83	110.88
94.91	81.74	2.78	86.13	239.26
97.69	90.53	0.46	91.36	42.30
98.15	92.19	0.93	94.04	87.08
99.07	95.90	0.93	97.95	90.69
100.00	100.00			
Total				2929.94

The area under the LORENZ curve is:

$$\sum_{i=1}^{i=11} (f_{i+1} \uparrow - f_i \uparrow) \frac{(q_i \uparrow + q_{i+1} \uparrow)}{2} = 2{,}929.94$$

It follows that:

$$\begin{aligned} \text{GINI Index} &= 100\% - \left(\frac{2{,}929.94}{5{,}000}\right) x100 \\ &= 100\% - 58.59\% \\ &= 41.40\% \end{aligned}$$

10-e: Let's calculate the GINI index by the definite integral calculus.

We will consider the following key $(f_i \uparrow, q_i \uparrow)$ points in the previous table:

(0,0)) – (21,4) – (57,27) – (75, 46) – (91, 73) – (100,100)

Following the computation, the interpolation polynomial is obtained as:

$$y(x) = \frac{427251329}{6793659348840000} x^5 - \frac{92505640501}{6793659348840000} x^4 + \frac{109984820587}{104517836136000} x^3$$

$$-\frac{58358161910161}{2264553116280000}x^2+\frac{411346371641}{1078358626800}x$$

Evaluating the area between the $q_i \uparrow$ curve and the $f_i \uparrow$ axis in the interval [0 ; 100)

$$\text{Area} = \int_0^{100} y\,dx = \frac{427251329}{6793659348840000}\left(\frac{10^{12}}{6}\right) - \frac{92505640501}{6793659348840000}\left(\frac{10^{10}}{5}\right) + \frac{109984820587}{104517836136000}\left(\frac{10^8}{4}\right) - \frac{58358161910161}{2264553116280000}\left(\frac{10^6}{3}\right) + \frac{411346371641}{1078358626800}\left(\frac{10^4}{2}\right)$$

$$= 2{,}873.54$$

Let's calculate the GINI concentration index by the interpolation method:

$$\text{GINI Index} = 100\% - \left(\frac{\int_0^{100} f(x)\,dx}{5{,}000}\right) \text{x}100$$

$$= 100\% - \left(\frac{2{,}873.54}{5{,}000}\right) \text{x}100$$

$$= 42.53\%$$

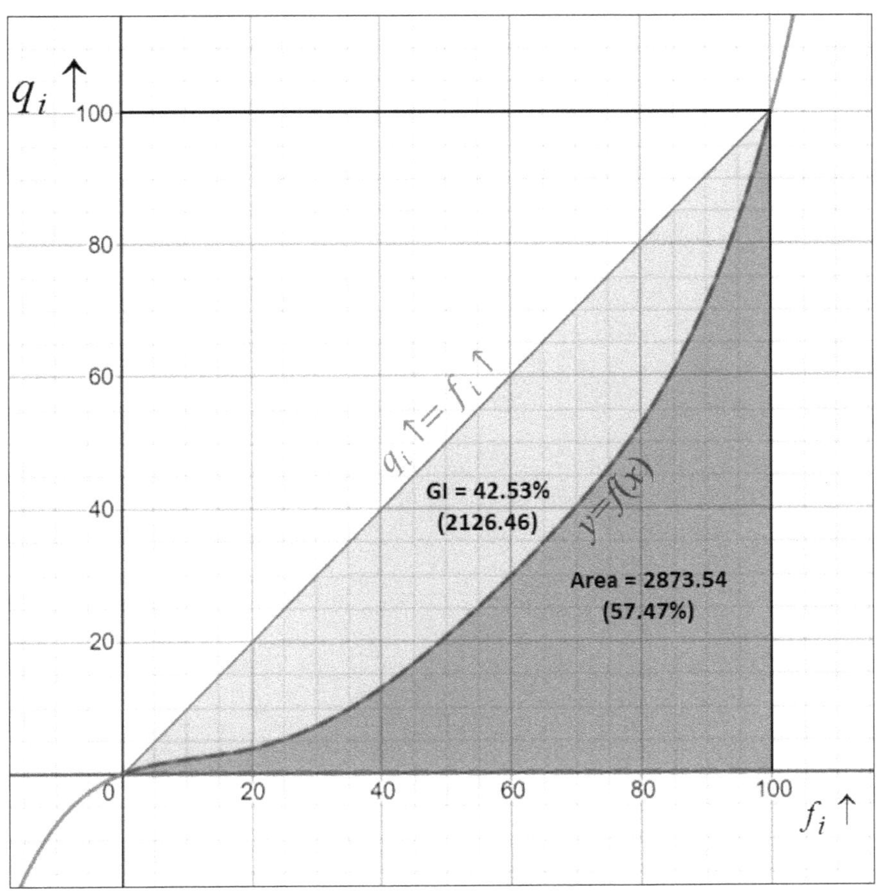

INTERPRETATION OF THE RESULTS

10-a. The medial population of the municipalities in the country is defined as the population at which the sum of the populations of the least populated municipalities is equal to the sum of the populations of the most populated municipalities. For the country under consideration, this median population is 194,160 people. Thus, 50% of the total population of the country lives in municipalities with a population equal to or greater than this median value, while the remaining 50% live in municipalities with a population less than this value. This result shows a strong concentration of the population in the most populated municipalities, while the majority of municipalities have a relatively low population.

10-b. A closer look at the distribution of population across municipalities in the country reveals a striking inequality: 77.5% of the least populated municipalities, accounting for over three-quarters of the total, represent only 50% of the national population. In other words, the other half of the population lives in just 22.5% of the most populated municipalities. This imbalance becomes even more pronounced when applying the Pareto principle, which states that 80% of effects are caused by 20% of the causes: here, approximately one-fifth of the national population living in the most populated municipalities represents the four-fifths of the national population living in the least populated municipalities. This concentration of population in a small number of municipalities poses challenges in terms of urban planning, provision of public services, and territorial development.

10-c/d. The GINI index of the distribution of the population in the country is 42.53%, indicating that disparities in the populations of different municipalities in the country are significant. This means that some municipalities have a significantly higher population than most others, while others have a much lower population. This value of the GINI index justifies the discrepancies between the modal, mean, and median populations found in **question 4**, and is consistent with the Coefficient of Variation of 82.7% found in **question 6-d**, as well as the shape coefficients observed in **question 9**.

9.7.4.2 Study of the Area

1-a: To count this series, we will first sort the municipalities in increasing order of area.

Nbr	Area	Nbr	Area	Nbr	Area
1	2.00	73	809.50	145	15397.00
2	2.90	74	1076.80	146	15498.00
3	2.90	75	1480.00	147	15634.00
4	3.20	76	1727.00	148	15770.00
5	3.40	77	1747.00	149	15786.00
6	3.90	78	1800.00	150	16015.00
7	4.00	79	1836.00	151	16055.00
8	4.70	80	1960.00	152	16201.00
9	4.80	81	2021.00	153	16797.00
10	4.90	82	2100.00	154	16898.00
11	5.00	83	2397.00	155	17328.00
12	5.30	84	3090.00	156	17411.00
13	5.60	85	3148.00	157	17494.00
14	6.60	86	3270.00	158	17682.00
15	6.80	87	3371.00	159	17861.00
16	8.50	88	3620.00	160	18096.00
17	10.00	89	4265.00	161	18098.00
18	10.30	90	4680.00	162	18126.00
19	11.40	91	4734.00	163	18773.00
20	11.40	92	5221.00	164	18926.00
21	12.80	93	5289.00	165	19073.00
22	14.00	94	5318.00	166	19187.00
23	14.20	95	5707.00	167	19264.00
24	14.30	96	5726.00	168	19513.00
25	15.00	97	6740.00	169	19718.00
26	16.60	98	6749.00	170	19805.00
27	18.00	99	6784.00	171	19865.00
28	18.00	100	7091.00	172	19996.00
29	18.50	101	7204.00	173	20155.00
30	20.00	102	7484.00	174	20503.00
31	20.00	103	7948.80	175	21181.00
32	20.40	104	7968.00	176	21239.00
33	23.00	105	8184.00	177	21675.00
34	23.20	106	8190.00	178	22436.00
35	23.70	107	8450.00	179	22466.00
36	23.90	108	8507.00	180	22480.00
37	24.00	109	8601.00	181	22567.00
38	25.10	110	8605.00	182	22673.00
39	27.00	111	8608.00	183	22909.00
40	27.10	112	8730.00	184	23081.00
41	28.50	113	8961.00	185	23475.00
42	34.20	114	9128.00	186	24184.00
43	36.00	115	10305.00	187	24350.00
44	36.10	116	10530.00	188	24430.00
45	39.00	117	10559.00	189	24598.00
46	40.60	118	10736.00	190	24700.00
47	44.00	119	11172.00	191	24953.00
48	53.30	120	11488.00	192	24963.00
49	59.90	121	11747.00	193	25074.00
50	69.70	122	12054.00	194	25175.00
51	76.90	123	12059.00	195	25216.00
52	81.70	124	12070.00	196	25417.00
53	88.70	125	12229.00	197	25490.00
54	90.00	126	12833.00	198	25894.00
55	102.10	127	12848.00	199	26217.00
56	112.00	128	12881.00	200	26346.00
57	142.50	129	13108.00	201	26648.00
58	153.00	130	13223.00	202	26665.00
59	185.10	131	13277.00	203	26676.00
60	192.20	132	13400.00	204	27910.00
61	197.50	133	13403.00	205	30363.00
62	221.80	134	13404.00	206	30512.00
63	230.20	135	13434.00	207	32446.00
64	274.90	136	13842.00	208	34198.00
65	281.00	137	14222.00	209	34704.00
66	291.60	138	14327.00	210	36385.00
67	300.00	139	14373.00	211	36783.00
68	358.90	140	14542.00	212	38075.00
69	405.00	141	14572.00	213	40214.00
70	641.40	142	14702.00	214	41824.00
71	753.40	143	14733.00	215	42196.00
72	768.00	144	15250.00	216	47087.00

1-b: Splitting and representing the series

As the study in this series is relatively identical to the first one, which is why some passages will not be elaborated on because the comments and formulas used in the first part remain valid. Similarly, by grouping the municipalities in classes of amplitude 5,000 km^2 we have the following statistical table:

Area	Number of Municipalities
[0;5000)	91
[5000;10000)	23
[10000;15000)	29
[15000;20000)	29
[20000;25000)	20
[25000;30000)	12
[30000;35000)	5
[35000;40000)	3
[40000;45000)	3
[45000;50000)	1
Total	216

2: Plotting the histogram and the polygon of frequencies of the distribution:

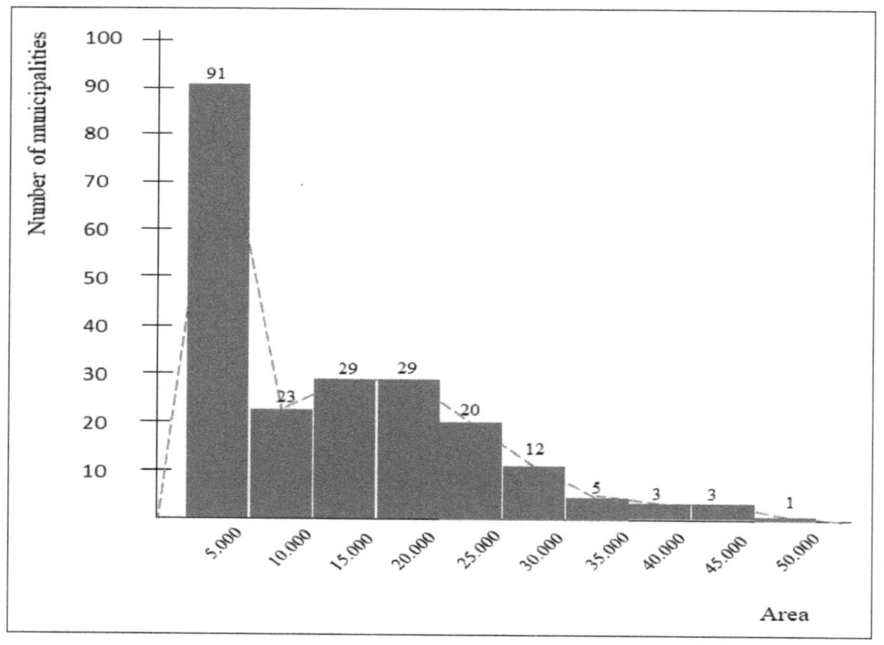

As with the first series, the plot of the histogram presents no difficulty because all classes are of equal amplitude.

The frequency polygon is obtained by joining the mid-stages of the histogram. The polygon is closed by taking the middle of the fictitious rectangles lying to the right and left of the histogram.

3: Let's define the distribution function and then draw the curves of the cumulative relative frequencies:

3-1: The distribution function associated with this statistical series is that which in any number x associates the number of municipalities with a strictly less area than x (continuous character).

Indeed

if

$x < 0$	$F(x) \leq 0$
$x < 0$	$F(x) \leq 0$
$x \epsilon [0\ ;\ 5000]$	$F(x) = 91$
$x \epsilon [5000\ ;\ 10000]$	$F(x) = 91 + 23 = 114$
$x \epsilon [10000\ ;\ 15000]$	$F(x) = 91 + 23 + 29 = 143$
$x \epsilon [15000\ ;\ 20000]$	$F(x) = 91 + 23 + 29 + 29 = 172$
$x \epsilon [20000\ ;\ 25000]$	$F(x) = 91 + 23 + 29 + 29 + 20 = 192$
$x \epsilon [25000\ ;\ 30000]$	$F(x) = 91 + 23 + 29 + 29 + 20 + 12 = 204$
$x \epsilon [30000\ ;\ 35000]$	$F(x) = 91 + 23 + 29 + 29 + 20 + 12 + 5 = 209$
$x \epsilon [35000\ ;\ 40000]$	$F(x) = 91 + 23 + 29 + 29 + 20 + 12 + 5 + 3 = 212$
$x \epsilon [40000\ ;\ 45000]$	$F(x) = 91 + 23 + 29 + 29 + 20 + 12 + 5 + 3 + 3 = 215$
$x \epsilon [45000\ ;\ 50000]$	$F(x) = 91 + 23 + 29 + 29 + 20 + 12 + 35 + 3 + 3 + 1 = 216$

3-2: Plotting the cumulative relative frequency curves.

Solving this exercise requires the following table:

Area	Number of municipalities	f_i	$f_i \uparrow$	$f_i \downarrow$
[0;5000)	91	42.13	42.13	100.00
[5000;10000)	23	10.65	52.78	57.87
[10000;15000)	29	13.43	66.20	47.22
[15000;20000)	29	13.43	79.63	33.80
[20000;25000)	20	9.26	88.89	20.37
[25000;30000)	12	5.56	94.44	11.11
[30000;35000)	5	2.31	96.76	5.56
[35000;40000)	3	1.39	98.15	3.24
[40000;45000)	3	1.39	99.54	1.85
[45000;50000)	1	0.46	100.00	0.46
Total	216	100.00		

The increasing and decreasing cumulative relative frequency curves are:

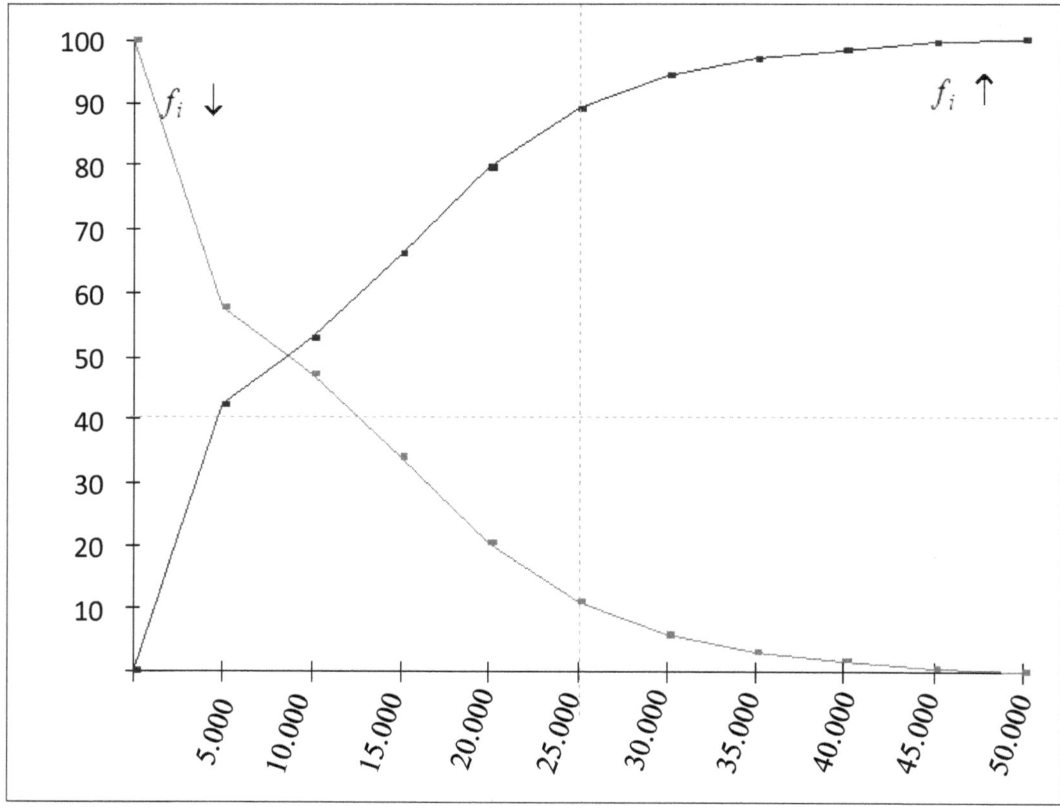

4: Finding central trend characteristics:

The following table shows the data required to solve this exercise:

Area	x_i	n_i	$n_i x_i$	f_i	$f_i \uparrow$
[0;5000)	2,500	91	227,500	42.13	42.13
[5000;10000)	7,500	23	172,500	10.65	52.78
[10000;15000)	12,500	29	362,500	13.43	66.20
[15000;20000)	17,500	29	507,500	13.43	79.63
[20000;25000)	22,500	20	450,000	9.26	88.89
[25000;30000)	27,500	12	330,000	5.56	94.44
[30000;35000)	32,500	5	162,500	2.31	96.76
[35000;40000)	37,500	3	112,500	1.39	98.15
[40000;45000)	42,500	3	127,500	1.39	99.54
[45000;50000)	47,500	1	47,500	0.46	100.00
Total	Total	216	2,500,000	100.00	

4-a: Finding the mode

$$\text{Mode } = 2.500 \text{ Km}^2$$

$$\text{Modal class} = [0\ ;\ 5000)$$

4-b: Finding the mean

$$\bar{x} = \frac{1}{n}\sum_{i=1}^{i=10} n_i x_i = \frac{2{,}500{,}000}{216}$$

$$= 11{,}574 \text{ Km}^2$$

4-c: Finding the median

$$\tilde{x} = a_i + (a_{i+1} - a_i)\frac{(50 - F_i)}{(F_{i+1} - F_i)}$$

$$= 5{,}000 + 5{,}000\frac{(50 - 42.13)}{(52.78 - 42.13)}$$

$$= 8{,}694.84 \text{ Km}^2$$

5: Finding the percentage of municipalities with a below-average area:

We know that the average $\bar{x} = 11{,}574 \text{ Km}^2$

Let P be the percentage to find, and we use the linear interpolation formula using the data pair $(a_i, f_i \uparrow)$:

$$\bar{x} = a_i + (a_{i+1} - a_i)\frac{(P_{<\bar{x}} - F_i)}{(F_{i+1} - F_i)}$$

$$11{,}574.07 = 10{,}000 + 5{,}000\frac{(P_{<\bar{x}} - 52.78)}{(66.20 - 52.78)}$$

$$= 57\%$$

INTERPRETATION OF THE RESULTS

4-a. The modal area is 2,500 Km^2, and the modal class is [0; 5,000[, which means that the majority of the communes in this country have an area around 2,500 Km^2. This information helps us to understand that most communes in the country are relatively small, with an area between 0 and 5,000 Km^2.

4-b. The average area of all the municipalities in this country is 11,574 km2, meaning that if the entire national territory were distributed equally among all the municipalities, the area of each municipality would be 11,574 km2. This information shows that the distribution of land area among the municipalities in this country is very unequal, with some municipalities having a very large area while others are very small. This inequality can have consequences on the management of natural resources, territorial planning, and economic development of the country.

4-c. The median area is 8,694.84 km2, which means that by arranging all the 216 municipalities of this country in ascending order of their areas (see counting at the beginning of the exercise), the median area corresponds to the municipality located in the middle of the ranking. In other words, it is the area where there are as many municipalities smaller than it as there are larger municipalities. This information allows us to understand that half of the municipalities in the country have an area smaller than 8,694.84 km2 and the other half has a larger area. This shows that the distribution of land area among the municipalities is very unequal.

5. There are 57% of municipalities with an area smaller than the average and 43% of municipalities with an area larger than the average. This information complements that of question **4-b** by showing that the majority of municipalities in the country have an area smaller than the average. This can have consequences on public policies regarding territorial planning and economic development.

6: Finding the characteristics of dispersion.

The following table shows the data required to solve this exercise:

Area	x_i	n_i	$x_i n_i$	$x_i n_i^2$
[0;5000)	2,500	91	227,500	568,750,000
[5000;10000)	7,500	23	172,500	1,293,750,000
[10000;15000)	12,500	29	362,500	4,531,250,000
[15000;20000)	17,500	29	507,500	8,881,250,000
[20000;25000)	22,500	20	450,000	10,125,000,000
[25000;30000)	27,500	12	330,000	9,075,000,000
[30000;35000)	32,500	5	162,500	5,281,250,000
[35000;40000)	37,500	3	112,500	4,218,750,000
[40000;45000)	42,500	3	127,500	5,418,750,000
[45000;50000)	47,500	1	47,500	2,256,250,000
Total	Total	216	2,500,000	51,650,000,000

6-a: The interquartile distance of the distribution is:

$$Q_1 = a_i + (a_{i+1} - a_i)\frac{(25 - F_i)}{(F_{i+1} - F_i)}$$

$$= 0 + (5{,}000 - 0)\frac{(25 - 0)}{(42.13 - 0)}$$

$$= 2{,}967$$

$$Q_3 = a_i + (a_{i+1} - a_i)\frac{(75 - F_i)}{(F_{i+1} - F_i)}$$

$$= 15{,}000 + (20{,}000 - 15{,}000)\frac{(75 - 66.2)}{(79.63 - 66.2)}$$

$$= 18{,}276.24$$

$$\text{Distance Interquartile} = Q_3 - Q_1$$
$$= 18{,}276.24 - 2{,}967$$
$$= 15{,}309.24 \text{ Km}^2$$

6-b: Finding the variance

$$V(x) = \frac{1}{n}\sum_{i=1}^{i=10} n_i x_i^2 - \bar{x}^2$$
$$= \frac{51{,}650{,}000{,}000}{216} - (11{,}574.07)^2$$
$$= 105{,}161{,}179.70$$

6-c: Finding the standard deviation

$$\sigma = \sqrt{V(x)}$$
$$= \sqrt{105{,}161{,}179.70}$$
$$= 10{,}254.81$$

6-d: Finding the coefficient of variation

$$\text{CV} = \frac{\sigma}{|\bar{x}|}$$
$$= \frac{10{,}254.81}{|11{,}574\,|} \text{ x } 100$$
$$= 88.6\%$$

7: Finding the percentage of the area occupied by:

7-a: by 50% of the smallest municipalities.

Let $P_{<50\%}$ be the percentage to find and use linear interpolation formula on $(q_i \uparrow, f_i \uparrow)$:

$$P_{<50\%} = Q_i + (Q_{i+1} - Q_i)\frac{(50 - F_i)}{(F_{i+1} - F_i)}$$
$$= 9.10 + (16 - 9.10)\frac{(50 - 42.13)}{(52.78 - 42.13)}$$
$$= 14.2\%$$

7-b: 25% of the smallest municipalities.

Let $P_{<25\%}$ be the percentage to find and use linear interpolation formula using the data pair $(q_i \uparrow, f_i \uparrow)$:

$$P_{<25\%} = Q_i + (Q_{i+1} - Q_i)\frac{(25 - F_i)}{(F_{i+1} - F_i)}$$

$$= 0 + 9.10\frac{(25 - 0)}{(42.13 - 0)}$$

$$= 4.4\%$$

7-c: 10% of the largest municipalities.

Let $P_{>10}\%$ be the percentage of the area occupied by 10% of the largest municipalities. This percentage can be obtained by subtracting by 100% the percentage of the area occupied by 90% of the smallest municipalities denoted by $P_{<90\%}$. We will then apply and use the linear interpolation formula using the data pair $(q_i \uparrow, f_i \uparrow)$:

$$P_{>10\%} = 100\% - P_{<90\%}$$

$$= 100\% - \left[Q_i + (Q_{i+1} - Q_i)\frac{(90 - F_i)}{(F_{i+1} - F_i)}\right]$$

$$= 100\% - \left[68.80 + (82 - 68.80)\frac{(90 - 88.89)}{(94.44 - 88.89)}\right]$$

$$= 100 - 71.44$$

$$= 28.56\%$$

8: Finding the percentage of municipalities having an area:

8-a: Lower than average minus the standard deviation:

$$\bar{x} - \sigma = 11{,}574 - 10{,}254.81$$
$$= 1{,}319.19 \text{ Km}^2$$

Let $P_{<\bar{x}-\sigma}$ be the percentage of municipalities with a higher-than-average area minus the standard deviation and use linear interpolation formula using the data pair: $(a_i, f_i \uparrow)$

$$\bar{x} - \sigma = a_i + (a_{i+1} - a_i)\frac{(P_{<\bar{x}-\sigma} - F_i)}{(F_{i+1} - F_i)}$$

$$1{,}319.1 = 0 + (5{,}000 - 0)\frac{(P_{<\bar{x}-\sigma} - 0)}{(42.13 - 0)}$$

$$P_{<\bar{x}-\sigma} = 8.53\%$$

$$P_{<\bar{x}-\sigma} = 11.12\%$$

8-b: Lower than average plus the standard deviation

$$\bar{x} + \sigma = 11{,}574 + 10{,}254.81$$
$$= 21{,}828.81 \text{ Km}^2$$

Let $P_{<\bar{x}+\sigma}$ be the percentage to find and use linear interpolation formula using the data pair $(a_i, f_i \uparrow)$

$$\bar{x} + \sigma = a_i + (a_{i+1} - a_i)\frac{(P_{<\bar{x}+\sigma} - F_i)}{(F_{i+1} - F_i)}$$

$$21{,}828.81 = 20{,}000 + (25{,}000 - 20{,}000)\frac{(P_{<\bar{x}+\sigma} - 79.63)}{(88.89 - 79.63)}$$

$$P_{<\bar{x}+\sigma} = 83.02\%$$

8-c: Between the average minus the standard deviation and the average plus the standard deviation:

$$P_{<\bar{x}+\sigma} - P_{<\bar{x}-\sigma} = 83.01\% - 11.12\%$$
$$= 71.90\%$$

8-d: Larger than the Paris region that has 12012 Km^2.

Let $P_{>\text{Paris}}$ be the percentage of municipalities with a larger area than the Paris region. This percentage can be obtained by subtracting from the 100% the percentage of municipalities with an area less than that of the Paris region denoted by $P_{<\text{Paris}}$. We will then apply and use the linear interpolation formula using the data pair $(a_i \uparrow, f_i \uparrow)$:

$$P_{>Paris} = 100\% - P_{<Paris}$$

Finding $P_{<Paris}$

$$P_{>\text{Paris}} = a_i + (a_{i+1} - a)\frac{(P_{<Paris} - F_i)}{(F_{i+1} - F_i)}$$

$$12{,}012 = 10{,}000 + (15{,}000 - 10{,}000)\frac{(P_{<Paris} - 52.78)}{(66.20 - 52.78)}$$

$$P_{<\text{Paris}} = 58.18\%$$

$$P_{>\text{Paris}} = 100\% - 58.18\%$$

$$= 41.82\%$$

41.82% of municipalities in the DRC are larger than the Paris region.

INTERPRETATION OF THE RESULTS

6-a. The interquartile distance of the distribution, which is 15,309.24 Km^2, that is, the difference between the value of the 3rd quartile (75th percentile) and that of the 1st quartile (25th percentile) of the distribution is 15,309.24 Km^2. This means that there are relatively minor variations in the areas of the communes located in the central half of the distribution. In other words, the sizes of most communes are quite similar.

6-d. The coefficient of variation of 88.6% confirms the existence of significant differences in the sizes of different communes in the country. This means that some communes are much larger than others, and the distribution of commune sizes is not uniform.

7-a. The fact that the 50% smallest communes in the country only occupy 14.2% of the total land area, highlights the stark inequality in the distribution of land area among communes. This means that some communes have much larger land areas than others, and the distribution is heavily skewed towards a few large communes. This could have implications for the management of natural resources, territorial planning, and economic development in the country, as some communes may have more resources and opportunities than others.

7-b. It can be observed that 25% of the smallest communes represent only 4.4% of the total land area of the country. This shows that the majority of the land area is concentrated in medium to large-sized communes.

7-c. The indicator that the land area of the largest 10% of communes represents approximately 29% of the total land area of the country is very revealing. It shows that the distribution of land area is heavily dominated by a small number of very large communes.

8-c. Approximately 72% of communes have a land area that falls between the mean minus the standard deviation and the mean plus the standard deviation, that is, in the interval of [1,319, 21,828] Km^2. This shows that there is a relative concentration of commune size around the mean. This relative concentration may reflect some coherence in territorial planning and natural resource management.

8-d. The fact that 41.82% of communes in this country have a land area greater than that of the Paris region is important information. This shows how vast and diverse this country is, with very different regions in terms of size. This diversity may present challenges in governance and development planning but may also offer opportunities for balanced and inclusive regional development.

9: Finding characteristics of shape:

The following table shows the data required to solve this exercise:

x_i	n_i	$n_i x_i$	$n_i x_i^2$	$n_i(x_i - \bar{x})^3$	$n_i(x_i - \bar{x})^4$
2,500	91	227500	568750000	-67990518467713	616951000910732000
7,500	23	172500	1293750000	-1555301529238	6336413637638050
12,500	29	362500	4531250000	23021134990	21315865731097
17,500	29	507500	8881250000	6034852410710	35762088359761500
22,500	20	450000	10125000000	26085835492557	285011906307567000
27,500	12	330000	9075000000	48472488949855	771969268460657000
32,500	5	162500	5281250000	45816726362851	958757422037442000
37,500	3	112500	4218750000	52278616064624	1355371527601370000
42,500	3	127500	5418750000	88733862978205	2744176873585210000
47,500	1	47500	2256250000	46368592186151	1665834608169110000
	216	2,500,000	51,650,000,000	244,268,175,582,990	8,440,192,424,935,220,000

9-a: The YULE's coefficient of skewness is:

$$C_Y = \frac{Q_1 - 2Q_2 + Q_3}{Q_3 - Q_1}$$

$$= \frac{2{,}967 - 2(8{,}694.84) + 18{,}276.24}{18{,}276.24 - 2{,}967}$$

$$= 0.251714651$$

9-b: The PEARSON's coefficient of skewness is:

$$\beta_1 = \frac{(\bar{x} - \text{Mode})}{\sigma}$$

$$= \frac{(11{,}574 - 2{,}500)}{10{,}254.81}$$

$$= 0.8848$$

9-c: The FISHER's coefficient of skewness is:

$$\mu_3 = \frac{\sum_{i=1}^{i=10} n_i (x_i - \bar{x})^3}{\sum_{i=1}^{i=10} n_i} = \frac{244{,}268{,}175{,}582{,}9900}{216}$$

$$= 1{,}130{,}871{,}183{,}254.59$$

$$\gamma_1 = \frac{\mu_3}{\sigma^3} = \frac{1{,}130{,}871{,}183{,}254.59}{(10{,}254.81)^3}$$

$$= 1.05$$

9-d: The FISHER's coefficient of kurtosis is:

$$\mu_4 = \frac{\sum_{i=1}^{i=10} n_i (x_i - \bar{x})^4}{\sum_{i=1}^{i=10} n_i} = \frac{8{,}440{,}192{,}424{,}935{,}220{,}000}{216}$$

$$= 390{,}749{,}64{,}930{,}255{,}700.00$$

$$\gamma_2 = \frac{\mu_4}{\sigma^4} - 3 = \frac{39{,}074{,}964{,}930{,}255{,}700.00}{(10{,}254.81)^4} - 3$$

$$= 0.53$$

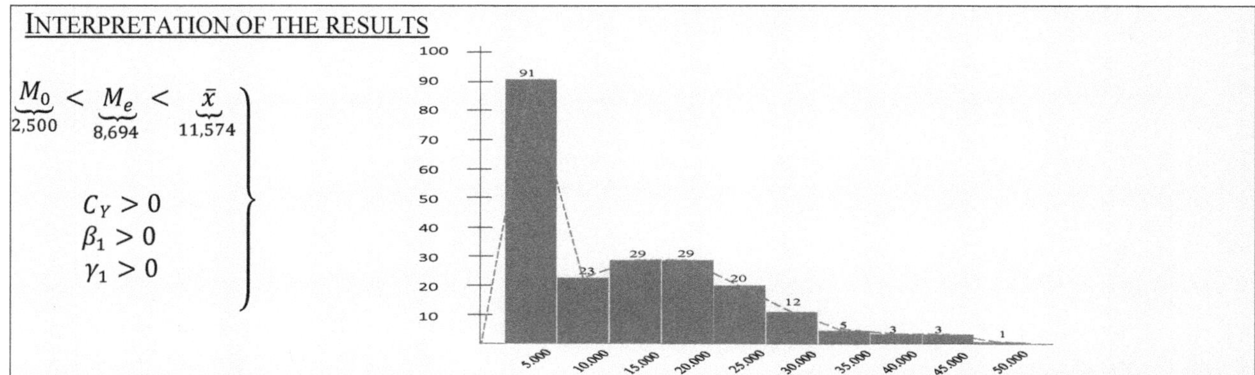

The distribution exhibits a right-skewed pattern, meaning that the tail of the distribution extends further to the right than to the left. This implies that the majority of observed values are concentrated in the lower end of the distribution, with relatively small land areas for most communes. In other words, the distribution is biased towards small land areas, which is confirmed by a strong concentration of values in the left part of the distribution.

$\gamma_2 > 0$, this indicates that the curve of this distribution is sharper than that of the normal distribution. This shape is largely attributed to the significant number of municipalities with areas less than 10,000 km2, which causes the distribution to rapidly increase to 91 municipalities before sharply declining as area size increases.

10: Let's calculate the characteristics of concentration

10-a: To determine the medial, we use linear interpolation formula using the data pair$(q_i, f_i \uparrow)$ using the table below:

Area	x_i	n_i	$x_i n_i$	f_i	q_i	$f_i \uparrow$	$q_i \uparrow$
[0;5000 [	2,500	91	227500	42.13	9.10	42.13	9.10
[5000;10000 [	7,500	23	172500	10.65	6.90	52.78	16.00
[10000;15000 [	12,500	29	362500	13.43	14.50	66.20	30.50
[15000;20000 [	17,500	29	507500	13.43	20.30	79.63	50.80
[20000;25000 [	22,500	20	450000	9.26	18.00	88.89	68.80
[25000;30000 [	27,500	12	330000	5.56	13.20	94.44	82.00
[30000;35000 [	32,500	5	162500	2.31	6.50	96.76	88.50
[35000;40000 [	37,500	3	112500	1.39	4.50	98.15	93.00
[40000;45000 [	42,500	3	127500	1.39	5.10	99.54	98.10
[45000;50000 [	47,500	1	47500	0.46	1.90	100.00	100.00
Total	Total	216	2500000	100.00	100.00		

$$\tilde{\tilde{M}}_e = a_i + (a_{i+1} - a_i)\frac{(50 - q_i)}{(Q_{i+1} - Q_i)}$$

$$= 15{,}000 + 5{,}000\frac{(50 - 30.5)}{(50.8 - 30.5)}$$

$$= 19{,}802.96$$

10-b: Finding the percentage of municipalities with an area less than the medial:

Let $P_{<\text{medial}}$ be the percentage to find and use linear interpolation formula using the data pair ($f_i \uparrow$ $, q_i \uparrow$):

$$P_{<\text{medial}} = f_i + (f_{i+1} - f_i)\frac{(50 - Q_i)}{(Q_{i+1} - Q_i)}$$

$$= 66.20 + (79.63 - 66.20)\frac{(50 - 30.50)}{(50.80 - 30.50)}$$

$$= 79.1\%$$

79.1% of the smaller municipalities represent 50% of the national area. That is, 20.9% of the largest municipalities also account for 50% of the national area.

10-c: Plotting the LORENZ curve

The plot of the LORENZ curve uses the previous table:

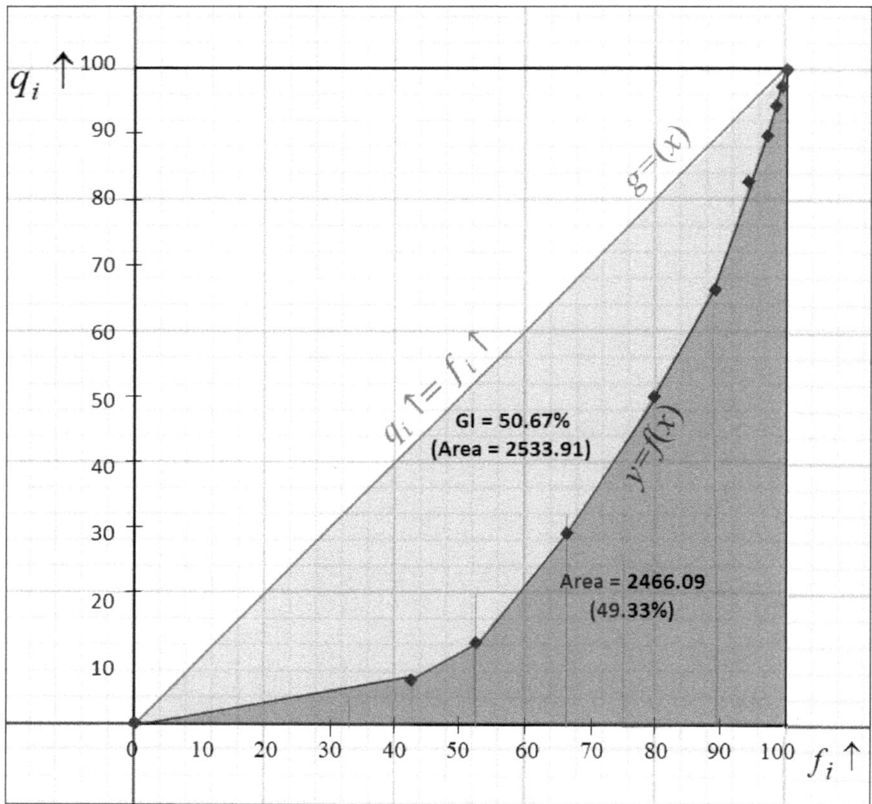

10-d: Let's calculate the GINI index by the Trapezoid method

Resolving the remainder of this exercise requires the columns in the following table:

$f_i \uparrow$	$q_i \uparrow$	$H_i = f_{i+1}\uparrow - f_i\uparrow$	$B_i = \frac{q_i\uparrow + q_{i+1}\uparrow}{2}$	Trapezoid Area (T_i) $T_i = B_i \text{x} H_i$
42.13	9.10	10.65	12.55	133.63
52.78	16.00	13.43	23.25	312.15
66.20	30.50	13.43	40.65	545.76
79.63	50.80	9.26	59.80	553.70
88.89	68.80	5.56	75.40	418.89
94.44	82.00	2.31	85.25	197.34
96.76	88.50	1.39	90.75	126.04
98.15	93.00	1.39	95.55	132.71
99.54	98.10	0.46	99.05	45.86
100.00	100.00			
Total				2466.09

The area under the LORENZ curve is:

$$\sum_{i=1}^{i=10}(f_{i+1}\uparrow - f_i\uparrow)\frac{(q_i\uparrow + q_{i+1}\uparrow)}{2} = 2{,}466.09$$

It follows that:

$$\text{GINI Index} = 100\% - \left(\frac{2{,}466.09}{5{,}000}\right)\text{x}100$$
$$= 100\% - 49.32\%$$
$$= 50.67\%$$

10-e: Let's calculate the GINI index by the definite integral calculus.

We will consider the following key $(f_i\uparrow, q_i\uparrow)$ points in the previous table:

(0.0)) - (42.9) - (52.16) -(66.31) - (94, 82) -(100,100)

After calculation, the polynomial interpolation is:

$$y(x) = \frac{7681}{133598424960}x^5 - \frac{336337}{22266404160}x^4 + \frac{185987}{122342880}x^3 - \frac{955671551}{16699803120}x^2 + \frac{243915887}{278330052}x$$

Evaluating the area between the $q_i\uparrow$ curve and the $f_i\uparrow$ axis in the interval[0 ; 100]

$$\text{Area} = \int_0^{100} y\,dx = \frac{7681}{133598424960}\left(\frac{10^{12}}{6}\right) - \frac{336337}{22266404160}\left(\frac{10^{10}}{5}\right) + \frac{185987}{122342880}\left(\frac{10^{8}}{4}\right) - \frac{955671551}{16699803120}\left(\frac{10^{6}}{3}\right) + \frac{243915887}{278330052}\left(\frac{10^{4}}{2}\right)$$

$$= 2{,}683.48$$

Let's calculate the GINI concentration curve and index by the interpolation method:

$$\text{GINI Index} = 100\% - \left(\frac{\int_0^{100} f(x)\,dx}{5{,}000}\right)\text{x}100$$
$$= 100\% - \left(\frac{2{,}683.48}{5{,}000}\right)\text{x}100$$

$$= 46.33\%$$

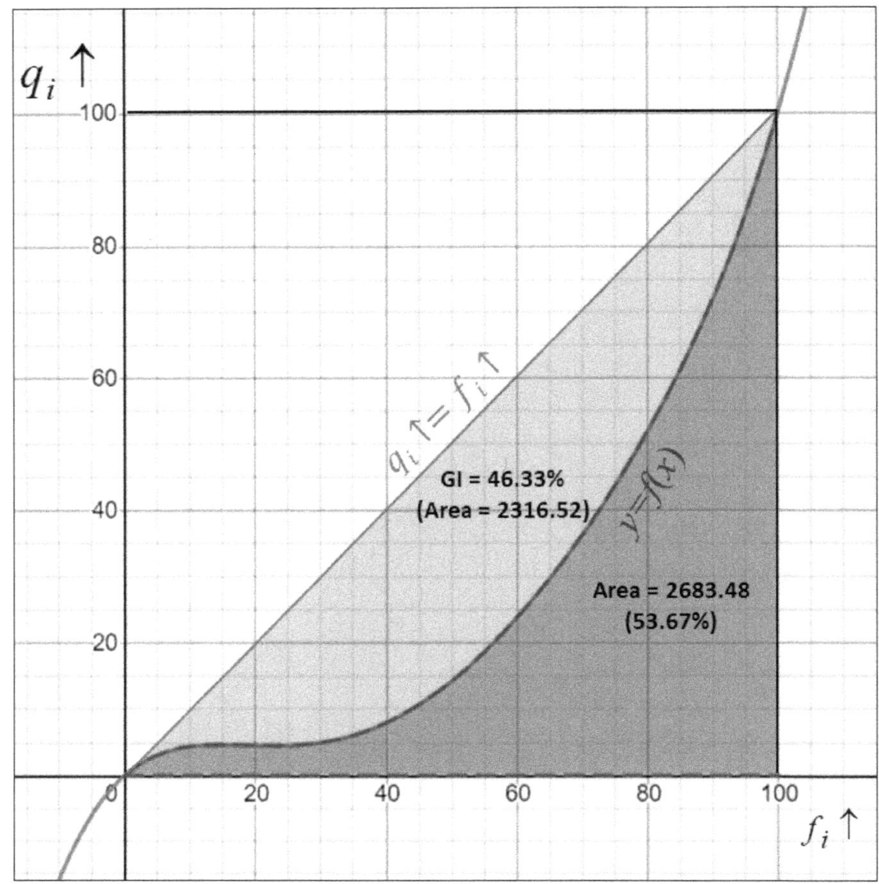

INTERPRETATION OF THE RESULTS

10-a. The medial area of the municipalities of this country is the area such that the sum of the areas of the smaller municipalities than the medial value (19,802.96 Km2) is equivalent to the sum of the areas of the larger municipalities. This means that half of the municipalities have an area larger than this value, and the other half have a smaller area.

10-b. 79.1% of the smallest municipalities represent 50% of the national surface. That is to say, 20.9% of the largest municipalities also represent 50% of the national surface. So we have reached the Pareto's law (law of 80-20), approximately one-fifth of the national area representing the largest municipalities is equivalent to four-fifths of the national area representing the smallest municipalities.

10-d/e. The GINI index of the distribution is 50.67%, indicating significant disparities in the area of different municipalities in the country. This value of the GINI index justifies the disparities between the modal, the average and the median areas found in **question 4** while corroborating the 86.6% of the Coefficient of Variation found in **question 6.d** and the values of the shape coefficients found to **question 9**. These results suggest that the distribution of municipality areas in the country is highly unequal.

9.7.4.3 Study of area and population (two-dimensional series)

1-a: The frequency Contingency Table is:

Area(Y) / *Population (X)*	[0; 5000)	[5000;10000)	[10000;15000)	[15000;20000)	[20000;25000)	[25000;30000)	[30000;35000)	[35000;40000)	[40000;45000)	[45000;50000)	Total
[0;60000 [	34	3	2	1	1	1	1	1	0	1	45
[60000;120000 [	32	5	**8**	14	8	6	1	2	2	0	78
[120000;180000 [	15	2	4	3	9	3	2	0	1	0	39
[180000;240000 [	2	6	7	5	2	0	1	0	0	0	23
[240000;300000 [	2	1	5	3	0	1	0	0	0	0	12
[300000;360000 [	**4**	0	0	1	0	0	0	0	0	0	5
[360000;420000 [	1	2	0	0	0	0	0	0	0	0	3
[420000;480000 [	1	2	2	1	0	0	0	0	0	0	6
[480000;540000 [	0	0	0	0	0	1	0	0	0	0	1
[540000;600000 [	0	2	0	0	0	0	0	0	0	0	2
[600000;660000 [	0	0	1	1	0	0	0	0	0	0	2
Total	91	23	29	29	20	12	5	3	3	1	216

INTERPRETATION OF THE RESULTS

For example, the data shows:

- 4 municipalities with a population of 300,000 to 360,000 people have an average area of less than 5,000 km2, while 8 municipalities with a population of 60,000 to 120,000 people have an average area of 10,000 to 15,000 km2. This indicates a significant difference in population density between these two groups of municipalities.
- There are no municipalities with an area between 20,000 and 25,000 km2 and a population of 240,000 to 300,000 people. This may be due to specific geographical or economic factors in the region or simply to a distribution of population and space that does not favor the creation of municipalities with these characteristics. By examining population and area data for municipalities, different patterns of distribution can be identified, for example:

In addition, we can read:

1. $n_{2*} = \sum_{j=1}^{j=10} n_{2j} = 32 + 5 + 8 + 14 + 8 + 6 + 1 + 2 + 2 + 0 = 78$

This means that there are 78 municipalities in this country with a population ranging from 60,000 to 120,000 people, regardless of their area.

2. $n_{*3} = \sum_{i=1}^{i=11} n_{i3} = 2 + 8 + 4 + 7 + 5 + 0 + 0 + 2 + 0 + 0 + 1 = 29$

This means that there are 29 municipalities in the DRC with an area ranging from 15,000 to 20,000 km2, regardless of their population.

1-b: The Relative Frequency Contingency Table is:

Area(Y) / Population (X)	[0; 5000)	[5000;10000)	[10000;15000)	[15000;20000)	[20000;25000)	[25000;30000)	[30000;35000)	[35000;40000)	[40000;45000)	[45000;50000)	Total
[0;60000 [	15.74	1.39	0.93	0.46	0.46	0.46	0.46	0.46	0.00	**0.46**	20.83
[60000;120000 [	14.81	2.31	3.70	6.48	3.70	2.78	0.46	0.93	0.93	0.00	36.11
[120000;180000 [	6.94	0.93	1.85	1.39	4.17	1.39	0.93	0.00	0.46	0.00	18.06
[180000;240000 [	0.93	**2.78**	3.24	2.31	0.93	0.00	0.46	0.00	0.00	0.00	10.65
[240000;300000 [	0.93	0.46	2.31	1.39	0.00	0.46	0.00	0.00	0.00	0.00	5.56
[300000;360000 [	1.85	0.00	0.00	0.46	0.00	0.00	0.00	0.00	0.00	0.00	2.31
[360000;420000 [	0.46	0.93	0.00	0.00	0.00	0.00	0.00	0.00	0.00	0.00	1.39
[420000;480000 [	0.46	0.93	0.93	0.46	0.00	0.00	0.00	0.00	0.00	0.00	2.78
[480000;540000 [	0.00	0.00	0.00	0.00	0.00	0.46	0.00	0.00	0.00	0.00	0.46
[540000;600000 [	0.00	0.93	0.00	0.00	0.00	0.00	0.00	0.00	0.00	0.00	0.93
[600000;660000 [	0.00	0.00	0.46	0.46	0.00	0.00	0.00	0.00	0.00	0.00	0.93
Total	42.13	10.65	13.43	13.43	9.26	5.56	2.31	1.39	1.39	0.46	100.00

INTERPRETATION OF THE RESULTS

The data shows that among the municipalities in the country:

- 2.78% have an area ranging from 5,000 to 10,000 Km^2 with a population between 180,000 and 240,000 people. Additionally,
- 0.46% of municipalities have an area of 45,000 to 50,000 Km^2 with a population of less than 60,000 people.

In addition, we can read:

1. $f_{11*} = \sum_{j=1}^{j=10} f_{7j} = 32 + 5 + 8 + 14 + 8 + 6 + 1 + 2 + 2 + 0 = 78$

This indicates that there are 78 municipalities in the country with a population ranging from 60,000 to 120,000 people, irrespective of their area.

2. $n_{*3} = \sum_{i=1}^{i=11} n_{*5} = 2 + 8 + 4 + 7 + 5 + 0 + 0 + 2 + 0 + 0 + 1 = 29$

This indicates that there are 78 municipalities in this country with an area ranging from 15,000 to 20,000 km2, irrespective of their population size.

2-a:

Distribution marginale des Populations (X)	
Population	Number of Municipalities
[0;60000)	45
[60000;120000)	78
[120000;180000)	39
[180000;240000)	23
[240000;300000)	12
[300000;360000)	5
[360000;420000)	3
[420000;480000)	6
[480000;540000)	1
[540000;600000)	2
[600000;660000)	2
Total	216

2-b:

Distribution marginale des superficies (Y)	
Area	Number of Municipalities
[0;5000)	91
[5000;10000)	23
[10000;15000)	29
[15000;20000)	29
[20000;25000)	20
[25000;30000)	12
[30000;35000)	5
[35000;40000)	3
[40000;45000)	3
[45000;50000)	1
Total	216

3: The Conditional law of the area, knowing that the population is between 60,000 and 120,000, is:

Area	[60 000;120 000)
[0;5000)	32
[5000;10000)	5
[10000;15000)	8
[15000;20000)	14
[20000;25000)	8
[25000;30000)	6
[30000;35000)	1
[35000;40000)	2
[40000;45000)	2
[45000;50000)	0
Total	78

INTERPRETATION OF THE RESULTS

The conditional law of the surface area given that the population is between 60,000 and 120,000 people allows for a better understanding of the distribution of the surface areas of municipalities in this population category.

Indeed, by fixing the population between 60,000 and 120,000 people, it can be observed that a large majority of municipalities have a surface area less than 5,000 km2, which results in a higher population density in these urban areas. However, there are also a significant number of municipalities with a surface area between 5,000 and 10,000 km2, which can be characterized by geographical factors such as rural or mountainous areas, where inhabitants are more dispersed.

On the other hand, there are no municipalities with a surface area between 45,000 and 50,000 km2 in this population category. This observation can be explained by the fact that municipalities of this size are very rare, and they are often located in sparsely populated areas.

These results allow for a better understanding of the distribution of the surface areas of municipalities in this population category.

4: Conditional law of the population knowing that the area is between 10.000 and 15.000 Km^2

Population	[10 000;15 000)
[0;60000)	2
[60000;120000)	8
[120000;180000)	4
[180000;240000)	7
[240000;300000)	5
[300000;360000)	0
[360000;420000)	0
[420000;480000)	2
[480000;540000)	0
[540000;600000)	0
[600000;660000)	1
Total	29

INTERPRETATION OF THE RESULTS

The conditional law of population given that the area is between 10,000 and 15,000 km2 allows for a better understanding of the distribution of population within this category of municipal area.

Indeed, by fixing the area between 10,000 and 15,000 km2, it is observed that the majority of municipalities have a population between 180,000 and 300,000 people, which may be related to socio-economic factors such as the presence of important economic activities in these areas.

On the other hand, it is noted that there are only two municipalities with a population of less than 60,000 people, which may be due to the fact that these municipal areas are more associated with higher population densities.

Similarly, the presence of two municipalities with a population between 420,000 and 480,000 people may be explained by the presence of large urban areas within these municipal areas.

Finally, the absence of municipalities with a population between 300,000 and 360,000 people in this area category may be due to their rarity throughout the country, as they represent only 2.31% of municipalities in general.

These results provide a better understanding of the distribution of population within this category of municipal areas.

5: The table to help calculate the marginal characteristics of the population (X):

Population	x_i	n_i	$n_i x_i$	$n_i x_i^2$
[0;60000)	30,000	45	1,350,000	40,500,000,000.00
[60000;120000)	90,000	78	7,020,000	631,800,000,000.00
[120000;180000)	150,000	39	5,850,000	877,500,000,000.00
[180000;240000)	210,000	23	4,830,000	1,014,300,000,000.00
[240000;300000)	270,000	12	3,240,000	874,800,000,000.00
[300000;360000)	330,000	5	1,650,000	544,500,000,000.00
[360000;420000)	390,000	3	1,170,000	456,300,000,000.00
[420000;480000)	450,000	6	2,700,000	1,215,000,000,000.00
[480000;540000)	510,000	1	510,000	260,100,000,000.00
[540000;600000)	570,000	2	1,140,000	649,800,000,000.00
[600000;660000)	630,000	2	1,260,000	793,800,000,000.00
Total	Total	216	30,720,000	7,358,400,000,000.00

5-a: Finding the marginal average of x:

$$\bar{\bar{x}} = \frac{1}{n}\sum_{i=1}^{i=11} n_i x_i = \frac{30,720,000}{216}$$

$$= 142,222.2$$

5-b: Finding the marginal variance of x:

$$V(x) = \frac{1}{n}\sum_{i=1}^{i=11} n_i x_{i^2} - \bar{x}^2$$

$$= \frac{7,358,400,000,000}{216} - (142,222.2)^2$$

$$= 13,839,506,173$$

INTERPRETATION OF THE RESULTS

It is clear that the marginal mean of the population is equal to 142,222.2 and the marginal variance of the population is equal to 13,839,506,173. These values match the corresponding values obtained in **Exercise 35** of **Section 7.7.1**, which confirms the accuracy of the analysis.

It's important to note that the marginal mean and variance provide an overview of the distribution of the population across all observations. However, it is also crucial to examine the distribution of the population for each group separately to get a more detailed understanding of the data.

Moreover, it is worth mentioning that the variance of the population is relatively large compared to its mean, which could indicate that there is a significant variation in the data. This variability may be due to a variety of factors, including measurement error, sampling variability, or differences in the characteristics of the groups being studied.

6: In the same way, let's determine the marginal characteristics of the area (Y):

Area	y_i	n_i	$n_i y_i$	$n_i y_i^2$
[0;5000)	2,500	91	227,500	568,750,000
[5000;10000)	7,500	23	172,500	1,293,750,000
[10000;15000)	12,500	29	362,500	4,531,250,000
[15000;20000)	17,500	29	507,500	8,881,250,000
[20000;25000)	22,500	20	450,000	10,125,000,000
[25000;30000)	27,500	12	330,000	9,075,000,000
[30000;35000)	32,500	5	162,500	5,281,250,000
[35000;40000)	37,500	3	112,500	4,218,750,000
[40000;45000)	42,500	3	127,500	5,418,750,000
[45000;50000)	47,500	1	47,500	2,256,250,000
Total	Total	216	2,500,000	51,650,000,000

6-a: Finding the marginal average of y:

$$\bar{\bar{y}} = \frac{1}{n}\sum_{i=1}^{i=10} n_i y_i = \frac{2,500,000}{216}$$

$$= 11,574.07$$

6-b: Finding the marginal variance of y:

$$V(y) = \frac{1}{n}\sum_{i=1}^{i=10} n_i y_{i^2} - y^2$$

$$= \frac{51{,}650{,}000{,}000}{216} - (11{,}574)^2$$

$$= 105{,}162{,}894$$

INTERPRETATION OF THE RESULTS

Similarly, as for the population, these data suggest that the marginal mean of surface area is 11,574.07 and the marginal variance is 105,162,894. These values correspond to the results obtained in **Exercise 35** of **Section 7.7**, confirming the accuracy of the analysis.

It is important to emphasize that the marginal mean and variance provide an overview of the distribution of surface area across all observations. However, it is also crucial to examine the surface area distribution for each group separately to obtain a more detailed understanding of the data.

Furthermore, it is worth mentioning that the variance of surface area is relatively large compared to its mean, which may indicate significant variation in the data. This variability could be due to various factors, including differences in the characteristics of the groups being studied.

7: Let us verify that the marginal means are equal to the means of the conditional means weighted by the marginal proportions of the link variable.

We will verify it for the variable X and this will also be valid for the variable Y.

We know that the marginal mean of X calculated above was 142,222.2 and let's verify that the average of the weighted conditional means equals the same value.

Note: *The data in this table are expressed in thousands of populations*

xi (*x1000[*	[0; 5000)		[5000;10000)		[10000;15000)		[15000;20000)		[20000;25000)		[25000;30000)		[30000;35000)		[35000;40000)		[40000;45000)		[45000;50000)	
x_i	n_1	n_1x_i	n_2	n_2x_i	n_3	n_3x_i	n_4	n_4x_i	n_5	n_5x_i	n_6	n_6x_i	n_7	n_7x_i	n_8	n_8x_i	n_9	n_9x_i	n_{10}	$n_{10}x_i$
30	34	1020	3	90	2	60	1	30	1	30	1	30	1	30	1	30	0	0	1	30
90	32	2880	5	450	8	720	14	1260	8	720	6	540	1	90	2	180	2	180	0	0
150	15	2250	2	300	4	600	3	450	9	1350	3	450	2	300	0	0	1	150	0	0
210	2	420	6	1260	7	1470	5	1050	2	420	0	0	1	210	0	0	0	0	0	0
270	2	540	1	270	5	1350	3	810	0	0	1	270	0	0	0	0	0	0	0	0
330	4	1320	0	0	0	0	1	330	0	0	0	0	0	0	0	0	0	0	0	0
390	1	390	2	780	0	0	0	0	0	0	0	0	0	0	0	0	0	0	0	0
450	1	450	2	900	2	900	1	450	0	0	0	0	0	0	0	0	0	0	0	0
510	0	0	0	0	0	0	0	0	0	0	1	510	0	0	0	0	0	0	0	0
570	0	0	2	1140	0	0	0	0	0	0	0	0	0	0	0	0	0	0	0	0
630	0	0	0	0	1	630	1	630	0	0	0	0	0	0	0	0	0	0	0	0
	91	9270	23	5190	29	5730	29	5010	20	2520	12	1800	5	630	3	210	3	330	1	30
$f_{.j}$	42.13%		10.65%		13.43%		13.43%		9.26%		5.56%		2.31%		1.39%		1.39%		0.46%	
$\bar{x}_j$	102		226		198		173		126		150		126		70		110		30	
$f_{.j}\bar{x}_j$	43		24		27		23		12		8		3		1		2		0	
	$\bar{\bar{x}}$= 142.222																			

INTERPRETATION OF THE RESULTS

From the above table, we can read, for example, that the average population is 142,222 people:

- The municipalities with less than 5,000 people have an average size of 102 km² and represent 42.13% of all municipalities. Their contribution to the overall average is 43%.
- The municipalities with between 5,000 and 10,000 people have an average size of 226 km² and represent 10.65% of all municipalities. Their contribution to the overall average is 24%.

Thus, we can easily observe that the marginal mean is equal to the mean of the conditional means weighted by the marginal frequencies of the linking variable.
Indeed,

$$\bar{\bar{x}} = \frac{(102 \text{ x } 42.13) + (226\text{x}10.65) + (198\text{x}13.43)}{100} + \frac{(173\text{x}13.43) + (126\text{x}9.26) + (150\text{x}5.56) + (126\text{x}2.31)}{100} + \frac{(70\text{x}1.39) + (110\text{x}1.39) + (30\text{x}0.46)}{100} = 142\,222$$

8: Let's consider the raw data on the area and population of municipalities:

8-a: Let's create a two-dimension statistical table representing the area and population

Nbr	Sup.	Pop.
1	2	26581
2	2.9	74708
3	2.9	49173
4	3.2	74447
5	3.4	82303
6	3.9	49297
7	4	88732
8	4.7	69147
9	4.8	75644
10	4.9	104904
11	5	74888
12	5.3	113968
13	5.6	108939
14	6.6	160719
15	6.8	97214
16	8.5	55645
17	10	64274
18	10.3	120016
19	11.4	157010
20	11.4	50345
21	12.8	337368
22	14	138413
23	14.2	55212
24	14.3	148363
25	15	75070
26	16.6	117774
27	18	35093
28	18	41357
29	18.5	67378
30	20	34405
31	20	50198
32	20.4	96950
33	23	48718
34	23.2	126589
35	23.7	159775
36	23.9	62622
37	24	65635
38	25.1	11824
39	27	54958
40	27.1	128197
41	28.5	32815
42	34.2	64230
43	36	22648
44	36.1	126074
45	39	44743
46	40.6	121836
47	44	42612
48	53.3	27805
49	59.9	31662
50	69.7	158080
51	76.9	353209
52	81.7	252181
53	88.7	20785
54	90	16249
55	102.1	20078
56	112	12988
57	142.5	123425
58	153	41593
59	185.1	61659
60	192.2	17360
61	197.5	97281
62	221.8	50756
63	230.2	28957
64	274.9	75632
65	281	92247
66	291.6	6093
67	300	98097
68	358.9	52820
69	405	115659
70	641.4	35780
71	753.4	12273
72	768	64664

Nbr	Sup.	Pop.
73	809.5	62298
74	1076.8	28963
75	1480	63575
76	1727	74227
77	1747	125085
78	1800	365675
79	1836	127120
80	1960	340597
81	2021	191635
82	2100	21811
83	2397	86960
84	3090	244900
85	3148	320022
86	3270	183709
87	3371	54899
88	3620	122782
89	4265	102633
90	4680	92956
91	4734	478450
92	5221	422919
93	5289	479064
94	5318	88154
95	5707	226811
96	5726	227801
97	6740	396062
98	6749	99583
99	6784	140825
100	7091	23445
101	7204	60138
102	7484	604210
103	7948.8	52676
104	7968	234291
105	8184	551137
106	8190	197675
107	8450	111133
108	8507	386121
109	8601	178573
110	8605	215679
111	8608	45824
112	8730	295107
113	8961	187593
114	9128	99419
115	10305	227268
116	10530	58366
117	10559	92568
118	10736	35760
119	11172	215895
120	11488	439079
121	11747	265237
122	12054	258568
123	12059	88492
124	12070	85640
125	12229	77438
126	12833	122012
127	12848	220854
128	12881	266863
129	13108	158258
130	13223	139748
131	13277	119627
132	13400	80850
133	13403	184633
134	13404	555124
135	13434	221932
136	13842	119993
137	14222	242629
138	14327	428962
139	14373	188112
140	14542	171600
141	14572	291310
142	14702	119993
143	14733	214404
144	15250	98788

Nbr	Sup.	Pop.
145	15397	112436
146	15498	359450
147	15634	167258
148	15770	245548
149	15786	204843
150	16015	109269
151	16055	52907
152	16201	246959
153	16797	69718
154	16898	214231
155	17328	90255
156	17411	203243
157	17494	111806
158	17682	84808
159	17861	99607
160	18096	619827
161	18098	74972
162	18126	268960
163	18773	232528
164	18926	477458
165	19073	65927
166	19187	114839
167	19264	92597
168	19513	112580
169	19718	157760
170	19805	60448
171	19865	217054
172	19996	140875
173	20155	231440
174	20503	179480
175	21181	149164
176	21239	84484
177	21675	84363
178	22436	116871
179	22466	123366
180	22480	136463
181	22567	137746
182	22673	131765
183	22909	112233
184	23081	95809
185	23475	137065
186	24184	150788
187	24350	213230
188	24430	91226
189	24598	98246
190	24700	82844
191	24953	56017
192	24963	161174
193	25074	92693
194	25175	56695
195	25216	173948
196	25417	93434
197	25490	169955
198	25894	79052
199	26217	530257
200	26346	76056
201	26648	278346
202	26665	110411
203	26676	105887
204	27910	149605
205	30363	114081
206	30512	180164
207	32446	122499
208	34198	171453
209	34704	48862
210	36385	47566
211	36783	84031
212	38075	116538
213	40214	171008
214	41824	85590
215	42196	119637
216	47087	59646

Solving the rest of this exercise requires the data in the following table:

x_i	y_i	x_iy_i	x_i^2	y_i^2
2	26581	53162	4	706549561
3	74708	216653	8	5581285264
3	49173	142602	8	2417983929
3	74447	238230	10	5542355809
3	82303	279830	12	6773783809
4	49297	192258	15	2430194209
4	88732	354928	16	7873367824
5	69147	324991	22	4781307609
5	75644	363091	23	5722014736
5	104904	514030	24	11004849216
5	74888	374440	25	5608212544
5	113968	604030	28	12988705024
6	108939	610058	31	11867705721
7	160719	1060745	44	25830596961
7	97214	661055	46	9450561796
9	55645	472983	72	3096366025
10	64274	642740	100	4131147076
10	120016	1236165	106	14403840256
11	157010	1789914	130	24652140100
11	50345	573933	130	2534619025
13	337368	4318310	164	113817167424
14	138413	1937782	196	19158158569
14	55212	784010	202	3048364944
14	148363	2121591	204	22011579769
15	75070	1126050	225	5635504900
17	117774	1955048	276	13870715076
18	35093	631674	324	1231518649
18	41357	744426	324	1710401449
19	67378	1246493	342	4539794884
20	34405	688100	400	1183704025
20	50198	1003960	400	2519839204
20	96950	1977780	416	9399302500
23	48718	1120514	529	2373443524
23	126589	2936865	538	16024774921
24	159775	3786668	562	25528050625
24	62622	1496666	571	3921514884
24	65635	1575240	576	4307953225
25	11824	296782	630	139806976
27	54958	1483866	729	3020381764
27	128197	3474139	734	16434470809
29	32815	935228	812	1076824225
34	64230	2196666	1170	4125492900
36	22648	815328	1296	512931904
36	126074	4551271	1303	15894653476
39	44743	1744977	1521	2001936049
41	121836	4946542	1648	14844010896
44	42612	1874928	1936	1815782544
53	27805	1482007	2841	773118025
60	31662	1896554	3588	1002482244
70	158080	11018176	4858	24989286400
77	353209	27161772	5914	124756597681
82	252181	20603188	6675	63595256761
89	20785	1843630	7868	432016225
90	16249	1462410	8100	264030001
102	20078	2049964	10424	403126084
112	12988	1454656	12544	168688144
143	123425	17588063	20306	15233730625
153	41593	6363729	23409	1729977649
185	61659	11413081	34262	3801832281
192	17360	3336592	36941	301369600
198	97281	19212998	39006	9463592961
222	50756	11257681	49195	2576171536
230	28957	6665901	52992	838507849
275	75632	20791237	75570	5720199424
281	92247	25921407	78961	8509509009
292	6093	1776719	85031	37124649
300	98097	29429100	90000	9623021409
359	52820	18957098	128809	2789952400
405	115659	46841895	164025	13377004281
641	35780	22949292	411394	1280208400
753	12273	9246478	567612	150626529
768	64664	49661952	589824	4181432896
810	62298	50430231	655290	3881040804

x_i	y_i	$x_i y_i$	x_i^2	y_i^2
1077	28963	31187358	1159498	838855369
1480	63575	94091000	2190400	4041780625
1727	74227	128190029	2982529	5509647529
1747	125085	218523495	3052009	15646257225
1800	365675	658215000	3240000	133718205625
1836	127120	233392320	3370896	16159494400
1960	340597	667570120	3841600	116006316409
2021	191635	387294335	4084441	36723973225
2100	21811	45803100	4410000	475719721
2397	86960	208443120	5745609	7562041600
3090	244900	756741000	9548100	59976010000
3148	320022	1007429256	9909904	102414080484
3270	183709	600728430	10692900	33748996681
3371	54899	185064529	11363641	3013900201
3620	122782	444470840	13104400	15075419524
4265	102633	437729745	18190225	10533532689
4680	92956	435034080	21902400	8640817936
4734	478450	2264982300	22410756	228914402500
5221	422919	2208060099	27258841	178860480561
5289	479064	2533769496	27973521	229502316096
5318	88154	468802972	28281124	7771127716
5707	226811	1294410377	32569849	51443229721
5726	227801	1304388526	32787076	51893295601
6740	396062	2669457880	45427600	156865107844
6749	99583	672085667	45549001	9916773889
6784	140825	955356800	46022656	19831680625
7091	23445	166248495	50282281	549668025
7204	60138	433234152	51897616	3616579044
7484	604210	4521907640	56010256	365069724100
7949	52676	418710989	63183421	2774760976
7968	234291	1866830688	63489024	54892272681
8184	551137	4510505208	66977856	303751992769
8190	197675	1618958250	67076100	39075405625
8450	111133	939073850	71402500	12350543689
8507	386121	3284731347	72369049	149089426641
8601	178573	1535906373	73977201	31888316329
8605	215679	1855917795	74046025	46517431041
8608	45824	394452992	74097664	2099838976
8730	295107	2576284110	76212900	87088141449
8961	187593	1681020873	80299521	35191133649
9128	99419	907496632	83320384	9884137561
10305	227268	2341996740	106193025	51650743824
10530	58366	614593980	110880900	3406589956
10559	92568	977425512	111492481	8568834624
10736	35760	383919360	115261696	1278777600
11172	215895	2411978940	124813584	46610651025
11488	439079	5044139552	131974144	192790368241
11747	265237	3115739039	137992009	70350666169
12054	258568	3116778672	145298916	66857410624
12059	88492	1067125028	145419481	7830834064
12070	85640	1033674800	145684900	7334209600
12229	77438	946989302	149548441	5996643844
12833	122012	1565779996	164685889	14886928144
12848	220854	2837532192	165071104	48776489316
12881	266863	3437462303	165920161	71215860769
13108	158258	2074445864	171819664	25045594564
13223	139748	1847887804	174847729	19529503504
13277	119627	1588287679	176278729	14310619129
13400	80850	1083390000	179560000	6536722500
13403	184633	2474636099	179640409	34089344689
13404	555124	7440882096	179667216	308162655376
13434	221932	2981434488	180472356	49253812624
13842	119993	1660943106	191600964	14398320049
14222	242629	3450669638	202265284	58868831641
14327	428962	6145738574	205262929	184008397444
14373	188112	2703733776	206583129	35386124544
14542	171600	2495407200	211469764	29446560000
14572	291310	4244969320	212343184	84861516100
14702	119993	1764137086	216148804	14398320049
14733	214404	3158814132	217061289	45969075216
15250	98788	1506517000	232562500	9759068944
15397	112436	1731177092	237067609	12641854096
15498	359450	5570756100	240188004	129204302500
15634	167258	2614911572	244421956	27975238564

x_i	y_i	$x_i y_i$	x_i^2	y_i^2
15770	245548	3872291960	248692900	60293820304
15786	204843	3233651598	249197796	41960654649
16015	109269	1749943035	256480225	11939714361
16055	52907	849421885	257763025	2799150649
16201	246959	4000982759	262472401	60988747681
16797	69718	1171053246	282139209	4860599524
16898	214231	3620075438	285542404	45894921361
17328	90255	1563938640	300259584	8145965025
17411	203243	3538663873	303142921	41307717049
17494	111806	1955934164	306040036	12500581636
17682	84808	1499575056	312653124	7192396864
17861	99607	1779080627	319015321	9921554449
18096	619827	11216389392	327465216	384185509929
18098	74972	1356843256	327537604	5620800784
18126	268960	4875168960	328551876	72339481600
18773	232528	4365248144	352425529	54069270784
18926	477458	9036370108	358193476	227966141764
19073	65927	1257425671	363779329	4346369329
19187	114839	2203415893	368140969	13187995921
19264	92597	1783788608	371101696	8574204409
19513	112580	2196773540	380757169	12674256400
19718	157760	3110711680	388799524	24888217600
19805	60448	1197172640	392238025	3653960704
19865	217054	4311777710	394618225	47112438916
19996	140875	2816936500	399840016	19845765625
20155	231440	4664673200	406224025	53564473600
20503	179480	3679878440	420373009	32213070400
21181	149164	3159442684	448634761	22249898896
21239	84484	1794355676	451095121	7137546256
21675	84363	1828568025	469805625	7117115769
22436	116871	2622117756	503374096	13658830641
22466	123366	2771540556	504721156	15219169956
22480	136463	3067688240	505350400	18622150369
22567	137746	3108513982	509269489	18973960516
22673	131765	2987507845	514064929	17362015225
22909	112233	2571145797	524822281	12596246289
23081	95809	2211367529	532732561	9179364481
23475	137065	3217600875	551075625	18786814225
24184	150788	3646656992	584865856	22737020944
24350	213230	5192150500	592922500	45467032900
24430	91226	2228651180	596824900	8322183076
24598	98246	2416655108	605061604	9652276516
24700	82844	2046246800	610090000	6863128336
24953	56017	1397792201	622652209	3137904289
24963	161174	4023386562	623151369	25977058276
25074	92693	2324184282	628705476	8591992249
25175	56695	1427296625	633780625	3214323025
25216	173948	4386272768	635846656	30257906704
25417	93434	2374811978	646023889	8729912356
25490	169955	4332152950	649740100	28884702025
25894	79052	2046972488	670499236	6249218704
26217	530257	13901747769	687331089	281172486049
26346	76056	2003771376	694111716	5784515136
26648	278346	7417364208	710115904	77476495716
26665	110411	2944109315	711022225	12190588921
26676	105887	2824641612	711608976	11212056769
27910	149605	4175475550	778968100	22381656025
30363	114081	3463841403	921911769	13014474561
30512	180164	5497163968	930982144	32459066896
32446	122499	3974602554	1052742916	15006005001
34198	171453	5863349694	1169503204	29396131209
34704	48862	1695706848	1204367616	2387495044
36385	47566	1730688910	1323868225	2262524356
36783	84031	3090912273	1352989089	7061208961
38075	116538	4437184350	1449705625	13581105444
40214	171008	6876915712	1617165796	29243736064
41824	85590	3579716160	1749246976	7325648100
42196	119637	5048202852	1780502416	14313011769
47087	59646	2808551202	2217185569	3557645316
2327355	30898035	368963879040	50886277801	7348610680837

8-b: Let's calculate the variance of the area of the municipalities:

$$\bar{\bar{x}} = \frac{1}{n}\sum_{i=1}^{i=216} x_i = \frac{2,327,355}{216} = 10,775$$

$$V(X) = \frac{1}{216}\sum_{i=1}^{i=216} x_i^2 - \bar{\bar{x}}^2 = \frac{50,886,277,801}{216} - (10,775)^2$$

$$= 235,584,619 - (10,775)^2$$

$$= 119,488,514$$

8-c: Let's calculate the variance of the population of the municipalities:

$$\bar{\bar{y}} = \frac{1}{n}\sum_{y=1}^{y=216} y_i = \frac{30,898,035}{216} = 143,046$$

$$V(Y) = \frac{1}{216}\sum_{i=1}^{i=216} y_i^2 - \bar{\bar{y}}^2 = \frac{7,348,610,680,837}{216} - (143,046)^2$$

$$= 34,021,345,745 - (143,046)^2$$

$$= 13,559,056,503$$

8-d: Let's calculate the covariance of the population and area of the municipalities:

$$\mathrm{COV}(X,Y) = \frac{1}{n}\sum_{i=1}^{i=216} x_i y_i - \bar{\bar{x}}\bar{\bar{y}}$$

$$= \frac{368,963,879,040}{216} - (143,046)(10,775)$$

$$= 1,708,166,107 - 1,541,295,589$$

$$= 166,870,518$$

8-e: Let's calculate the equation of the regression line of the surface area with respect to the population of municipalities:

$$a = \frac{COV(x,y)}{V(x)} = \frac{166,870,518}{119,488,514}$$

$$= 1,3965$$

Let's calculate the regression line of the surface area with respect to the population of municipalities:

$$y = 1.3965x+b$$

The mean point $(\bar{\bar{x}}, \bar{\bar{y}})$ is on this line:

$$143\,046 = 1.3965(10{,}775) + b$$
$$b = 127{,}998.71$$

The equation of the linear regression line is therefore:

$$y = 1{,}3965x + 127{,}998.71$$

8-f: Let's calculate the linear correlation coefficient.:

$$\rho = \frac{\text{COV}(X,Y)}{\sqrt{V(X)V(Y)}} = \frac{166{,}870{,}518}{\sqrt{119{,}488{,}514 \text{ x } 13{,}559{,}056{,}503}}$$

$$= \frac{166{,}870{,}518}{1{,}272{,}851{,}724.148}$$

$$= 0.13110$$

RESULT INTERPRETATION

By examining this correlation coefficient very close to zero, we can conclude that in this country, the size of a municipality is not a determining factor for the size of its population. In other words, the transition from a small municipality to a larger one does not necessarily lead to an increase in population. This result indicates the absence of a mathematical relationship between the increase in the surface area of a municipality and that of its population.

However, it is important to note that in general, there are factors that may influence the size of a municipality's population in a country, such as geographical conditions, the socio-economic appeal of the region, land use policies, the presence of infrastructure and public services, as well as proximity to urban or industrial areas.

For example, some rural regions may have a low population density due to rural exodus and migration towards more attractive urban areas in terms of employment and services.

CHAPTER 10

Socio-Economic Study on net monthly salaries in full-time equivalent (FTEQ) in the private sector for the year 2018 in France

10.1 Origin of Data

This chapter presents real data that have been made publicly available by the National Institute of Statistics and Economic Studies (INSEE) of France through their website, https://www.insee.fr/fr/statistiques/4990766. The authors chose these data because they are comprehensive, complete, and representative of the entire statistical population they are characterizing.

The data are real, as they are based on the full-time equivalent salary (FTEQ) of all French individuals working in the private sector in 2018. The FTEQ salary is a salary that is converted to full-time for the entire year, regardless of the actual volume of work. For example, if an individual worked for nine months at 60% and earned a total of 4,500 euros, their FTEQ salary would be calculated as $\frac{9000}{0.75*0.6} = 20.000$ euros per year.

The data are also complete, as they include all positions, including part-time ones, that are taken into account proportionally to their actual volume of work. In the example above, the individual's EQTP would be 0.75 * 0.6 = 0.45.

Finally, the data are exhaustively inclusive and representative of the entire statistical population of all French individuals working in the private sector in France, which is 16,516,417 workers. This is a complete census of the statistical population, not a statistical sample.

The raw data was gathered, organized, analyzed, and interpreted as part of this socio-economic study. Therefore, any assumptions, results, and interpretations presented in this book are solely those of the authors.

10.2 Presentation of the Study

This study is divided into three main parts: an analysis of basic statistical decision-making parameters, an analysis of parameters related to taxation, and an analysis of socio-economic parameters.

In the first part, we examine and interpret all the key parameters of central tendency, dispersion, shape, and concentration.

The second part of the study focuses on parameters related to taxation. We evaluate the percentage of French people working in the private sector with a full-time equivalent salary (FTEQ) who are not subject to taxes and those who are subject to taxes, including a component of taxable salary that is subject to the highest tax bracket of 45%.

The last part of our study focuses on socio-economic aspects, in which we compared the numbers to different concepts such as the Minimum Wage (SMIC), poverty, wealth, and middle class to draw conclusions.

For the four aforementioned concepts, we used definitions established by official organizations such as INSEE, the Observatory of Inequalities, the Center for Research on the Study and Observation of Living Conditions (CREDOC), and the Organization for Economic Cooperation and Development (OCDE). It should be noted that while their definitions are perfectly aligned when it comes to the concepts of poverty and wealth, there is no universal definition for the concept of "middle class".

However, the discrepancies between the numbers used to define this class are not significant. Therefore, we used the numerical results obtained from the two definitions of the middle class chosen, those of CREDOC and OCDE, to draw our conclusions.

The following table shows the distribution of net monthly wages in FTEQ for the year 2018 in France. This distribution is analyzed through a series of questions aimed at understanding the central tendency, dispersion, shape, and concentration of the data. Additionally, decision-making data is calculated to gain insights into the percentage of French people working in the private sector with a certain salary range.

Salary Bracket (Classes)	Class Center (x_i)	Number of Employees (n_i)
[0,1200)	600	895447
[1200,1300)	1,250	966112
[1300,1400)	1,350	1141747
[1400,1500)	1,450	1293283
[1500,1600)	1,550	1230465
[1600,1700)	1,650	1106458
[1700,1800)	1,750	983950
[1800,1900)	1,850	887476
[1900,2000)	1,950	785072
[2000,2100)	2,050	702178
[2100,2200)	2,150	618695
[2200,2300)	2,250	544494
[2300,2400)	2,350	486429
[2400,2500)	2,450	430428
[2500,2600)	2,550	382365
[2600,2700)	2,650	331462
[2700,2800)	2,750	296115
[2800,2900)	2,850	265910
[2900,3000)	2,950	245152
[3000,3100)	3,050	224543
[3100,3200)	3,150	201448
[3200,3300)	3,250	182529
[3300,3400)	3,350	165202
[3400,3500)	3,450	149852
[3500,3600)	3,550	137060
[3600,3700)	3,650	124374
[3700,3800)	3,750	113153
[3800,3900)	3,850	103644
[3900,4000)	3,950	96278
[4000,4100)	4,050	88372
[4100,4200)	4,150	80701
[4200,4300)	4,250	74793
[4300,4400)	4,350	68269
[4400,4500)	4,450	63161
[4500,4600)	4,550	58615
[4600,4700)	4,650	54404
[4700,4800)	4,750	50555
[4800,4900)	4,850	45986
[4900,5000)	4,950	43679
Total	5050	272721

Salary Bracket (Classes)	Class Center (x_i)	Number of Employees (n_i)
[5000,5100)	5,050	40535
[5100,5200)	5,150	37540
[5200,5300)	5,250	35400
[5300,5400)	5,350	33073
[5400,5500)	5,450	30563
[5500,5600)	5,550	28839
[5600,5700)	5,650	26751
[5700,5800)	5,750	25303
[5800,5900)	5,850	23474
[5900,6000)	5,950	22477
[6000,6100)	6,050	21449
[6100,6200)	6,150	20119
[6200,6300)	6,250	18840
[6300,6400)	6,350	17654
[6400,6500)	6,450	16374
[6500,6600)	6,550	15562
[6600,6700)	6,650	14955
[6700,6800)	6,750	14049
[6800,6900)	6,850	13166
[6900,7000)	6,950	12526
[7000,7100)	7,050	11798
[7100,7200)	7,150	11113
[7200,7300)	7,250	10746
[7300,7400)	7,350	10029
[7400,7500)	7,450	9716
[7500,7600)	7,550	9184
[7600,7700)	7,650	8585
[7700,7800)	7,750	8181
[7800,7900)	7,850	7855
[7900,8000)	7,950	7724
[8000,8100)	8,050	7386
[8100,8200)	8,150	6894
[8200,8300)	8,250	6555
[8300,8400)	8,350	6346
[8400,8500)	8,450	6048
[8500,8600)	8,550	5758
[8600,8700)	8,650	5663
[8700,+8700)	17,440	188331
Total		188331

Salary Bracket (Classes)	Class Center (x_i)	Number of Employees (n_i)
[0,1200)	600	895447
[1200,1300)	1250	966112
[1300,1400)	1350	1141747
[1400,1500)	1450	1293283
[1500,1600)	1550	1230465
[1600,1700)	1650	1106458
[1700,1800)	1750	983950
[1800,1900)	1850	887476
[1900,2000)	1950	785072
[2000,2100)	2050	702178
[2100,2200)	2150	618695
[2200,2300)	2250	544494
[2300,2400)	2350	486429
[2400,2500)	2450	430428
[2500,2600)	2550	382365
[2600,2700)	2650	331462
[2700,2800)	2750	296115
[2800,2900)	2850	265910
[2900,3000)	2950	245152
[3000,3100)	3050	224543
[3100,3200)	3150	201448
[3200,3300)	3250	182529
[3300,3400)	3350	165202
[3400,3500)	3450	149852
[3500,3600)	3550	137060
[3600,3700)	3650	124374
[3700,3800)	3750	113153
[3800,3900)	3850	103644
[3900,4000)	3950	96278
[4000,4100)	4050	88372
[4100,4200)	4150	80701
[4200,4300)	4250	74793
[4300,4400)	4350	68269
[4400,4500)	4450	63161
[4500,4600)	4550	58615
[4600,4700)	4650	54404
[4700,4800)	4750	50555
[4800,4900)	4850	45986
[4900,5000)	4950	43679
Total	**118400**	**272721**

Salary Bracket (Classes)	Class Center (x_i)	Number of Employees (n_i)
[5000,5100)	5050	40535
[5100,5200)	5150	37540
[5200,5300)	5250	35400
[5300,5400)	5350	33073
[5400,5500)	5450	30563
[5500,5600)	5550	28839
[5600,5700)	5650	26751
[5700,5800)	5750	25303
[5800,5900)	5850	23474
[5900,6000)	5950	22477
[6000,6100)	6050	21449
[6100,6200)	6150	20119
[6200,6300)	6250	18840
[6300,6400)	6350	17654
[6400,6500)	6450	16374
[6500,6600)	6550	15562
[6600,6700)	6650	14955
[6700,6800)	6750	14049
[6800,6900)	6850	13166
[6900,7000)	6950	12526
[7000,7100)	7050	11798
[7100,7200)	7150	11113
[7200,7300)	7250	10746
[7300,7400)	7350	10029
[7400,7500)	7450	9716
[7500,7600)	7550	9184
[7600,7700)	7650	8585
[7700,7800)	7750	8181
[7800,7900)	7850	7855
[7900,8000)	7950	7724
[8000,8100)	8050	7386
[8100,8200)	8150	6894
[8200,8300)	8250	6555
[8300,8400)	8350	6346
[8400,8500)	8450	6048
[8500,8600)	8550	5758
[8600,8700)	8650	5663
[8700,+8700)	17440	188331
Total	**270890**	**188331**

10.3 Questions

10.3.1 Analysis of basic decision-making parameters

1. Draw a histogram to represent the distribution of net monthly wages in FTEQ for the year 2018 in France.

2. Draw the curves of the cumulative ascending and descending proportions representing the number of employees based on the salary.

3. Calculate the characteristics of the central tendency of the distribution:

a. What is the modal monthly salary in the private sector in 2018 in France?
b. What is the average monthly salary in the private sector in 2018 in France?
c. What is the median monthly salary in the private sector in 2018 in France?
a. Give an interpretation of the order of magnitude of the elements of this statistical series

4 Calculate the characteristics of dispersion of the distribution:
a. What is the interquartile range?
b. What is the variance?
c. What is the standard deviation?
d. What is the coefficient of variation?
e. Give an interpretation of how are wages spread out from the average.

5. Calculate the shape characteristics of the distribution:
a. What is the YULE's coefficient of skewness ?
b. What is the PEARSON's coefficient of skewness?
c. What is the FISHER's coefficient of skewness?
d. What is the FISHER's coefficient of skewness?
e. What is the shape of the frequency curves of this distribution?

6. Calculate the concentration characteristics of the distribution:
a. What is the median?
b. What is the GINI index by the graphical method (trapezoid method)?
c. What is the GINI index by the analytical method (integral method)?
d. How is the wealth created by companies working in the private sector redistributed to its employees in France? In other words, what is the interpretation of the homogeneity or heterogeneity of wages in the private sector in France?

7. Determine some decision-making data by calculating the percentage of French people working in the private sector and having a salary:
a. Less than the mode.

b. Below average.
c. Less than the median.
d. Between the mean minus the standard deviation and the mean plus the standard deviation.
e. What is the interpretation of the percentage of French people working in the private sector in 2018 with a salary lower than these central measures and dispersion?

10.3.2 Analysis of tax parameters

Income tax rates in France (Income 2018)	
Levels	
Up to €9,964	0%
From €9,964 to €27,519	14%
From 27,519 to 73,779 euros	30%
From 73,779 to 156,244 euros	41%
Above €156,244	45%

Taxing Wages 2018 – OECD

Note: In the following of this study, we will consider the following assumptions:

1. We assume that every individual belonging to the population of French people working in the private sector in 2018 and having a monthly net salary in EQTP, has only this source of income. Social benefits are not taken into account.
2. The term "French people working in the private sector" refers to any person residing in France, legally entitled to work, and working in the private sector.

8. In France, the income tax threshold for monthly net salary in EQTP is 9,964 euros gross annual, which translates to 830.33 euros net monthly in 2018. This means that private sector workers whose salary does not exceed this threshold are not subject to income tax. In other words, their gross salary is equal to their net salary.
Calculate the percentage of French private sector workers in 2018 who had a monthly net salary in EQTP below the gross income tax threshold.

9. Calculate the percentage of French people working in the private sector whose salary was subject to a 45% income tax, i.e., those who have a salary above the last tax bracket set at 156,244 euros per year, which is 13,020.33 euros per month (in 2018).
10. Provide an interpretation.

10.3.3 Analysis of Socio-Economic Parameters

Definitions

1. **SMIC**: The Minimum Interprofessional Growth Wage (SMIC) is the minimum wage applied by companies in France for anyone aged 18 at least. In 2018, the amount of the gross SMIC in France was 1498.47 euros per month, or 17,981.64 euros per year. (Decree No. 2017-1719 of 20 December 2017 on increasing the minimum growth wage).

2. **Poverty**: According to INSEE and the Observatory of Inequalities, an individual (or a household) is considered poor when they live in a household whose standard of living is below the poverty line. In France and Europe, this threshold is most often set at 60% of the median standard of living. In France, the poverty rate was 14.8% in 2018.

3. **Wealth:** According to the Observatory of Inequalities, the wealth threshold is double the median standard of living.

4. **Middle Class**: There is no official definition of the middle class.
   - According to CREDOC and the Observatory of Inequalities, the middle class represents the population between the poorest 30% and the richest 20%, which is 50% of the population.
   - The ranges of percentages above applied to INSEE figures on the entire French population would include people whose disposable income (income and social assistance received from which we subtract taxes directly paid to the tax authorities) would be between 1390 euros and 2568 euros per month in 2018.
   - According to the OECD, the middle class is represented by people with incomes between 75% and 200% of the median income.

11. Calculate the percentage of French people working in the private sector in 2018and having a net monthly salary in FTEQ lower than the SMIC.

12. Calculate the percentage of French people working in the private sector in 2018 who earned a monthly net salary in EQTP below the poverty line as defined by INSEE and the Observatory of Inequalities. We assume here that the employee has only one source of income and that social benefits are not taken into account. This percentage will allow us to assess the prevalence of poverty among private sector workers in France.

13. Calculate the percentage of French people working in the private sector in 2018 who earned a monthly net salary in EQTP, considered to be above the wealth threshold according to the Observatory of Inequalities.

14. Calculate the percentage of French people working in the private sector in 2018 and having a net monthly salary in FTEQ that allows them to be classified as middle class by using the definition of the INSEE.

15. Calculate the percentage of French workers in the private sector in 2018 who have a monthly net salary in EQTP that allows them to be classified as middle class according to the OECD definition. Here we assume that the employee has only one source of income, and social benefits are not taken into account.

16. If we only consider the population of French workers in the private sector in 2018 with a monthly net salary in EQTP:

a. What would be the lower and upper limits of the middle class according to the definitions of INSEE, CREDOC, and the Observatory of Inequalities?
b. What would be the proportions of this middle class in terms of percentages?

17. Drawing conclusions from these results, what interpretation can we give of the socio-economic situation of French workers in the private sector in 2018 with a monthly net salary in EQTP?

10.3 Solutions

10.4.1 Analysis of basic decision-making parameters

The following table shows the data required to solve this exercise:

Classes	x_i	n_i	$n_i x_i$	$n_i x_i^2$	f_i	$f_i \uparrow$	$f_i \downarrow$
[0,1200)	600	895447	537268200	322360920000	5.42	5.42	100.00
[1200,1300)	1250	966112	1207640000	1509550000000	5.85	11.27	94.58
[1300,1400)	1350	1141747	1541358450	2080833907500	6.91	18.18	88.73
[1400,1500)	1450	1293283	1875260350	2719127507500	7.83	26.01	81.82
[1500,1600)	1550	1230465	1907220750	2956192162500	7.45	33.46	73.99
[1600,1700)	1650	1106458	1825655700	3012331905000	6.70	40.16	66.54
[1700,1800)	1750	983950	1721912500	3013346875000	5.96	46.12	59.84
[1800,1900)	1850	887476	1641830600	3037386610000	5.37	51.49	53.88
[1900,2000)	1950	785072	1530890400	2985236280000	4.75	56.25	48.51
[2000,2100)	2050	702178	1439464900	2950903045000	4.25	60.50	43.75
[2100,2200)	2150	618695	1330194250	2859917637500	3.75	64.24	39.50
[2200,2300)	2250	544494	1225111500	2756500875000	3.30	67.54	35.76
[2300,2400)	2350	486429	1143108150	2686304152500	2.95	70.49	32.46
[2400,2500)	2450	430428	1054548600	2583644070000	2.61	73.09	29.51
[2500,2600)	2550	382365	975030750	2486328412500	2.32	75.41	26.91
[2600,2700)	2650	331462	878374300	2327691895000	2.01	77.41	24.59
[2700,2800)	2750	296115	814316250	2239369687500	1.79	79.21	22.59
[2800,2900)	2850	265910	757843500	2159853975000	1.61	80.82	20.79
[2900,3000)	2950	245152	723198400	2133435280000	1.48	82.30	19.18
[3000,3100)	3050	224543	684856150	2088811257500	1.36	83.66	17.70
[3100,3200)	3150	201448	634561200	1998867780000	1.22	84.88	16.34
[3200,3300)	3250	182529	593219250	1927962562500	1.11	85.99	15.12
[3300,3400)	3350	165202	553426700	1853979445000	1.00	86.99	14.01
[3400,3500)	3450	149852	516989400	1783613430000	0.91	87.89	13.01
[3500,3600)	3550	137060	486563000	1727298650000	0.83	88.72	12.11
[3600,3700)	3650	124374	453965100	1656972615000	0.75	89.48	11.28
[3700,3800)	3750	113153	424323750	1591214062500	0.69	90.16	10.52
[3800,3900)	3850	103644	399029400	1536263190000	0.63	90.79	9.84
[3900,4000)	3950	96278	380298100	1502177495000	0.58	91.37	9.21
[4000,4100)	4050	88372	357906600	1449521730000	0.54	91.91	8.63
[4100,4200)	4150	80701	334909150	1389872972500	0.49	92.40	8.09
[4200,4300)	4250	74793	317870250	1350948562500	0.45	92.85	7.60
[4300,4400)	4350	68269	296970150	1291820152500	0.41	93.26	7.15
[4400,4500)	4450	63161	281066450	1250745702500	0.38	93.64	6.74
[4500,4600)	4550	58615	266698250	1213477037500	0.35	94.00	6.36
[4600,4700)	4650	54404	252978600	1176350490000	0.33	94.33	6.00
[4700,4800)	4750	50555	240136250	1140647187500	0.31	94.63	5.67
[4800,4900)	4850	45986	223032100	1081705685000	0.28	94.91	5.37
[4900,5000)	4950	43679	216211050	1070244697500	0.26	95.18	5.09
[5000,5100)	5050	40535	204701750	1033743837500	0.25	95.42	4.82
[5100,5200)	5150	37540	193331000	995654650000	0.23	95.65	4.58
[5200,5300)	5250	35400	185850000	975712500000	0.21	95.86	4.35
[5300,5400)	5350	33073	176940550	946631942500	0.20	96.06	4.14
[5400,5500)	5450	30563	166568350	907797507500	0.19	96.25	3.94
[5500,5600)	5550	28839	160056450	888313297500	0.17	96.42	3.75
[5600,5700)	5650	26751	151143150	853958797500	0.16	96.59	3.58
[5700,5800)	5750	25303	145492250	836580437500	0.15	96.74	3.41
[5800,5900)	5850	23474	137322900	803338965000	0.14	96.88	3.26
[5900,6000)	5950	22477	133738150	795741992500	0.14	97.02	3.12
[6000,6100)	6050	21449	129766450	785087022500	0.13	97.15	2.98
[6100,6200)	6150	20119	123731850	760950877500	0.12	97.27	2.85
[6200,6300)	6250	18840	117750000	735937500000	0.11	97.38	2.73
[6300,6400)	6350	17654	112102900	711853415000	0.11	97.49	2.62
[6400,6500)	6450	16374	105612300	681199335000	0.10	97.59	2.51
[6500,6600)	6550	15562	101931100	667648705000	0.09	97.68	2.41
[6600,6700)	6650	14955	99450750	661347487500	0.09	97.77	2.32
[6700,6800)	6750	14049	94830750	640107562500	0.09	97.86	2.23
[6800,6900)	6850	13166	90187100	617781635000	0.08	97.94	2.14
[6900,7000)	6950	12526	87055700	605037115000	0.08	98.01	2.06
[7000,7100)	7050	11798	83175900	586390095000	0.07	98.09	1.99
[7100,7200)	7150	11113	79457950	568124342500	0.07	98.15	1.91
[7200,7300)	7250	10746	77908500	564836625000	0.07	98.22	1.85
[7300,7400)	7350	10029	73713150	541791652500	0.06	98.28	1.78
[7400,7500)	7450	9716	72384200	539262290000	0.06	98.34	1.72
[7500,7600)	7550	9184	69339200	523510960000	0.06	98.39	1.66
[7600,7700)	7650	8585	65675250	502415662500	0.05	98.45	1.61
[7700,7800)	7750	8181	63402750	491371312500	0.05	98.50	1.55
[7800,7900)	7850	7855	61661750	484044737500	0.05	98.54	1.50
[7900,8000)	7950	7724	61405800	488176110000	0.05	98.59	1.46
[8000,8100)	8050	7386	59457300	478631265000	0.04	98.63	1.41
[8100,8200)	8150	6894	56186100	457916715000	0.04	98.68	1.37
[8200,8300)	8250	6555	54078750	446149687500	0.04	98.72	1.32
[8300,8400)	8350	6346	52989100	442458985000	0.04	98.75	1.28
[8400,8500)	8450	6048	51105600	431842320000	0.04	98.79	1.25
[8500,8600)	8550	5758	49230900	420924195000	0.03	98.83	1.21
[8600,8700)	8650	5663	48984950	423719817500	0.03	98.86	1.17
[8700,+8700)	17440	188331	3284492640	57281551641600	1.14	100.00	1.14
Total		**16516417**	**39127451690**	**158480352899100**	**100**		

1: The histogram of the distribution is:

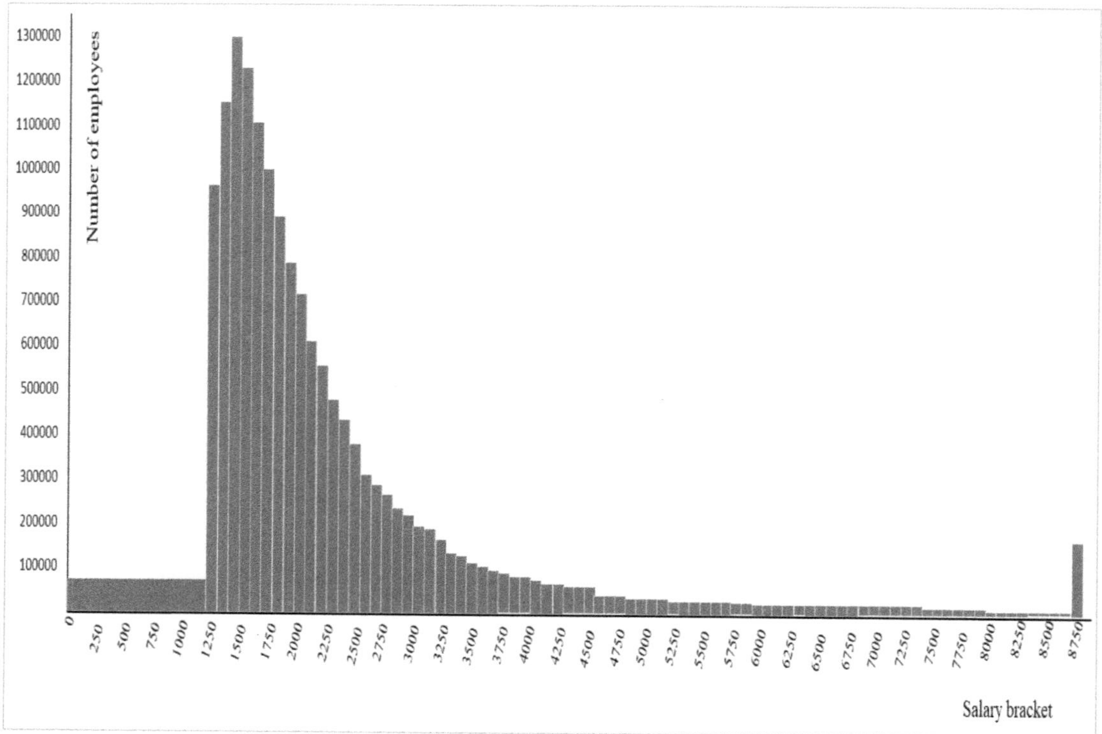

2: Plotting the curves of the relative cumulative increasing and decreasing frequencies:

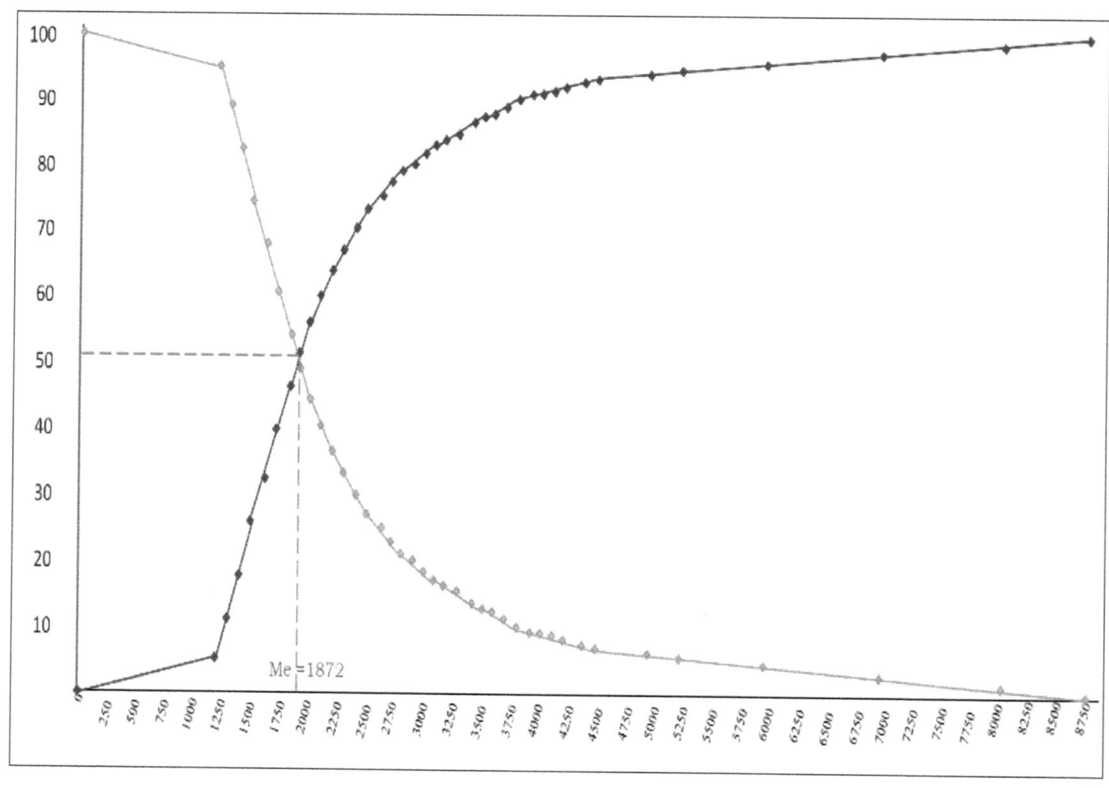

3: Determining the characteristics of central tendency:

3-a: The modal class of the distribution is [1400 ; 1500), meaning the modal monthly salary after tax in FTEQ in the private sector in 2018 in France is 1,450 euros.

3-b: The average monthly salary after tax in FTEQ in the private sector in 2018 in France is:

$$\bar{x} = \frac{1}{n}\sum_{i=1}^{i=77} n_i x_i = \frac{39{,}127{,}372{,}040}{16{,}516{,}390}$$

$$= 2{,}369.00 \text{ euros}$$

3-c: The median monthly salary after tax in FTEQ in the private sector in 2018 in France is:

$$M_e = a_i + (a_{i+1} - a_i)\frac{(50 - F_i)}{(F_{i+1} - F_i)}$$

$$= 1.800 + (1.900 - 1.800)\frac{(50 - 46.12)}{(51.49 - 46.12)}$$

$$= 1.872.25 \text{ euros}$$

3-d:

INTERPRETATION OF THE RESULTS

Considering the net monthly FTEQ salary in the private sector in France in 2018:

3.a. The most frequent salary class is between 1,400 and 1,500 euros. This means that the majority of private sector workers in France in this social category earn a monthly net salary in EQTP between these two values. This information can be useful in understanding the distribution of salaries and the potential need for public policies aimed at improving the living conditions of the most disadvantaged individuals in the private sector.

3-b. The average monthly net salary in EQTP in the private sector in 2018 was 2,369.00 euros. If salaries were distributed equally, each private sector worker would earn this amount each month. This information can be useful in assessing the gap between the highest and lowest salaries in the private sector in France.

3-c. The median salary is 1,872.25 euros, which means that there are as many private sector workers who earn a monthly net salary in EQTP below this amount as there are workers who earn more. This indicates that approximately half of private sector workers in France (8,258,208 people) have a monthly net salary below 1,872.25 euros in EQTP. This information can be useful in better understanding salary inequalities in the private sector in France.

4: Determining the characteristics of dispersion

4-a: The interquartile distance of the distribution is:

$$Q_1 = a_i + (a_{i+1} - a_i)\frac{(25 - F_i)}{(F_{i+1} - F_i)}$$

$$= 1{,}400 + (1{,}500 - 1{,}400)\frac{(25 - 18.18)}{(26.01 - 18.18)}$$

$$= 1{,}487.10 \text{ euros}$$

$$Q_3 = a_i + (a_{i+1} - a_i)\frac{(75 - F_i)}{(F_{i+1} - F_i)}$$

$$= 2{,}500 + (2{,}600 - 2{,}500)\frac{(75 - 73.09)}{(75.41 - 73.09)}$$

$$= 2{,}582.33 \text{ euros}$$

$$\text{Distance Interquartile} = Q_3 - Q_1$$

$$= 2{,}582.3 - 1{,}487.1$$

$$= 1{,}095.23 \text{ euros}$$

4-b: The variance of the distribution is:

$$V(x) = \frac{1}{n}\sum_{i=1}^{i=77} n_i x_i^2 - \bar{x}^2$$

$$= \frac{158{,}480{,}352{,}899{,}100}{16{,}516{,}417} - (2{,}369)^2$$

$$= 3{,}983{,}161.82$$

4-c: The standard deviation of the distribution is:

$$\sigma = \sqrt{V(x)}$$

$$= \sqrt{3,983,144.65}$$

$$= 1,995.79 \text{ euros}$$

4-d: The coefficient of variation of the distribution is:

$$\text{CV} = \frac{\sigma}{|\bar{x}|}\text{x}100 = \frac{1,995.78}{2,369.00}\text{x}100$$

$$= 84.25\%$$

4-e:

<u>INTERPRETATION OF THE RESULTS</u>

When considering the monthly net salary in EQTP in the private sector in France in 2018,

4-a. The interquartile range of the salary distribution of 1,095.23 euros indicates that there are large differences in salaries in the central half of the distribution, which confirms the presumption of a high salary disparity.

4-c The standard deviation of the salary distribution of 1,995.78 euros also shows that there is a great divergence in salaries.

4-d The coefficient of variation of 84.24% is a relatively high value but does not exceed 100%. This curve should therefore be not very narrow but high.

All these characteristics confirm the existence of a significant dispersion of salaries in the private sector in France in 2018.

5: Calculation of shape characteristics

The following table shows the data required to solve this exercise:

x_i	n_i	$n_i x_i$	$n_i(x_i - \bar{x})^3$	$n_i(x_i - \bar{x})^4$	f_i	q_i	$f_i \uparrow$	$f_i \downarrow$
600	895447	537268200.00	-4957081416096720	876909497798456000	5.42	1.37	5.42	1.37
1250	966112	1207640000.00	-1353698516172020	151479354224481000	5.85	3.09	11.27	4.46
1350	1141747	1541358450.00	-1208083803251210	123104177079297000	6.91	3.94	18.18	8.40
1450	1293283	1875260350.00	-1003795484099120	92249168530236000	7.83	4.79	26.01	13.19
1550	1230465	1907220750.00	-675968925286852	55362099794582000	7.45	4.87	33.46	18.07
1650	1106458	1825655700.00	-411271075713569	29570539292587500	6.70	4.67	40.16	22.73
1750	983950	1721912500.00	-233374069877372	14445939445779200	5.96	4.40	46.12	27.13
1850	887476	1641830600.00	-124070285765589	6439292765388260	5.37	4.20	51.49	31.33
1950	785072	1530890400.00	-57751440154767	2419806258146370	4.75	3.91	56.25	35.24
2050	702178	1439464900.00	-22794709371830	727159484451243	4.25	3.68	60.50	38.92
2150	618695	1330194250.00	-6498759971223	142325197005707	3.75	3.40	64.24	42.32
2250	544494	1225111500.00	-917642742603	10920280976824	3.30	3.13	67.54	45.45
2350	486429	1143108150.00	-3338324780	63440261119	2.95	2.92	70.49	48.37
2450	430428	1054548600.00	228716404924	18525200463646	2.61	2.70	73.09	51.07
2550	382365	975030750.00	2267189317932	410353055535773	2.32	2.49	75.41	53.56
2650	331462	878374300.00	7354208084418	2066505837212110	2.01	2.24	77.41	55.80
2750	296115	814316250.00	16376570144360	6239413914480380	1.79	2.08	79.21	57.89
2850	265910	757843500.00	29591030464985	14233178484732000	1.61	1.94	80.82	59.82
2950	245152	723198400.00	48079032117449	27933743533884200	1.48	1.85	82.30	61.67
3050	224543	684856150.00	70914317504045	48292393392049800	1.36	1.75	83.66	63.42
3150	201448	634561200.00	95964370739120	74947825996039900	1.22	1.62	84.88	65.04
3250	182529	593219250.00	124811396860010	109958388608068000	1.11	1.52	85.99	66.56
3350	165202	553426700.00	155961539289868	152997705202246000	1.00	1.41	86.99	67.97
3450	149852	516989400.00	189293307834515	204625380211354000	0.91	1.32	87.89	69.29
3550	137060	486563000.00	225764901268366	266627530752134000	0.83	1.24	88.72	70.54
3650	124374	453965100.00	261440766188827	334904674635897000	0.75	1.16	89.48	71.70
3750	113153	424323750.00	298018820628891	411562911962892000	0.69	1.08	90.16	72.78
3850	103644	399029400.00	336671345861658	498609043908846000	0.63	1.02	90.79	73.80
3950	96278	380298100.00	380469357699749	601520676589120000	0.58	0.97	91.37	74.77
4050	88372	357906600.00	419773498801831	705637731205094000	0.54	0.91	91.91	75.69
4150	80701	334909150.00	455898355097579	811953319315722000	0.49	0.86	92.40	76.55
4250	74793	317870250.00	497765544742951	936295186919311000	0.45	0.81	92.85	77.36
4350	68269	296970150.00	530731159534088	1051376504904290000	0.41	0.76	93.26	78.12
4450	63161	281066450.00	569197482451381	1184497899536290000	0.38	0.72	93.64	78.84
4550	58615	266698250.00	608098038508925	1326259619657980000	0.35	0.68	94.00	79.52
4650	54404	252978600.00	645661096181794	1472750622019720000	0.33	0.65	94.33	80.16
4750	50555	240136250.00	682402044244569	1624796795911800000	0.31	0.61	94.63	80.78
4850	45986	223032100.00	702269853727717	1742328963709310000	0.28	0.57	94.91	81.35
4950	43679	216211050.00	750991242063539	1938305675924090000	0.26	0.55	95.18	81.90
5050	40535	204701750.00	781121981234162	2094185202723310000	0.25	0.52	95.42	82.42
5150	37540	193331000.00	807412629144061	2245411597468120000	0.23	0.49	95.65	82.92
5250	35400	185850000.00	846508647556889	2438788347837040000	0.21	0.47	95.86	83.39
5350	33073	176940550.00	876108585122394	2611676519274310000	0.20	0.45	96.06	83.84
5450	30563	166568350.00	893860085853511	2753979687249060000	0.19	0.43	96.25	84.27
5550	28839	160056450.00	928260180543609	2952792272457840000	0.17	0.41	96.42	84.68
5650	26751	151143150.00	944837831853220	3100009504420270000	0.16	0.39	96.59	85.07
5750	25303	145492250.00	977926313772851	3306365325140320000	0.15	0.37	96.74	85.44
5850	23474	137322900.00	990142756217365	3446683348423050000	0.14	0.35	96.88	85.79
5950	22477	133738150.00	1032167050043530	3696186468038190000	0.14	0.34	97.02	86.13
6050	21449	129766450.00	1069801542317870	3937935602804710000	0.13	0.33	97.15	86.46
6150	20119	123731850.00	1087489756221050	4111794829743540000	0.12	0.32	97.27	86.78
6250	18840	117750000.00	1101312607311950	4274190240387650000	0.11	0.30	97.38	87.08
6350	17654	112102900.00	1113828868154410	4434148690202930000	0.11	0.29	97.49	87.37
6450	16374	105612300.00	1112893008678570	4541712337886820000	0.10	0.27	97.59	87.64
6550	15562	101931100.00	1137377899147050	4755372877127220000	0.09	0.26	97.68	87.90
6650	14955	99450750.00	1173332225337570	5023031007248860000	0.09	0.25	97.77	88.15
6750	14049	94830750.00	1181310583763150	5175317389150110000	0.09	0.24	97.86	88.39
6850	13166	90187100.00	1184616096371380	5308260437552430000	0.08	0.23	97.94	88.62
6950	12526	87055700.00	1204182336961450	5516354924470240000	0.08	0.22	98.01	88.85
7050	11798	83175900.00	1210105701572470	5664500406458080000	0.07	0.21	98.09	89.06
7150	11113	79457950.00	1214469349022280	5806373559269190000	0.07	0.20	98.15	89.26
7250	10746	77908500.00	1249603671274620	6099310993840190000	0.07	0.20	98.22	89.46
7350	10029	73713150.00	1239385209783010	6173373241285850000	0.06	0.19	98.28	89.65
7450	9716	72384200.00	1274483338312510	6475645226208720000	0.06	0.18	98.34	89.83
7550	9184	69339200.00	1277237735618930	6617364082509000000	0.06	0.18	98.39	90.01
7650	8585	65675250.00	1264409960735470	6677344423369320000	0.05	0.17	98.45	90.18
7750	8181	63402750.00	1274660394565800	6858942966760170000	0.05	0.16	98.50	90.34
7850	7855	61661750.00	1293375811812710	7088988140366050000	0.05	0.16	98.54	90.50
7950	7724	61405800.00	1342695357113530	7493577925252100000	0.05	0.16	98.59	90.65
8050	7386	59457300.00	1354200096942310	7693205846264410000	0.04	0.15	98.63	90.81
8150	6894	56186100.00	1331923774299720	7699846515439270000	0.04	0.14	98.68	90.95
8250	6555	54078750.00	1333292486943520	7841088286970390000	0.04	0.14	98.72	91.09
8350	6346	52989100.00	1357752734301610	8120714186526570000	0.04	0.14	98.75	91.22
8450	6048	51105600.00	1359990944941410	8270100010751320000	0.04	0.13	98.79	91.35
8550	5758	49230900.00	1359712618416910	8404378770005510000	0.03	0.13	98.83	91.48
8650	5663	48984950.00	1403240758091580	8813750119499250000	0.03	0.13	98.86	91.61
17440	188331	3284492640.00	644685212758782000	971604850665097000000	1.14	8.39	100.00	100.00
Total	**16516417**	**39127451690**	**685320929407072000**	**994264896139670000000**				

5-a: Finding the YULE's coefficient of skewness:

$$C_Y = \frac{Q_1 - 2Q_2 + Q_3}{Q_3 - Q_1}$$

$$= \frac{1{,}487.1 - 2(1{,}872.25) + 2{,}582.33}{2{,}582.3 - 1{,}487.1}$$

$$= 0.2966774102$$

5-b: Finding the PEARSON's coefficient of skewness:

$$\beta_1 = \frac{(\bar{x} - \text{Mode})}{\sigma}$$

$$= \frac{(2{,}369 - 1{,}450)}{1{,}995.78}$$

$$= 0.4604692879$$

5-c: Finding the FISHER's coefficient of skewness:

$$\mu_3 = \frac{\sum_{i=1}^{i=77} n_i (x_i - \bar{x})^3}{\sum_{i=1}^{i=77} n_i} = \frac{685{,}320{,}929{,}407{,}072{,}000}{16{,}516{,}417}$$

$$= 41{,}493{,}317{,}189.01$$

$$\gamma_1 = \frac{\mu_3}{\sigma^3} = \frac{475{,}825{,}529{,}395.70}{(1{,}995.78)^3}$$

$$= 5.2196$$

5-d: Finding the FISHER's coefficient of kurtosis:

$$\mu_4 = \frac{\sum_{i=1}^{i=77} n_i (x_i - \bar{x})^4}{\sum_{i=1}^{i=77} n_i} = \frac{9{,}942{,}648{,}961{,}396{,}700{,}000{,}000}{16{,}516{,}417}$$

$$= 601{,}985{,}827{,}882{,}445.69$$

$$\gamma_2 = \frac{\mu_4}{\sigma^4} - 3 = \frac{601{,}985{,}827{,}882{,}445.69}{(1{,}995.78)^4} - 3$$

$$= 34.9432$$

5-e:

<u>INTERPRETATION OF THE RESULTS</u>

In 2018, considering the net monthly FTEQ salary in the private sector in France,

$$\left.\begin{array}{c} \underbrace{M_0}_{1{,}450} < \underbrace{M_e}_{1{,}872} < \underbrace{\bar{x}}_{2{,}369} \\ C_Y > 0 \\ \beta_1 > 0 \\ \gamma_1 > 0 \end{array}\right\}$$

The curve of wage distribution is characterized by right-skewness, with a long tail of high values on the right side of the curve. This skewness indicates a strong concentration of low salaries in this social category. Indeed, in France, the majority of people working in the private sector in 2018 have a low net EQTP salary that is below the average of 2,369.00 euros, which is confirmed by the high coefficient of variation of 84.24% found in **question 4.d**.

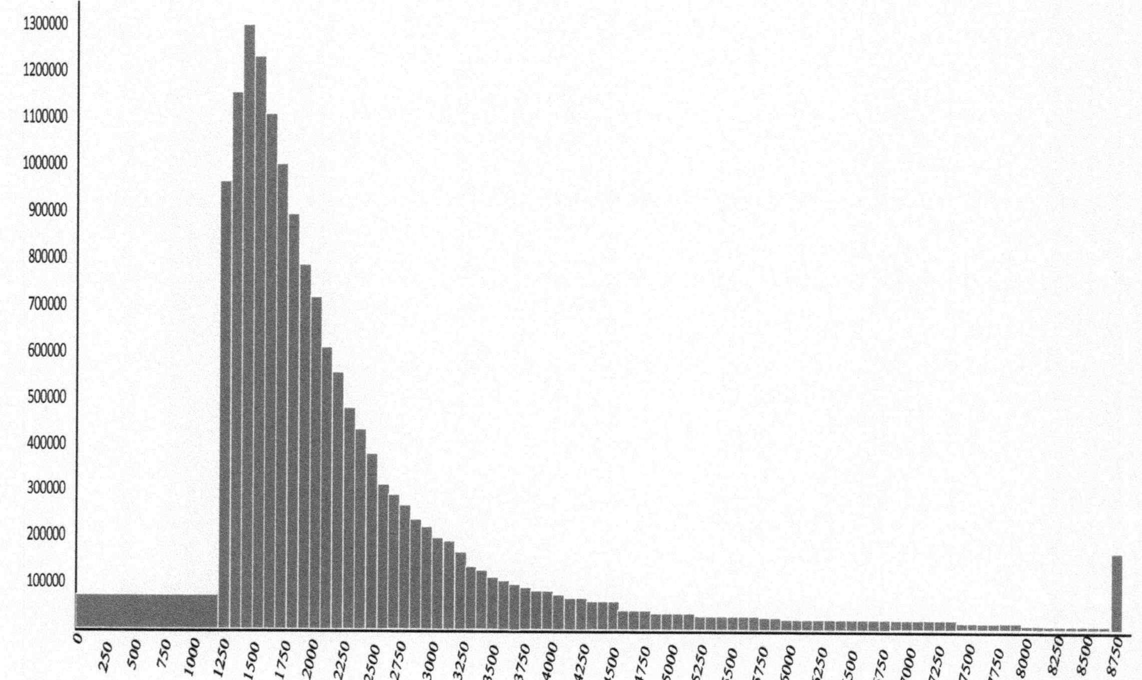

$\gamma_2 > 0$, the curve of this distribution is sharper than that of the normal distribution. This concentration of low salaries quickly grows the distribution up to a peak of 1,293,283 employees, after which it declines rapidly as salaries become higher.

6: Calculation of concentration characteristics:

The following table shows the data required to solve this exercise:

Classes	$f_i \uparrow$	$q_i \uparrow$	$H_i = f_{i+1}\uparrow - f_i\uparrow$	$B_i = \frac{q_i\uparrow + q_{i+1}\uparrow}{2}$	$T_i = B_i \text{x} H_i$
[0 , 1200 [	5.42	1.37	5.85	2.92	17.06
[1200 , 1300 [	11.27	4.46	6.91	6.43	44.44
[1300 , 1400 [	18.18	8.40	7.83	10.80	84.53
[1400 , 1500 [	26.01	13.19	7.45	15.63	116.43
[1500 , 1600 [	33.46	18.07	6.70	20.40	136.66
[1600 , 1700 [	40.16	22.73	5.96	24.93	148.53
[1700 , 1800 [	46.12	27.13	5.37	29.23	157.07
[1800 , 1900 [	51.49	31.33	4.75	33.29	158.21
[1900 , 2000 [	56.25	35.24	4.25	37.08	157.65
[2000 , 2100 [	60.50	38.92	3.75	40.62	152.16
[2100 , 2200 [	64.24	42.32	3.30	43.89	144.68
[2200 , 2300 [	67.54	45.45	2.95	46.91	138.16
[2300 , 2400 [	70.49	48.37	2.61	49.72	129.57
[2400 , 2500 [	73.09	51.07	2.32	52.31	121.11
[2500 , 2600 [	75.41	53.56	2.01	54.68	109.74
[2600 , 2700 [	77.41	55.80	1.79	56.85	101.91
[2700 , 2800 [	79.21	57.89	1.61	58.85	94.75
[2800 , 2900 [	80.82	59.82	1.48	60.75	90.17
[2900 , 3000 [	82.30	61.67	1.36	62.55	85.03
[3000 , 3100 [	83.66	63.42	1.22	64.23	78.34
[3100 , 3200 [	84.88	65.04	1.11	65.80	72.72
[3200 , 3300 [	85.99	66.56	1.00	67.27	67.28
[3300 , 3400 [	86.99	67.97	0.91	68.63	62.27
[3400 , 3500 [	87.89	69.29	0.83	69.92	58.02
[3500 , 3600 [	88.72	70.54	0.75	71.12	53.55
[3600 , 3700 [	89.48	71.70	0.69	72.24	49.49
[3700 , 3800 [	90.16	72.78	0.63	73.29	45.99
[3800 , 3900 [	90.79	73.80	0.58	74.29	43.30
[3900 , 4000 [	91.37	74.77	0.54	75.23	40.25
[4000 , 4100 [	91.91	75.69	0.49	76.12	37.19
[4100 , 4200 [	92.40	76.55	0.45	76.95	34.85
[4200 , 4300 [	92.85	77.36	0.41	77.74	32.13
[4300 , 4400 [	93.26	78.12	0.38	78.48	30.01
[4400 , 4500 [	93.64	78.84	0.35	79.18	28.10
[4500 , 4600 [	94.00	79.52	0.33	79.84	26.30
[4600 , 4700 [	94.33	80.16	0.31	80.47	24.63
[4700 , 4800 [	94.63	80.78	0.28	81.06	22.57
[4800 , 4900 [	94.91	81.35	0.26	81.62	21.59
[4900 , 5000 [	95.18	81.90	0.25	82.16	20.16
[5000 , 5100 [	95.42	82.42	0.23	82.67	18.79
[5100 , 5200 [	95.65	82.92	0.21	83.15	17.82
[5200 , 5300 [	95.86	83.39	0.20	83.62	16.74
[5300 , 5400 [	96.06	83.84	0.19	84.06	15.55
[5400 , 5500 [	96.25	84.27	0.17	84.47	14.75
[5500 , 5600 [	96.42	84.68	0.16	84.87	13.75
[5600 , 5700 [	96.59	85.07	0.15	85.25	13.06
[5700 , 5800 [	96.74	85.44	0.14	85.61	12.17
[5800 , 5900 [	96.88	85.79	0.14	85.96	11.70
[5900 , 6000 [	97.02	86.13	0.13	86.30	11.21
[6000 , 6100 [	97.15	86.46	0.12	86.62	10.55
[6100 , 6200 [	97.27	86.78	0.11	86.93	9.92
[6200 , 6300 [	97.38	87.08	0.11	87.22	9.32
[6300 , 6400 [	97.49	87.37	0.10	87.50	8.67
[6400 , 6500 [	97.59	87.64	0.09	87.77	8.27
[6500 , 6600 [	97.68	87.90	0.09	88.02	7.97
[6600 , 6700 [	97.77	88.15	0.09	88.27	7.51
[6700 , 6800 [	97.86	88.39	0.08	88.51	7.06
[6800 , 6900 [	97.94	88.62	0.08	88.73	6.73
[6900 , 7000 [	98.01	88.85	0.07	88.95	6.35
[7000 , 7100 [	98.09	89.06	0.07	89.16	6.00
[7100 , 7200 [	98.15	89.26	0.07	89.36	5.81
[7200 , 7300 [	98.22	89.46	0.06	89.55	5.44
[7300 , 7400 [	98.28	89.65	0.06	89.74	5.28
[7400 , 7500 [	98.34	89.83	0.06	89.92	5.00
[7500 , 7600 [	98.39	90.01	0.05	90.09	4.68
[7600 , 7700 [	98.45	90.18	0.05	90.26	4.47
[7700 , 7800 [	98.50	90.34	0.05	90.42	4.30
[7800 , 7900 [	98.54	90.50	0.05	90.58	4.24
[7900 , 8000 [	98.59	90.65	0.04	90.73	4.06
[8000 , 8100 [	98.63	90.81	0.04	90.88	3.79
[8100 , 8200 [	98.68	90.95	0.04	91.02	3.61
[8200 , 8300 [	98.72	91.09	0.04	91.16	3.50
[8300 , 8400 [	98.75	91.22	0.04	91.29	3.34
[8400 , 8500 [	98.79	91.35	0.03	91.42	3.19
[8500 , 8600 [	98.83	91.48	0.03	91.54	3.14
[8600 , 8700 [	98.86	91.61	1.14	95.80	109.24
[8700 , +8700 [	100.00	100.00			
Total					3443.61

6-a: Calculation of the medial:

$$\overset{\approx}{M}_e = a_i + (a_{i+1} - a_i)\frac{(50 - Q_i)}{(Q_{i+1} - Q_i)}$$

$$= 2.400 + (2.500 - 2.400)\frac{(50 - 48.37)}{(51.07 - 48.37)}$$

$$= 2{,}460.37 \text{ euros}$$

6-b: Calculation of the GINI index by the trapezoid method:

The area under the LORENZ curve is:

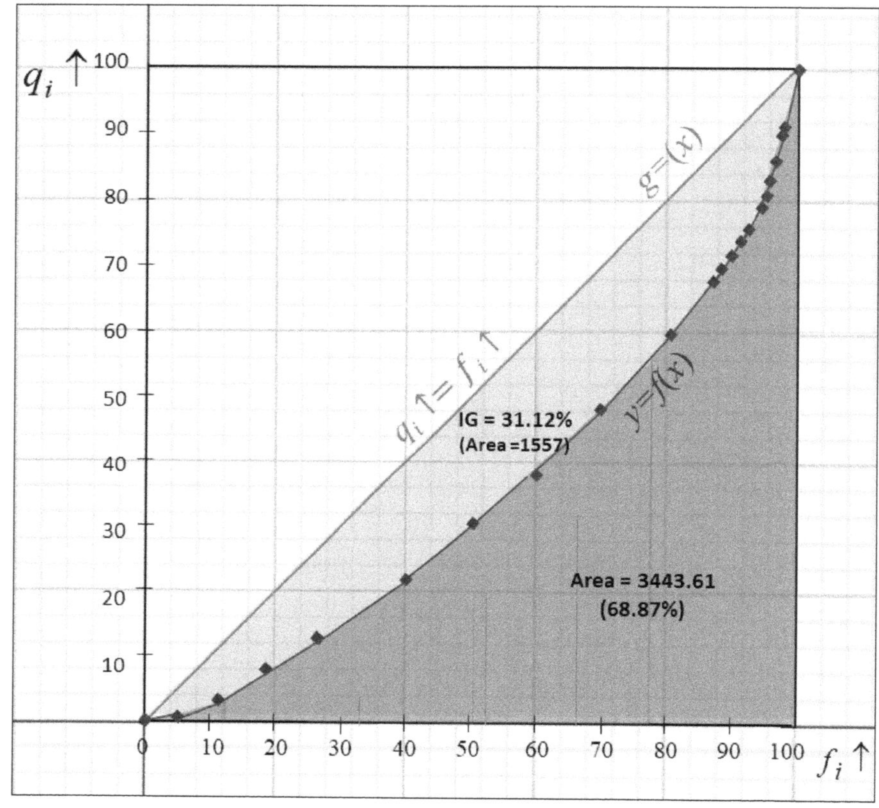

$$\sum_{i=1}^{n}(f_{i+1}\uparrow - f_i\uparrow)\frac{(q_i\uparrow + q_{i+1}\uparrow)}{2} = 3{,}443.61$$

It follows that:

$$\text{GINI Index} = 100\% - \left(\frac{3{,}443.61}{5{,}000}\right)\text{x}100$$

$$= 100\% - 68.87\%$$

$$= 31.13\%$$

6-c: Calculation of the GINI index by integrals:

We will consider the following key points: $(f_i \uparrow, q_i \uparrow)$

$$(0,0)-(5,1)-(11,4)-(19,8)-(26,13)-(40,23)-(64,42)-(80,60)-(85,65)-(94,80)-(98,90)-(100,100)$$

Following the computation, the interpolation polynomial is obtained as:

$$\begin{aligned} y(x) = {} & \frac{1147933087137362614546 3}{881446470379196522392000499650560000000}x^{11} - \frac{870227061491365879151983}{1291007456615994906533738105548800000000}x^{10} \\ & + \frac{3447882019201462598935520 17}{2323813421908790831760728589987840000 00}x^{9} - \frac{116706675668199638081783657 57}{6455037283079974532668690527744000000}x^{8} \\ & + \frac{30913503142851511525318698508369}{232381342190879083176072858998784000000}x^{7} - \frac{55390675391734574444317436974 3}{9156081252595708556976865996800000 0}x^{6} \\ & + \frac{1388126916042066983851573321792643}{829933364967425297057403067852800000 0}x^{5} - \frac{34937115748494060848608844934313}{1317354547567341741360957250560000 0}x^{4} \\ & + \frac{6075919554797551855152790121129737}{290476677738598853970091073748480000}x^{3} - \frac{185820939072321693696129736 91}{4715836705932184784240714880 00}x^{2} \\ & + \frac{259011464743970846795545178351}{219453534005809573741140234288 0}x \end{aligned}$$

Let's evaluate the area between the curve $q_i \uparrow$ and the $f_i \uparrow$ axes in the interval $[0\,;100]$:

$$\begin{aligned} \text{Area} = \int_0^{100} y\,dx = {} & \frac{1147933087137362614546 3}{881446470379196522392000499650560000000}\left(\frac{10^{24}}{12}\right) \\ & - \frac{870227061491365879151983}{1291007456615994906533738105548800000000}\left(\frac{10^{22}}{11}\right) \\ & + \frac{3447882019201462598935520 17}{2323813421908790831760728589987840000 00}\left(\frac{10^{20}}{10}\right) \\ & - \frac{116706675668199638081783657 57}{6455037283079974532668690527744000000}\left(\frac{10^{18}}{9}\right) \\ & + \frac{30913503142851511525318698508369}{232381342190879083176072858998784000000}\left(\frac{10^{16}}{8}\right) \\ & - \frac{55390675391734574444317436974 3}{9156081252595708556976865996800000 0}\left(\frac{10^{14}}{7}\right) \\ & + \frac{1388126916042066983851573321792643}{829933364967425297057403067852800000 0}\left(\frac{x^{12}}{6}\right) \\ & - \frac{34937115748494060848608844934313}{1317354547567341741360957250560000 0}\left(\frac{10^{10}}{5}\right) \\ & + \frac{6075919554797551855152790121129737}{290476677738598853970091073748480000}\left(\frac{10^{8}}{4}\right) - \frac{185820939072321693696129736 91}{4715836705932184784240714880 00}\left(\frac{10^{6}}{3}\right) \\ & + \frac{259011464743970846795545178351}{219453534005809573741140234288 0}\left(\frac{10^{4}}{2}\right) \end{aligned}$$

$$\text{Area} = 3.421$$

It follows that:

$$\text{GINI Index} = 100\% - \left(\frac{\int_0^{100} f(x)\,dx}{5{,}000}\right) \text{x} 100$$

$$= 100\% - \left(\frac{3{,}421}{5{,}000}\right) \text{x} 100$$

$$= 100\% - 68.42\%$$

$$= 31.58\%$$

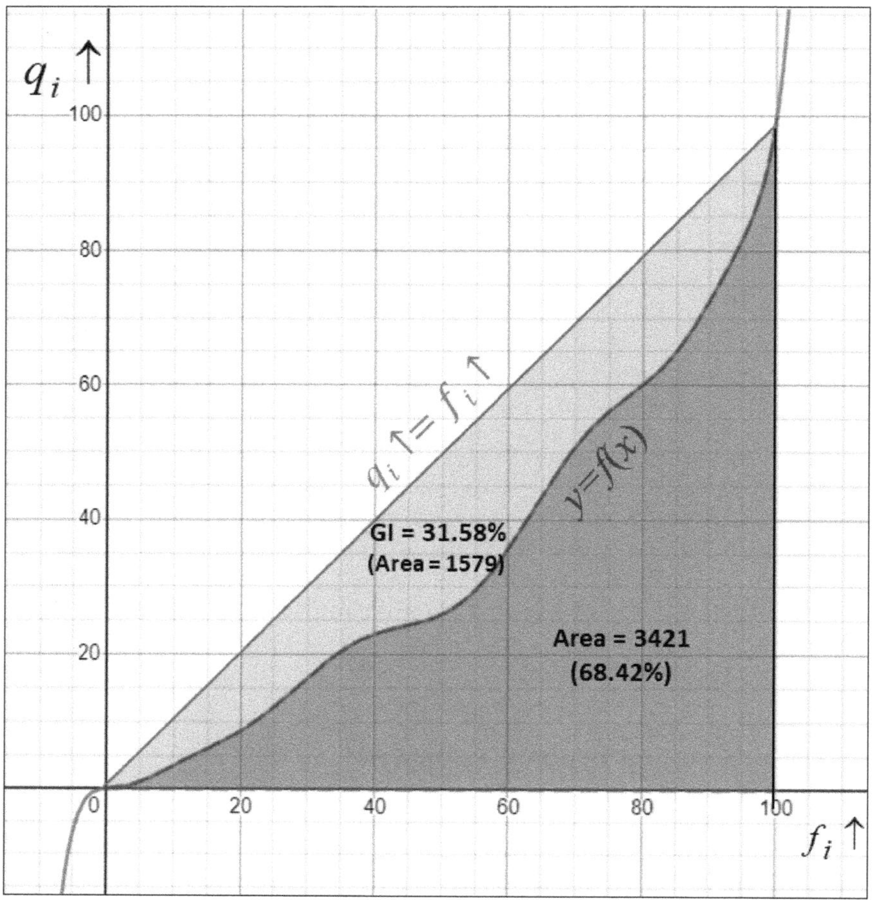

6-d:

<u>INTERPRETATION OF THE RESULTS</u>

Considering the net monthly FTEQ salary in the private sector in France in 2018,

6.a. People earning less than €2,460.37 share 50% of the wage bill, while those earning more than €2,460.37 also share 50% of the wage bill. This means that half of the private sector employees in France earn less than €2,460.37 net per month, while the other half earn more.

6-b/c. The GINI index being around 31%, indicates that inequality exists but is not excessively pronounced. This means that there is a notable difference between the highest and lowest salaries,

However, it is important to note that the GINI index does not take into account fringe benefits such as social benefits and bonuses, which can have a significant impact on workers' overall income.

Note that the GINI index for all of France (all sectors combined) was 29.8% in 2018.

7-a: Finding the percentage of French people working in the private sector and having a salary lower than the mode, i.e., lower than 1,450 euros:

Let's $P_{<\text{mode}}$ be the percentage to find, and we use the linear interpolation formula using the data pair $(a_i, f_i \uparrow)$:

$$Mode = a_i + (a_{i+1} - a_i)\frac{(P_{<\text{mode}} - F_i)}{(F_{i+1} - F_i)}$$

$$1{,}450 = 1{,}400 + (1{,}500 - 1{,}400)\frac{(P_{<\text{mode}} - 18.18)}{(26.01 - 18.18)}$$

$$P_{<\text{mode}} = 22.1\%$$

22.1% of French people working in the private sector, meaning 3,650,128 people have a lower salary than 1,450 euros (which represents the modal wage).

7-b: Finding the percentage of French people working in the private sector and with a salary below the average, i.e., lower than 2,369 euros:

Let $P_{<\bar{x}}$ be the percentage to find, and we use the linear interpolation formula using the data pair $(a_i, f_i \uparrow)$:

$$\bar{x} = a_i + (a_{i+1} - a_i)\frac{(P_{<\bar{x}} - F_i)}{(F_{i+1} - F_i)}$$

$$2{,}369 = 2{,}300 + (2{,}400 - 2{,}300)\frac{(P_{<\bar{x}} - 67.54)}{(70.49 - 67.54)}$$

$$P_{<\bar{x}} = 69.58\%$$

69.57% of French people in the private sector, meaning 11,490,471 people have a salary lower than 2,369 euros (which represents the average wage).

7-c: Finding the percentage of French people working in the private sector and having a salary lower than the medial, i.e., lower than 2,460.37 euros:

Let $P_{<\text{medial}}$ pe be the percentage to find, we use the linear interpolation formula using the data pair $(a_i, f_i \uparrow)$ as proceeded above:

$$Médiale = a_i + (a_{i+1} - a_i)\frac{(P_{<\bar{x}} - F_i)}{(F_{i+1} - F_i)}$$

$$2{,}460.37 = 2{,}400 + (2{,}500 - 2{,}400)\frac{(P_{<\text{medial}} - 70.43)}{(73.09 - 70.43)}$$

$$P_{<\text{medial}} = 72.06\%$$

72.06% of French people working in the private sector, meaning 11,900,078 people have a lower salary than 2,460.37 euros (which represents the median wage).

7-d: Finding the percentage of French people working in the private sector and with a salary between the average minus the standard deviation and the average plus the standard deviation.

7-d-1: The difference between the average plus the standard deviation and the average minus the standard deviation:

$$\bar{x} - \sigma = 2{,}369 \; - 1{,}995.79$$
$$= 373.21 \text{ euros}$$
$$\bar{x} + \sigma = 2{,}369 \; +1{,}995.79$$
$$= 4{,}364.79 \text{ euros}$$

7-d-2: Percentage of French people working in the private sector with a lower-than-average salary minus the standard deviation is:

Let $P_{<\bar{x}-\sigma}$ be the percentage to find, and we use the linear interpolation formula using the data pair $(a_i, f_i \uparrow)$:

$$\bar{x} - \sigma = a_i + (a_{i+1} - a_i)\frac{(P_{<\bar{x}-\sigma} - F_i)}{(F_{i+1} - F_i)}$$
$$373.21 = 0 + (1{,}200 - 0)\frac{(P_{<\bar{x}-\sigma} - 0)}{(5.42 - 0)}$$
$$P_{<\bar{x}-\sigma} = 1.69\%$$

7-d-3: Percentage of French people working in the private sector with a lower-than-average salary plus the standard deviation:

Let $P_{<\bar{x}+\sigma}$ be the percentage to find, and we use the linear interpolation formula using the data pair $(a_i, f_i \uparrow)$:

$$\bar{x} + \sigma = a_i + (a_{i+1} - a_i)\frac{(P_{<\bar{x}+\sigma} - F_i)}{(F_{i+1} - F_i)}$$
$$4{,}364.79 \; = 4{,}300 + (4{,}400 - 4{,}300)\frac{(P_{<\bar{x}+\sigma} - 92.85)}{(93.26 - 92.85)}$$
$$P_{<\bar{x}+\sigma} = 93.12\%$$

The percentage of French people working in the private sector, with a salary between 373.21 and 4,364.79 euros is:

$$93.12\% - 1.69\,\% = 91.43\%$$

Which represents 15,100,960 of French people working in the private sector.

7-e: interpretation of the percentage of French people working in the private sector in 2018 with a salary lower than these central measures and dispersion

INTERPRETATION OF THE RESULTS

In 2018, considering the net monthly FTEQ salary in the private sector in France,

7-a. 22.1% of French people working in the private sector, that is 3,650,128 people, have a net monthly FTEQ salary of less than 1,450 euros; that is to say, they have a lower salary than the salary earned by the majority of French people working in the private sector

7-b. 69.58% of French people working in the private sector, that is 11,490,471 people, have a net monthly salary in full time equivalent below 2,369.00 euros; that is to say, their salary is lower than the average salary in the private sector.

7-c. Approximately 72.06% of French private sector workers, or nearly 11.9 million people, have a monthly net salary in EQTP lower than 2,460.37 euros. This category of workers shares 50% of the total wage bill, while the remaining 27.94%, or about 4.6 million people, with a salary above 2,460.37 euros share the other 50% of the wage bill. In other words, nearly 11.9 million workers share the same wage bill as approximately 4.6 million people.

7-d. Approximately 91.43% of French private sector workers, or more than 15.1 million people, have a monthly net salary in EQTP between €373.21 and €4,364.79. This figure corresponds to the salary range that falls between the private sector average salary minus the standard deviation and the average salary plus the standard deviation. This means that the vast majority of private sector workers earn salaries that fall within a relatively narrow range, which may explain the relatively low level of inequality in the distribution of salaries in the French private sector in 2018.

It is clear that this distribution does not follow the normal distribution, as a salary distribution in a population that follows the normal distribution would have a more balanced distribution of salaries around the mean, with a lower proportion of people having very low or very high salaries. The proportion of people in this range in this distribution is therefore higher than what would be expected from a normal distribution.

This large proportion of workers in this range may suggest a concentration of salaries, probably due to economic, social, and cultural factors that influence the distribution of salaries in the French private sector. There are several possible reasons to explain why the distribution of salaries in the French private sector in 2018 does not follow a normal distribution. This can be explained by various factors such as the wage policy of companies, competition in the job market for mid-wage jobs, collective bargaining negotiations, etc.

- First, the labor market is not perfectly competitive, and salaries are not entirely determined by market forces. There may be factors such as wage negotiations, collective agreements, or minimum wage scales that can influence salary levels in a particular sector. In addition, some employers may choose to pay their workers higher wages than those on the market to retain talented or motivated workers.
- Furthermore, there may be differences in salaries based on sectors of activity, levels of education or professional experience, which can explain the observed salary disparities in the distribution.

As for passing judgment on the good or bad side of having so many people in this range, it is difficult to answer this question categorically because it depends on each person's point of view.

- On the one hand, the fact that the vast majority of private sector workers in France have a salary between the average salary minus the standard deviation and the average salary plus the standard deviation can be seen as a good thing because it indicates some stability and equity in salaries within the company or sector. This also means that the majority of private sector workers in France are not living in great financial insecurity.
- On the other hand, this can be seen as a bad thing because it may reflect a lack of diversity in salaries and an unequal distribution of wealth. For example, if a company only offers salaries in this range, it could indicate that it does not offer enough opportunities for growth and professional development for workers. Additionally, this can also highlight a difficulty for workers to move out of this salary range, even if they work hard and have experience, which could cause frustration and a stagnation of motivation.

10.4.2 Analysis of Tax-related Parameters

8: Let's calculate the percentage of private sector workers in France who have a monthly net salary in EQTP that is not subject to income tax, meaning their annual gross salary does not exceed the taxable threshold of 9,964 euros (830.33 euros gross monthly in 2018). At this threshold, the gross salary is equal to the net salary, both of which are equivalent to 830.33 euros per month. This is because, below this amount, the tax rate is set at zero percent.

Let $P_{<830}$ be the percentage to find, and we use the linear interpolation formula using the data pair $(a_i, f_i \uparrow)$:

$$Tax\_threshold_{2018} = a_i + (a_{i+1} - a_i)\frac{(P_{<830} - F_i)}{(F_{i+1} - F_i)}$$

$$830.33 = 0 + (1{,}200 - 0)\frac{(P_{<830} - 0)}{(5.42 - 0)}$$

$$P_{<830} = 3.75\%$$

3.75% of French people working in the private sector have a salary not subject to taxes, which represents 619,366 people.

9: Let's calculate the percentage of French workers in the private sector who pay 45% of taxes on their income, meaning they have a salary exceeding the highest tax bracket set at 156,244 euros gross per year, or 13,020.33 euros gross per month (in 2018).

First, we need to convert this amount to the net salary, which will be the basis of our calculations.

The net salary in FTEQ is:

$$13{,}020.33 - 13{,}020.33 \times 45\% = 7{,}161.18 \text{ euros.}$$

Let's first calculate the percentage of French people working in the private sector, having a salary of less than 7,161.18 euros, then we will subtract them from 100%.

Let $P_{<7161}$ be the percentage to find, and we use the linear interpolation formula using the data pair $(a_i, f_i \uparrow)$ as proceeded above:

$$\text{Net\_Salary\_FTEQ} = a_i + (a_{i+1} - a_i)\frac{(P_{<7161} - F_i)}{(F_{i+1} - F_i)}$$

$$7{,}161 = 7{,}100 + (7{,}200 - 7{,}100)\frac{(P_{<7161} - 98.09)}{(98.15 - 98.09)}$$

$$P_{<7161} = 98.12\%$$

So, the number of French people working in the private sector and paying 45% tax is:

$$100\% - 98.12\% = 1.88\%$$

Which represents 310,509 people.

10:

INTERPRETATION OF RESULTS
Considering the net monthly FTEQ salary in the private sector in France in 2018,
8-9. A small fraction of the population working in the private sector did not pay income tax. This corresponds to 3.75% of this category of French people whose monthly net salaries in EQTP do not exceed a certain threshold. This means that they are not subject to tax laws and do not pay income tax. However, 1.88% of private sector workers are subject to income tax, and a portion of their income is taxed at a rate of 45%. It is important not to confuse these two groups with workers whose salaries exceed the tax threshold but who may be exempt from taxes for various reasons, such as social status or the number of dependent children.

10.4.3 Analysis of Socio-economic parameters

11: Let's calculate the percentage of French workers in the private sector with a monthly net salary in EQTP lower than the Minimum Interprofessional Growth Wage (SMIC), which was 1,498.47 euros gross per month in France in 2018.

Since the SMIC is higher than the first tax bracket, we will first convert this amount to the net salary, which will be the basis of our calculation. According to the tax scale applied by the French tax administration in 2018, the tax rate for the first tax bracket was 14% (see table above).

The net SMIC in FTEQ is:

$$1{,}498.47 - 1{,}498.47 \times 14\% = 1{,}288.68 \text{ euros.}$$

Let us use $P_{<smic}$ the percentage of French people working in the private sector and having a net monthly salary in FTEQ lower than the SMIC, and we use the linear interpolation formula using the data pair $(a_i, f_i \uparrow)$:

$$\text{Smic} = a_i + (a_{i+1} - a_i)\frac{(P_{<smic} - F_i)}{(F_{i+1} - F_i)}$$

$$1{,}288.68 = 1{,}200 + (1{,}300 - 1{,}200)\frac{(P_{<smic} - 5.42)}{(11.27 - 5.42)}$$

$$P_{<smic} = 10.6\%$$

10.6% of French people working in the private sector, that is 1 750 740 people have a net monthly salary in FTEQ below the SMIC.

12: Let's calculate the percentage of French people working in the private sector in 2018 who earned a monthly net salary in EQTP below the poverty line as defined by INSEE and the Observatory of Inequalities. We assume here that the employee has only one source of income and that social benefits are not taken into account. This percentage will allow us to assess the prevalence of poverty among private sector workers in France.

The poverty line being set at 60% below the median, so it is below 1,872.25 x 60% = 1,123.35 euros.

Let P_{poverty} be the percentage of French people working in the private sector, and living below the poverty line, i.e., having a salary of less than 1,123.53 euros, and we use the linear interpolation formula using the data pair $(a_i, f_i \uparrow)$ as proceeded above:

$$\text{Poverty\_Treshold} = a_i + (a_{i+1} - a_i)\frac{(P_{\text{poverty}} - F_i)}{(F_{i+1} - F_i)}$$

$$1{,}123.35 = 1{,}100 + (1{,}200 - 1{,}100)\frac{(P_{\text{poverty}} - 0)}{(5.42 - 0)}$$

$$P_{<\text{poverty}} = 1.27\%$$

Thus, 1.27% of French people working in the private sector with a net monthly salary in FTEQ in 2018 life below the poverty line in the sense of the definition of INSEE and the Observatory of inequalities, which represents 209,759 people.

13: Let's calculate the percentage of French workers in the private sector who are considered to be above the wealth threshold according to the Observatory of Inequalities.

The wealth threshold corresponds to twice the median standard of living, so it is above 1,872.25 x 2 = 3,744.5 euros.

Let's calculate the percentage below the wealth threshold and subtract it from 100%.

Let P_{wealth} be the percentage to find, and we use the linear interpolation formula using the data pair $(a_i, f_i \uparrow)$ as proceeded above:

$$\text{Weath\_Treshold} = a_i + (a_{i+1} - a_i)\frac{(P_{\text{wealth}} - F_i)}{(F_{i+1} - F_i)}$$

$$3{,}744.5 = 3{,}700 + (3{,}800 - 3{,}700)\frac{(P_{\text{wealth}} - 89.48)}{(90.16 - 89.48)}$$

$$P_{<\text{wealth}} = 89.78\%$$

Thus, the percentage of French people working in the private sector with a net monthly salary in FTEQ in 2018 considered to be above the wealth threshold in the sense of the Observatory of inequalities is:

$$100\% - 89.78\% = 10.22\%$$

Which represents 1,687,978 people.

14: The percentage of French people working in the private sector and having a net monthly salary in FTEQ in 2018 that allows them to be classified as middle class by using the definition of the INSEE.

According to INSEE, the middle class for the year 2018 contains all people whose disposable income (in other words all income, including social benefits) is between 1390 euros and 2568 euros per month.

14-a: Let's calculate the percentage of French people working in the private sector with a salary of less than 1,390 euros (below the middle class)

Let $P_{<1390}$ be the percentage to find, and we use the linear interpolation formula using the data pair $(a_i, f_i \uparrow)$:

$$\lim_{inf\_INSEE}(\text{Middle Class}) = a_i + (a_{i+1} - a_i)\frac{(P_{<1390} - F_i)}{(F_{i+1} - F_i)}$$

$$1{,}390 = 1{,}300 + (1{,}400 - 1{,}300)\frac{(P_{<1390} - 11.27)}{(18.18 - 11.27)}$$

$$P_{<1390} = 17.49\%$$

2,888,721 French people working in the private sector have a salary below the middle class in the sense of INSEE.

14-b: Let's calculate the percentage of French workers in the private sector who have a salary lower than 2,568 euros, which is below the upper limit of the middle class.

Let $P_{<2568}$ be the percentage to find, and we use the linear interpolation formula using the data pair $(a_i, f_i \uparrow)$:

$$\lim_{sup\_INSEE}(\text{Middle Class}) = a_i + (a_{i+1} - a_i)\frac{(P_{<2568} - F_i)}{(F_{i+1} - F_i)}$$

$$2{,}568 = 2{,}500 + (2{,}600 - 2{,}500)\frac{(P_{<2568} - 73.09)}{(75.41 - 73.09)}$$

$$P_{<2568} = 74.67\%$$

12,332,809 French people working in the private sector have a salary below the upper limit of the middle class in the sense of INSEE.

The percentage of French people working in the private sector, with a salary that allows them to position themselves in the middle class in the direction of INSEE is:

$$74.67\% - 17.49\% = 57.18\%$$

Thus, 9,444,087 French people working in the private sector have a salary classifying them in the middle class in the sense of INSEE according to the INSEE definition.

15: Let's calculate the percentage of French workers in the private sector in 2018 who have a monthly net salary in EQTP that allows them to be classified as middle class according to the OECD definition. Here we assume that the employee has only one source of income and social benefits are not taken into account.

According to the OECD, the middle class is represented by people with an income between 75% and 200% of the median income.

75% of median income = 1,872.25 x 75% = 1,404.19 euros

Let's calculate the percentage of French people working in the private sector with a salary of less than 1,404.19 euros.

Let $P_{<1404}$ be the percentage to find, and we use the linear interpolation formula using the data pair $(a_i, f_i \uparrow)$:

$$\lim_{inf\_OCDE}(\text{Middle Class}) = a_i + (a_{i+1} - a_i)\frac{(P_{<1404} - F_i)}{(F_{i+1} - F_i)}$$

$$1{,}404.19 = 1{,}400 + (1{,}500 - 1{,}400)\frac{(P_{<1404} - 18.18)}{(26.01 - 18.18)}$$

$$P_{<1404} = 18.51\%$$

3,057,189 French people working in the private sector have a salary below the lower limit of the middle class according to the OECD definition.

200% of median income = 3,744.5 euros. The percentage of French people working in the private sector with a salary below this amount has already been calculated in question **14**; it is equivalent to 89.78%

The percentage of French people working in the private sector, with a salary that allows them to position themselves in the middle class in the direction of the OECD IS:

$$89.78\% - 18.51\% = 71.27\%$$

Thus, 11,771,250 French people working in the private sector have a salary classifying them in the middle class according to the OECD definition.

16: We will consider only the population of French workers in the private sector in 2018 with a monthly net salary in EQTP. The definition of the middle class varies according to INSEE, CREDOC, and the Observatory of Inequalities. According to these organizations, the middle class corresponds to the population situated between the 30% poorest and the 20% richest, representing 50% of the population.

16-a: Let's calculate the deciles to find the lower and upper bounds.

Let's use the formula:

$$D_j = a_i + (a_{i+1} - a_i)\frac{(10{*}j - F_i)}{(F_{i+1} - F_i)}$$

$$D_1 = 1{,}200 + (1{,}300 - 1{,}200)\frac{(10 - 5.42)}{(11.27 - 5.42)}$$

$$= 1{,}278.29 \text{ euros}$$

$$D_2 = 1{,}400 + (1{,}500 - 1{,}400)\frac{(20 - 18.18)}{(26.01 - 18.18)}$$

$$= 1{,}423.24 \text{ euros}$$

$$D_3 = 1{,}500 + (1{,}600 - 1{,}500)\frac{(30 - 26.01)}{(33.46 - 16.01)}$$

$$= 1{,}522.87 \text{ euros}$$

$$D_4 = 1{,}600 + (1{,}700 - 1{,}600)\frac{(40 - 33.46)}{(40.16 - 33.46)}$$

$$= \mathbf{1{,}697.61 \text{ euros}}$$

$$D_5 = 1{,}800 + (1{,}900 - 1{,}800)\frac{(50 - 46.12)}{(51.49 - 46.12)}$$

$$= 1{,}872.25 \text{ euros}$$

$$D_6 = 2{,}000 + (2{,}100 - 2{,}000)\frac{(60 - 56.25)}{(60.50 - 56.25)}$$

$$= 2{,}088.24 \text{ euros}$$

$$D_7 = 2{,}300 + (2{,}400 - 2{,}300)\frac{(70 - 67.54)}{(70.49 - 67.54)}$$

$$= 2{,}383.39 \text{ euros}$$

$$D_8 = 2{,}800 + (2{,}900 - 2{,}800)\frac{(80 - 79.21)}{(80.82 - 79.21)}$$

$$= \mathbf{2{,}849.07 \text{ euros}}$$

$$D_9 = 3{,}700 + (3{,}800 - 3{,}700)\frac{(90 - 89.48)}{(90.16 - 89.48)}$$

$$= 3{,}776.47 \text{ euros}$$

The lower limit is: 1,697.61 euros.

The upper limit is: 2,849.07 euros.

The middle class composed of French people working in the private sector and having a net monthly salary in FTEQ in 2018 ranges from 1,697.61 to 2,849.07 euros.

16-b: Let's calculate the number of French people who would be in this middle class:

16-b-1: Let's calculate the percentage of French people working in the private sector with a salary below 1,697.61 euros (below the middle class):

Let $P_{<1697}$ be the percentage of French people working in the private sector, having a salary lower than euros, and we use the linear interpolation formula using the data pair $(a_i , f_i \uparrow)$:

$$\lim_{inf\_ALL}(\text{Middle Class}) = a_i + (a_{i+1} - a_i)\frac{(P_{<1697} - F_i)}{(F_{i+1} - F_i)}$$

$$1{,}697.61 = 1{,}600 + (1{,}700 - 1{,}600)\frac{(P_{<1697} - 33.46)}{(40.16 - 33.46)}$$

$$P_{<1697} = 40\%$$

Which represents 6,605,567 people.

16-b-2: Let's calculate the percentage of French people working in the private sector, with a salary below 2,849.07 euros (below the upper limit of the middle class).

Let $P_{<2849.07}$ be the percentage of French people working in the private sector, having a salary lower than euros, and we use the linear interpolation formula using the data pair $(a_i , f_i \uparrow)$:

$$\lim_{sup\_ALL}(\text{Middle Class}) = a_i + (a_{i+1} - a_i)\frac{(P_{<2849.07} - F_i)}{(F_{i+1} - F_i)}$$

$$2{,}849.07 = 2{,}800 + (2{,}900 - 2{,}800)\frac{(P_{<2849.07} - 79.21)}{(80.82 - 79.21)}$$

$$P_{<2849.07} = 80\%$$

Which represents 13,213,134 people.

The percentage of French people working in the private sector, with a salary allowing them to position themselves in the middle class in the direction of CREDOC is:

$$80\% - 40\% = 40\%$$

Which represents 6,605,567 people.

1,278.29	1,423.24	1,522.86	1,697.61	1,872.25	2,088	2,383.38	2,849.06	3,776.47	
D1	D2	D3	D4	D5	D6	D7	D8	D9	D10
			40%				80%		
					40% middle class				

INTERPRETATION OF THE RESULTS

Considering the net monthly FTEQ salary in the private sector in France in 2018,

11: 10.6% of French people working in the private sector, that is 1,750,740 people, have a net monthly salary in FTEQ lower than the minimum wage.

12. 1.27% of French people working in the private sector, that is 209,759 people live below the poverty line according to the definition of INSEE and the observatory of inequalities.

13. 10.22% of French people working in the private sector, that is 1,687,978 people with a net monthly salary in FTEQ in 2018 have a salary above the wealth threshold according to the Observatory of Inequalities.

14. 54.18% of French people working in the private sector, that is 9,444,087 people and have having a net monthly salary in FTEQ in 2018 allowing them to be classified as middle class according to the INSEE definition.

15. 71.27% of French people working in the private sector, that is 11,771,250 people and having a net monthly salary in FTEQ in 2018 allowing them to be classified as middle class according to the definition of the OECD.

16. If we only consider the population of French workers in the private sector in 2018 with a monthly net salary in EQTP in 2018, the middle class according to the definition of the INSEE, the CREDOC and the Observatory of inequalities, is represented by the population located between the poorest 30% and the richest 20%, i.e., 50% of the population.

This middle class will go from 1,697.61 euros to 2,849.07 euros and would include around 40% of the French, which represents 6,605,567 people.

CHAPTER 11

Case Studies Leading to Optimized Decision-Making Through Data Analysis

11.1 Presentation of Case Studies

The purpose of these studies is to demonstrate how data analysis and interpretation can lead to the maximization of positive outcomes (gain, profit, satisfaction, well-being, etc.) while minimizing negative outcomes (loss, dissatisfaction, risk, etc.), ultimately resulting in optimized decision-making.

The four case studies presented illustrate this approach. The first case study is independent of the other three, which have similarities in terms of data processing and analysis but differ in terms of interpretation.

These case studies have been carefully selected for their relevance and ability to illustrate the data-driven decision-making method. It is important to note that these examples are not taken from real situations but have been created to demonstrate how data analysis and interpretation can be used to optimize decision-making.

We hope that these case studies will enable readers to better understand how data analysis can be used to optimize decision-making and maximize positive outcomes while minimizing negative outcomes.

The four case studies presented in this chapter are the following:

1. The first case study involves an international institution that wishes to collaborate with a non-profit organization (NPO) over a long period of time. The goal is to select an NPO capable of working with rigor, integrity, and impartiality, while also being able to measure and evaluate weights without using a balance or weighing scale. Three NPOs have been selected for on-site evaluation to determine the most suitable one for this collaboration.

2. The second case is that of a specialty fabric store. This store is facing a high demand from its customers for fabrics of various sizes and shapes. To meet this demand, the store wants to automate its fabric cutting system, replacing manual cutting with a machine. However, it is important to choose the machine that is best suited to the store's needs in terms of speed, precision, cut quality, and cost. To do so, three machines from different manufacturers were presented and subjected to validation tests to verify the manufacturers' claims. The results of these tests were analyzed to determine the most performant machine for the store.

3. The third case, based on the same data as the previous one, concerns a biological laboratory searching for a solution to administer medication to patients with a dosage that is limited to a certain maximum volume. Three medication injection guns were proposed by different manufacturers. To determine the most suitable solution, these guns were subjected to rigorous validation tests. The tests focused on the accuracy of the injection, and a careful evaluation of the test results allowed the biological laboratory to select the most appropriate medication injection gun for their patients.

4. The fourth case, like the previous one, is built on the same data as the second case and concerns a foundry specializing in the production of very specific metals, which require a precise and rigorous manufacturing process. To ensure the quality and efficiency of their production, the foundry is looking to automate their metal mixing process by finding a software that meets strict criteria. The software must be able to detect the complete melting of the metal, measure the net weight of the melted metal, calculate the amount of additive components to be added based on the weight of the metal, and activate the mixing process before the alloy cools and hardens. Additionally, the software must be asymptotically superior, meaning its response time must be strictly less than a pre-defined maximum time. Three software developers were asked to propose solutions, which were subjected to validation tests to determine the software that best suits the needs of the foundry.

In each of these cases, the statistical results were forwarded to a specialized team of statisticians for further analysis. This team conducted thorough data analysis, rigorous testing, and interpreted the results to identify the strengths and weaknesses of each proposed solution. Ultimately, this team drew accurate and relevant conclusions to help stakeholders make informed decisions based on the data.

11.2 First Case Study

11.2.1 Presentation of the Study

An international institution wishes to collaborate with a Non-Profit Organization (NPO) for a long-term partnership. After various stages of interviews, three NPOs, referred to as A, B, and C, have been selected to perform field tests that will demonstrate their integrity, rigor, selflessness, accuracy, and firmness. They are also expected to have excellent measurement and evaluation skills as their task is to estimate the weight of rice to be distributed without the use of a scale.

The three NPOs were sent to three different refugee camps, each with 60 families. Their mission was to distribute rice fairly and without favoritism to each of the 60 families in each camp for which they were responsible. For this purpose, bags of rice were given to the three NPOs: Association A received 237 kg, Association B received 233 kg, and Association C received 90 kg. They were therefore required to demonstrate absolute rigor, to not be influenced or swayed, and above all to be incorruptible, as each family was to receive exactly one-sixtieth of the quantity of rice given to them.

After the rice distribution, inspectors from the institution checked the portions given to each family in each camp. The data was collected in the following three tables.

ASBL **A** : Camp A		
Quantité de riz reçu par famille (en kg)		
6	4	5
3	6	3
3	4	4
6	4	2
5	4	3
4	7	4
3	5	3
3	5	5
4	2	2
4	3	5
4	4	4
5	2	4
5	4	5
1	3	3
2	5	4
4	5	1
5	6	5
3	2	5
3	5	3
7	4	3

ASBL **B** : Camp B		
Quantité de riz reçu par famille (en kg)		
3	6	4
4	4	1
2	3	2
1	6	1
6	4	2
5	5	6
8	4	2
5	3	6
5	4	1
3	3	2
4	4	2
3	1	3
3	4	2
4	4	2
4	6	2
7	4	8
3	8	4
5	6	7
5	3	0
5	5	4

ASBL **C** : Camp C		
Quantité de riz reçu par famille (en kg)		
1	3	0
0	0	2
2	0	1
3	1	0
1	3	3
0	2	1
1	1	1
0	2	2
1	2	0
3	1	2
3	2	1
1	3	1
2	0	1
2	0	0
4	1	2
2	3	3
0	1	3
2	5	0
1	2	3
0	2	1

Based on your evaluation, which non-profit organization would you recommend to this institution to be hired?

Note: To streamline the reading of the solutions, we will adopt the following notation: each response pertaining to a camp will be preceded by a letter indicating the corresponding camp.

11.2.2 Questions

11.2.2.1 Analysis of central tendency measures for decision-making

For each Camp,

1. Count the data and give the table of distribution of the number of families in relation to the quantity of rice received.
2. Draw the bar graph.
3. Calculate the central tendency characteristics of the distribution:
    a. The mode.
    b. The average.
    c. The median.
4. Give an interpretation of the order of magnitude of the elements of this statistical series.

11.2.2.2 Analysis of measures of dispersion for decision-making

Calculate and compare the following data for the three camps.

5. Determine the characteristics of dispersion.
    d. The Interquartile range
    e. The Standard deviation
    f. The Variance
    g. The Coefficient of Variation
6. Determine the percentage of families that received a lower
    h. The majority of the families
    i. The average amount of rice.
7. Determine the percentage of families that received a quantity of rice lower than the average minus the standard deviation.
8. Determine the percentage of families that received a quantity of rice lower than the average plus the standard deviation.
9. Determine the percentage of families receiving a quantity of rice between the mean and the mean minus the standard deviation.
10. Determine the percentage of families receiving a quantity of rice between the mean and the mean plus the standard deviation.
11. Determine the percentage of families receiving a quantity of rice in the range of the mean minus the standard deviation and the mean plus the standard deviation.
12. Determine the percentage of families that have received nothing.
13. Interpret the results then conclude.

11.2.2.3 Analysis of shape measures for decision-making

To complete the argument, you are asked to calculate the shape characteristics for each camp.

14. Determine the FISHER's coefficient of skewness.
15. Determine the FISHER's coefficient of kurtosis.

11.2.2.4 Analysis of concentration measures for decision-making

To refine and consolidate the decision, it is requested to calculate their concentration characteristics:

16. Determine the medial.
17. Determine The percentage of families with a rice ratio lower than the medial.
18. Determine the GINI index by the graphical method (trapezoidal method).
19. Determine the GINI index by the analytical method (integral method).

11.2.3 Solutions

11.2.3.1 Analysis of central tendency measures for decision-making

A: Camp A

A-1: The distribution table of the number of families with respect to the amount of rice received is:

Quantity of received rice (Kg)	Number of families
1	2
2	6
3	14
4	17
5	15
6	4
7	2
Total	60

A-2: The distribution bar diagram is:

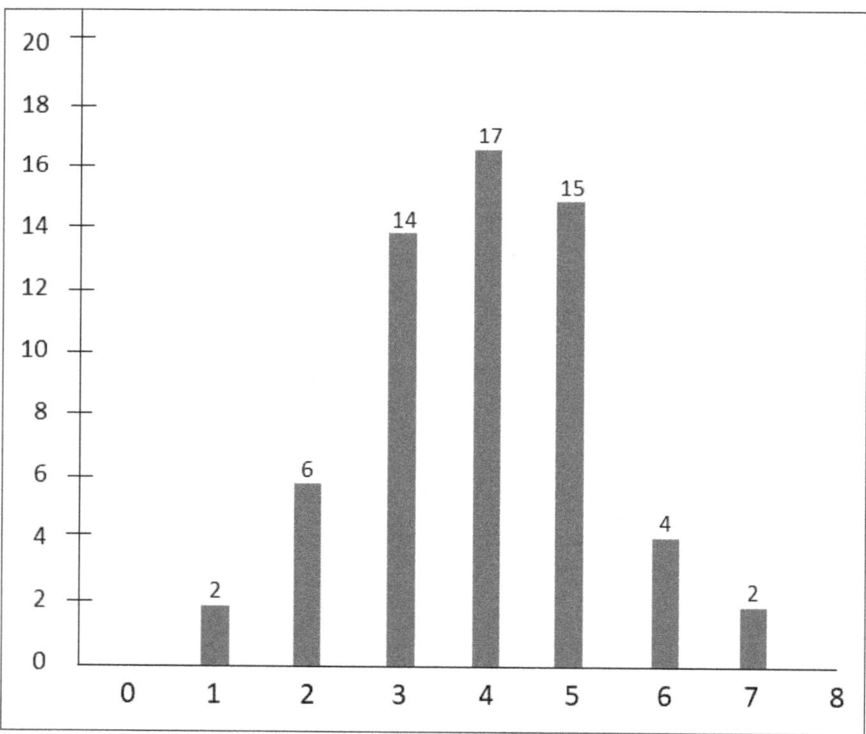

The following table shows the data required to solve this exercise:

x_i	n_i	$n_i x_i$	$n_i x_i^2$	$n_i(x_i - \bar{x})$	$n_i(x_i - \bar{x})^3$	$n_i(x_i - \bar{x})^4$	f_i	$f_i\uparrow$
1	2	2	2	-5.90	-51.34	151.47	3.33	3.33
2	6	12	24	-11.70	-44.49	86.75	10.00	13.33
3	14	42	126	-13.30	-12.00	11.40	23.33	36.67
4	17	68	272	0.85	0.00	0.00	28.33	65.00
5	15	75	375	15.75	17.36	18.23	25.00	90.00
6	4	24	144	8.20	34.46	70.64	6.67	96.67
7	2	14	98	6.10	56.75	173.07	3.33	100.00
	60	237	1041	0.00	0.73	511.57		

A-3: The characteristics of central tendency are:

A-3-a: The mode of camp A is 4 Kg.

A-3-b: The average of camp A is:

$$\bar{x} = \frac{1}{n}\sum_{i=1}^{i=7} n_i x_i = \frac{237}{60}$$
$$= 3.95 \text{ Kg}$$

A-3-c: The median of camp A is:

$$M_e = 3 + (4-3)\frac{(50 - 36.67)}{(65 - 36.67)}$$
$$= 3.47\text{Kg}$$

B: Camp B

B-1: The distribution table of the number of families with respect to the amount of rice received is:

Quantity of received rice (Kg)	Number of families
0	1
1	5
2	9
3	10
4	15
5	8
6	7
7	2
8	3
Total	60

B-2: The distribution bar diagram is:

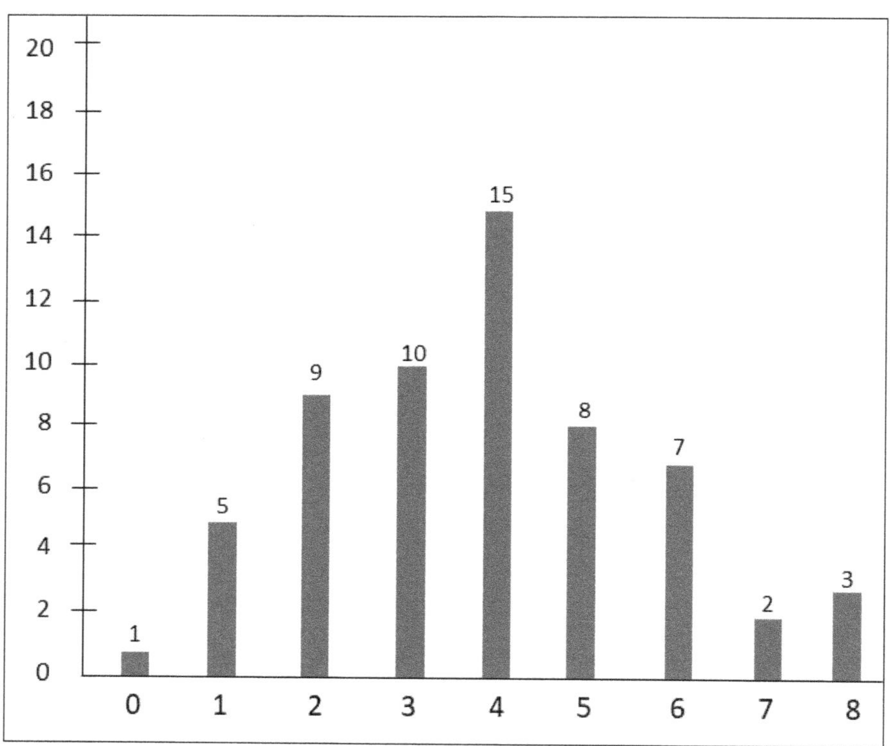

The following table shows the data required to solve this exercise:

x_i	n_i	n_ix_i	$n_ix_i^2$	$n_i(x_i-\bar{x})$	$n_i(x_i-\bar{x})^3$	$n_i(x_i-\bar{x})^4$	f_i	$f_i\uparrow$
0	1	0	0	-3.88	-61.63	227.41	1.67	1.67
1	5	5	5	-14.42	-128.36	345.58	8.33	10.00
2	9	18	36	-16.95	-66.73	113.23	15.00	25.00
3	10	30	90	-8.83	-8.57	6.09	16.67	41.67
4	15	60	240	1.75	0.00	0.00	25.00	66.67
5	8	40	200	8.93	9.26	12.44	13.33	80.00
6	7	42	252	14.82	60.31	140.51	11.67	91.67
7	2	14	98	6.23	56.75	188.71	3.33	95.00
8	3	24	192	12.35	199.29	861.60	5.00	100.00
	60	233	1113	0.00	60.30	1895.57		

B-3: The characteristics of Central tendency are:

B-3-a: The mode of camp B is 4 Kg.

B-3-b: The average of camp B is:

$$\bar{x} = \frac{1}{n}\sum_{i=1}^{i=8} n_i x_i = \frac{233}{60}$$
$$= 3.8833 \text{ Kg}$$

B-3-c: The median of camp B is:

$$M_e = 3 + (4-3)\frac{(50-41.67)}{(66.67-41.67)}$$
$$= 3.33 \text{ Kg}$$

C: Camp C

C-1: The distribution table of the number of families with respect to the amount of rice received is:

Quantity of received rice (Kg)	Number of families
0	14
1	18
2	15
3	11
4	1
5	1
Total	60

C-2: The distribution stick diagram is:

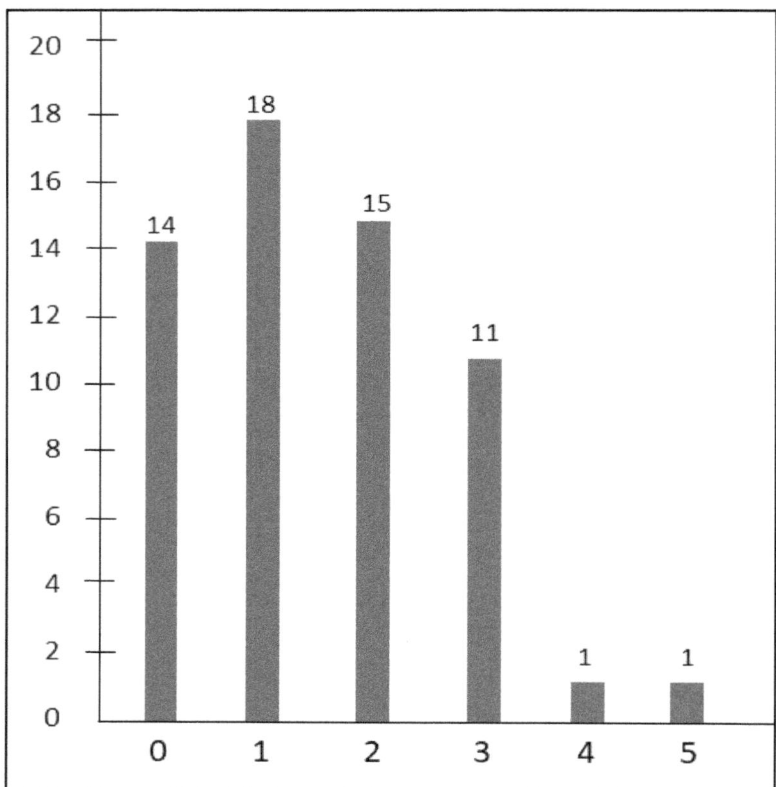

The following table shows the data required to solve this exercise:

x_i	n_i	n_ix_i	$n_ix_i^2$	$n_i(x_i-\bar{x})$	$n_i(x_i-\bar{x})^3$	$n_i(x_i-\bar{x})^4$	f_i	$f_i\uparrow$
0	14	0	0	-21.00	-862.82	70.88	23.33	23.33
1	18	18	18	-9.00	-462.10	1.13	30.00	53.33
2	15	30	60	7.50	-111.22	0.94	25.00	78.33
3	11	33	99	16.50	-9.43	55.69	18.33	96.67
4	1	4	16	2.50	0.00	39.06	1.67	98.33
5	1	5	25	3.50	1.16	150.06	1.67	100.00
	60	90	218	0.00	-1444.42	317.75		

C-3: The characteristics of central tendency are:

C-3-a: The mode of camp C is 1 Kg.

C-3-b: The average of camp B is:

$$\bar{x} = \frac{1}{n}\sum_{i=1}^{i=6} n_i x_i = \frac{90}{60}$$

$$= 1.5 \text{ Kg}$$

C-3.c: The median interval of camp B is [2 ; 3] and the median is:

$$M_e = 0 + (1-0)\frac{(50-23.33)}{(53.33-23.33)}$$
$$= 0.889\text{Kg}$$

4: Interpretation of the result

INTERPRETATION OF THE RESULTS

	Camp A	Camp B	Camp C
	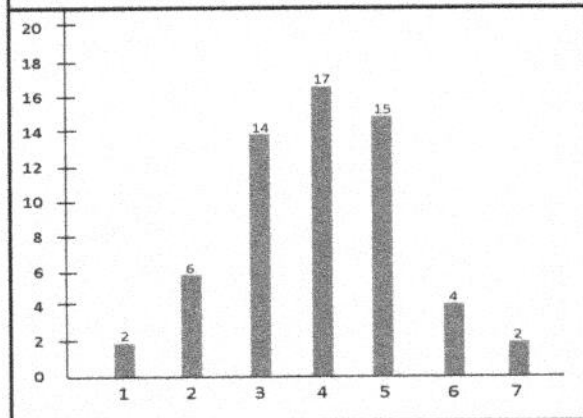	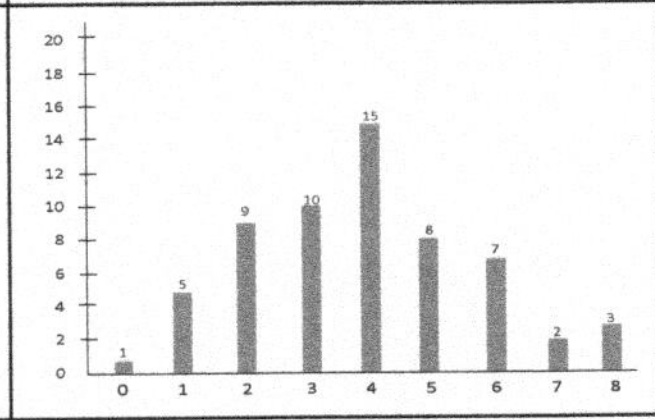	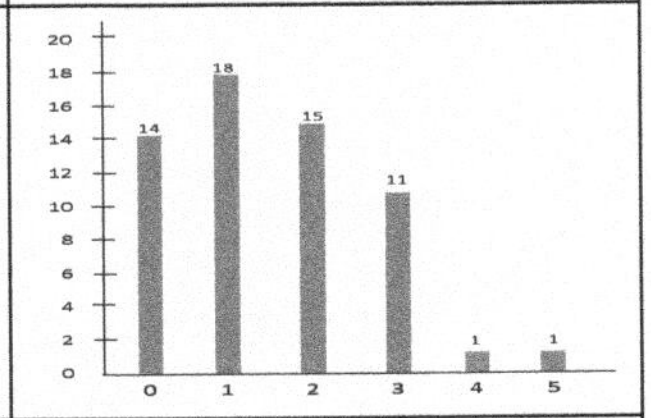
Mode (Kg)	4	4	1
% of total quantity	1.687763713%	1.7167381974%	1.111%
Average (Kg)	3.95	3.88	1.5
% of total quantity	1.666666%	1.665236	1.666666
Median (Kg)	3.47	3.33	0.889
% of total quantity	1.687763713%	1.7167381974%	0.987777%

Analysis

Camp A

- Firstly, the majority of families (more than half) received 4 kg of rice. This distribution shows some homogeneity, but it could also reflect complacency or an imprecise measurement by the NGO responsible for the distribution.
- Secondly, the average amount of rice received by each family in this camp was 3.95 kg, which represents 1.66% of the total quantity. If the distribution had been perfectly fair, each family would have received this amount of rice. Despite the variation in the quantities distributed, the difference between the amount received and the fair amount is very small.
- Thirdly, there are as many families who received a quantity lower than 3.47 kg of rice as those who received more. In other words, if families were ranked according to the amount of rice received, the middle family would have received 3.47 kg of rice.

Camp B:

- Firstly, the majority of families (more than half) received 4 kg of rice. Like Camp A, this distribution shows some homogeneity, but it could also reflect complacency or an imprecise measurement by the NGO responsible for the distribution.
- Secondly, the average amount of rice received by each family in this camp was 3.88 kg. If the distribution had been perfectly fair, each family would have received this amount of rice. Despite the variation in the quantities distributed, the difference between the amount received and the fair amount is relatively small.
- Thirdly, the distribution of the quantities of rice received shows that there are as many families who received a quantity lower than 3.33 kg as those who received more. In other words, the middle family would have received 3.33 kg of rice.

Camp C:

- Firstly, the majority of families received 1 kg of rice, which is a much lower quantity than in the other two camps.

- Secondly, the average amount of rice received was 1.5 kg, indicating that some families received a higher quantity of rice, but most did not. Moreover, if the distribution had been fair, each family should have received 1.5 kg of rice, but this was not the case in reality.
- Thirdly, the fact that the middle family would have received only 0.889 kg of rice. Additionally, there are as many families who received a quantity lower than 0.889 kg of rice as those who received more. In other words, if families were ranked according to the amount of rice received, the middle family would have received 0.889 kg of rice. This clearly shows that the quantity of rice distributed was very low in this camp compared to the other two.

Conclusion:

The variation in the quantities of rice distributed in the three camps can be attributed to the measuring and evaluating capacity of the three NGOs. The results of the rice distribution in Camp A show some homogeneity in the amount of rice received by families. Comparing these central tendency characteristics, rice seems to be better distributed in Camp A due to the proximity of its characteristics, followed by Camp B.

While Camp A seems to have a better distribution than the others, it is important not to jump to conclusions at this stage. It is essential to consider how the values are dispersed in these distributions. For example, it is possible that Camp A has a more dense distribution in the mean range, but that the values are more dispersed around the mean than in the other camps. This is why it is important to take into account the degree of dispersion of values in each distribution. A large deviation between values can indicate an unequal distribution of resources, even if the measures of central tendency indicate otherwise. So, this is the subject of the next step.

11.2.3.2 Analysis of measures of dispersion for decision-making

A: Camp A

A-5: Determining the characteristics of dispersion:

A-5-a: Let's calculate the Interquartile Range:

$$Q_1 = a_i + (a_{i+1} - a_i)\frac{(25 - F_i)}{(F_{i+1} - F_i)}$$
$$= 2 + (3 - 2)\frac{(25 - 13,33)}{(36,67 - 13,33)}$$
$$= 2,5 \text{ Kg}$$
$$Q_3 = a_i + (a_{i+1} - a_i)\frac{(75 - F_i)}{(F_{i+1} - F_i)}$$
$$= 4 + (5 - 4)\frac{(75 - 65)}{(90 - 65)}$$
$$= 4,4 \text{ Kg}$$

$$\text{Interquartile Range} =_3 - Q_1$$
$$= 4,4 - 2,5$$
$$= 1,9 \text{ Kg}$$

A-5-b: Let's calculate the Variance:

$$V(x) = \frac{1}{n}\sum_{i=1}^{i=7} n_i x_i^2 - \bar{x}^2$$
$$= \frac{1.041}{60} - (3.95)^2$$
$$= 1.75$$

A-5-c: Let's calculate the standard deviation:

$$\sigma = \sqrt{V(x)}$$
$$= \sqrt{1.75}$$
$$= 1.32$$

A-5-d: Let's calculate the coefficient of variation:

$$\text{CV} = \frac{\sigma}{|\bar{x}|}\text{x}100$$
$$= \frac{1.32}{|3.95|}\text{x}100$$
$$= 33.42\%$$

A-6-a: Let's calculate the percentage of families that received a lower-than-majority quantity of rice:

$$Mode = 4 \text{ Kg}$$

From the immediate reading of the table below, this percentage is 65%.

x_i	n_i	$n_i x_i$	$n_i x_i^2$	$n_i(x_i - \bar{x})$	$n_i(x_i - \bar{x})^3$	$n_i(x_i - \bar{x})^4$	f_i	$f_i\uparrow$
1	2	2	2	-5.90	-51.34	151.47	3.33	3.33
2	6	12	24	-11.70	-44.49	86.75	10.00	13.33
3	14	42	126	-13.30	-12.00	11.40	23.33	36.67
4	17	68	272	0.85	0.00	0.00	28.33	65.00
5	15	75	375	15.75	17.36	18.23	25.00	90.00
6	4	24	144	8.20	34.46	70.64	6.67	96.67
7	2	14	98	6.10	56.75	173.07	3.33	100.00
	60	237	1041	0.00	0.73	511.57		

A-6-b: Let's calculate the percentage of families that received a lower-than-average quantity of rice:

$$\bar{x} = 3.95 \text{ Kg}$$

Let $P_{<\bar{x}}$ be this percentage, we use the linear interpolation formula using the data pair $(a_i, f_i \uparrow)$:

$$\bar{x} = a_i + (a_{i+1} - a_i)\frac{(P_{<\bar{x}} - F_i)}{(F_{i+1} - F_i)}$$

$$3.95 = 3 + (4 - 3)\frac{(P_{<\bar{x}} - 36.67)}{(65 - 36.67)}$$

$$P_{<\bar{x}} = 63.6\%$$

A-7: Calculation of the percentage of families that received less than $\bar{x} - \sigma$ quantity of rice:

$$\bar{x} - \sigma = 3.95 - 1.32$$
$$= 2.63 \text{ Kg}$$

Let $P_{<\bar{x}-\sigma}$ be this percentage, we use the linear interpolation formula using the data pair $(a_i, f_i \uparrow)$:

$$\bar{x} - \sigma = a_i + (a_{i+1} - a_i)\frac{(P_{<\bar{x}-\sigma} - F_i)}{(F_{i+1} - F_i)}$$

$$2.63 = 2 + (3 - 2)\frac{(P_{<\bar{x}-\sigma} - 13.33)}{(36.67 - 13.33)}$$

$$P_{<\bar{x}-\sigma} = 28.03\ \%$$

A-8: Calculation of the percentage of families receiving less than $\bar{x} + \sigma$ quantity of rice:

$$\bar{x} + \sigma = 3.95+1.32$$
$$= 5.27 \text{ Kg}$$

Let $P_{<\bar{x}+\sigma}$ be this percentage, we use the linear interpolation formula using the data pair $(a_i, f_i \uparrow)$:

$$\bar{x} + \sigma = a_i + (a_{i+1} - a_i)\frac{(P_{<\bar{x}+\sigma} - F_i)}{(F_{i+1} - F_i)}$$

$$5.27 = 5 + (6 - 5)\frac{(P_{<\bar{x}+\sigma} - 90)}{(96.67 - 90)}$$

$$P_{<\bar{x}+\sigma} = 91.8\%$$

A-9: Calculation of the percentage of families that received a quantity of rice within the range of the mean, i.e., 3.95 Kg and the mean minus the standard deviation, i.e., 2.63 Kg.

This percentage is:

$$63.6\% - 28.034\% = 35.57\%$$

A-10: Calculation of the percentage of families that received a quantity of rice within the range of the mean, i.e., 3.95 Kg and the mean plus the standard deviation, i.e., 5.27 Kg.

This percentage is:

$$91.8\% \ - \ 63.6\% = 28.2\%$$

A-11: Calculation of the percentage of families that received a quantity of rice in the interval of the mean minus the standard deviation, i.e., 2.63 Kg and the mean plus the standard deviation, i.e., 5.27 Kg.

This percentage is:

$$91.8\% - 28.04\% = 63.77\%$$

A-12: No family received anything.

B: Camp B

B-5: Determining the characteristics of dispersion:

B-5-a: Let's calculate the Interquartile Range:

$$Q_1 = 2 \text{ Kg}$$

$$\begin{aligned} Q_3 &= a_i + (a_{i+1} - a_i)\frac{(75 - F_i)}{(F_{i+1} - F_i)} \\ &= 4 + (5 - 4)\frac{(75 - 66{,}67)}{(80 - 66{,}67)} \\ &= 4{,}62 \text{ Kg} \end{aligned}$$

$$\begin{aligned} \text{Interquartile Range} &= Q_3 - Q_1 \\ &= 4{,}62 - 2 \\ &= 2{,}62 \text{ Kg} \end{aligned}$$

B-5-b: Let's calculate the Variance:

$$\begin{aligned} V(x) &= \frac{1}{n}\sum_{i=1}^{i=8} n_i x_i^2 - \bar{x}^2 \\ &= \frac{1113}{60} - (3.88)^2 \\ &= 3.50 \end{aligned}$$

B-5-c: Let's calculate the standard deviation:

$$\begin{aligned} \sigma &= \sqrt{V(x)} \\ &= \sqrt{3.47} \\ &= 1.87 \text{ Kg} \end{aligned}$$

B-5-d: Let's calculate the coefficient of variation:

$$CV = \frac{\sigma}{|\bar{x}|} x100$$

$$= \frac{1.87}{|3.8833|} x100$$

$$= 48\%$$

B-6-a: Let's calculate the percentage of families that received a lower-than-majority quantity of rice:

$$Mode = 4 \text{ Kg}$$

From the immediate reading of the table below, this percentage is 66.67%.

x_i	n_i	$n_i x_i$	$n_i x_i^2$	$n_i(x_i - \bar{x})$	$n_i(x_i - \bar{x})^3$	$n_i(x_i - \bar{x})^4$	f_i	$f_i\uparrow$
0	1	0	0	-3.88	-61.63	227.41	1.67	1.67
1	5	5	5	-14.42	-128.36	345.58	8.33	10.00
2	9	18	36	-16.95	-66.73	113.23	15.00	25.00
3	10	30	90	-8.83	-8.57	6.09	16.67	41.67
4	15	60	240	1.75	0.00	0.00	25.00	66.67
5	8	40	200	8.93	9.26	12.44	13.33	80.00
6	7	42	252	14.82	60.31	140.51	11.67	91.67
7	2	14	98	6.23	56.75	188.71	3.33	95.00
8	3	24	192	12.35	199.29	861.60	5.00	100.00
	60	233	1113	0.00	60.30	1895.57		

B-6-b: Let's calculate of the percentage of families that received below-average rice:

$$\bar{x} = 3.88 \text{ Kg}$$

Let $P_{<\bar{x}}$ be this percentage, we use the linear interpolation formula using the data pair $(a_i, f_i \uparrow)$:

$$\bar{x} = a_i + (a_{i+1} - a_i)\frac{(P_{<\bar{x}} - F_i)}{(F_{i+1} - F_i)}$$

$$3.88 = 3 + (4 - 3)\frac{(P_{<\bar{x}} - 41.67)}{(66.67 - 41.67)}$$

$$P_{<\bar{x}} = 63.67\%$$

B-7: Calculation of the percentage of families that received less than $\bar{x} - \sigma$ quantity of rice:

$$\bar{x} - \sigma = 3.88 - 1.87$$

$$= 2.01 \text{ Kg}$$

Let $P_{<\bar{x}-\sigma}$ be this percentage, we use the linear interpolation formula using the data pair $(a_i, f_i \uparrow)$:

$$\bar{x} - \sigma = a_i + (a_{i+1} - a_i)\frac{(P_{<\bar{x}-\sigma} - F_i)}{(F_{i+1} - F_i)}$$
$$2.01 = 2 + (3 - 2)\frac{(P_{<\bar{x}-\sigma} - 25)}{(41.67 - 25)}$$
$$P_{<\bar{x}-\sigma} = 25.17\ \%$$

B-8: Calculation of the percentage of families receiving less than $\bar{x} + \sigma$ quantity of rice:

$$\bar{x} + \sigma = 3.88 + 1.87$$
$$= 5.75\ \text{Kg}$$

Let $P_{<\bar{x}+\sigma}$ be this percentage, we use the linear interpolation formula using the data pair $(a_i, f_i \uparrow)$:

$$\bar{x} + \sigma = a_i + (a_{i+1} - a_i)\frac{(P_{<\bar{x}+\sigma} - F_i)}{(F_{i+1} - F_i)}$$
$$5.75 = 5 + (6 - 5)\frac{(P_{<\bar{x}+\sigma} - 80)}{(91.67 - 80)}$$
$$P_{<\bar{x}+\sigma} = 88.75\%$$

B-9: Calculation of the percentage of families that received a quantity of rice within the range of the mean, i.e., 3.88 Kg and the mean minus the standard deviation, i.e., 2.01 Kg.

This percentage is:

$$63.67\% - 25.17\% = 38.5\%$$

B-10: Calculation of the percentage of families that received a quantity of rice within the range of the mean, i.e., 3.88 Kg and the mean plus the standard deviation, i.e., 5.75 Kg.

This percentage is:

$$88.75\% - 63.67\% = 25.08\%$$

B-11: Calculation of the percentage of families that received a quantity of rice in the range of the mean minus the standard deviation, i.e., 2.01 Kg and the mean plus the standard deviation, i.e., 5.75 Kg.

This percentage is:

$$88.75\% - 25.17\% = 63.58\%$$

B-12: Only one out of 60 families received nothing, or 1.67% of families received nothing.

C: Camp C

C-5: Determining the characteristics of dispersion:

C-5-a: Let's calculate the Interquartile Range:

$$Q_1 = a_i + (a_{i+1} - a_i)\frac{(25 - F_i)}{(F_{i+1} - F_i)}$$
$$= 0 + (1 - 0)\frac{(25 - 23{,}33)}{(53{,}33 - 23{,}33)}$$
$$= 0{,}055 \text{ Kg}$$
$$Q_3 = a_i + (a_{i+1} - a_i)\frac{(75 - F_i)}{(F_{i+1} - F_i)}$$
$$= 1 + (2 - 1)\frac{(75 - 53{,}33)}{(78{,}33 - 53{,}33)}$$
$$= 1{,}866 \text{ Kg}$$
$$\text{Interquartile Range} = Q_3 - Q_1$$
$$= 1{,}866 - 0{,}055$$
$$= 1{,}811 \text{ Kg}$$

C-5-b: Let's calculate the Variance:

$$V(x) = \frac{1}{n}\sum_{i=1}^{i=6} n_i x_i^2 - \bar{x}^2$$
$$= \frac{218}{60} - (1.5)^2$$
$$= 1.38$$

C-5-c: Let's calculate the standard deviation:

$$\sigma = \sqrt{V(x)}$$
$$= \sqrt{1.38}$$
$$= 1.176 \text{ Kg}$$

C-5-d: Let's calculate the coefficient of variation:

$$\text{CV} = \frac{\sigma}{|\bar{x}|}\text{x}100$$
$$= \frac{1.176}{|1.5|}\text{x}100$$
$$= 78.4\%$$

C-6-a: Let's calculate the percentage of families that received a lower-than-majority quantity of rice:

$$Mode = 1 \text{ Kg}$$

From the immediate reading of the table below, this percentage is 53.33%.

x_i	n_i	$n_i x_i$	$n_i x_i^2$	$n_i(x_i - \bar{x})$	$n_i(x_i - \bar{x})^3$	$n_i(x_i - \bar{x})^4$	f_i	$f_i\uparrow$
0	14	0	0	-21.00	-862.82	70.88	23.33	23.33
1	18	18	18	-9.00	-462.10	1.13	30.00	53.33
2	15	30	60	7.50	-111.22	0.94	25.00	78.33
3	11	33	99	16.50	-9.43	55.69	18.33	96.67
4	1	4	16	2.50	0.00	39.06	1.67	98.33
5	1	5	25	3.50	1.16	150.06	1.67	100.00
	60	90	218	0.00	-1444.42	317.75		

C-6-b: Let's calculate the percentage of families that received a lower-than-average quantity of rice

$$\bar{x} = 1.5 \text{ Kg}$$

Let $P_{<\bar{x}}$ be this percentage, we use the linear interpolation formula using the data pair $(a_i, f_i \uparrow)$:

$$\bar{x} = a_i + (a_{i+1} - a_i)\frac{(P_{<\bar{x}} - F_i)}{(F_{i+1} - F_i)}$$

$$1.5 = 1 + (2 - 1)\frac{(P_{<\bar{x}} - 53.33)}{(78.33 - 53.33)}$$

$$P_{<\bar{x}} = 65.83\%$$

C-7: Calculation of the percentage of families that received less than $\bar{x} - \sigma$ quantity of rice:

$$\bar{x} - \sigma = 1.5 - 1.176$$
$$= 0.324 \text{ Kg}$$

Let $P_{<\bar{x}-\sigma}$ be this percentage, we use the linear interpolation formula using the data pair $(a_i, f_i \uparrow)$:

$$\bar{x} - \sigma = a_i + (a_{i+1} - a_i)\frac{(P_{<\bar{x}-\sigma} - F_i)}{(F_{i+1} - F_i)}$$

$$0.324 = 0 + (1 - 0)\frac{(P_{<\bar{x}-\sigma} - 23.33)}{(53.33 - 23.33)}$$

$$P_{<\bar{x}-\sigma} = 33.05\ \%$$

C-8: Calculation of the percentage of families receiving less than $\bar{x} + \sigma$ quantity of rice:

$$\bar{x} + \sigma = 1.5 + 1.176 = 2.676 \text{ Kg}$$

Let $P_{<\bar{x}+\sigma}$ be this percentage, we use the linear interpolation formula using the data pair $(a_i , f_i \uparrow)$:

$$\bar{x} + \sigma = a_i + (a_{i+1} - a_i)\frac{(P_{<\bar{x}+\sigma} - F_i)}{(F_{i+1} - F_i)}$$

$$2.676 = 2 + (3 - 2)\frac{(P_{<\bar{x}+\sigma} - 78.33)}{(96.67 - 78.33)}$$

$$P_{<\bar{x}+\sigma} = 90.73\%$$

C-9: Calculation of the percentage of families that received a quantity of rice within the range of the mean, i.e., 1.5 Kg and the mean minus the standard deviation, i.e., 0.324 Kg.

This percentage is:

$$65.83\% - 33.05\% = 32.78\%$$

C-10: Calculation of the percentage of families that received a quantity of rice within the range of the mean, i.e., 1.5 Kg and the mean plus the standard deviation, i.e., 2.676 Kg.

This percentage is:

$$90.73\% \; - \; 65.83\% = 24.90\%$$

C-11: Calculation of the percentage of families that received a quantity of rice in the interval of the mean minus the standard deviation, i.e., 0.324 Kg, and the mean plus the standard deviation, i.e., 2.676 Kg.

This percentage is:

$$90.73\% - 33.05\% = 57.68\%$$

C-12: 14 out of 60 families received nothing, i.e., 23.3%

INTERPRETATION OF RESULTS, ANALYSIS, AND DECISION

	Camp A	Camp B	Camp C
Mode (Kg)	4	4	1
Average (Kg)	3.95	3.88	1.5
Median (Kg)	3.47	3.33	0.889
% Pop. < Majority	65%	66.67%	53.33%
% Pop. < Average	63.6%	63.67%	65.83.2%

NPO **A**

- 4 kg of rice provided to a majority of families representing 65% of the population. However, these families received a quantity of rice that represents 1.6877% of the total amount of rice provided by ASBL A. This percentage is slightly above the average of 1.66666% that could have been allocated to each family if they had all received an equal amount of rice.

- Additionally, the median family, the one in the middle of the distribution of families, received only 1.4641% of the total amount of rice provided, which is 87.8% of the average. This means that half of the families received an amount of rice lower than 1.4641% and the other half received a higher amount. In other words, half of the families received only 87.8% of what they should have received, and the other half received 112.2% of what they should have received.
- It is also important to note that 63.6% of the population received a quantity of rice lower than the average, once again highlighting that the distribution was not fair for everyone.

NPO **B**

- 4 kg of rice provided to a majority of families representing 66.67% of the population. However, these families received a quantity of rice that represents 1.66523% of the total amount of rice provided by ASBL A. This percentage is slightly above the average of 1.66666% that could have been allocated to each family if they had all received an equal amount of rice.
- Additionally, the median family, the one in the middle of the distribution of families, received only 1.4291% of the total amount of rice provided, which is 85.8% of the average. This means that half of the families received an amount of rice lower than 1.4291% and the other half received a higher amount. In other words, half of the families received only 85.8% of what they should have received, and the other half received 114.2% of what they should have received.
- It is also important to note that 63.67% of the population received a quantity of rice lower than the average, once again highlighting that the distribution was not fair for everyone.

NPO **C**

- The majority of families representing 53.3% of the population received only 1 kg of rice each, which is 1.1111% of the total amount of rice provided by the ASBL. However, this percentage is significantly lower than the average of 1.66666% that could have been allocated to each family if they had all received an equal amount of rice.
- Additionally, the median family, the one in the middle of the distribution of families, received only 0.987% of the total amount of rice provided, which is 59.2% of the average. This means that half of the families received an amount of rice lower than 0.987% and the other half received a higher amount. In other words, half of the families received only 59.2% of what they should have received, and the other half received 140.8% of what they should have received.
- Moreover, the median family received 0.9877% of the total rice, which represents 59.2% of the average. This means that the rice distribution is not fair and that some families received more than others.
  - It is also important to note that 65.83% of the population received a quantity of rice lower than the average, once again highlighting that the distribution was not fair for everyone.

	A	B	C
CV	33.6%	48%	78.4%

The distribution of rice in Camp A has a lower coefficient of variation than the other two camps. This indicates that the distribution is moderately dispersed, meaning that the amount of rice provided to each family is relatively close to the average.

% in $[\bar{x}, \bar{x} - \sigma]$	38.5%	38.5%	32.78%
% in $[\bar{x}, \bar{x} + \sigma]$	28.20%	25.08%	24.90%
% in $[\bar{x} - \sigma, \bar{x} + \sigma]$	63.58%	63.58%	57.68%

Camp A stands out from the other two camps in terms of population density. Indeed, it exhibits a significant population within the range of one standard deviation above and one standard deviation below the mean, indicating a high concentration of individuals around the mean. Camp B closely follows Camp A in terms of population but has a greater dispersion around the mean. As for Camp C, its population density is noticeably lower. Furthermore, the distribution of Camp A more closely resembles a normal distribution than those of the other two camps, suggesting that the data is more regular and predictable, resulting in a more homogeneous and equitable distribution among different groups.

Conclusion

Based on these results, we can conclude that the distribution of rice in camp A is more homogeneous and regular than in the other two camps. The proximity of the mode, mean, and median values indicates a balanced distribution of rice among the families. Moreover, the low coefficient of variation testifies to a less significant dispersion around the mean.

Furthermore, the obtained data suggest that the population of this camp is more concentrated in the interval of the mean plus or minus the standard deviation, which means that most families received a similar amount of rice. In comparison, the distributions of the other two camps show a greater dispersion, with values further away from the mean.

Therefore, these results indicate that the distribution of rice in Camp A is better than the other two camps. However, although the distribution appears to be the best at this level, we will further examine the shape of the distribution curve, which will clarify the shape of the curve and provide insight into the concentration of values distributed around the central values.

11.2.3.3 Analysis of shape measures for decision-making

To analyze the shape characteristics of the distributions, we will only calculate the two FISHER's coefficient as they are more appropriate in this context. Indeed, rounding errors in the calculation of other coefficients may distort the data on the shape of the curve.

A: Camp **A**

A-14: Let's calculate the coefficients of skewness:

A-14 -a: Calculation of the YULE's coefficient of skewness:

$$C_Y = \frac{Q_1 - 2Q_2 + Q_3}{Q_3 - Q_1}$$

$$= \frac{2.5 - 2(3.95) + 4.4}{4.4 - 2.5}$$

$$= -0.52$$

A-14 -b: Calculation of the PEARSON's coefficient of skewness:

$$\beta_1 = \frac{(\bar{x} - \text{Mode})}{\sigma}$$

$$= \frac{(3.95 - 4)}{1.32}$$

$$= -0.0378$$

A-14 -c: Calculation of the FISHER's coefficient of skewness:

$$\mu_3 = \frac{\sum_{i=1}^{i=7} n_i (x_i - \bar{x})^3}{\sum_{i=1}^{i=7} n_i} = \frac{0.73}{60}$$

$$= 0.01$$

$$\gamma_1 = \frac{\mu_3}{\sigma^3} = \frac{0.01}{(1.32)^3}$$

$$= 0.00434$$

A-15: Let's calculate the FISHER's coefficient of kurtosis:

$$\mu_4 = \frac{\sum_{i=1}^{i=7} n_i (x_i - \bar{x})^4}{\sum_{i=1}^{i=7} n_i} = \frac{511.57}{60}$$

$$= 8.526$$

$$\gamma_2 = \frac{\mu_4}{\sigma^4} - 3 = \frac{8.526}{(1.32)^4} - 3$$

$$= -0.1916$$

B: Camp **B**

B-14: Let's calculate the coefficients of skewness:

B-14 -a: Calculation of the YULE's coefficient of skewness:

$$C_Y = \frac{Q_1 - 2Q_2 + Q_3}{Q_3 - Q_1}$$

$$= \frac{2 - 2(3.88) + 4.62}{4.62 - 2}$$

$$= -0.435$$

B-14 -b: Calculation of the PEARSON's coefficient of skewness:

$$\beta_1 = \frac{(\bar{x} - \text{Mode})}{\sigma}$$

$$= \frac{(3.98 - 4)}{1.86}$$

$$= -0.010$$

B-14 -c: Calculation of the:

$$\mu_3 = \frac{\sum_{i=1}^{i=8} n_i (x_i - \bar{x})^3}{\sum_{i=1}^{i=8} n_i} = \frac{101.96}{60}$$

$$= 1.699$$

$$\gamma_1 = \frac{\mu_3}{\sigma^3} = \frac{1.699}{(1.86)^3}$$

$$= 0.251$$

B-15: Let's calculate the FISHER's coefficient of kurtosis:

$$\mu_4 = \frac{\sum_{i=1}^{i=6} n_i (x_i - \bar{x})^4}{\sum_{i=1}^{i=6} n_i} = \frac{1{,}895.57}{60}$$

$$= 31.59$$

$$\gamma_2 = \frac{\mu_4}{\sigma^4} - 3 = \frac{31.59}{(1.86)^4} - 3$$

$$= -0.36$$

C: Camp **C**

C-14: Let's calculate the coefficients of skewness:

C-14 -a: Calculation of the YULE's coefficient of skewness:

$$C_Y = \frac{Q_1 - 2Q_2 + Q_3}{Q_3 - Q_1}$$

$$= \frac{0.055 - 2(1.5) + 1.866}{1.866 - 0.055}$$

$$= -0.5958$$

C-14 -b: Calculation of the PEARSON's coefficient of skewness

$$\beta_1 = \frac{(\bar{x} - \text{Mode})}{\sigma}$$

$$= \frac{(1.5 - 1)}{1.176}$$

$$= 0.4251$$

C-14 -c: Calculation of the FISHER's coefficient of skewness:

$$\mu_3 = \frac{\sum_{i=1}^{i=6} n_i (x_i - \bar{x})^3}{\sum_{i=1}^{i=6} n_i} = \frac{48}{60}$$

$$= 0.8$$

$$\gamma_1 = \frac{\mu_3}{\sigma^3} = \frac{0.8}{(1.176)^3}$$

$$= 0.4917$$

C-15: Let's calculate the FISHER's coefficient of kurtosis:

$$\mu_4 = \frac{\sum_{i=1}^{i=6} n_i (x_i - \bar{x})^4}{\sum_{i=1}^{i=6} n_i} = \frac{317.75}{60}$$

$$= 5.295$$

$$\gamma_2 = \frac{\mu_4}{\sigma^4} - 3 = \frac{5.295}{(1.176)^4} - 3$$

$$= -0.231$$

INTERPRETATION OF RESULTS, ANALYSIS, AND DECISION

	Camp A	Camp B	Camp C
	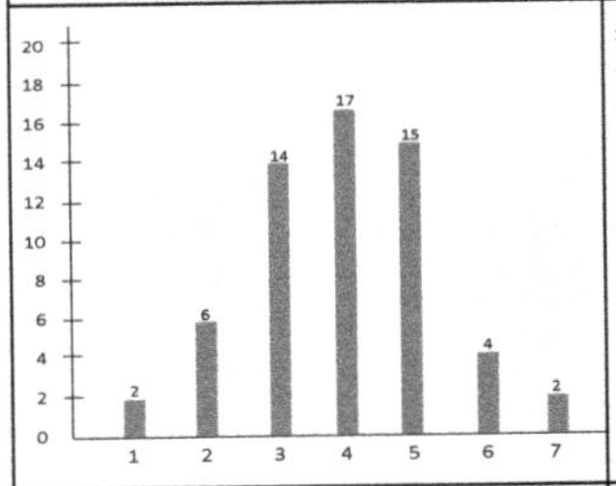	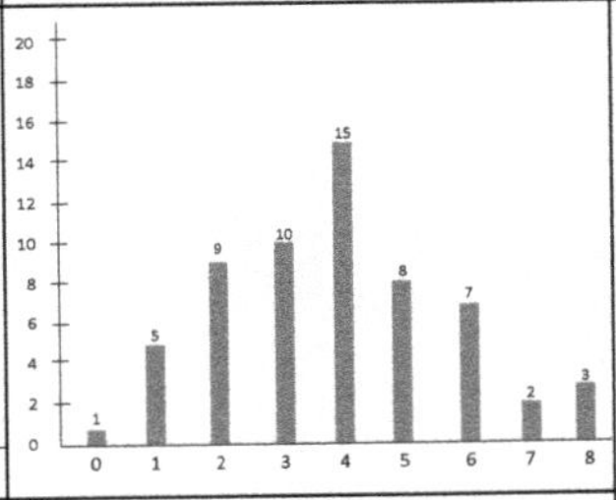	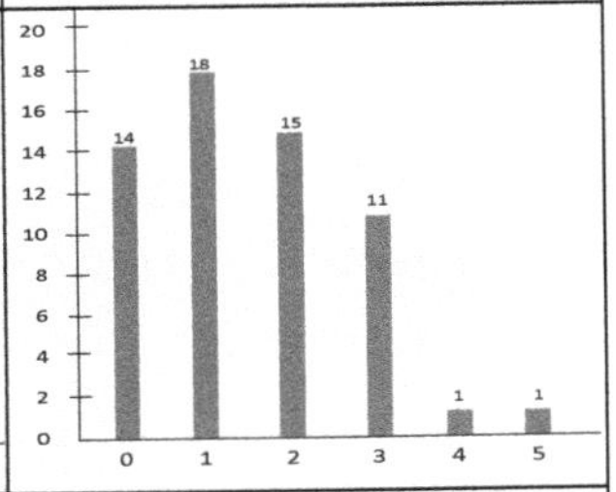
% dans $[\bar{x} - \sigma, \bar{x}]$	35.57%	38.50%	32.78%
% dans $[\bar{x}, \bar{x} + \sigma]$	28.20%	25.08%	24.90%
% dans $[\bar{x} - \sigma, \bar{x} + \sigma]$	63.77%	63.58%	57.68%
	$C_y < 0$	$C_y < 0$	$C_y < 0$
	$\beta_1 < 0$	$\beta_1 < 0$	$\beta_1 > 0$
	$\gamma_1 > 0$	$\gamma_1 > 0$	$\gamma_1 > 0$
	$\gamma_2 < 0$	$\gamma_2 < 0$	$\gamma_2 < 0$

Analysis

As for the skewness coefficients, their proximity to zero indicates that the data is relatively symmetric around their mean, but the rounding has caused oscillations around zero, leading to slightly positive or negative values.

Indeed, based on three out of four coefficients, distributions A and B are left-skewed with a tail of distribution stretched towards lower values, while distribution C is right-skewed. The distribution curves of A and B indicate a high concentration of high values, which means there is a strong prevalence of families receiving large amounts of rice. In contrast, distribution C is right-skewed, indicating a high concentration of families with a low rice ratio.

Finally, all three distribution curves are flatter than that of the normal distribution. This means that the data is less dispersed than in a normal distribution, which may be due to a high concentration of values at a specific location. These two arguments reinforce the conclusions established in the previous section regarding the choice to be made.

11.2.3.4 Analysis of concentration measures for decision-making

A: Camp A

The following table shows the data required to solve this exercise:

x_i	n_i	$n_i x_i$	f_i	q_i	$f_i\uparrow$	$q_i\uparrow$	$H_i=f_{i+1}\uparrow - f_i\uparrow$	$B_i = \frac{q_i\uparrow + q_{i+1}\uparrow}{2}$	Trapezoid Area (T_i) $T_i=B_i x H_i$
1	2	2.00	3.33	0.84	3.33	0.84	10.00	3.38	33.76
2	6	12.00	10.00	5.06	13.33	5.91	23.33	14.77	344.59
3	14	42.00	23.33	17.72	36.67	23.63	28.33	37.97	1075.95
4	17	68.00	28.33	28.69	65.00	52.32	25.00	68.14	1703.59
5	15	75.00	25.00	31.65	90.00	83.97	6.67	89.03	593.53
6	4	24.00	6.67	10.13	96.67	94.09	3.33	97.05	323.49
7	2	14.00	3.33	5.91	100.00	100.00			
	60	237							4074.89

A-16: The medial a of this distribution is:

$$\overset{\approx}{M_e} = a_i + (a_{i+1} - a_i)\frac{(50 - Q_i)}{(Q_{i+1} - Q_i)}$$

$$= 3 + (4 - 3)\frac{(50 - 23.63)}{(52.32 - 23.63)}$$

$$= 3.92 \text{ Kg}$$

A-17: Determining the percentage of families with a rice ratio below the medial:

Let $P_{<\text{medial}}$ be this percentage, let's use the linear interpolation formula using the data pair $(f_i \uparrow, q_i \uparrow)$:

$$P_{<\text{medial}} = a_i + (a_{i+1} - a_i)\frac{(50 - Q_i)}{(Q_{i+1} - Q_i)}$$

$$= 36.67 + (65 - 36.67)\frac{(50 - 23.63)}{(52.32 - 23.63)}$$

$$= 62.71\%$$

A-18: Let's calculate the GINI index by trapezoids.

The area under the LORENZ curve is:

$$\sum_{i=1}^{i=7}(f_{i+1}\uparrow - f_i\uparrow)\frac{(q_i\uparrow + q_{i+1}\uparrow)}{2} = 4{,}074.89$$

It follows that:

$$\begin{aligned}\text{GINI Index} &= 100\% - \left(\frac{4{,}074.89}{5{,}000}\right)\text{x}100 \\ &= 100\% - 81.5\% \\ &= 18.5\%\end{aligned}$$

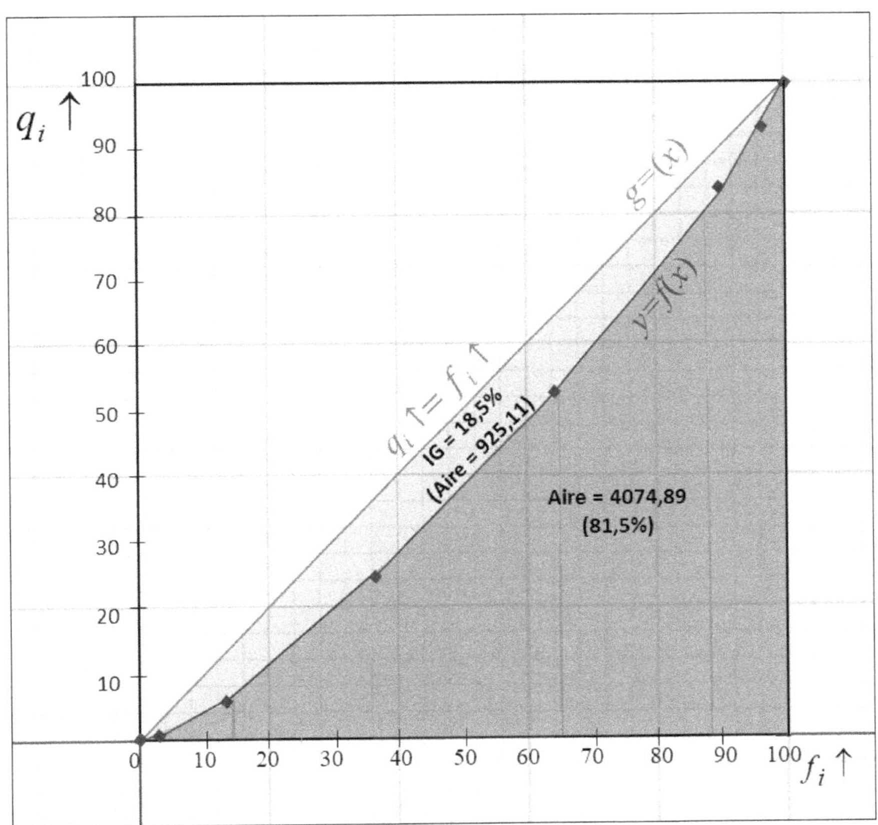

A-19: Let's calculate the GINI index by the use of the defined integrals.

We will consider the following $(f_i\uparrow, q_i\uparrow)$key points:

(0,0) – (13, 6) – (37, 24) – (65,52) – (90,84) – (100,100)

Following the computation, the interpolation polynomial is obtained as:

$$y(x)=\frac{69933430066500}{1398668601330000}x^5+\frac{32808767}{69933430066500}x^4-\frac{146913606641}{1398668601330000}x^3+\frac{1686736536899}{139866860133000}x^2+\frac{1280993991}{3984810830}x$$

Let's evaluate the area between the curve $q_i\uparrow$ and the $f_i\uparrow$ axes in the interval [0 ; 100]:

$$\text{Area}=\int_0^{100} y\,dx=\frac{69933430066500}{1398668601330000}\left(\frac{10^{12}}{6}\right)+\frac{32808767}{69933430066500}\left(\frac{10^{10}}{5}\right)-\frac{146913606641}{1398668601330000}\left(\frac{10^{8}}{4}\right)$$
$$+\frac{1686736536899}{139866860133000}\left(\frac{10^{6}}{3}\right)+\frac{1280993991}{3984810830}\left(\frac{10^{4}}{2}\right)$$

$$\text{Area}=4\ 029{,}22$$

It follows that:

$$\text{GINI Index}=100\%-\left(\frac{\int_0^{100} f(x)\,dx}{5{,}000}\right)\text{x}100$$
$$=100\%-\left(\frac{4{,}092.22}{5{,}000}\right)\text{x}100$$
$$=100\%-81.84\%$$
$$=18.16\%$$

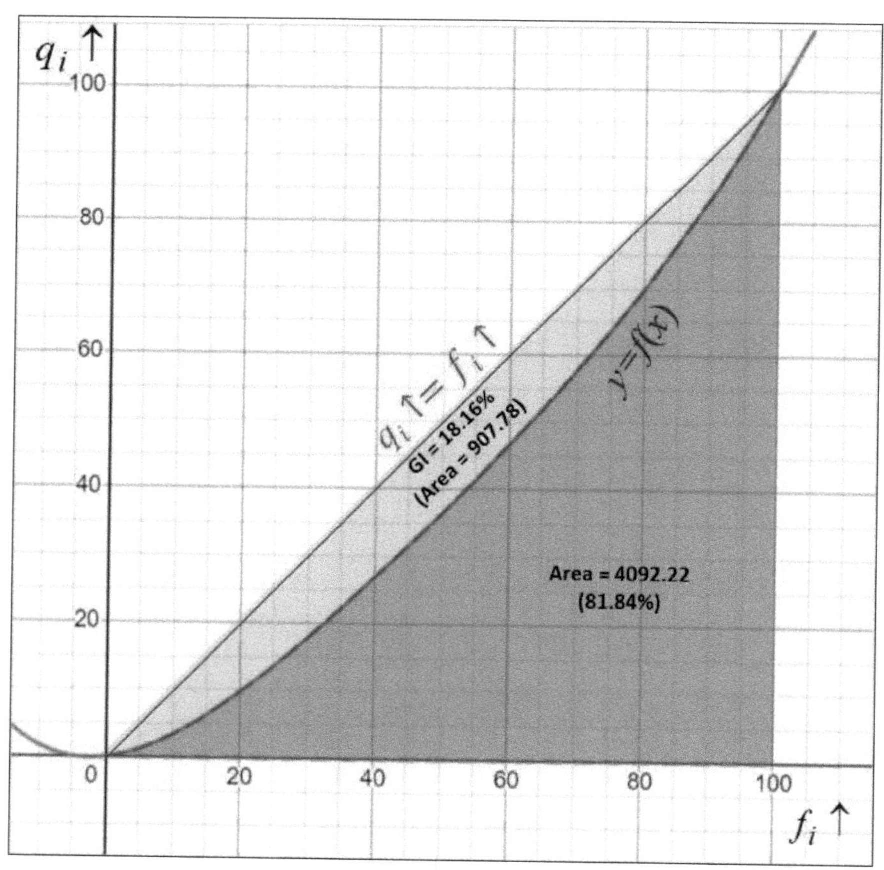

B: Camp B

The following table shows the data required to solve this exercise:

x_i	n_i	$n_i x_i$	f_i	q_i	$f_i\uparrow$	$q_i\uparrow$	$H_i = f_{i+1}\uparrow - f_i\uparrow$	$B_i = \frac{q_i\uparrow + q_{i+1}\uparrow}{2}$	Trapezoid Area (T_i) $T_i = B_i \text{x} H_i$
0	1	0.00	1.67	0.00	1.67	0.00	8.33	1.07	8.94
1	5	5.00	8.33	2.15	10.00	2.15	15.00	6.01	90.13
2	9	18.00	15.00	7.73	25.00	9.87	16.67	16.31	271.82
3	10	30.00	16.67	12.88	41.67	22.75	25.00	35.62	890.56
4	15	60.00	25.00	25.75	66.67	48.50	13.33	57.08	761.09
5	8	40.00	13.33	17.17	80.00	65.67	11.67	74.68	871.24
6	7	42.00	11.67	18.03	91.67	83.69	3.33	86.70	288.98
7	2	14.00	3.33	6.01	95.00	89.70	5.00	94.85	474.25
8	3	24.00	5.00	10.30	100.00	100.00			
	60	233							3657.01

B-16: The medial of this distribution is:

$$\tilde{\tilde{M}}_e = a_i + (a_{i+1} - a_i)\frac{(50 - Q_i)}{(Q_{i+1} - Q_i)}$$

$$= 4 + (5 - 4)\frac{(50 - 48.5)}{(65.67 - 48.5)}$$

$$= 4.09 \text{ Kg}$$

B-17: Determining the percentage of families with a rice ratio below the medial:

Let $P_{<\text{medial}}$ be this percentage, let's use the linear interpolation formula using the data pair $(f_i\uparrow, q_i\uparrow)$:

$$P_{<\text{medial}} = 66.67 + (80 - 66.67)\frac{(50 - 48.50)}{(65.67 - 48.50)}$$

$$= 67.83\%$$

B-18: Let's calculate the GINI index by trapezoids.

The area under the LORENZ curve is:

$$\sum_{i=1}^{i=8}(f_{i+1}\uparrow - f_i\uparrow)\frac{(q_i\uparrow + q_{i+1}\uparrow)}{2} = 3{,}657.01$$

It follows that:

$$\begin{aligned}\text{G{\small INI} Index} &= 100\% - \left(\frac{3{,}657.01}{5{,}000}\right)\text{x}100 \\ &= 100\% - 73.14\% \\ &= 26.86\%\end{aligned}$$

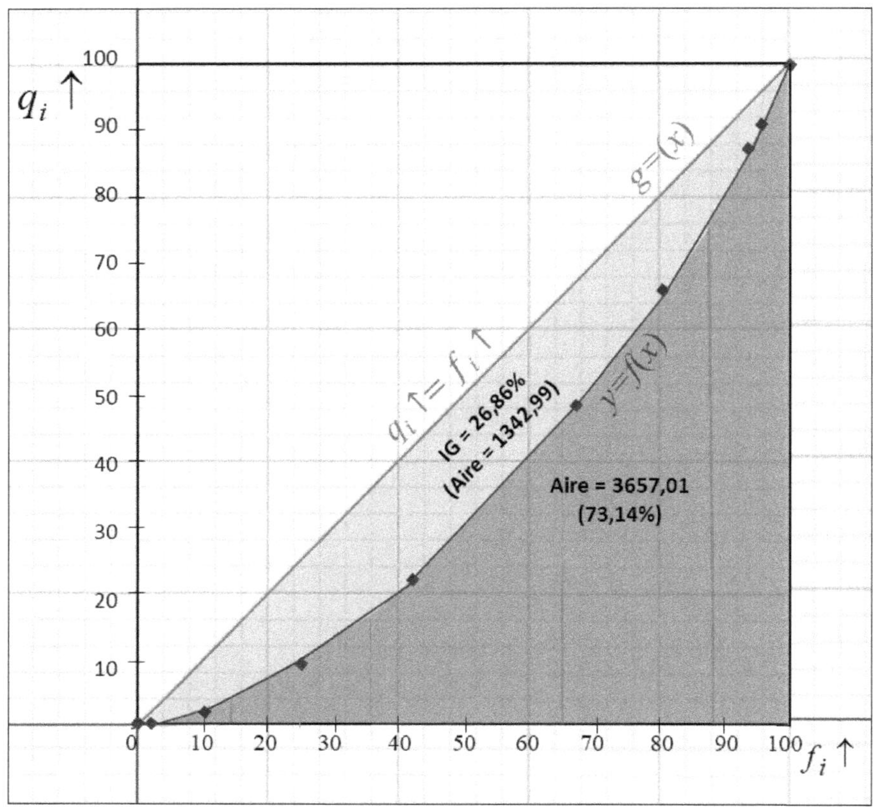

B-19: Calculation of the GINI index by the use of defined integrals

We will consider the following $(f_i \uparrow, q_i \uparrow)$key points:

(0.0) – (10, 2) – (25, 10) – (42.23) – (66.48) – (80.65) – (91.83) – (100,100)

Following the computation, the interpolation polynomial is obtained as:

$$\begin{aligned}y(x) = &-\frac{10624240813}{3222504015546816000000}x^7 + \frac{915285849983}{805626003886704000000}x^6 - \frac{32805771817237}{2148336010364544000 0}x^5 \\ &+ \frac{372611316218227}{366193638130320000 0}x^4 - \frac{8167916230023661}{2301788582533440000}x^3 + \frac{6291436260931673}{89514000431856000}x^2 \\ &- \frac{1369322114287}{5812597430640}x\end{aligned}$$

Let's evaluate the area between the curve $q_i \uparrow$ and the $f_i \uparrow$ axes in the interval [0 ; 100]:

$$\text{Area} = \int_0^{100} y\,dx = -\frac{10624240813}{3222504015546816000000}\left(\frac{10^{16}}{8}\right) + \frac{915285849983}{80562600388670400000}\left(\frac{10^{14}}{7}\right) - \frac{32805771817237}{2148336010364544000}\left(\frac{10^{12}}{6}\right)$$
$$+ \frac{372611316218227}{36619363813032000000}\left(\frac{10^{10}}{5}\right) - \frac{8167916230023661}{2301788582533440000}\left(\frac{10^{8}}{4}\right) + \frac{6291436260931673}{89514000431856000}\left(\frac{10^{6}}{3}\right)$$
$$- \frac{1369322114287}{5812597430640}\left(\frac{10^{4}}{2}\right)$$

Aire = 3,628.7

It follows that:

$$\text{GINI Index} = 100\% - \left(\frac{\int_0^{100} f(x)\,dx}{5{,}000}\right) \text{x}100$$
$$= 100\% - \left(\frac{3{,}628.7}{5{,}000}\right) \text{x}100$$
$$= 100\% - 72.57\%$$
$$= 27.43\%$$

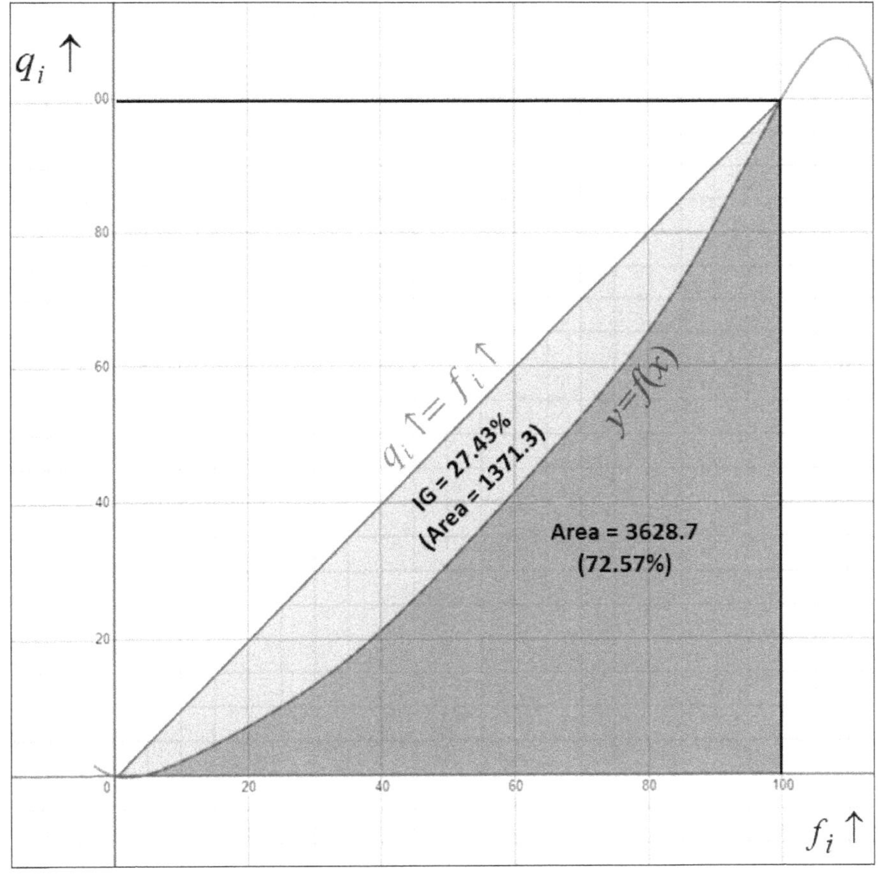

B: Camp C

The following table shows the data required to solve this exercise:

x_i	n_i	$n_i x_i$	f_i	q_i	$f_i\uparrow$	$q_i\uparrow$	$H_i = f_{i+1}\uparrow - f_i\uparrow$	$B_i = \frac{q_i\uparrow + q_{i+1}\uparrow}{2}$	Trapezoid Area (T_i) $T_i = B_i \text{x} H_i$
0	14	0.00	23.33	0.00	23.33	0.00	30.00	10.00	300.00
1	18	18.00	30.00	20.00	53.33	20.00	25.00	36.67	916.67
2	15	30.00	25.00	33.33	78.33	53.33	18.33	71.67	1313.89
3	11	33.00	18.33	36.67	96.67	90.00	1.67	92.22	153.70
4	1	4.00	1.67	4.44	98.33	94.44	1.67	97.22	162.04
5	1	5.00	1.67	5.56	100.00	100.00			
	60	90							2846.30

C-16: The medial of this distribution is:

$$\overset{\approx}{M}_e = a_i + (a_{i+1} - a_i)\frac{(50 - Q_i)}{(Q_{i+1} - Q_i)}$$

$$= 1 + (2 - 1)\frac{(50 - 20)}{(53.33 - 20)}$$

$$= 1.9 \text{ Kg}$$

C-17: Determining the percentage of families with a rice ratio below the medial:

Let $P_{<\text{médial}}$ be this percentage, let's use the linear interpolation formula using the data pair $(f_i \uparrow, q_i \uparrow)$:

$$P_{<\text{medial}} = 53.33 + (78.33 - 53.33)\frac{(50 - 20)}{(53.33 - 20)}$$

$$= 75.83\%$$

C-18: Let's calculate the GINI index by trapezoids.

The area under the LORENZ curve is:

$$\sum_{i=1}^{i=6}(f_{i+1} \uparrow - f_i \uparrow)\frac{(q_i \uparrow + q_{i+1} \uparrow)}{2} = 2{,}846.30$$

It follows that:

$$\text{GINI Index} = 100\% - \left(\frac{2{,}846.30}{5{,}000}\right) \text{x} 100$$
$$= 100\% - 56.93\%$$
$$= 43.07\%$$

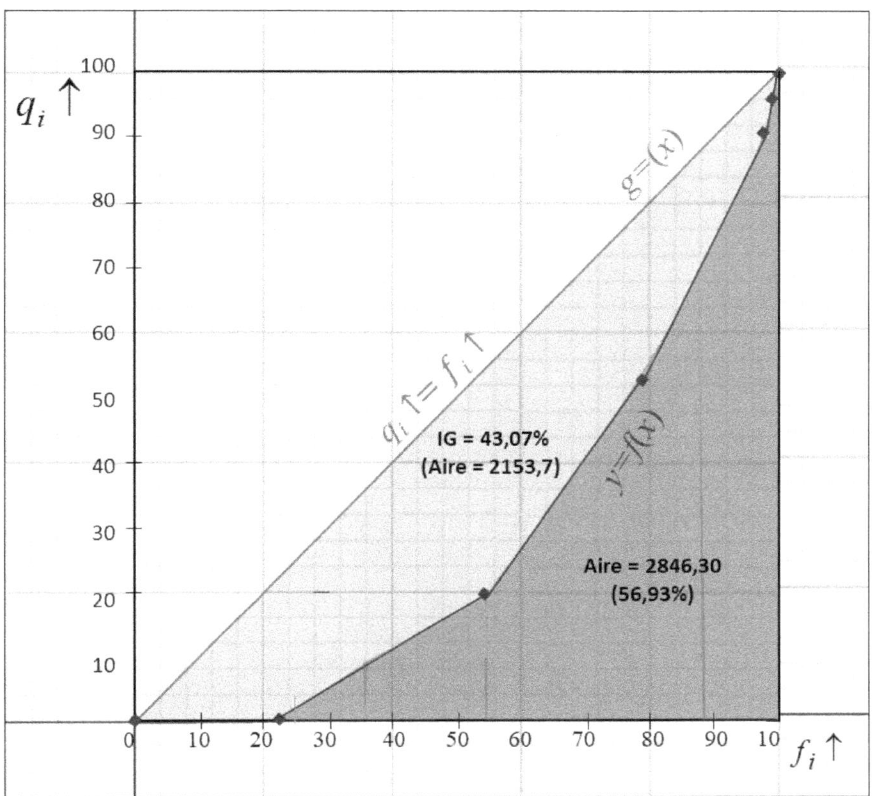

C-19: Let's calculate of the GINI index by the calculation of defined integrals:

We will consider the following $(f_i \uparrow, q_i \uparrow)$ key points:

(0,0) – (23, 0) – (53, 20) – (78,53) – (97,90) – (100,100)

Following the computation, the interpolation polynomial is obtained as:

$$y(x) = -\frac{24232847}{2797337202660000}x^5 + \frac{441394159}{139866860133000}x^4 - \frac{1086175391407}{2797337202660000}x^3 + \frac{6912336570673}{279733720266000}x^2 + \frac{974803607}{7969621660}x$$

Let's evaluate the area between the curve $q_i \uparrow$ and the $f_i \uparrow$ axes in the interval [0 ; 100]:

$$\text{Area} = \int_0^{100} y\,dx = -\frac{24232847}{2797337202660000}\left(\frac{10^{12}}{6}\right) + \frac{441394159}{139866860133000}\left(\frac{10^{10}}{5}\right) - \frac{1086175391407}{2797337202660000}\left(\frac{10^{8}}{4}\right)$$
$$+ \frac{6912336570673}{279733720266000}\left(\frac{10^{6}}{3}\right) + \frac{974803607}{7969621660}\left(\frac{10^{4}}{2}\right)$$

$$\text{Area} = 2{,}716{,}07$$

It follows that:

$$\text{GINI index} = 100\% - \left(\frac{\int_0^{100} f(x)\,dx}{5.000}\right) \text{x } 100\%$$

$$= 100\% - \left(\frac{2{,}716.07}{5{,}000}\right) \text{x } 100\%$$

$$= 100\% - 54.32\%$$

$$= 45.68\%$$

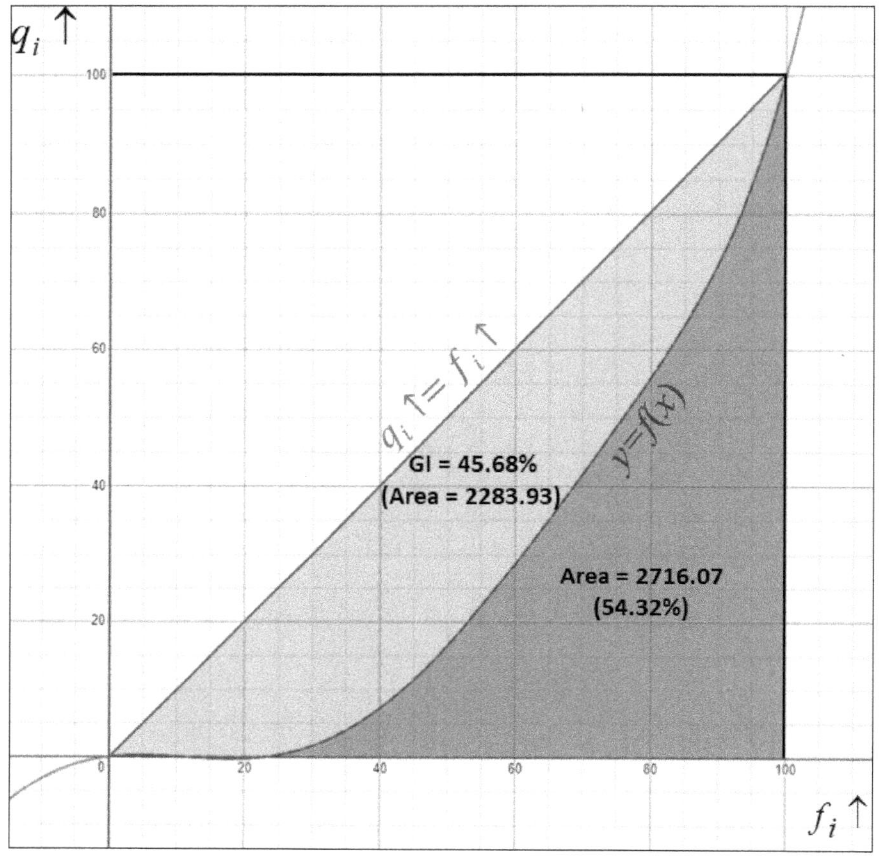

I INTERPRETATION OF RESULTS, ANALYSIS, AND DECISION

	Camp A	Camp B	Camp C
Medial	3.92 Kg	4.09 Kg	1.9 Kg
% Family that received median < ratio	62.71%	67.83%	75.83%
GINI index	18.5	26.86	43.07

Analysis

NPO **A**

- The total amount of rice distributed in portions that are each less than 2.11 kg is equivalent to the total amount distributed in portions that are each greater than 2.11 kg. This means that the families who received less than 2.11 kg of rice received in total the same amount as those who received more.
- The percentage of families who received portions of less than 2.11 kg of rice is 62.70%. This indicates that about 37.30% of families received the same amount of rice as 62.70% of families. This suggests that almost two-thirds of families received relatively small quantities of rice, while the remaining third received higher quantities.
- The GINI index for camp A is 18.5, which suggests a relatively equal distribution at 81.5%.

NPO **B**

- The total amount of rice distributed in portions that are each less than 4.08 kg is equivalent to the total amount distributed in portions that are each greater than 4.08 kg. This means that the families who received less than 4.08 kg of rice received in total the same amount as those who received more.
- The percentage of families who received portions of less than 4.08 kg of rice is 67.83%. This indicates that about 32.17% of families received the same amount of rice as 67.83% of families. This suggests that almost two-thirds of families received relatively small quantities of rice, while the remaining third received higher quantities.
- The GINI index for camp B is 26.86, which suggests a relatively equal distribution at 73.14%.

NPO **C**

- The total amount of rice distributed in portions that are each less than 1.9 kg is equivalent to the total amount distributed in portions that are each greater than 1.9 kg.
- This means that the families who received less than 1.9 kg of rice received in total the same amount as those who received more.
- The percentage of families who received portions of less than 1.9 kg of rice is 75.83%. This indicates that about 24.17% of families received the same amount of rice as 75.83% of families. This suggests that almost three-quarters of families received relatively small quantities of rice, while the remaining quarter received higher quantities.
- The GINI index for camp C is 43.07, which suggests a relatively unequal distribution at only 56.93%.

Conclusion

- The rice distribution proposed by NPO **C** clearly has major flaws. Firstly, it is very poorly done in that one quarter of families received as much rice as the remaining three quarters of families. This imbalanced distribution is unfair and inefficient in meeting established needs. Moreover, the GINI index of this distribution is excessively high, indicating that the rice distribution was highly unequal. The equity rate of the distribution is only 56.93%, which is very low and confirms that this distribution should be disqualified.

In comparison, the distribution of NPO **B** is moderately unequal, with an equity rate of 73.14%, placing it in second position. However, it is important to note that this rate is still below average, showing that this distribution is far from ideal.

- Finally, the distribution of NPO **A** is by far the best, for several reasons. Firstly, it meets the criteria mentioned in the previous questions and interpretations. But in addition to that, it also has the best GINI index of all the proposed distributions, with an equity rate of 81.5%. This shows that the rice distribution is much more fair and equitable and is more effective in meeting the requirements of this international institution.

In summary, it is clear that ASBL A is by far the best option for rice distribution.

11.3 Second Case Study

11.3.1 Presentation of the Study

A store specializing in the sale of fabrics is facing a high demand from its customers for fabrics of various sizes and shapes. To meet this demand, the store wants to automate its fabric cutting system, replacing manual cutting with a machine. However, it is important to choose the machine that is best suited to the store's needs in terms of precision and cut quality. Three machine manufacturers responded to the call for tenders by presenting their machines. For a wise decision, the technical team tested them, and the following tables give the results of one hundred individual tests carried out. These tables give the number of tests as a function of the cut length.

Machine A	
Cut (Cm)	Freq.
[98 , 98,5)	2
[98,5 , 99)	7
[99 , 99,5)	21
[99,5 , 100)	53
[100 , 100,5)	87
[100,5 , 101)	30

Machine B	7
Cut (Cm)	Freq.
[98 , 98.5)	4
[98.5 , 99)	9
[99 , 99.5)	31
[99.5 , 100)	47
[100 , 100.5)	57
[100.5 , 101)	52

Machine C	
Cut (Cm)	Freq.
[98 , 98,5)	9
[98,5 , 99)	11
[99 , 99,5)	38
[99,5 , 100)	90
[100 , 100,5)	45
[100,5 , 101)	7

The validation team is faced with a significant challenge: minimizing the gap around a meter when cutting fabric. It is crucial not to waste fabric by cutting pieces that are too large, but it is equally important not to provide fabric pieces that are smaller than what customers have paid for. In this situation, it is essential to find the machine that will maintain a minimal gap while satisfying customers.

As the leader of the technical team, you must choose the machine that will best suit this task. To make an informed decision, it is important to perform calculations to determine which machine is the most accurate and effective.

Note: Note: In order to facilitate the reading of the solutions, we will use the following notation: each answer related to a machine will be preceded by a letter indicating the corresponding machine.

11.3.2 Questions

11.3.2.1 Analysis of central tendency measures for decision-making

For each of the machines,

1. Plot its histogram.

2. Plot its curve of cumulative relative frequencies.
3. Calculate the characteristics of central tendency of the distribution:
    a. The mode.
    b. The average.
    c. The median.
4. Give an interpretation of the order of magnitude of the elements of this statistical series.

11.3.2.2 Analysis of measures of dispersion for decision-making

To verify the accuracy of the statements made by the three suppliers, it is required to calculate and compare the following:

5. The measures of dispersion for the three machines:
    a. The interquartile range.
    b. The standard deviation.
    c. The coefficient of variation.
6. The percentage of their one-meter fabrics that is less than average in length.
7. The percentage of their one-meter fabrics that is less than average in length minus the standard deviation.
8. The percentage of their one-meter fabrics that is less than average in length plus the standard deviation.
9. The percentage of their one-meter fabrics that is in length in the range of the mean and mean minus the standard deviation.
10. The percentage of their one-meter fabrics that is in length in the range of the mean and the mean plus the standard deviation.
11. The percentage of their one-meter fabrics whose length is in the range of the mean plus the standard deviation and the mean minus the standard deviation.
12. The percentage of their one-meter fabrics that is less than one meter in length (the percentage of possible dissatisfied).
13. Length and percentage of dosage wasted over the 200 tests.

11.3.2.3 Analysis of shape measures for decision-making

To refine and consolidate the decision, it is requested to calculate their shape characteristics:

14. The YULE's coefficient of skewness.
15. The PEARSON's coefficient of skewness.
16. The FISHER's coefficient of skewness.
17. The FISHER's coefficient of kurtosis.

11.3.3 Solutions

11.3.3.1 Analysis of central tendency measures for decision-making

A: Machine A

A-1: The histogram of the machine **A** is:

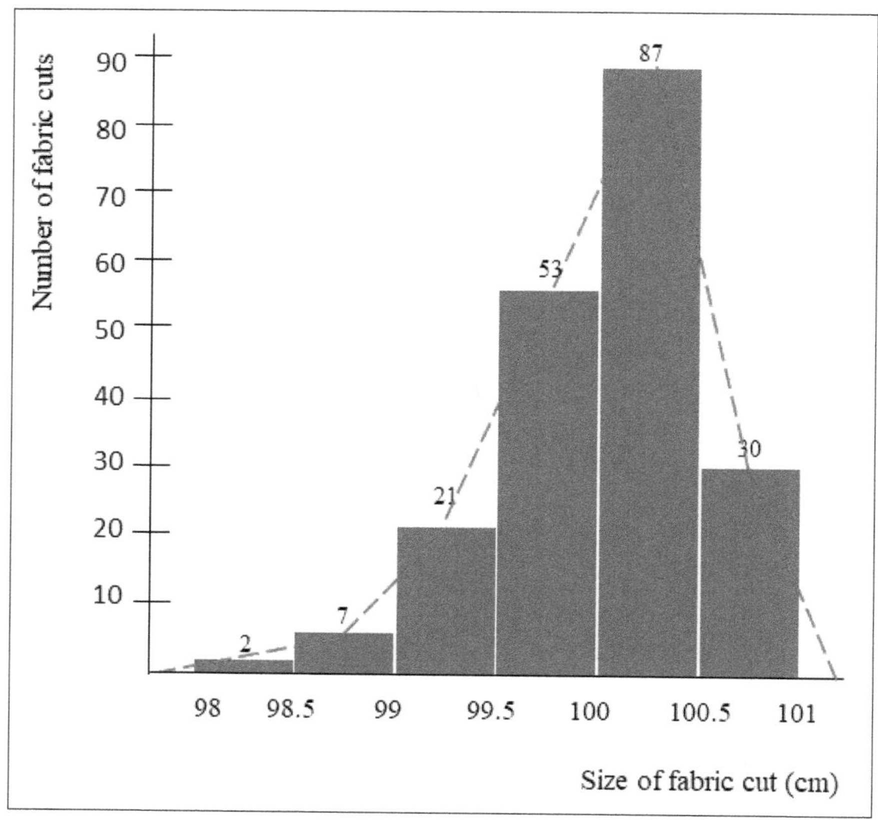

Solving this exercise requires the following table:

Machine A							
Fabric cuts	x_i	n_i	$n_i x_i$	$n_i x_i^2$	f_i	$f_i \uparrow$	$f_i \downarrow$
[98 , 98,5)	98.25	2	196.50	19,306.13	1.00	1.00	100.00
[98,5 , 99)	98.75	7	691.25	68,260.94	3.50	4.50	99.00
[99 , 99,5)	99.25	21	2,084.25	206,861.81	10.50	15.00	95.50
[99,5 , 100)	99.75	53	5,286.75	527,353.31	26.50	41.50	85.00
[100 , 100,5)	100.25	87	8,721.75	874,355.44	43.50	85.00	58.50
[100,5 , 101)	100.75	30	3,022.50	304,516.88	15.00	100.00	15.00
	Total	200	20,003.00	2,000,654.50	100.00		

A-2: The cumulative relative frequency curve of machine **A** is:

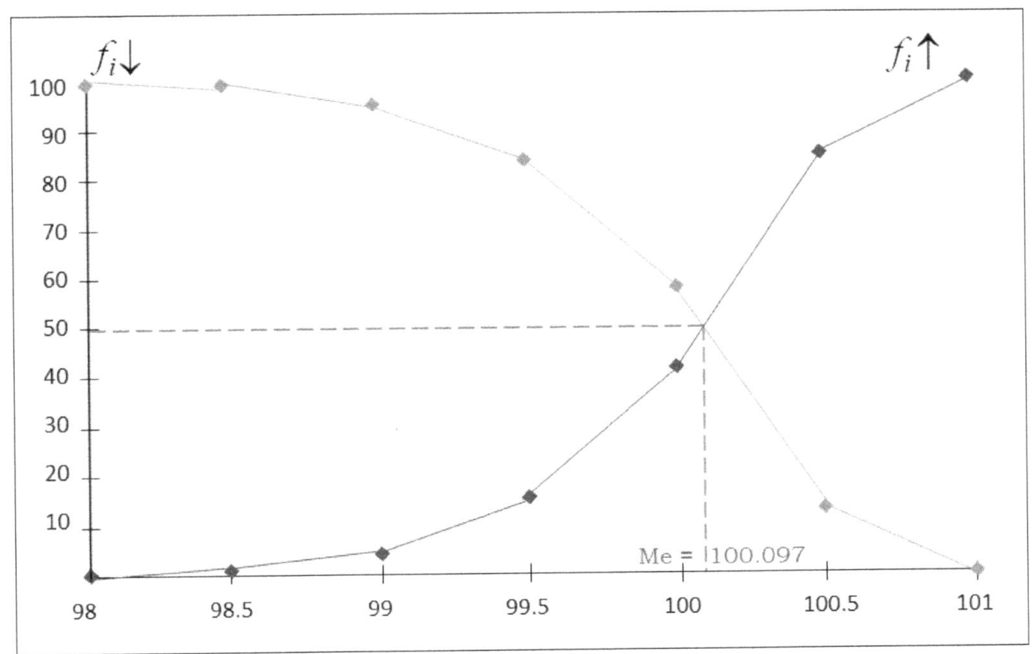

A-3: Determining the characteristics of central tendency:

A-3-a: The mode of machine **A** is 100.25 cm.

A-3-b: The average of machine **A** is:

$$\bar{x} = \frac{1}{n}\sum_{i=1}^{i=6} n_i x_i = \frac{20{,}003}{200}$$

$$= 100.015 \text{ cm}$$

A-3-c: The median of machine **A** is:

$$Me = a_i + (a_{i+1} - a_i)\frac{(50 - F_i)}{(F_{i+1} - F_i)}$$

$$= 100 + (100.5 - 100)\frac{(50 - 41.50)}{(85 - 41.50)}$$

$$= 100.097701149425 \text{ cm}$$

B: Machine B

B-1: The histogram of the machine **B** is:

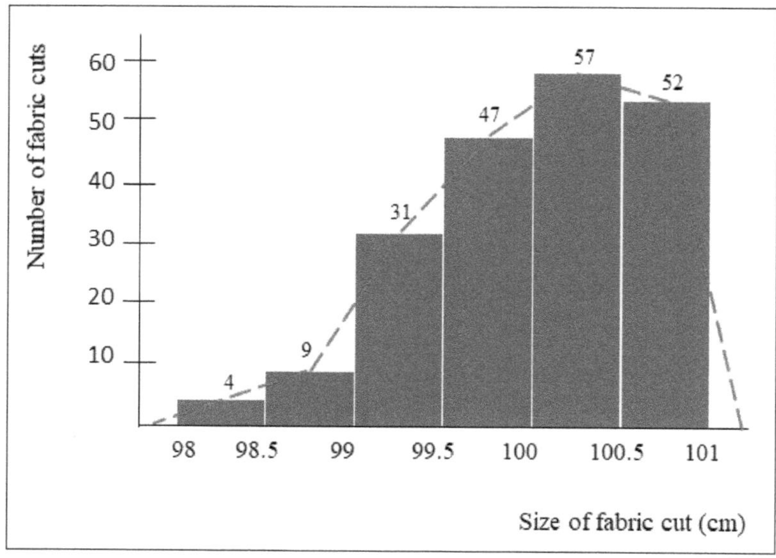

Resolving this part of the exercise requires the following table:

Machine B							
Fabric cuts	x_i	n_i	$n_i x_i$	$n_i x_i^2$	f_i	$f_i \uparrow$	$f_i \downarrow$
[98 , 98,5)	98.25	4	393.00	38,612.25	2.00	2.00	100.00
[98,5 , 99)	98.75	9	888.75	87,764.06	4.50	6.50	98.00
[99 , 99,5)	99.25	31	3,076.75	305,367.44	15.50	22.00	93.50
[99,5 , 100)	99.75	47	4,688.25	467,652.94	23.50	45.50	78.00
[100 , 100,5)	100.25	57	5,714.25	572,853.56	28.50	74.00	54.50
[100,5 , 101)	100.75	52	5,239.00	527,829.25	26.00	100.00	26.00
	Total	200	20,000.00	2,000,079.50	100.00		

B-2: The cumulative relative frequency curve of the machine **B** is:

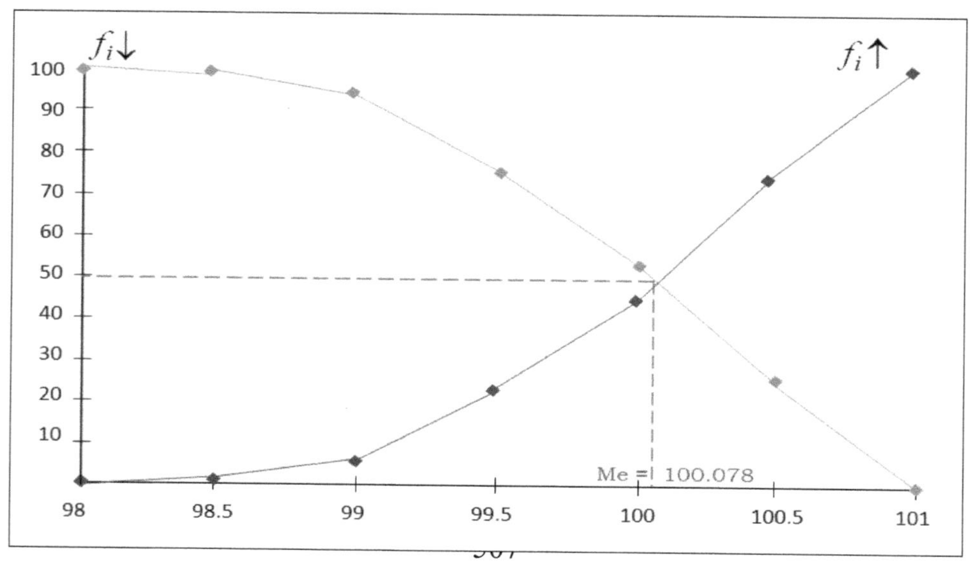

B-3: Determining the characteristics of central tendency:

B-3-a: The mode of machine **B** is 100.25 cm.

B-3-b: The average of machine **B** is:

$$\bar{x} = \frac{1}{n}\sum_{i=1}^{i=6} n_i x_i = \frac{20{,}000}{200}$$
$$= 100 \text{ cm}$$

B-3-c: The median of machine **B** is:

$$Me = a_i + (a_{i+1} - a_i)\frac{(50 - F_i)}{(F_{i+1} - F_i)}$$
$$= 100 + (100.5 - 100)\frac{(50 - 45.5)}{(74 - 45.5)}$$
$$= 100.078947368421 \text{ cm}$$

C: Machine C

C-1: The histogram of the machine **C** is:

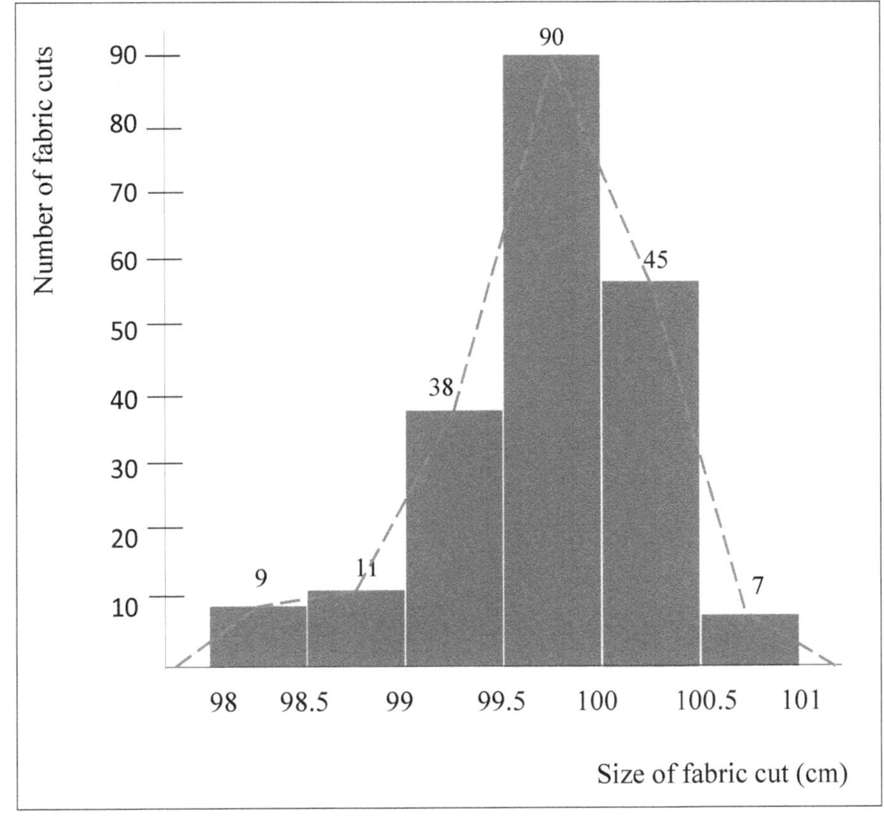

The following table shows the data required to solve this exercise:

Machine C							
Fabric cuts	x_i	n_i	$n_i x_i$	$n_i x_i^2$	f_i	$f_i \uparrow$	$f_i \downarrow$
[98 , 98,5)	98.25	4	393.00	38,612.25	2.00	2.00	100.00
[98,5 , 99)	98.75	9	888.75	87,764.06	4.50	6.50	98.00
[99 , 99,5)	99.25	31	3,076.75	305,367.44	15.50	22.00	93.50
[99,5 , 100)	99.75	47	4,688.25	467,652.94	23.50	45.50	78.00
[100 , 100,5)	100.25	57	5,714.25	572,853.56	28.50	74.00	54.50
[100,5 , 101)	100.75	52	5,239.00	527,829.25	26.00	100.00	26.00
	Total	200	20,000.00	2,000,079.50	100.00		

C-2: The cumulative relative frequency curve of the machine **C** is:

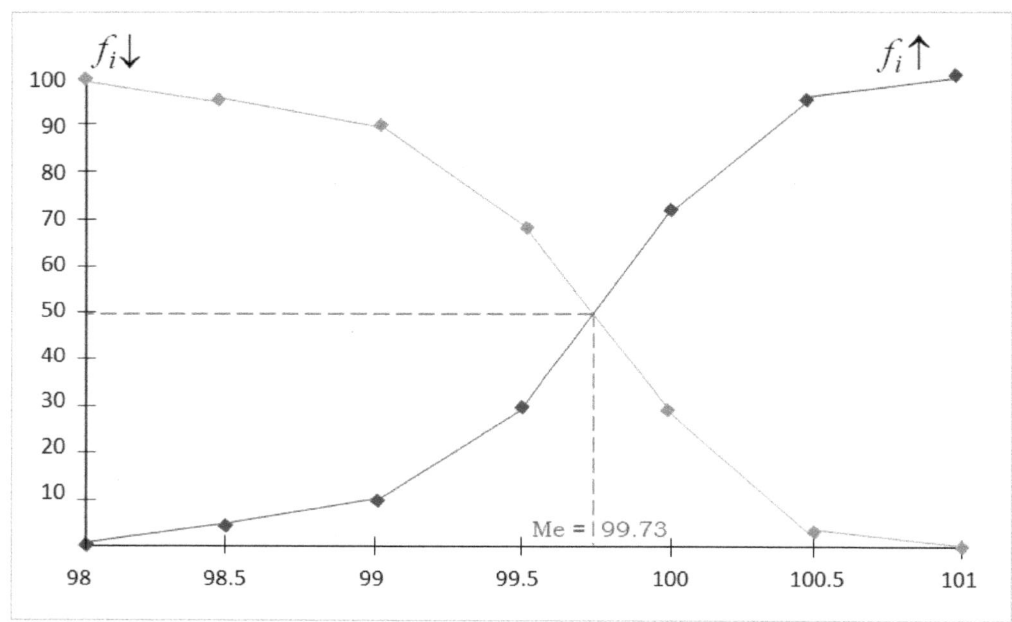

C-3: Determining the characteristics of central tendency:

C-3-a: The mode of machine **C** is 99.75 cm.

C-3-b: The average of machine **C** is:

$$\bar{x} = \frac{1}{n}\sum_{i=1}^{i=6} n_i x_i = \frac{19{,}936}{200}$$

$$= 99.68 \text{ cm}$$

C-3-c: The median of machine **C** is:

$$Me = a_i + (a_{i+1} - a_i)\frac{(50 - F_i)}{(F_{i+1} - F_i)}$$

$$= 99.5 + (100 - 99.5)\frac{(50 - 29)}{(74 - 29)}$$

$$= 99.733 \text{ cm}$$

4: Interpretation

<u>INTERPRETATION OF RESULTS, ANALYSIS AND DECISION</u>

Comparing the three parameters, it is observed that:

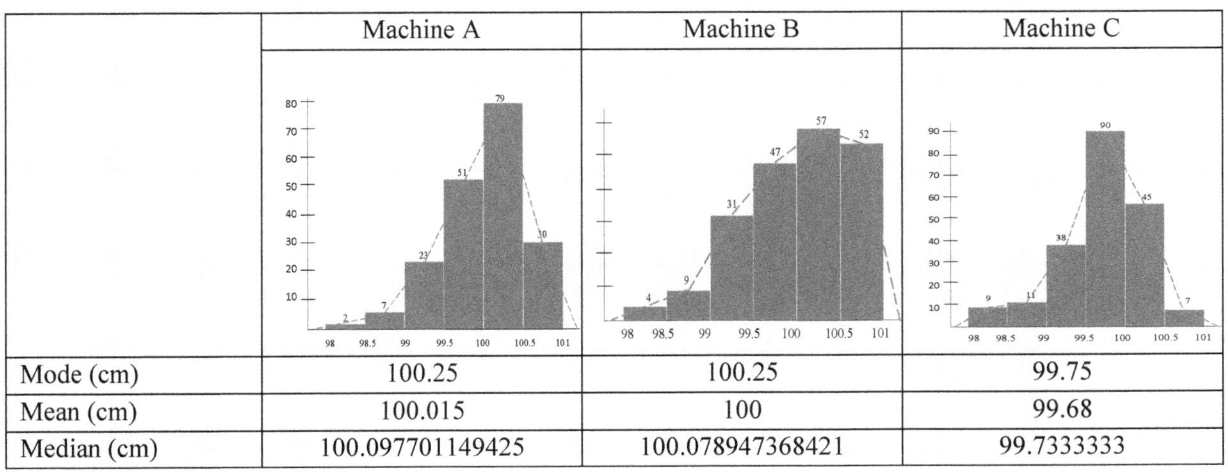

	Machine A	Machine B	Machine C
Mode (cm)	100.25	100.25	99.75
Mean (cm)	100.015	100	99.68
Median (cm)	100.097701149425	100.078947368421	99.7333333

Analysis

The quality control of the three fabric cutting machines revealed the following information:

- The measures of central tendency of the three machines are almost aligned, with very close means, medians, and modes.
- Additionally, the majority of fabrics cut by the three machines have a length of about one meter, with a margin of error of plus or minus 25 centimeters.
- Finally, there are as many fabrics cut by the three machines that have a length above their respective median as those that have a length below.

However, there is a notable difference in the results of machine B, which has an exact average of 100 centimeters for all cuts made. This distinctive value could indicate a higher precision or consistency of machine B compared to the others.

Conclusion
Given the data presented above, it is interesting to note that machine **B** appears to have an average cut length of 100 centimeters, which could make it more attractive than the other two machines. However, this information alone is not sufficient to make a decision. Indeed, the mean only gives an approximate idea of the central tendency of the length distribution of cuts and does not allow for the evaluation of the dispersion or symmetry of the distributions.

It is therefore necessary to continue the analysis by calculating other measures that will give an idea of the variability of cut lengths for each machine, and thus compare their level of precision and reliability.

11.3.3.2 Analysis of measures of dispersion for decision-making

A: Machine A

A-5: Characteristics of dispersion

Resolving this exercise requires the columns in the following table:

Machine A							
Fabric cuts	x_i	n_i	$n_i x_i$	$n_i x_i^2$	f_i	$f_i \uparrow$	$f_i \downarrow$
[98 , 98,5)	98.25	2	196.50	19,306.13	1.00	1.00	100.00
[98,5 , 99)	98.75	7	691.25	68,260.94	3.50	4.50	99.00
[99 , 99,5)	99.25	21	2,084.25	206,861.81	10.50	15.00	95.50
[99,5 , 100)	99.75	53	5,286.75	527,353.31	26.50	41.50	85.00
[100 , 100,5)	100.25	87	8,721.75	874,355.44	43.50	85.00	58.50
[100,5 , 101)	100.75	30	3,022.50	304,516.88	15.00	100.00	15.00
	Total	200	20,003.00	2,000,654.50	100.00		

A-5-1: The interquartile deviation of machine **A** is:

$$Q_1 = a_i + (a_{i+1} - a_i)\frac{(25 - F_i)}{(F_{i+1} - F_i)}$$
$$= 99.5 + (100 - 99.5)\frac{(25 - 15)}{(41.5 - 15)}$$
$$= 99.688679245283 \text{ cm}$$

$$Q_3 = a_i + (a_{i+1} - a_i)\frac{(75 - F_i)}{(F_{i+1} - F_i)}$$
$$= 100 + (100.5 - 100)\frac{(75 - 41.5)}{(85 - 41.5)}$$
$$= 100.385057471264 \text{ cm}$$

$$\text{Interquartile Range} = Q_3 - Q_1$$
$$= 100.385057471264 - 99.688679245283$$
$$= 0.696378225981363 \text{ centimeters}$$

A-5-2: The variance of machine **A** is:

$$V(x) = \frac{1}{n}\sum_{i=1}^{i=6} n_i x_i^2 - \bar{x}^2$$
$$= \frac{2{,}000{,}654.50}{20{,}003} - (100.015)^2$$
$$= 0.2864583$$

A-5-3: The standard deviation of machine **A** is:

$$\sigma = \sqrt{V(x)}$$
$$= \sqrt{0.272274999999354}$$
$$= 0.52179977002616 \text{ cm}$$

A-5-4: The coefficient of variation of machine **A** is:

$$\text{CV} = \frac{\sigma}{|\bar{x}|}$$
$$= \frac{0.52179977002616}{100.015} \text{x} 100$$
$$= 0.52\%$$

A-6: Let's calculate the percentage of one-meter fabric items that are less than average in length.

$$\bar{x} = 100.015 \text{ cm}$$

Let $P_{<\bar{x}}$ be this percentage, and we use the linear interpolation formula using the data pair $(a_i, f_i \uparrow)$:

$$\bar{x} = a_i + (a_{i+1} - a_i)\frac{(P_{<\bar{x}} - F_i)}{(F_{i+1} - F_i)}$$

$$100.015 = 100 + (100.5 - 100)\frac{(P_{<\bar{x}} - 41.5)}{(85 - 41.5)}$$

$$P_{<\bar{x}} = 42.805\%$$

A-7: Calculate the percentage of one-meter fabric items whose length is less than $\bar{x} - \sigma$:

$$\begin{aligned}\bar{x} - \sigma &= 100.015 - 0.52179977002616 \\ &= 99.4932002299738 \text{ cm}\end{aligned}$$

Let $P_{<\bar{x}-\sigma}$ be this percentage, and we use the linear interpolation formula using the data pair $(a_i, f_i \uparrow)$:

$$\bar{x} - \sigma = a_i + (a_{i+1} - a_i)\frac{(P_{<\bar{x}-\sigma} - F_i)}{(F_{i+1} - F_i)}$$

$$99.4932002299738 = 99 + (99.5 - 99)\frac{(P_{<\bar{x}-\sigma} - 4.5)}{(15 - 4.5)}$$

$$P_{<\bar{x}-\sigma} = 14.8572\%$$

A-8: Calculate the percentage of one-meter fabric items whose length is less than $\bar{x} + \sigma$:

$$\begin{aligned}\bar{x} + \sigma &= 100.015+0.52179977002616 \\ &= 100.53679977002616 \text{ cm}\end{aligned}$$

Let $P_{<\bar{x}+\sigma}$ be this percentage, and we use the linear interpolation formula using the data pair $(a_i, f_i \uparrow)$:

$$\bar{x} + \sigma = a_i + (a_{i+1} - a_i)\frac{(P_{<\bar{x}+\sigma} - F_i)}{(F_{i+1} - F_i)}$$

$$100.53679977002616 = 100.5 + (101 - 100.5)\frac{(P_{<\bar{x}+\sigma} - 85)}{(100 - 85)}$$

$$P_{<\bar{x}+\sigma} = 86.1037\%$$

A-9: Let's calculate the percentage of one-meter fabric items whose length is within the range of the mean, i.e., 100.015 cm and the mean minus the standard deviation, i.e., 99.4932002299738 cm.

This percentage is:

$$42.805\% - 14.8572\% = 27.9478\%$$

A-10: Let's calculate the percentage of one-meter fabric items whose length is within the range of the mean, i.e., 100.015 cm and the mean plus the standard deviation, i.e., 100.53679977002616 cm.

This percentage is:

$$86.1037\% \ - \ 42.805\% = 43.2987\%$$

A-11: Let's calculate the percentage of one-meter fabric items whose length is in the mean range minus the standard deviation, which is 99.4932002299738 cm and the mean plus the standard deviation, which is 100.53679977002616 cm.

This percentage is:

$$43.2987\% - 27.9478\% = 71.2465\%$$

This result can be obtained by:

$$86{,}1037\% \ - \ 14{,}8572\% = 71{,}2465\%$$

A.12: Let's calculate the percentage of one-meter fabric items less than 1 meter in length.

This percentage can be obtained directly by intersecting the column of cumulative increasing frequencies with the row corresponding to the interval [99,5 ; 100[in the distribution table. This percentage is 41.5%.

$$P_{<meter} = 41.5\%$$

A.13: Determining the length and percentage of fabric wasted over the 200 tests:

- Ideal length of a fabric: 100 cm.
- Number of pieces of fabric cut (n_i)= 200.
- Ideal length of total cut fabrics = 200 x 100 cm = 20,000 cm.
- Total length of cut fabrics = 20.003 cm.
- Length wasted fabrics = 20,003 cm – 20,000 cm = 3 cm.
- Percentage of wasted fabric = $\frac{3}{20{,}000} = 0.015\%$.

B: Machine B

B-5: Characteristics of dispersion

The following table shows the data required to solve this exercise:

Machine B							
Fabric cuts	x_i	n_i	$n_i x_i$	$n_i x_i^2$	f_i	$f_i \uparrow$	$f_i \downarrow$
[98 , 98,5)	98.25	4	393.00	38,612.25	2.00	2.00	100.00
[98,5 , 99)	98.75	9	888.75	87,764.06	4.50	6.50	98.00
[99 , 99,5)	99.25	31	3,076.75	305,367.44	15.50	22.00	93.50
[99,5 , 100)	99.75	47	4,688.25	467,652.94	23.50	45.50	78.00
[100 , 100,5)	100.25	57	5,714.25	572,853.56	28.50	74.00	54.50
[100,5 , 101)	100.75	52	5,239.00	527,829.25	26.00	100.00	26.00
	Total	200	20,000.00	2,000,079.50	100.00		

B-5-1: The interquartile deviation of machine **B** is:

$$Q_1 = a_i + (a_{i+1} - a_i)\frac{(25 - F_i)}{(F_{i+1} - F_i)}$$

$$= 99.5 + (100 - 99.5)\frac{(25 - 22)}{(45.5 - 22)}$$

$$= 99.563829787234 \text{ cm}$$

$$Q_3 = a_i + (a_{i+1} - a_i)\frac{(75 - F_i)}{(F_{i+1} - F_i)}$$

$$= 100.5 + (101 - 100.5)\frac{(75 - 74)}{(100 - 74)}$$

$$= 100.51923076923 \text{ cm}$$

$$\text{Interquartile deviation} = Q_3 - Q_1$$

$$= 100.51923076923 - 99.563829787234$$

$$= 0.955400981996732 \text{ cm}$$

B-5-2: The variance of machine **B** is:

$$V(x) = \frac{1}{n}\sum_{i=1}^{i=6} n_i x_i^2 - \bar{x}^2$$

$$= \frac{2{,}000{,}079.5}{200} - (100)^2$$

$$= 0.397499999999127$$

B-5-3: The standard deviation of machine **B** is:

$$\sigma = \sqrt{V(x)}$$
$$= \sqrt{0.397499999999127}$$
$$= 0.630476010645232 \text{ cm}$$

B-5-4: The coefficient of variation of machine **B** is:

$$\text{CV} = \frac{\sigma}{|\bar{x}|}$$
$$= \frac{0.630476010645232}{100} \text{x} 100$$
$$= 0.630476010645232\%$$

B-6: Let's calculate the percentage of one-meter fabric items that are less than average in length.

This percentage can be obtained directly from the distribution table on the increasing cumulative frequency column which corresponds to the interval [99,5 ; 100[this value is 45.5%.

$$P_{<\bar{x}} = 41.5\%$$

B-7: Let's calculate the percentage of one-meter fabric items whose length is less than $\bar{x} - \sigma$

$$\bar{x} - \sigma = 100 - 0.630476010645232$$
$$= 99.369523989354768 \text{ cm}$$

Let $P_{<\bar{x}-\sigma}$ be this percentage, and we use the linear interpolation formula using the data pair $(a_i, f_i \uparrow)$:

$$\bar{x} - \sigma = a_i + (a_{i+1} - a_i)\frac{(P_{<\bar{x}-\sigma} - F_i)}{(F_{i+1} - F_i)}$$
$$99.369523989354768 = 99 + (99.5 - 99)\frac{(P_{<\bar{x}-\sigma} - 6.5)}{(22 - 6.5)}$$
$$P_{<\bar{x}-\sigma} = 17.95512\%$$

B-8: Let's calculate the percentage of one-meter fabric items whose length is less than $\bar{x} + \sigma$

$$\bar{x} + \sigma = 100 + 0.630476010645232$$
$$= 100.630476010645232 \text{ cm}$$

Let $P_{<\bar{x}+\sigma}$ be this percentage, and we use the linear interpolation formula using the data pair $(a_i, f_i \uparrow)$ as proceeded above:

$$\bar{x} + \sigma = a_i + (a_{i+1} - a_i)\frac{(P_{<\bar{x}+\sigma} - F_i)}{(F_{i+1} - F_i)}$$

$$100.630476010645232 = 100.5 + (101 - 100.5)\frac{(P_{<\bar{x}+\sigma} - 74)}{(100 - 74)}$$

$$P_{<\bar{x}+\sigma} = 80.78444\%$$

B-9: Let's calculate the percentage of one-meter fabric items whose length is within the range of the mean, which is 100 cm and the mean minus the standard deviation, which is 99.369523989354768 cm.

This percentage is:

$$41.5\% - 17.95512\% = 23.54488\%$$

B-10: Let's calculate the percentage of one-meter fabric items whose length is within the range of the mean, i.e., 100 cm and the mean plus the standard deviation, i.e.,100.630476010645232 cm.

This percentage is:

$$80.78444\% \; - \; 41.5\% = 39.28444\%$$

B-11: Let's calculate the percentage of one-meter fabric items whose length is in the range of the mean minus the standard deviation, i.e., 99.369523989354768 cm and the mean plus the standard deviation, i.e., 100.630476010645232 cm.

This percentage is:

$$80.78444\ \% - 17.95512\% = 62.82932\%$$

B-12: Let's calculate the percentage of one-meter fabric items less than 1 meter in length.

This percentage can be obtained directly by intersecting the column of cumulative increasing frequencies with the row corresponding to the interval [99,5 ; 100[in the distribution table. This percentage is 45.5%.

$$P_{<meter} = 45.5\%$$

B-13: Length and percentage of fabric wasted:

- Ideal length of a fabric: 100 cm.
- Number of pieces of fabric cut (n_i)= 200.
- Ideal length of total cut fabrics = 200 x 100 cm = 20,000 cm.
- Total length of cut fabrics = 20,000 cm.

- Length wasted fabrics = 20,000 cm – 20,000 cm = 0 cm.
- Percentage of wasted fabric = $\frac{0}{20{,}000} = 0\%$.

C: Machine C

C-5: Characteristics of dispersion

The following table shows the data required to solve this exercise:

Machine C							
Fabric cuts	x_i	n_i	$n_i x_i$	$n_i x_i^2$	f_i	$f_i \uparrow$	$f_i \downarrow$
[98 , 98,5)	98.25	4	393.00	38,612.25	2.00	2.00	100.00
[98,5 , 99)	98.75	9	888.75	87,764.06	4.50	6.50	98.00
[99 , 99,5)	99.25	31	3,076.75	305,367.44	15.50	22.00	93.50
[99,5 , 100)	99.75	47	4,688.25	467,652.94	23.50	45.50	78.00
[100 , 100,5)	100.25	57	5,714.25	572,853.56	28.50	74.00	54.50
[100,5 , 101)	100.75	52	5,239.00	527,829.25	26.00	100.00	26.00
	Total	200	20,000.00	2,000,079.50	100.00		

C-5-1: The interquartile deviation of machine **C** is:

$$Q_1 = a_i + (a_{i+1} - a_i)\frac{(25 - F_i)}{(F_{i+1} - F_i)}$$
$$= 99 + (99.5 - 99)\frac{(25 - 10)}{(29 - 10)}$$
$$= 99.3947368421053 \text{ cm}$$
$$Q_3 = a_i + (a_{i+1} - a_i)\frac{(75 - F_i)}{(F_{i+1} - F_i)}$$
$$= 100 + (100.5 - 100)\frac{(75 - 74)}{(96.5 - 74)}$$
$$= 100.022222222222 \text{ cm}$$
$$\text{The interquartile range } = Q_3 - Q_1$$
$$= 100.022222222222 - 99.3947368421053$$
$$= 0.627485380116966 \text{ cm}$$

C-5-2: The variance of machine **C** is equal to:

$$V(x) = \frac{1}{n}\sum_{i=1}^{i=6} n_i x_i^2 - \bar{x}^2$$
$$= \frac{1{,}987{,}278.5}{200} - (99.68)^2$$
$$= 0.290099999998347$$

C-5-3: The standard deviation of machine **C** is:

$$\sigma = \sqrt{V(x)}$$
$$= \sqrt{0.290099999998347}$$
$$= 0.538609320378275 \text{ cm}$$

C-5-4: The coefficient of variation of machine **C** is:

$$CV = \frac{\sigma}{|\bar{x}|}$$
$$= \frac{0.538609320378275}{99.68} \text{x}100$$
$$= 0.5403384032687\%$$

C-6: Let's calculate the percentage of one-meter fabric items whose length is less than average.

$$\bar{x} = 99.68 \text{ cm}$$

Let $P_{<\bar{x}}$ be this percentage, we use the linear interpolation formula on the couple $(a_i, f_i \uparrow)$:

$$\bar{x} = a_i + (a_{i+1} - a_i)\frac{(P_{<-\sigma} - F_i)}{(F_{i+1} - F_i)}$$
$$99.68 = 99.5 + (100 - 99.5)\frac{(P_{<\bar{x}} - 29)}{(74 - 29)}$$
$$P_{<\bar{x}} = 45.2\%$$

C-7: Let's calculate the percentage of one-meter fabric items whose length is less than $\bar{x} - \sigma$

$$\bar{x} - \sigma = 99.68 - 0{,}538609320378275$$
$$= 99.141390679621725 \text{ cm}$$

Let $P_{<\bar{x}-\sigma}$ be this percentage, and we use the linear interpolation formula using the data pair $(a_i, f_i \uparrow)$ as proceeded above:

$$\bar{x} - \sigma = a_i + (a_{i+1} - a_i)\frac{(P_{<\bar{x}-\sigma} - F_i)}{(F_{i+1} - F_i)}$$
$$99.141390679621725 = 99 + (99.5 - 99)\frac{(P_{<\bar{x}-\sigma} - 10)}{(29 - 10)}$$
$$P_{<\bar{x}-\sigma} = 15.37282\%$$

C-8: Let's calculate the percentage of one-meter fabric items whose length is less than $\bar{x} + \sigma$

$$\bar{x} + \sigma = 99.68 + 0{,}538609320378275$$
$$= 100.218609320378275 \text{ cm}$$

Let $P_{<\bar{x}+\sigma}$ be this percentage, and we use the linear interpolation formula using the data pair $(a_i, f_i \uparrow)$ as proceeded above:

$$\bar{x} + \sigma = a_i + (a_{i+1} - a_i)\frac{(P_{<\bar{x}-\sigma} - F_i)}{(F_{i+1} - F_i)}$$

$$100.218609320378275 = 100 + (100.5 - 100)\frac{(P_{<\bar{x}+\sigma} - 74)}{(96.5 - 74)}$$

$$P_{<\bar{x}+\sigma} = 83.837\%$$

C-9: Let's calculate the percentage of one-meter fabric items whose length is within the range of the mean, i.e., 99,68 cm and the mean minus the standard deviation, i.e., 99.141390679621725 cm.

This percentage is:

$$45.2\% - 15.37282\% = 29.82718\%$$

C-10: Let's calculate the percentage of one-meter fabric items whose length is within the range of the mean, i.e., 99.68 cm and the mean plus the standard deviation, i.e., 100.218609320378275 cm

This percentage is:

$$83.837\ \% - 45.2\% = 38.637\%$$

C-11: Let's calculate the percentage of one-meter fabric items whose length is in the range of the mean minus the standard deviation, i.e., 99.141390679621725 cm and the mean plus the standard deviation, i.e., 100.218609320378275 cm.

This percentage is:

$$29.82718\% + 38.637\% = 68.46418\%$$

The same result can be obtained by:

$$83.837\% - 15.37282\% = 68.46418\%$$

C-12: Let's calculate the percentage of one-meter fabric items whose length is less than 1 meter.

This percentage can be obtained directly by intersecting the column of cumulative increasing frequencies with the row corresponding to the interval [99,5 ; 100[in the distribution table. This percentage is 74%.

$$P_{<meter} = 74\%$$

C-13: Length and percentage of fabric wasted:

- Ideal length of a fabric: 100 cm.
- Number of pieces of fabric cut (n_i)= 200.
- Ideal length of total cut fabrics = 200 x 100 cm = 20,000 cm.
- Total length of cut fabrics = 19936 cm
- Length wasted fabrics = 19,936 cm – 20,000 cm = –64 cm
- Percentage of wasted fabric = $\frac{-64}{20,000}$ =- 3.2%

INTERPRETATION OF RESULTS, ANALYSIS AND DECISION

	Machine A	Machine B	Machine C
Mode (cm)	**100.25**	**100.25**	99.75
Mean (cm)	100.015	**100**	99.68
Median (cm)	100.0977	100.0789	99.7333
% Majority	43.5%	28.5%	45%
% Pop. < Mean	42.8%	45.5%	45.2%
% Pop. < Median	50%	50%	50%
% Pop. < 100 cm	41.5%	45.5%	74%

Machine **A:**

- The majority of the population, 43.5%, received a fabric that was 0.25 cm larger than the required size of 100 cm. Additionally, 42.8% of the population received fabric that was smaller than the average of 100.015 cm. There are as many customers who received fabric with a length greater than 100.0977 cm as those who received fabric with a length less than it. Finally, 41.5% of the population received fabric with a length less than the required meter.
- Analyzing these results, it is clear that Machine A's cutting system is evidently oriented towards customer satisfaction at the expense of the store. Although this machine can satisfy customers by offering them slightly larger fabric pieces than the required size, it can also cause losses for the store.

Machine **B:**

- The majority of the population, 28.5%, received fabric that was 0.25 cm larger than the required size of 100 cm, and 45.5% of the population received fabric that was smaller than the average of 100 cm. There are as many customers who received fabric with a length greater than 100.0789 cm as those who received fabric with a length less than it. 45.5% of the population received fabric with a length less than the required meter. This system is also oriented towards customer satisfaction but to a lesser degree than the previous one.
- It can be noted that 28.5% of the population received fabric whose length was 0.25 cm larger than the required size of 100 cm. However, this percentage is lower than that of Machine A. On the other hand,

45.5% of the population received fabric whose length was smaller than the average of 100 cm, a slightly higher percentage than that of Machine A.
- Regarding the accuracy of Machine B, it can be noted that there are as many customers who received fabric with a length greater than 100.0789 cm as those who received fabric with a length less than it. This indicates good precision of the machine compared to Machine A.
- Finally, it should be noted that 45.5% of the population received fabric whose length was less than the required meter. Although this percentage is higher than that of Machine A, it still indicates that Machine B is capable of producing fabrics of various sizes with reasonable precision.
- Overall, it can be said that Machine B is also oriented towards customer satisfaction, but to a lesser degree than Machine A. However, it offers good precision and is capable of producing fabrics of various sizes with reasonable precision, making it a viable option for the store.

Machine **C**:
- The majority of the population, 45%, received fabric that was 0.25 cm smaller than the required 100 cm size, meaning that almost half of the customers will not receive the amount of fabric they need. Additionally, 45.2% of the population received fabric that was smaller than the average of 99.68 cm, suggesting that the machine is prone to frequent measurement errors.
- The most concerning result is that 74% of the population received fabric with a length less than one meter. This means that Machine C cannot produce fabric pieces of the required length for the majority of customers, which will inevitably lead to returns from dissatisfied customers and financial losses for the store. Furthermore, although Machine C can also produce fabric whose length is greater than the required size, the percentage of customers who received such fabric is relatively low, indicating that the machine is not precise enough.
- It is therefore clear that Machine C is not suitable to meet the needs of the store as it preserves the interests of the store by creating customer dissatisfaction. The store must opt for a machine capable of producing precise fabric pieces of the required size while maintaining a high level of customer satisfaction

CV	0.52%	0.630%	0.540%

The coefficients of variation of less than 1% for each of the three machines show that the variance of their distributions is very low compared to their respective means. This implies that the difference between the results produced by each of these machines is very low, if not negligible. This also means that the repeatability of each machine is high and the precision of its results is reliable.
However, even though the results of each machine have a quasi-homogeneous distribution, it is important to consider them in the overall context of decision-making.

% in $[\bar{x}, \bar{x} - \sigma]$	27.9478%	23.54488%	29.82718%
% in $[\bar{x}, \bar{x} + \sigma]$	43.2987%	39.28444%	38.637%
% In $[\bar{x} - \sigma, \bar{x} + \sigma]$	71.2465%	62.82932%	68.46418%
% Population < 1 meter	41.5%	45.5%	74%

The majority of fabrics provided by machine **A** are located between the mean minus the standard deviation and the mean plus the standard deviation, which means it provides fabrics that are slightly shorter and slightly longer than average. On the other hand, machines **B** and **C** provide fabrics that are closer to the mean, with less spread in their distributions.
Despite this difference in fabric distribution, machine A manages to maintain a high level of customer satisfaction by providing few fabrics whose length is less than the required meter. Machine B closely follows in terms of customer satisfaction, but machine **C** leaves many customers dissatisfied with its 74% of fabrics that are inferior to the required size.

Length wasted fabrics	3 meters	0 meters	−64 meters
% Fabric wasted	0.015%	0%	−3.2%

Machine **C** remains with 3.2% of fabric at the end of the distribution, which is not normal because the customer seems aggrieved. This is not normal as it can result in a dissatisfied customer. It is crucial to consider customer satisfaction during fabric production.
Machine **A** wastes 0.015% tissue, although negligible can make up the difference.
Machine **B** does not waste any fabrics, nor does it save any on the total fabric cut either.

Conclusion

The recommendation for selecting the machine to use depends on the priority of the business: customer satisfaction, store satisfaction, or both.

- If the business prioritizes customer satisfaction, machine A is the best choice as it provides a large quantity of fabrics close to the average with few fabrics shorter than the required meter, thus limiting customer dissatisfaction.
- If the business aims to maximize store satisfaction, machine C is the recommended choice as it allows for more fabric savings than the other machines, but this may cause customer dissatisfaction due to the high percentage of fabrics shorter than the required meter.
- If the business wants to satisfy both the customer and the store, machine B is recommended as it neither saves nor wastes fabric but provides an average quality fabric and may not fully satisfy demanding customers.

11.3.3.3 Analysis of shape measures for decision-making

Determining the shape characteristics of

A: Machine **A**

Resolving this exercise requires the columns in the following table:

x_i	n_i	$n_i x_i$	$n_i(x_i-\bar{x})^3$	$n_i(x_i-\bar{x})^4$
98.25	2	196.50	-11.00	19.41
98.75	7	691.25	-14.17	17.93
99.25	21	2084.25	-9.40	7.19
99.75	53	5286.75	-0.99	0.26
100.25	87	8721.75	1.13	0.27
100.75	30	3022.50	11.91	8.76
	200	20003.00	-22.51365000	53.80854463

a: Let's calculate the YULE's coefficient of skewness:

$$C_Y = \frac{Q_1 - 2Q_2 + Q_3}{Q_3 - Q_1}$$

$$= \frac{99.68867924528 - 2(100.09770114) + 100.3850574712}{100.3850574712 - 99.68867924528}$$

$$= -0.17473$$

b: Let's calculate the PEARSON's coefficient of skewness:

$$\beta_1 = \frac{(\bar{x} - \text{Mode})}{\sigma}$$

$$= \frac{(100.015 - 99.75)}{0.52179977002616}$$

$$= -0.450364323$$

c: Let's calculate the FISHER's coefficient of skewness:

$$\mu_3 = \frac{\sum_{i=1}^{i=6} n_i (x_i - \bar{x})^3}{\sum_{i=1}^{i=6} n_i} = \frac{-22.51365}{200}$$

$$= -0.1125682497$$

$$\gamma_1 = \frac{\mu_3}{\sigma^3} = \frac{-0.1125682497}{(0.52179977)^3}$$

$$= -0.792327$$

d: Let's calculate the FISHER's coefficient of kurtosis:

$$\mu_4 = \frac{\sum_{i=1}^{i=6} n_i (x_i - \bar{x})^4}{\sum_{i=1}^{i=6} n_i} = \frac{53.80854463}{200}$$

$$= 0.26904278994$$

$$\gamma_2 = \frac{\mu_4}{\sigma^4} - 3 = \frac{0.26904278994}{(0.52179977)^4} - 3$$

$$= -0.6291576$$

B: Machine **B**

Resolving this exercise requires the columns in the following table:

x_i	n_i	$n_i x_i$	$n_i(x_i - \bar{x})^3$	$n_i(x_i - \bar{x})^4$
98.25	4	393.00	-21.44	37.52
98.75	9	888.75	-17.58	21.97
99.25	31	3076.75	-13.08	9.81
99.75	47	4688.25	-0.73	0.18
100.25	57	5714.25	0.89	0.22
100.75	52	5239.00	21.94	16.45
	200	20000.00	-30.00000000	86.15625000

a: Let's calculate the YULE's coefficient of skewness:

$$C_Y = \frac{Q_1 - 2Q_2 + Q_3}{Q_3 - Q_1}$$

$$= \frac{99.56382978723 - 2(100.07894736) + 100.5192307692}{100.5192307692 - 99.56382978723}$$

$$= -0.078327493$$

b: Let's calculate the PEARSON's coefficient of skewness:

$$\beta_1 = \frac{(\bar{x} - \text{Mode})}{\sigma}$$

$$= \frac{(100 - 100.25)}{0.630476}$$

$$= -0,3965257996$$

c: Let's calculate the FISHER's coefficient of skewness:

$$\mu_3 = \frac{\sum_{i=1}^{i=6} n_i(x_i - \bar{x})^3}{\sum_{i=1}^{i=6} n_i} = \frac{-30}{200}$$

$$= -0.15$$

$$\gamma_1 = \frac{\mu_3}{\sigma^3} = \frac{-0.15}{(0.630476)^3}$$

$$= -0.598529$$

d: Let's calculate the FISHER's coefficient of kurtosis:

$$\mu_4 = \frac{\sum_{i=1}^{i=6} n_i (x_i - \bar{x})^4}{\sum_{i=1}^{i=6} n_i} = \frac{86.15625}{200}$$

$$= -0.430781$$

$$\gamma_2 = \frac{\mu_4}{\sigma^4} - 3 = \frac{-0.430781}{(0.630476)^4} - 3$$

$$= -0.2736443$$

C: Machine **C**

Resolving this exercise requires the columns in the following table:

x_i	n_i	$n_i x_i$	$n_i(x_i - \bar{x})^3$	$n_i(x_i - \bar{x})^4$
98.25	9	884.25	-26.32	37.63
98.75	11	1086.25	-8.85	8.23
99.25	38	3771.50	-3.02	1.30
99.75	90	8977.50	0.03	0.00
100.25	45	4511.25	8.33	4.75
100.75	7	705.25	8.58	9.18
	200	19936.00	-21.24720000	61.09019400

a: Let's calculate the YULE's coefficient of skewness:

$$C_Y = \frac{Q_1 - 2Q_2 + Q_3}{Q_3 - Q_1}$$

$$= \frac{99.39473684210 - 2(99.7333) + 100.0222}{100.0222 - 99.39473684210}$$

$$= -0.079149122$$

b: Let's calculate the PEARSON's coefficient of skewness:

$$\beta_1 = \frac{(\bar{x} - \text{Mode})}{\sigma}$$

$$= \frac{(99.68 - 99.75)}{0.5386}$$

$$= -0.12996658$$

c: Let's calculate the FISHER's coefficient of skewness:

$$\mu_3 = \frac{\sum_{i=1}^{i=6} n_i (x_i - \bar{x})^3}{\sum_{i=1}^{i=6} n_i} = \frac{-21.2472}{200}$$

$$= -0.1062359998$$

$$\gamma_1 = \frac{\mu_3}{\sigma^3} = \frac{-0.1062359998}{(0.5386)^3}$$

$$= -0.679908$$

d: Let's calculate the FISHER's coefficient of kurtosis:

$$\mu_4 = \frac{\sum_{i=1}^{i=6} n_i (x_i - \bar{x})^4}{\sum_{i=1}^{i=6} n_i} = \frac{61.090194}{200}$$

$$= 0.30545091629$$

$$\gamma_2 = \frac{\mu_4}{\sigma^4} - 3 = \frac{0.30545091629}{(0.5386)^4} - 3$$

$$= 0.6294931$$

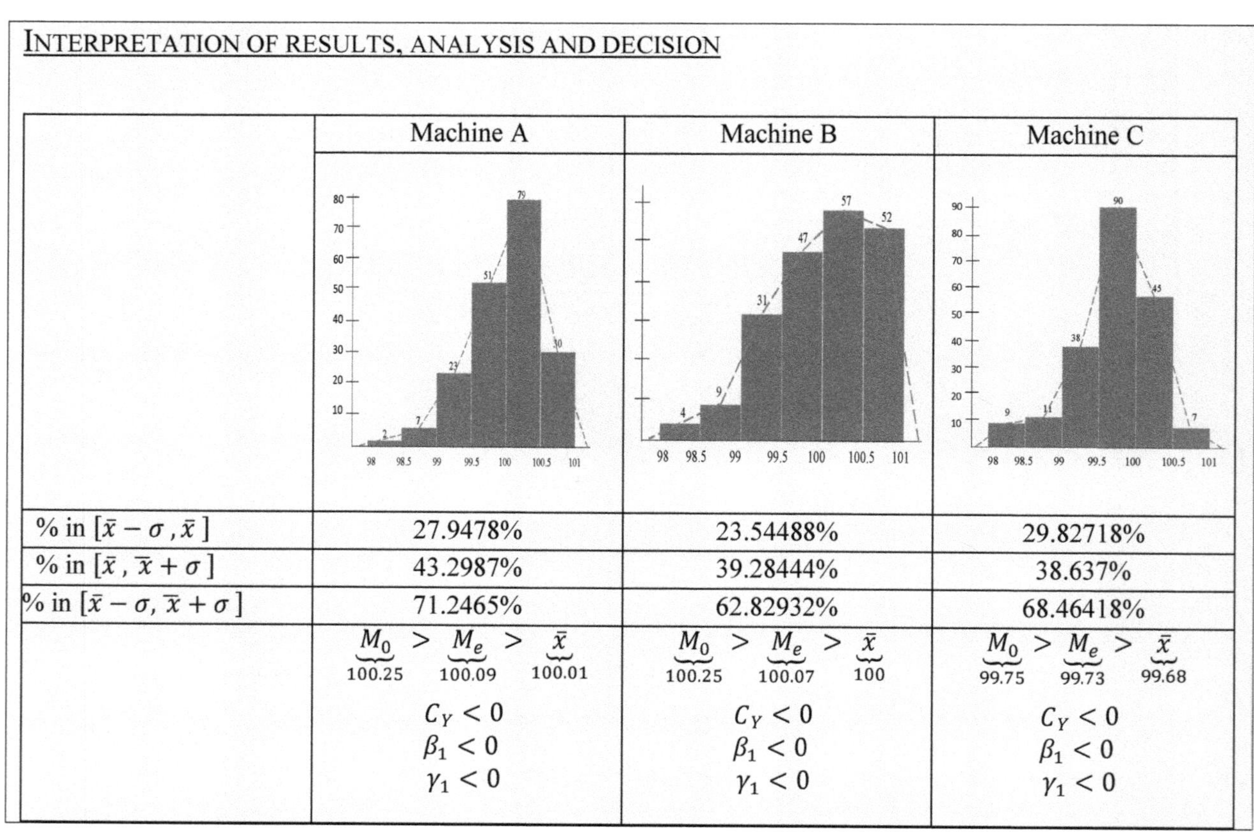

INTERPRETATION OF RESULTS, ANALYSIS AND DECISION

	Machine A	Machine B	Machine C
% in $[\bar{x} - \sigma, \bar{x}]$	27.9478%	23.54488%	29.82718%
% in $[\bar{x}, \bar{x} + \sigma]$	43.2987%	39.28444%	38.637%
% in $[\bar{x} - \sigma, \bar{x} + \sigma]$	71.2465%	62.82932%	68.46418%
	$\underbrace{M_0}_{100.25} > \underbrace{M_e}_{100.09} > \underbrace{\bar{x}}_{100.01}$ $C_Y < 0$ $\beta_1 < 0$ $\gamma_1 < 0$	$\underbrace{M_0}_{100.25} > \underbrace{M_e}_{100.07} > \underbrace{\bar{x}}_{100}$ $C_Y < 0$ $\beta_1 < 0$ $\gamma_1 < 0$	$\underbrace{M_0}_{99.75} > \underbrace{M_e}_{99.73} > \underbrace{\bar{x}}_{99.68}$ $C_Y < 0$ $\beta_1 < 0$ $\gamma_1 < 0$

	$\gamma_2 < 0$	$\gamma_2 < 0$	$\gamma_2 > 0$

Analysis

The three curves A, B and C all have asymmetric distributions on the right; their distribution tail is spread to the left. There is therefore a concentration of high values that is to say, there is a strong preponderance of number of dosages administered above the average.

It is easy to notice that for the three machines the population in the interval $[\bar{x}, \bar{x} + \sigma]$ is much higher than that in the interval.$[\bar{x} - \sigma, \bar{x}]$. This means there is a larger population that is higher than the average than one that is lower.

The distribution curves of the two machines A and B are more flattened than that of the normal distribution while the distribution curve of C is sharper than that of the normal distribution.

These three blocks of arguments do not affect the conclusions drawn in the previous part on the choice of machine.

11.4 Third Case Study

11.4.1 Presentation of the Study

A biological laboratory is searching for ways to speed up the procedure of vaccine administration via injection. Experts have proposed the use of a gun for serum injection to accelerate the process. Therefore, the laboratory has launched a call for tenders for a gun capable of respecting a strict dosage of 100 microliters. Any quantity below 99 microliters or above 100 microliters is ineffective or toxic, respectively. The toxicity increases as the administered dose deviates from the standard dosage. Three manufacturers of injection guns have responded to the call for tenders and submitted their guns for testing. The following tables present the results of two hundred individual tests conducted based on the administered dosage. To make an informed decision, the technical team must examine the results and determine which gun best meets the requirements.

Injector gun A	
Dosage (µl)	Frequency
[98 , 98,5)	2
[98,5 , 99)	7
[99 , 99,5)	21
[99,5 , 100)	53
[100 , 100,5)	87
[100,5 , 101)	30

Injector gun B	
Dosage (µl)	Frequency
[98 , 98,5)	4
[98,5 , 99)	9
[99 , 99,5)	31
[99,5 , 100)	47
[100 , 100,5)	57
[100,5 , 101)	52

Injector gun C	
Dosage (µl)	Frequency
[98 , 98,5)	9
[98,5 , 99)	11
[99 , 99,5)	38
[99,5 , 100)	90
[100 , 100,5)	45
[100,5 , 101)	7

Given the importance of minimizing the risk of toxicity to patients while ensuring that each vaccinated patient receives the required dose and is thus immunized, the validation team must make an informed decision regarding the choice of injection gun to use. Based on the results of the tests conducted, which of the three proposed injection guns is the best option to ensure the safety and efficacy of the vaccination process? This decision must be supported by verification calculations.

11.4.2 Questions

11.4.2.1 Analysis of central tendency measures for decision-making

For each of the injection guns,

1. Plot its histogram.
2. Plot its curve of cumulative frequencies.
3. Calculate the central tendency characteristics of the distribution:
   a. The mode.
   b. The average.
   c. The median.
4. Give an interpretation of the order of magnitude of the elements of this statistical series.

11.4.2.2 Analysis of measures of dispersion for decision-making

To verify the accuracy of the statements made by the three suppliers, it is required to calculate and compare the following measures of dispersion for the three-injection guns:

5. Their characteristics of dispersion:
    a. The interquartile range.
    b. The standard deviation.
    c. The coefficient of variation.
6. The percentage of dosages below average.
7. The percentage of dosages below average minus the standard deviation.
8. The percentage of dosages below average plus the standard deviation.
9. The percentage of dosages being in the range of the mean and the mean minus the standard deviation.
10. The percentage of dosages being in the range of the mean and the mean plus the standard deviation.
11. The percentage of dosages within the range of the mean plus the standard deviation and the mean minus the standard deviation.
12. The failure rates by calculating:
    13. The non-responsiveness rates.
    14. The toxicity rates.
    15. The failure rates.
13. The success rates using the items in question **12.**
14. The success rates using only the cumulative frequency curve.
15. The amount and percentage of serum administered in excess. of the admissible limit (theoretical quantity liable to toxicity) out of the 200 tests.

11.4.2.3 Analysis of shape measures for decision-making

To refine and consolidate the decision, it is asked to calculate their characteristics of shape.

16. The YULE's coefficient of skewness.
17. The PEARSON's coefficient of skewness.
18. The FISHER's coefficient of skewness.
19. The FISHER's coefficient of kurtosis.

11.4.3 Solutions

11.4.3.1 Analysis of central tendency measures for decision-making

A: Injection gun A

A-1: The histogram of injection gun **A** is:

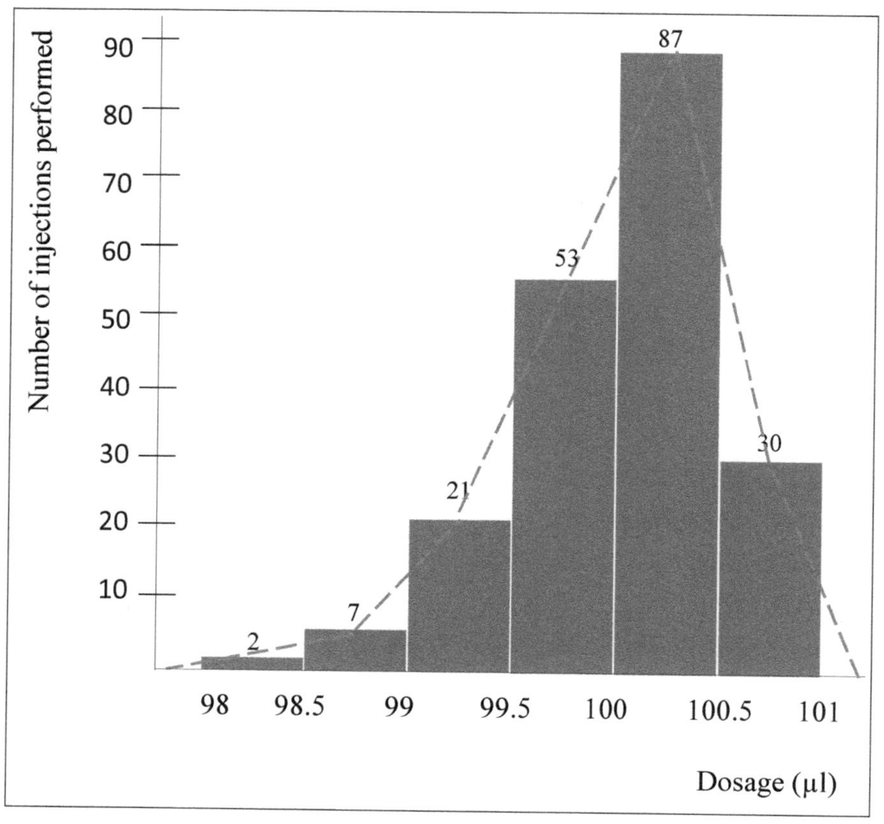

Solving this part of the exercise requires the following table:

Injection Gun A							
Dosage (μl)	x_i	n_i	$n_i x_i$	$n_i x_i^2$	f_i	$f_i \uparrow$	$f_i \downarrow$
[98 , 98,5 [	98.25	2	196.50	19,306.13	1.00	1.00	100.00
[98,5 , 99 [	98.75	7	691.25	68,260.94	3.50	4.50	99.00
[99 , 99,5 [	99.25	21	2,084.25	206,861.81	10.50	15.00	95.50
[99,5 , 100 [	99.75	53	5,286.75	527,353.31	26.50	41.50	85.00
[100 , 100,5 [	100.25	87	8,721.75	874,355.44	43.50	85.00	58.50
[100,5 , 101 [	100.75	30	3,022.50	304,516.88	15.00	100.00	15.00
	Total	200	20,003.00	2,000,654.50	100.00		

A-2: The cumulative relative frequency curve of injection gun **A** is:

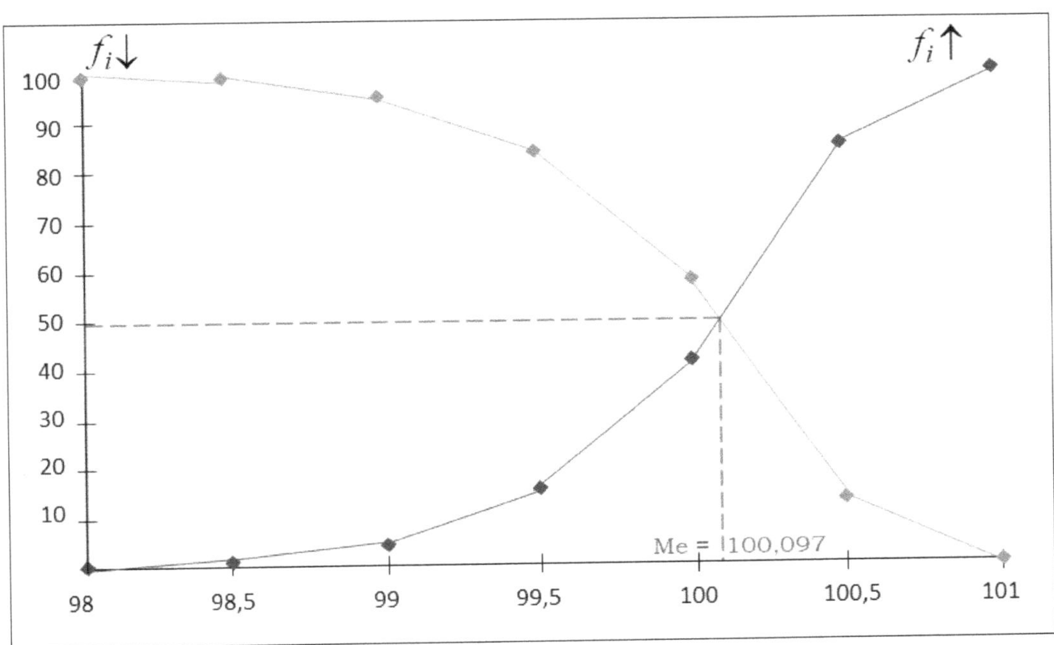

A-3: Determining the characteristics of central tendency:

A-3-a: The mode of injection gun **A** is 100.25 μl.

A-3-b: The average of injection gun **A** is:

$$\bar{x} = \frac{1}{n}\sum_{i=1}^{i=6} n_i x_i = \frac{20{,}003}{200}$$

$$= 100.015 \text{ μl}$$

A-3-c: The median of injection gun **A** is:

$$Me = a_i + (a_{i+1} - a_i)\frac{(50 - F_i)}{(F_{i+1} - F_i)}$$

$$= 100 + (100.5 - 100)\frac{(50 - 41.50)}{(85 - 41.50)}$$

$$= 100.097701149425 \text{ μl}$$

B: Injection gun B

B-1: The histogram of injection gun **B** is:

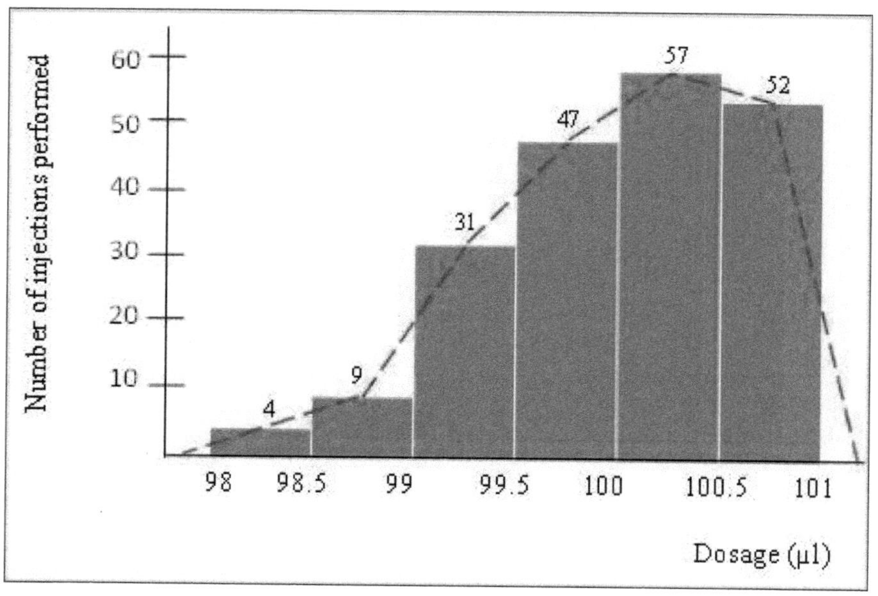

Resolving this part of the exercise requires the following table:

Injection Gun B							
Dosage (μl)	x_i	n_i	$n_i x_i$	$n_i x_i^2$	f_i	$f_i \uparrow$	$f_i \downarrow$
[98 , 98,5 [	98.25	4	393.00	38,612.25	2.00	2.00	100.00
[98,5 , 99 [	98.75	9	888.75	87,764.06	4.50	6.50	98.00
[99 , 99,5 [	99.25	31	3,076.75	305,367.44	15.50	22.00	93.50
[99,5 , 100 [	99.75	47	4,688.25	467,652.94	23.50	45.50	78.00
[100 , 100,5 [	100.25	57	5,714.25	572,853.56	28.50	74.00	54.50
[100,5 , 101 [	100.75	52	5,239.00	527,829.25	26.00	100.00	26.00
	Total	200	20,000.00	2,000,079.50	100.00		

B-2: The cumulative relative frequency curve of injection gun **B** is:

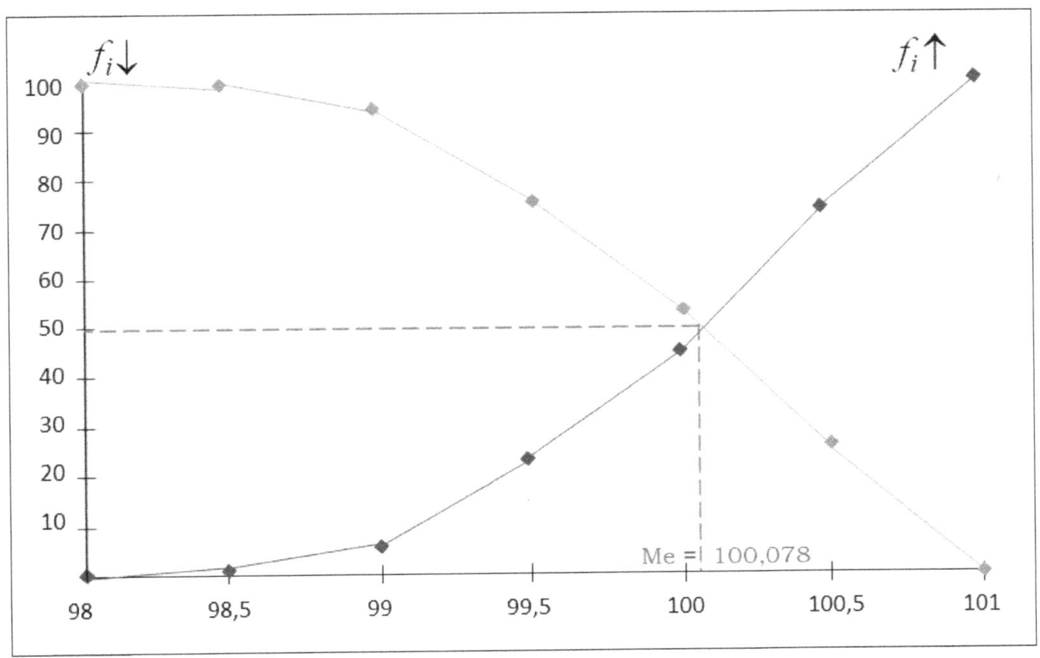

B-3: Determining the characteristics of central tendency:

B-3-a: The mode of injection gun **B** is 100.25 µl.

B-3-b: The average of injection gun **B** is:

$$\bar{x} = \frac{1}{n}\sum_{i=1}^{i=6} n_i x_i = \frac{20{,}000}{200}$$
$$= 100 \text{ µl}$$

B-3-c: The median of injection gun **B** is:

$$Me = a_i + (a_{i+1} - a_i)\frac{(50 - F_i)}{(F_{i+1} - F_i)}$$
$$= 100 + (100.5 - 100)\frac{(50 - 45.5)}{(74 - 45.5)}$$
$$= 100.078947368421 \text{ µl}$$

C: Injection gun C

C-1: The histogram of injection gun **C** is:

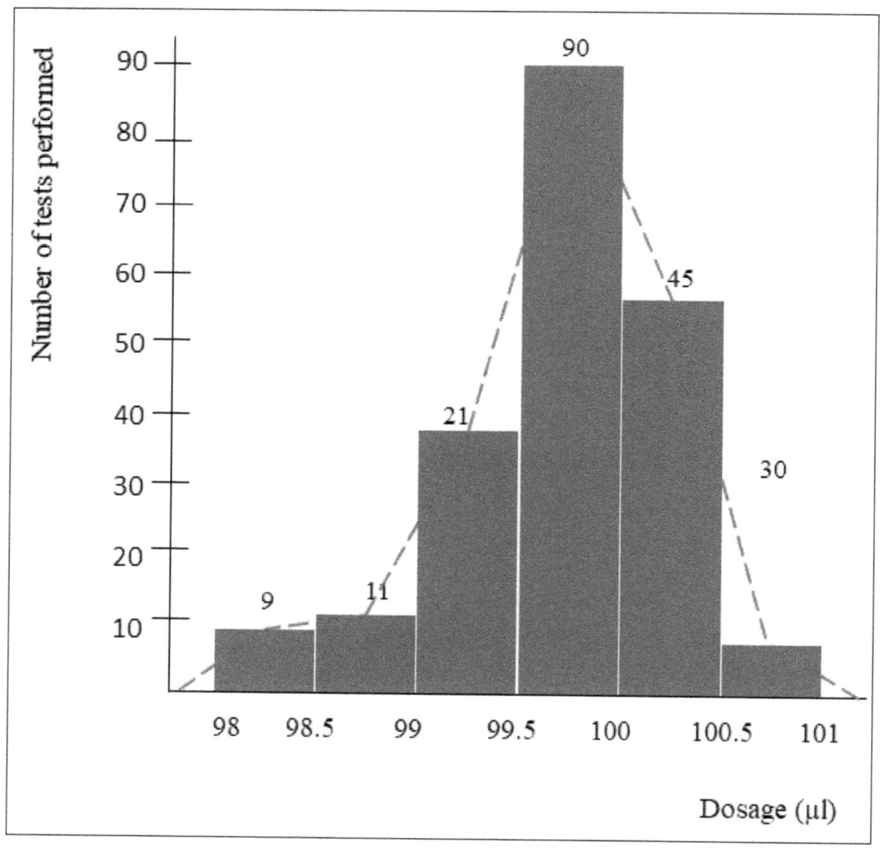

Resolving this part of the exercise requires the following table:

Injection Gun C							
Dosage (µl)	x_i	n_i	$n_i x_i$	$n_i x_i^2$	f_i	$f_i \uparrow$	$f_i \downarrow$
[98 , 98,5)	98.25	9	884.25	86,877.56	4.50	4.50	100.00
[98,5 , 99)	98.75	11	1,086.25	107,267.19	5.50	10.00	95.50
[99 , 99,5)	99.25	38	3,771.50	374,321.38	19.00	29.00	90.00
[99,5 , 100)	99.75	90	8,977.50	895,505.63	45.00	74.00	71.00
[100 , 100,5)	100.25	45	4,511.25	452,252.81	22.50	96.50	26.00
[100,5 , 101)	100.75	7	705.25	71,053.94	3.50	100.00	3.50
	Total	200	19,936.00	1,987,278.50	100.00		

C-2: The cumulative relative frequency curve of injection gun **C** is:

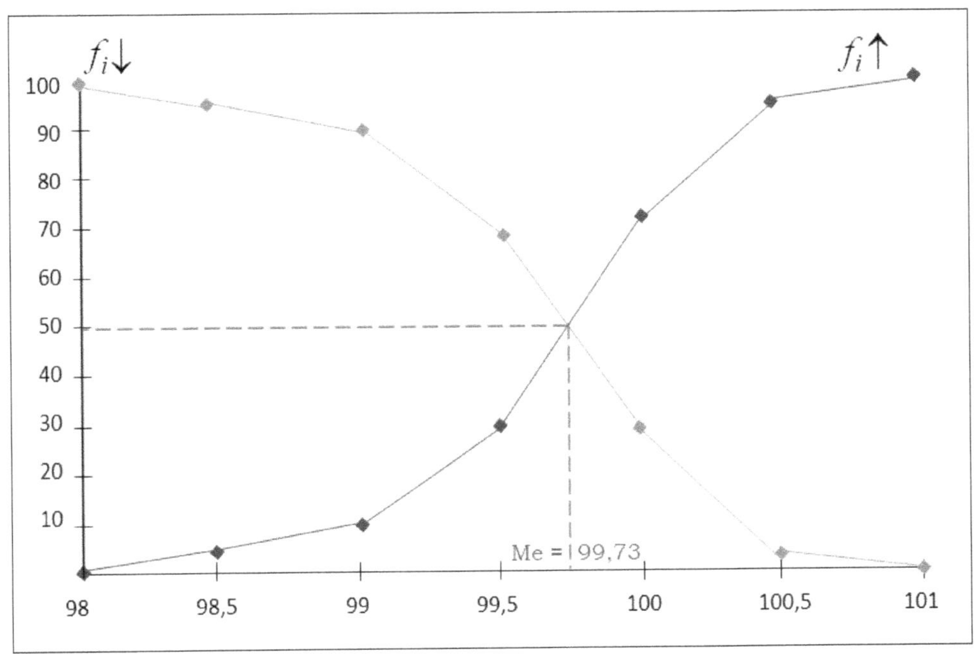

C-3 Determining the characteristics of central tendency:

C-3-a: The mode of injection gun **C** is 99.75 µl.

C-3-b: The average of injection gun **C** is:

$$\bar{x} = \frac{1}{n}\sum_{i=1}^{i=6} n_i x_i = \frac{19{,}936}{200}$$

$$= 99.68 \ \mu l$$

C-3-c: The median of injection gun **C** is:

$$Me = a_i + (a_{i+1} - a_i)\frac{(50 - F_i)}{(F_{i+1} - F_i)}$$

$$= 99.5 + (100 - 99.5)\frac{(50 - 29)}{(74 - 29)}$$

$$= 99.733 \ \mu l$$

4: Interpretation

INTERPRETATION OF RESULTS, ANALYSIS AND DECISION

By comparing the three parameters, we observe that:

	Injection gun A	Injection gun B	Injection gun C
	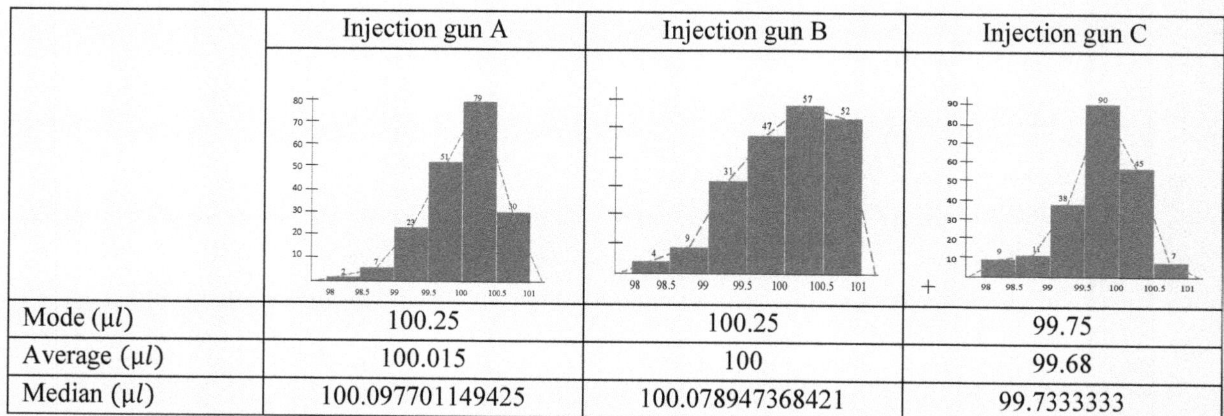		
Mode (µl)	100.25	100.25	99.75
Average (µl)	100.015	100	99.68
Median (µl)	100.097701149425	100.078947368421	99.7333333

Analysis

The quality control of three vaccine administration guns revealed the following information:

Gun **A**:

- The mode of Gun **A** is 100.25 µl, which means that the majority of the population is at risk of being intoxicated.
- However, the mean is slightly higher than 100 µl, which implies that a small proportion of the population could be slightly intoxicated.
- The median, on the other hand, is 100.0977 µl, which suggests that more than half of the population could be slightly intoxicated.

Gun **B**:

- The mode of Gun **B** is 100.25 µl, which means that the majority of the population is at risk of being intoxicated.
- The mean, however, is 100 µl, which implies that the average of the population will be perfectly immunized without any risk of toxicity.
- However, the median is 100.0789 µl, which suggests that more than half of the population could be slightly intoxicated.

Gun **C**:

- The mode of Gun **C** is 99.75 µl, which means that the majority of the population will be perfectly immunized without any risk of toxicity.
- The mean is 99.68 µl, which implies that the average of the population will be perfectly immunized without any risk of toxicity.
- The median is 99.733 µl, which suggests that more than half of the population will be perfectly immunized without any risk of toxicity.

Conclusion

The mode of gun **C** is 99.75 µl, which means that the majority of the population will be perfectly immunized without any risk of toxicity. Its mean is 99.68 µl, indicating that the average of the population will be perfectly immunized without any risk of toxicity. Its median, at 99.733 µl, also suggests that more than half of the population will be perfectly immunized without any risk of toxicity.

On the other hand, both gun **A** and gun **B** have modes of 100.25 µl, which means that the majority of the population risks being intoxicated. The mean of gun **A** is slightly higher than 100 µl, indicating a small proportion of the population may be slightly intoxicated. While the mean of gun **B** is 100 µl, implying that the average of the population will be perfectly immunized without any risk of toxicity, the median of gun **B** is 100.0789 µl, which suggests that more than half of the population could be slightly intoxicated.

Therefore, given the requirements for the gun and the data presented, gun **C** seems to be the best option. However, we will explore other characteristics to see if this trend holds.

11.4.3.2 Analysis of measures of dispersion for decision-making

A: Injection gun A

A-5: Characteristics of dispersion

The following table shows the data required to solve this exercise:

Injection Gun A							
Dosage (µl)	x_i	n_i	$n_i x_i$	$n_i x_i^2$	f_i	$f_i \uparrow$	$f_i \downarrow$
[98 , 98,5 [	98.25	2	196.50	19,306.13	1.00	1.00	100.00
[98,5 , 99 [	98.75	7	691.25	68,260.94	3.50	4.50	99.00
[99 , 99,5 [	99.25	21	2,084.25	206,861.81	10.50	15.00	95.50
[99,5 , 100 [	99.75	53	5,286.75	527,353.31	26.50	41.50	85.00
[100 , 100,5 [	100.25	87	8,721.75	874,355.44	43.50	85.00	58.50
[100,5 , 101 [	100.75	30	3,022.50	304,516.88	15.00	100.00	15.00
	Total	200	20,003.00	2,000,654.50	100.00		

A-5-1: The interquartile Range of injection gun **A** is:

$$Q_1 = a_i + (a_{i+1} - a_i)\frac{(25 - F_i)}{(F_{i+1} - F_i)}$$
$$= 99.5 + (100 - 99.5)\frac{(25 - 15)}{(41.5 - 15)}$$
$$= 99.688679245283 \text{ µl}$$
$$Q_3 = a_i + (a_{i+1} - a_i)\frac{(75 - F_i)}{(F_{i+1} - F_i)}$$
$$= 100 + (100.5 - 100)\frac{(75 - 41.5)}{(85 - 41.5)}$$
$$= 100.385057471264 \text{ µl}$$
$$\text{Interquartile Range} = Q_3 - Q_1$$
$$= 100.385057471264 - 99.688679245283$$
$$= 0.696378225981363 \text{ µl}$$

A-5-2: The variance of injection gun **A** is:

$$V(x) = \frac{1}{n}\sum_{i=1}^{i=6} n_i x_i^2 - \bar{x}^2$$
$$= \frac{2{,}000{,}654.50}{200} - (100.015)^2$$
$$= 0.2864583$$

A-5-3: The standard deviation of injection gun **A** is:

$$\sigma = \sqrt{V(x)}$$
$$= \sqrt{0.272274999999354}$$
$$= 0.52179977002616\ \mu\text{l}$$

A-5-4: The coefficient of variation of injection gun **A** is:

$$\text{CV} = \frac{\sigma}{|\bar{x}|}$$
$$= \frac{0.52179977002616}{100.015}\text{x}100$$
$$= 0.52\%$$

A-6: Let's calculate the percentage of dosages below the average:

$$\bar{x} = 100.015\ \mu\text{l}$$

Let $P_{<\bar{x}}$ be this percentage, we use the linear interpolation formula using the data pair $(a_i, f_i \uparrow)$:

$$\bar{x} = a_i + (a_{i+1} - a_i)\frac{(P_{<\bar{x}} - F_i)}{(F_{i+1} - F_i)}$$

$$100.015 = 100 + (100.5 - 100)\frac{(P_{<\bar{x}} - 41.5)}{(85 - 41.5)}$$

$$P_{<\bar{x}} = 42.805\%$$

A7: Let's calculate the percentage of dosages below $\bar{x} - \sigma$:

$$\bar{x} - \sigma = 100.015 - 0.52179977002616$$
$$= 99{,}4932002299738\ \mu\text{l}$$

Let $P_{<\bar{x}-\sigma}$ be this percentage, we use the linear interpolation formula using the data pair $(a_i, f_i \uparrow$ $)$:

$$\bar{x} - \sigma = a_i + (a_{i+1} - a_i)\frac{(P_{<\bar{x}-\sigma} - F_i)}{(F_{i+1} - F_i)}$$

$$99.4932002299738 = 99 + (99.5 - 99)\frac{(P_{<\bar{x}-\sigma} - 4.5)}{(15 - 4.5)}$$

$$P_{<\bar{x}-\sigma} = 14.8572\%$$

A-8: Let's calculate the percentage of dosages below $\bar{x} + \sigma$:

$$\bar{x} + \sigma = 100.015 + 0.52179977002616$$
$$= 100.53679977002616 \text{ µl}$$

Let $P_{<\bar{x}+\sigma}$ be this percentage, we use the linear interpolation formula using the data pair $(a_i, f_i \uparrow)$:

$$\bar{x} + \sigma = a_i + (a_{i+1} - a_i)\frac{(P_{<\bar{x}+\sigma} - F_i)}{(F_{i+1} - F_i)}$$

$$100.53679977002616 = 100.5 + (101 - 100.5)\frac{(P_{<\bar{x}+\sigma} - 85)}{(100 - 85)}$$

$$P_{<\bar{x}+\sigma} = 86.1037\%$$

A-9: Let's calculate the percentage of dosages between the mean, i.e., 100.015 cm and the mean minus the standard deviation, i.e., 99.4932002299738 µl.

This percentage is:

$$42.805\% - 14.8572\% = 27.9478\%$$

A-10: Let's calculate the percentage of dosages between the mean, i.e., 100.015 cm and the mean plus the standard deviation, i.e., 100.53679977002616 µl.

This percentage is:

$$86.1037\% \; - \; 42.805\% = 43.2987\%$$

A-11: Let's calculate the percentage of dosages between the mean minus the standard deviation, which is 99.4932002299738 µl and the mean plus the standard deviation, which is 100.53679977002616 µl.

This percentage is:

$$43.2987\% - 27.9478\% = 71.2465\%$$

The same result can be obtained with:

$$86.1037\% \ - \ 14.8572\% = 71.2465\%$$

A.12: Determining the failure rate of gun **A**

A-12-a: Let's first determine the non-responsiveness rate by calculating the percentage of the ineffective, but not toxic dosage, i.e., the percentage of the dosage strictly less than 99 µl.

This percentage can be obtained directly from the distribution table on the intersection of the column of increasing cumulative frequencies with the line of the interval $[98{,}5\,;99[$, this value is 4.5%.

A-12-b: We then calculate its toxicity rate by calculating the percentage of its toxic dosage, i.e., the percentage of the dosage greater than 100 µl.

This percentage can be obtained directly by intersecting the column of cumulative increasing frequencies with the row corresponding to the interval $[100\,;100.5[$ in the distribution table. This percentage is 58.5%.

A-12-c: Let's now calculate its failure rate:

$$\begin{aligned}\text{Failure Rate} &= \text{Non} - \text{responsiveness rate} + \text{Toxicity rate}\\ &= 4.5\% + 58.5\%\\ &= 63\%\end{aligned}$$

The failure rate of gun **A** is 63%

A-13: Deduce the success rate of gun **A**

$$\begin{aligned}\text{Success Rate} &= 100 - \text{Failure Rate}\\ &= 100\% - 63\%\\ &= 37\%\end{aligned}$$

The success rate of gun **A** is 37%

A-14: Determining the success rate of gun A using only the cumulative frequency curve.

We know that the success interval is the one that starts from the beginning of the effectiveness, that is to say 99 µl until the end of the effectiveness, which corresponds to the beginning of the toxicity that is 100 µl excluded.

A-14-a: We first need the percentage of the dosage strictly less than 99 µl.

As in point **A-12-a**, this percentage equivalent to 4.5% can be obtained directly from the distribution table on the intersection of the column of increasing cumulative frequencies with the line of the interval [98,5 ; 99[.

A-14-b: Next, we need the percentage of dosages strictly below 100 µl.

This percentage can also be obtained directly from the distribution table on the intersection of the column of increasing cumulative frequencies with the line of the interval [98,5 ; 99[this value is 41.5%.

A-14-c: The percentage of success is therefore equivalent to the difference of these two percentages:

$$41.5\% - 4.5\% = 37\%$$

The success rate of pistol **A** is 37%, so we have found the same value as in point **A.13.**

A-15: Let's calculate the amount and percentage of serum administered in excess of the admissible limit (theoretical quantity liable to toxicity) out of the 200 tests:

- Maximum ideal dosage of an injection: 100 µl.
- Number of injections administered (n_i)= 200.
- Maximum ideal total amount of serum administered = 200 x 100 µl = 20,000 µl.
- Total amount of serum administered =20,003 µl.
- Amount of serum in excess of the admissible limit = 20,003 µl – 20,000 µl = 3 µl.
- Percentage of serum in excess of the admissible limit = $\frac{3}{20,000} = 0.015\%$.

Injection gun B

B-5: Characteristics of dispersion

The following table shows the data required to solve this exercise:

Injection Gun B							
Dosage (µl)	x_i	n_i	$n_i x_i$	$n_i x_i^2$	f_i	$f_i \uparrow$	$f_i \downarrow$
[98 , 98,5 [	98.25	4	393.00	38,612.25	2.00	2.00	100.00
[98,5 , 99 [	98.75	9	888.75	87,764.06	4.50	6.50	98.00
[99 , 99,5 [	99.25	31	3,076.75	305,367.44	15.50	22.00	93.50
[99,5 , 100 [	99.75	47	4,688.25	467,652.94	23.50	45.50	78.00
[100 , 100,5 [	100.25	57	5,714.25	572,853.56	28.50	74.00	54.50
[100,5 , 101 [	100.75	52	5,239.00	527,829.25	26.00	100.00	26.00
	Total	200	20,000.00	2,000,079.50	100.00		

B-5-1: The interquartile range of injection gun **B** is:

$$Q_1 = a_i + (a_{i+1} - a_i)\frac{(25 - F_i)}{(F_{i+1} - F_i)}$$
$$= 99{,}5 + (100 - 99{,}5)\frac{(25 - 22)}{(45{,}5 - 22)}$$
$$= 99.563829787234 \text{ µl}$$
$$Q_3 = a_i + (a_{i+1} - a_i)\frac{(75 - F_i)}{(F_{i+1} - F_i)}$$
$$= 100.5 + (101 - 100.5)\frac{(75 - 74)}{(100 - 74)}$$
$$= 100.51923076923 \text{ µl}$$
$$\text{The interquartile range } = Q_3 - Q_1$$
$$= 100.51923076923 - 99.563829787234$$
$$= 0.955400981996732 \text{ µl}$$

B-5-2: The variance of injection gun **B** is:

$$V(x) = \frac{1}{n}\sum_{i=1}^{i=6} n_i x_i^2 - \bar{x}^2$$
$$= \frac{2{,}000{,}079.5}{200} - (100)^2$$
$$= 0.397499999999127$$

B-5-3: The standard deviation of injection gun **B** is:

$$\sigma = \sqrt{V(x)}$$
$$= \sqrt{0.397499999999127}$$
$$= 0.630476010645232 \text{ µl}$$

B-5-4: The coefficient of variation of injection gun **B** is:

$$\text{CV} = \frac{\sigma}{|\bar{x}|}$$
$$= \frac{0.630476010645232}{100} \text{x}100$$
$$= 0.630476010645232\%$$

B-6: Let's calculate the percentage of the dosages less than the average.

This percentage can be obtained directly from the table, it is:

$$P_{<\bar{x}} = 41.5\%$$

B-7: Let's calculate the percentage of dosages whose less than $\bar{x} - \sigma$

$$\bar{x} - \sigma = 100 - 0.630476010645232$$
$$= 99.369523989354768 \text{ µl}$$

Let $P_{<\bar{x}-\sigma}$ be this percentage, we use the linear interpolation formula using the data pair $(a_i, f_i \uparrow)$:

$$\bar{x} - \sigma = a_i + (a_{i+1} - a_i)\frac{(P_{<\bar{x}-\sigma} - F_i)}{(F_{i+1} - F_i)}$$
$$99.369523989354768 = 99 + (99.5 - 99)\frac{(P_{<\bar{x}-\sigma} - 6.5)}{(22 - 6.5)}$$

$$P_{<\bar{x}-\sigma} = 17.95512\%$$

B-8: Let's calculate the percentage of dosages less than $\bar{x} + \sigma$

$$\bar{x} + \sigma = 100 + 0.630476010645232$$
$$= 100.630476010645232 \text{ µl}$$

Let $P_{<\bar{x}+\sigma}$ be this percentage, we use the linear interpolation formula using the data pair $(a_i, f_i \uparrow)$ as proceeded above:

$$\bar{x} + \sigma = a_i + (a_{i+1} - a_i)\frac{(P_{<\bar{x}+\sigma} - F_i)}{(F_{i+1} - F_i)}$$
$$100.630476010645232 = 100.5 + (101 - 100.5)\frac{(P_{<\bar{x}+\sigma} - 74)}{(100 - 74)}$$
$$P_{<\bar{x}+\sigma} = 80.78444\%$$

B-9: Let's calculate the percentage of dosages between the mean, which is 100 µl and the mean minus the standard deviation, which is 99.369523989354768 µl.

This percentage is:

$$41.5\% - 17.95512\% = 23.54488\%$$

B-10: Let's calculate the percentage of dosages between the mean, i.e., 100 µl and the mean plus the standard deviation, i.e.,100.630476010645232 µl.

This percentage is:

$$80.78444\% \ - \ 41.5\% = 39.28444\%$$

B-11: Let's calculate the percentage of dosages between the mean minus the standard deviation, i.e., 99.369523989354768 µl and the mean plus the standard deviation, i.e., 100.630476010645232 µl.

This percentage is:

$$80.78444\ \% - 17.95512\% = 62.82932\%$$

B-12: Determining the failure rate of gun **B**

B-12-a: Let's first determine the non-responsiveness rate by calculating the percentage of the ineffective, but not toxic dosage, i.e., the percentage of the dosage strictly less than 99 µl.

This percentage can be obtained directly by intersecting the column of cumulative increasing frequencies with the row corresponding to the interval $[99{,}5\,;100[$ in the distribution table. This percentage is 6.5%.

B-12-b: We then calculate its toxicity rate by calculating the percentage of its toxic dosage, i.e., the percentage of the dosage greater than 100 µl.

This percentage can be obtained directly from the distribution table on the column of decreasing cumulative frequencies which corresponds to the interval $[100\,; 100{,}5[$ this value is 54.5%.

B-12-c: Let's now calculate its failure rate.

$$\begin{aligned}\text{Failure Rate} &= \text{Non responsiveness rate} + \text{Toxicity rate}\\ &= 6.5\% + 54.5\%\\ &= 61\%\end{aligned}$$

The failure rate of gun **B** is 61%

B-13: Deduce the success rate of gun **B**

$$\begin{aligned}\text{Success Rate} &= 100 - \text{Failure Rate}\\ &= 100\% - 61\%\\ &= 39\%\end{aligned}$$

The success rate of gun **B** is 39%

B-14: Determining the success rate of gun **B** using only the cumulative frequency curve.

We know that the success interval is the one that starts from the beginning of the effectiveness, that is to say 99 µl until the end of the effectiveness, which corresponds to the beginning of the toxicity that is 100 µl excluded.

B-14-a: We first need the percentage of the dosage strictly less than 99 µl.

As in point **B-12-a**, this percentage equivalent to 6.5% can be obtained directly from the distribution table on the intersection of the column of increasing cumulative frequencies with the line of the interval [98,5 ; 99[.

B-14-b: Next we need the percentage of dosages strictly below 100 µl.

This percentage can also be obtained directly from the distribution table on the intersection of the column of increasing cumulative frequencies with the line of the interval [98,5 ; 99[this value is 45.5%.

B-14-c: The percentage of success is therefore equivalent to the difference of these two percentages:

$$45.5\% - 6.5\% = 39\%$$

The success rate of pistol **B** is 39%, so we have found the same value as in point **B.13.**

B-15: Let's calculate the amount and percentage of serum administered in excess of the admissible limit (theoretical quantity liable to toxicity) out of the 200 tests:

- Maximum ideal dosage of an injection: 100 µl.
- Number of injections administered (n_i)= 200.
- Maximum ideal total amount of serum administered = 200 x 100 µl = 20,000 µl.
- Total amount of serum administered =20,000 µl.
- Amount of serum in excess of the admissible limit = 20,000 µl – 20,000 µl = 3 µl.
- Percentage of serum in excess of the admissible limit = $\frac{0}{20,000} = 0\%$.

C: Injection gun C

C-5: Characteristics of dispersion

The following table shows the data required to solve this exercise:

Injection Gun C							
Dosage (µl)	x_i	n_i	$n_i x_i$	$n_i x_i^2$	f_i	$f_i \uparrow$	$f_i \downarrow$
[98 , 98,5)	98.25	9	884.25	86,877.56	4.50	4.50	100.00
[98,5 , 99)	98.75	11	1,086.25	107,267.19	5.50	10.00	95.50
[99 , 99,5)	99.25	38	3,771.50	374,321.38	19.00	29.00	90.00
[99,5 , 100)	99.75	90	8,977.50	895,505.63	45.00	74.00	71.00
[100 , 100,5)	100.25	45	4,511.25	452,252.81	22.50	96.50	26.00
[100,5 , 101)	100.75	7	705.25	71,053.94	3.50	100.00	3.50
	Total	200	19,936.00	1,987,278.50	100.00		

C-5-1: The interquartile range of injection gun **C** is:

$$Q_1 = a_i + (a_{i+1} - a_i)\frac{(25 - F_i)}{(F_{i+1} - F_i)}$$

$$= 99 + (99.5 - 99)\frac{(25 - 10)}{(29 - 10)}$$

$$= 99.3947368421053 \text{ µl}$$

$$Q_3 = a_i + (a_{i+1} - a_i)\frac{(75 - F_i)}{(F_{i+1} - F_i)}$$

$$= 100 + (100.5 - 100)\frac{(75 - 74)}{(96.5 - 74)}$$

$$= 100.022222222222 \text{ µl}$$

$$\text{The interquartile range} = Q_3 - Q_1$$

$$= 100.022222222222 - 99.3947368421053$$

$$= 0.627485380116966 \text{ µl}$$

C-5-2: The variance of injection gun **C** is:

$$V(x) = \frac{1}{n}\sum_{i=1}^{i=6} n_i x_i^2 - \bar{x}^2$$

$$= \frac{1\,987\,278.5}{200} - (99.68)^2$$

$$= 0.290099999998347$$

C-5-3: The standard deviation of injection gun **C** is:

$$\sigma = \sqrt{V(x)}$$

$$= \sqrt{0.290099999998347}$$
$$= 0.538609320378275 \text{ µl}$$

C-5-4: The coefficient of variation of injection gun **C** is:

$$CV = \frac{\sigma}{|\bar{x}|}$$
$$= \frac{0.538609320378275}{99.68} \text{x100}$$
$$= 0.5403384032687\%$$

C-6: Let's calculate the percentage of dosages less than average.

$$\bar{x} = 99.68 \text{ µl}$$

Let $P_{<\bar{x}}$ be this percentage, we use the linear interpolation formula on the couple $(a_i, f_i \uparrow)$, in the same way that we used for the median:

$$\bar{x} = a_i + (a_{i+1} - a_i)\frac{(P_{<-\sigma} - F_i)}{(F_{i+1} - F_i)}$$
$$99.68 = 99.5 + (100 - 99.5)\frac{(P_{<\bar{x}} - 29)}{(74 - 29)}$$
$$P_{<\bar{x}} = 45.2\%$$

C-7: Let's calculate the percentage of dosages less than $\bar{x} - \sigma$

$$\bar{x} - \sigma = 99.68 - 0.538609320378275$$
$$= 99.141390679621725 \text{ µl}$$

Let $P_{<\bar{x}-\sigma}$ be this percentage, we use the linear interpolation formula using the data pair $(a_i, f_i \uparrow)$ as proceeded above:

$$\bar{x} - \sigma = a_i + (a_{i+1} - a_i)\frac{(P_{<\bar{x}-\sigma} - F_i)}{(F_{i+1} - F_i)}$$
$$99.141390679621725 = 99 + (99.5 - 99)\frac{(P_{<\bar{x}-\sigma} - 10)}{(29 - 10)}$$
$$P_{<\bar{x}-\sigma} = 15.37282\%$$

C-8: Let's calculate the percentage of dosage less than $\bar{x} + \sigma$

$$\bar{x} + \sigma = 99.68 + 0.538609320378275$$

$$= 100.218609320378275 \text{ µl}$$

Let $P_{<\bar{x}+\sigma}$ be this percentage, we use the linear interpolation formula using the data pair $(a_i, f_i \uparrow)$ as proceeded above:

$$\bar{x} + \sigma = a_i + (a_{i+1} - a_i)\frac{(P_{<\bar{x}-\sigma} - F_i)}{(F_{i+1} - F_i)}$$

$$100.218609320378275 = 100 + (100.5 - 100)\frac{(P_{<\bar{x}+\sigma} - 74)}{(96.5 - 74)}$$

$$P_{<\bar{x}+\sigma} = 83.837\%$$

C-9: Let's calculate the percentage of dosages between the mean, i.e., 99,68 µl and the mean minus the standard deviation, i.e., 99.141390679621725 µl.

This percentage is:

$$45.2\% - 15.37282\% = 29.82718\%$$

C-10: Let's calculate the percentage of dosages between the mean, i.e., 99.68 µl and the mean plus the standard deviation, or.100.218609320378275 µl.

This percentage is:

$$83.837\ \% \ - \ 45.2\% = 38.637\%$$

C-11: Let's calculate the percentage of dosages between the mean minus the standard deviation, i.e., 99.141390679621725 µl and the mean plus the standard deviation, i.e., 100.218609320378275 µl.

This percentage is:

$$29.82718\% + 38.637\% = 68.46418\%$$

C-12: Determining the failure rate of gun **C**

C-12-a: Let's first determine the non-responsiveness rate by calculating the percentage of the ineffective, but not toxic dosage, i.e., the percentage of the dosage strictly less than 99 µl.

This percentage can be obtained directly by intersecting the column of cumulative increasing frequencies with the row corresponding to the interval [99,5 ; 99[in the distribution table. This percentage is 10%.

C-12-b: We then calculate its toxicity rate by calculating the percentage of its toxic dosage, i.e., the percentage of the dosage greater than 100 μl.

This percentage can be obtained directly from the distribution table on the column of decreasing cumulative frequencies which corresponds to the interval [100 ; 100,5[this value is 26%.

C-12-c: Let's now calculate its failure rate.

$$\begin{aligned}\text{Failure Rate} &= \text{Non responsiveness rate} + \text{Toxicity rate}\\ &= 10\% + 26\%\\ &= 36\%\end{aligned}$$

The failure rate of gun **C** is 36%

C-13: Deduce the success rate of gun **C**

$$\begin{aligned}\text{Success Rate} &= 100 - \text{Failure Rate}\\ &= 100\% - 36\%\\ &= 64\%\end{aligned}$$

The success rate of gun **C** is 64%

C-14: Determining the success rate of gun **C** using only the cumulative frequency curve.

We know that the success interval is the one that starts from the beginning of the effectiveness, that is to say 99 μl until the end of the effectiveness, which corresponds to the beginning of the toxicity that is 100 μl excluded.

C-14-a: We first need the percentage of the dosage strictly less than 99 μl.

As in point **C-12-a**, this percentage equivalent to 10% can be obtained directly from the distribution table on the intersection of the column of increasing cumulative frequencies with the line of the interval [98,5 ; 99[.

C-14-b: Next, we need the percentage of dosages strictly below 100 μl.

This percentage can also be obtained directly from the distribution table on the intersection of the column of increasing cumulative frequencies with the line of the interval [98,5 ; 99[this value is 74%.

C-14-c: The percentage of success is therefore equivalent to the difference of these two percentages:

$$74\% - 10\% = 64\%$$

The success rate of pistol **C** is 64%, so we have found the same value as in point **C.13.**

C-15: Let's calculate the amount and percentage of serum administered in excess of the admissible limit (theoretical quantity liable to toxicity) out of the 200 tests:

- Maximum ideal dosage of an injection: 100 µl.
- Number of injections administered (n_i)= 200.
- Maximum ideal total amount of serum administered = 200 x 100 µl = 20,000 µl.
- Total amount of serum administered = 19 936 µl.

Amount of serum in excess of the admissible limit = 19 936 µl – 20,000 µl = –64 µl.

- Percentage of serum in excess of the admissible limit = $\frac{-64}{20\,000} = -0{,}32\%$.

INTERPRETATION OF RESULTS, ANALYSIS AND DECISION

Analysis

	Injection gun A	Injection gun B	Injection gun C
Mode (µl)	100.25	100.25	99.75
Average (µl)	100.015	100	99.68
Median (µl)	100.0977	100.0789	99.7333
% Pop.< Majority	43.5%	28.5%	45%
% Toxicity	58.5%	54.5%	26%
%Non-responsiveness	4.5%	6.5%	10%
%Failure	63%	61%	36%
%Efficiency	37%	39%	64%

Gun A

- 37% of the population will be perfectly immunized and not intoxicated while 63% will not be.

Gun B

- 39% of the population will be perfectly immunized and not intoxicated while 61% will not be.

Gun C

- 64% of the population will be perfectly immunized and not intoxicated while 36% will not be.

These results show that the three tested injection guns are not all equivalent in terms of efficiency and safety for vaccination. The vaccination effectiveness results vary significantly between the three guns, with a success rate ranging from 37% for gun A to 64% for gun C. Similarly, the toxicity rates also differ between the three guns, with a higher risk for guns A and B, which have a higher probability of administering an inadequate vaccine dose. It is important to note that the safety and effectiveness of vaccination are crucial for preventing the spread of infectious diseases and protecting public health. The results indicate that gun C appears to be the safest, most effective, and least toxic choice of the three for vaccination, but it is also important to consider the proportion of the population that will not be perfectly immunized with this gun.

CV	0.52%	0.630%	0.540%

The three guns each have a coefficient of variation of less than 1% and around 0.50%, so their distributions are almost homogeneous, and they do not have too much immediate influence on the decision.

% in $[\bar{x} - \sigma, \bar{x} + \sigma]$	71.2465%	62.82932%	68.46418%
% in [99µl;100µl]	70%	52%	67%
	0.015%	0%	-3.2%

Gun A

- Gun A has a 37% effectiveness rate with a 63% miss rate.
- Gun A has a high toxicity risk.

Gun B

- Gun B has a 39% effectiveness rate with a 61% miss rate.
- Gun B has a high toxicity risk.

Gun C

- Gun C has a 64% effectiveness rate with a 36% miss rate.
- Gun C has a low toxicity risk.

These results show that pistols A and B have relatively low effectiveness rates, with a high risk of toxicity for both. This means they are less effective in immunizing the population against infectious diseases and may pose risks to patients' health.

On the other hand, pistol C has a higher effectiveness rate and a lower risk of toxicity, making it the preferred choice for vaccination. However, it is important to note that even with pistol C, there is still a proportion of the population that will not be perfectly immunized.

Conclusion

Based on the data presented above, the gun **C** appears to be the most favorable option for vaccination compared to the other two gun. It has the highest percentage of the population that will be perfectly immunized and not intoxicated, with an efficacy rate of 64% and a low toxicity risk. Additionally, it has the second-highest percentage of individuals within the interval $[\bar{x} - \sigma, \bar{x} + \sigma]$, which indicates that the vaccine doses administered by this gun are more consistent and accurate than those administered by the other two.

It is important to note that the effectiveness and safety of vaccination are critical in preventing the spread of infectious diseases and protecting public health. The results indicate that the gun **C** is the safest, most effective, and least toxic option for vaccination. However, it is also important to consider the percentage of the population that will not be perfectly immunized with this gun, which is 36%. This means that some individuals may still be susceptible to the disease, and additional measures may be necessary to prevent its spread.

11.4.3.3 Analysis of shape measures for decision-making

Calculation of the characteristics of shape.

A: Injection gun **A**

The following table shows the data required to solve this exercise:

x_i	n_i	$n_i x_i$	$n_i(x_i - \bar{x})^3$	$n_i(x_i - \bar{x})^4$
98.25	2	196.50	-11.00	19.41
98.75	7	691.25	-14.17	17.93
99.25	21	2084.25	-9.40	7.19
99.75	53	5286.75	-0.99	0.26
100.25	87	8721.75	1.13	0.27
100.75	30	3022.50	11.91	8.76
	200	20003.00	-22.51365000	53.80854463

A-16: Let's calculate the YULE's coefficient of skewness:

$$C_Y = \frac{Q_1 - 2Q_2 + Q_3}{Q_3 - Q_1}$$

$$= \frac{99.68867924528 - 2(100.09770114) + 100.3850574712}{100.3850574712 - 99.68867924528}$$

$$= -0.17473$$

A-17: Let's calculate the PEARSON's coefficient of skewness:

$$\beta_1 = \frac{(\bar{x} - \text{ Mode})}{\sigma}$$

$$= \frac{(100.015 - 100.25)}{0.52179977}$$

$$= -0.450364323$$

A-18: Let's calculate the FISHER's coefficient of skewness:

$$\mu_3 = \frac{\sum_{i=1}^{i=6} n_i (x_i - \bar{x})^3}{\sum_{i=1}^{i=6} n_i} = \frac{-22.51365}{200}$$

$$= -0.1125682497$$

$$\gamma_1 = \frac{\mu_3}{\sigma^3} = \frac{-0.1125682497}{(0.52179977)^3}$$

$$= -0.792327$$

A-19: Let's calculate the FISHER's coefficient of kurtosis:

$$\mu_4 = \frac{\sum_{i=1}^{i=6} n_i (x_i - \bar{x})^4}{\sum_{i=1}^{i=6} n_i} = \frac{53.80854463}{200}$$

$$= 0.26904278994$$

$$\gamma_2 = \frac{\mu_4}{\sigma^4} - 3 = \frac{0.26904278994}{(0.52179977)^4} - 3$$

$$= -0.6291576$$

B: Injection gun **B**

The following table shows the data required to solve this exercise:

x_i	n_i	$n_i x_i$	$n_i(x_i - \bar{x})^3$	$n_i(x_i - \bar{x})^4$
98.25	4	393.00	-21.44	37.52
98.75	9	888.75	-17.58	21.97
99.25	31	3076.75	-13.08	9.81
99.75	47	4688.25	-0.73	0.18
100.25	57	5714.25	0.89	0.22
100.75	52	5239.00	21.94	16.45
	200	20000.00	-30.00000000	86.15625000

B-16: Let's calculate the YULE's coefficient of skewness:

$$C_Y = \frac{Q_1 - 2Q_2 + Q_3}{Q_3 - Q_1}$$
$$= \frac{99.56382978723 - 2(100.07894736) + 100.5192307692}{100.5192307692 - 99.56382978723}$$
$$= -0.078327493$$

B-17: Let's calculate the PEARSON's coefficient of skewness:

$$\beta_1 = \frac{(\bar{x} - \text{ Mode})}{\sigma}$$
$$= \frac{(100 - 100.25)}{0.630476}$$
$$= -0.078327493$$

B-18: Let's calculate the FISHER's coefficient of skewness:

$$\mu_3 = \frac{\sum_{i=1}^{i=6} n_i (x_i - \bar{x})^3}{\sum_{i=1}^{i=6} n_i} = \frac{-30}{200}$$
$$= -0.15$$

$$\gamma_1 = \frac{\mu_3}{\sigma^3} = \frac{-0.15}{(0.630476)^3}$$
$$= -0.598529$$

B-19: Let's calculate the FISHER's coefficient of kurtosis:

$$\mu_4 = \frac{\sum_{i=1}^{i=6} n_i(x_i - \bar{x})^4}{\sum_{i=1}^{i=6} n_i} = \frac{86.15625}{200}$$

$$= -0.430781$$

$$\gamma_2 = \frac{\mu_4}{\sigma^4} - 3 = \frac{-0.430781}{(0.630476)^4} - 3$$

$$= -0.2736443$$

C: Injection gun **C**

The following table shows the data required to solve this exercise:

x_i	n_i	$n_i x_i$	$n_i(x_i - \bar{x})^3$	$n_i(x_i - \bar{x})^4$
98.25	9	884.25	-26.32	37.63
98.75	11	1086.25	-8.85	8.23
99.25	38	3771.50	-3.02	1.30
99.75	90	8977.50	0.03	0.00
100.25	45	4511.25	8.33	4.75
100.75	7	705.25	8.58	9.18
	200	19936.00	-21.24720000	61.09019400

C-16: Let's calculate the YULE's coefficient of skewness:

$$C_Y = \frac{Q_1 - 2Q_2 + Q_3}{Q_3 - Q_1}$$

$$= \frac{99.39473684210 - 2(99.7333) + 100.0222}{100.0222 - 99.39473684210}$$

$$= -0.079149122$$

C-17: Let's calculate the PEARSON's coefficient of skewness:

$$\beta_1 = \frac{(\bar{x} - \text{Mode})}{\sigma}$$

$$= \frac{(99.78 - 99.75)}{0.5386}$$

$$= -0.129964331$$

C-18: Let's calculate the FISHER's coefficient of skewness:

$$\mu_3 = \frac{\sum_{i=1}^{i=6} n_i (x_i - \bar{x})^3}{\sum_{i=1}^{i=6} n_i} = \frac{-21.2472}{200}$$

$$= -0.1062359998$$

$$\gamma_1 = \frac{\mu_3}{\sigma^3} = \frac{-0.1062359998}{(0.5386)^3}$$

$$= -0.679908$$

C-19: Let's calculate the FISHER's coefficient of kurtosis:

$$\mu_4 = \frac{\sum_{i=1}^{i=6} n_i (x_i - \bar{x})^4}{\sum_{i=1}^{i=6} n_i} = \frac{61.090194}{200}$$

$$= 0.30545091629$$

$$\gamma_2 = \frac{\mu_4}{\sigma^4} - 3 = \frac{0.30545091629}{(0.5386)^4} - 3$$

$$= 0.6294931$$

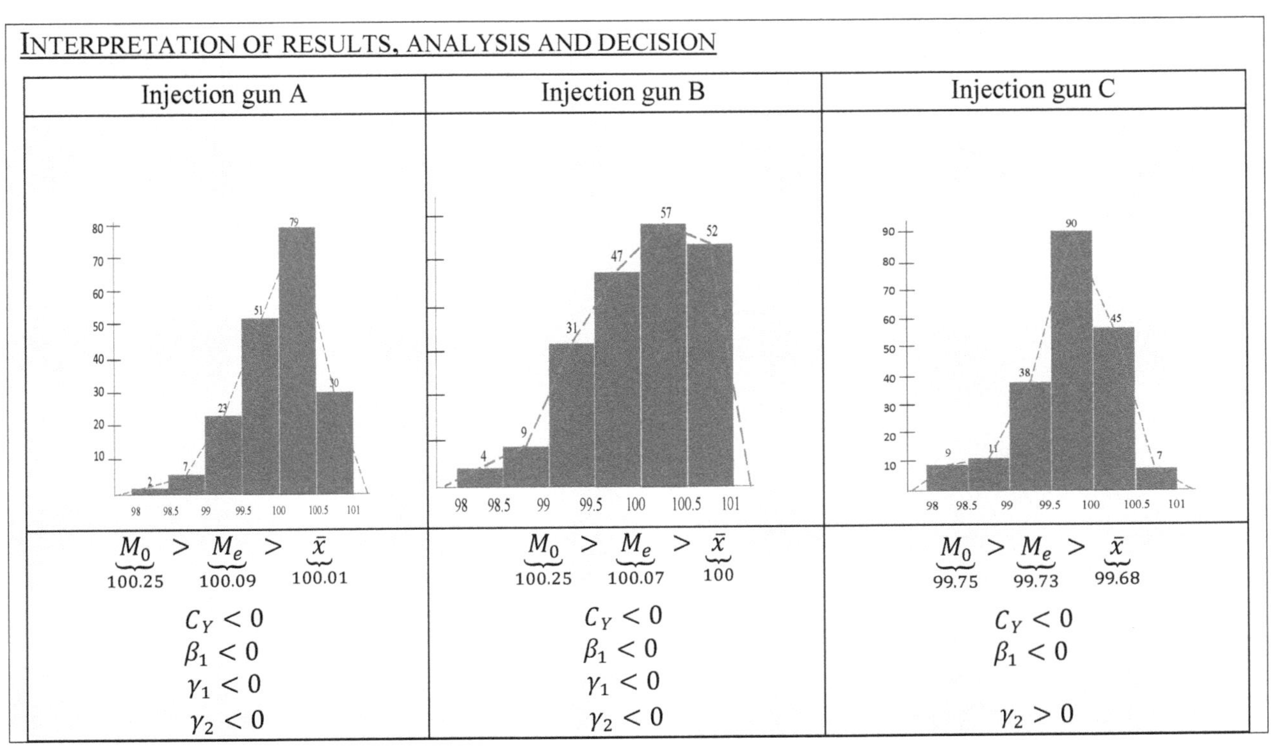

Injection gun A	Injection gun B	Injection gun C
$\underbrace{M_0}_{100.25} > \underbrace{M_e}_{100.09} > \underbrace{\bar{x}}_{100.01}$	$\underbrace{M_0}_{100.25} > \underbrace{M_e}_{100.07} > \underbrace{\bar{x}}_{100}$	$\underbrace{M_0}_{99.75} > \underbrace{M_e}_{99.73} > \underbrace{\bar{x}}_{99.68}$
$C_Y < 0$ $\beta_1 < 0$ $\gamma_1 < 0$ $\gamma_2 < 0$	$C_Y < 0$ $\beta_1 < 0$ $\gamma_1 < 0$ $\gamma_2 < 0$	$C_Y < 0$ $\beta_1 < 0$ $\gamma_2 > 0$

Analysis

The curves of the three injection guns **A** and **B** and **C** have asymmetric distributions on the right; their distribution tail is spread to the left, hence the preponderance of above-average dosages. This would lead to extreme risks of poisoning patients.

The distribution curves of guns **A** and **B** are more flattened than that of the normal distribution while the distribution curve of gun **C** is sharper than that of the normal distribution.

These two blocks of arguments do not affect the conclusions drawn in the second part, that is, gun **C** is more advantageous.

11.5 Fourth Case Study

11.5.1 Presentation of the Study

A foundry of very particular metals is looking for a software whose response time would be strictly less than 100 milliseconds (ms). This time is estimated to be the time needed for the software to detect the total melting of the metal, to obtain the net weight of the melted metal, to evaluate the quantity of additive components to be added according to the weight of the metal and to activate the mixture before cooling and hardening of the alloy.

To this end, it launches a call for tenders to select software capable of detecting when the metal has completely melted, triggering and carrying out all operations within the time allowed, that is to say in less than 100 ms. To this end, three software developers responded to the call for tenders by presenting their software. For a wise decision, the technical team tested them, and the following tables give the results of one hundred individual tests carried out.

Software A	
Response Time (ms)	Frequency
[98 , 98,5)	2
[98,5 , 99)	7
[99 , 99,5)	21
[99,5 , 100)	53
[100 , 100,5)	87
[100,5 , 101)	30

Software B	
Response Time (ms)	Frequency
[98 , 98,5)	4
[98,5 , 99)	9
[99 , 99,5)	31
[99,5 , 100)	47
[100 , 100,5)	57
[100,5 , 101)	52

Software C	
Response Time (ms)	Frequency
[98 , 98,5)	9
[98,5 , 99)	11
[99 , 99,5)	38
[99,5 , 100)	90
[100 , 100,5)	45
[100,5 , 101)	7

Since the validation team wants the asymptotically superior software or the fastest software to be used, which of this three software will receive the approval of the technical team you lead?
The decision must be accompanied by verifications by calculation.

11.5.2 Questions

11.5.2.1 Analysis of central tendency measures for decision-making

For each of the software,

1. Plot your histogram.
2. Plot its curve of cumulative frequencies.
3. Calculate the characteristics of central tendency of the distribution:
   a. The mode.
   b. The average.
   c. The median
4. Give an interpretation of the order of magnitude of the elements of this statistical series.

11.5.2.2 Analysis of measures of dispersion for decision-making

To verify the accuracy of the statements made by the three suppliers, it is required to calculate and compare the following measures of dispersion for the three software:

5. Their characteristics of dispersion
   a. The interquartile range.
   b. The standard deviation.
   c. The coefficient of variation.
6. The percentage of response times below average.
7. The percentage of response times below the mean minus the standard deviation.
8. The percentage of response times below the mean plus the standard deviation.
9. The percentage of response time within the range of the mean and the mean minus the standard deviation.
10. The percentage of response time within the range of the mean and the mean plus the standard deviation.
11. The percentage of response time within the range of the mean plus the standard deviation and the mean minus the standard deviation.
12. The percentage of response time less than 100 ms (the percentage of success cases).

11.5.2.3 Analysis of shape measures for decision-making

To refine and consolidate the decision, if possible, it is asked to calculate their characteristics of shape.

13. The YULE's coefficient of skewness.
14. The PEARSON's coefficient of skewness.
15. The FISHER's coefficient of skewness.
16. The FISHER's coefficient of kurtosis.

11.5.3 Solutions

11.5.3.1 Analysis of central tendency measures for decision-making

A: Software A

A-1: The histogram of software **A** is:

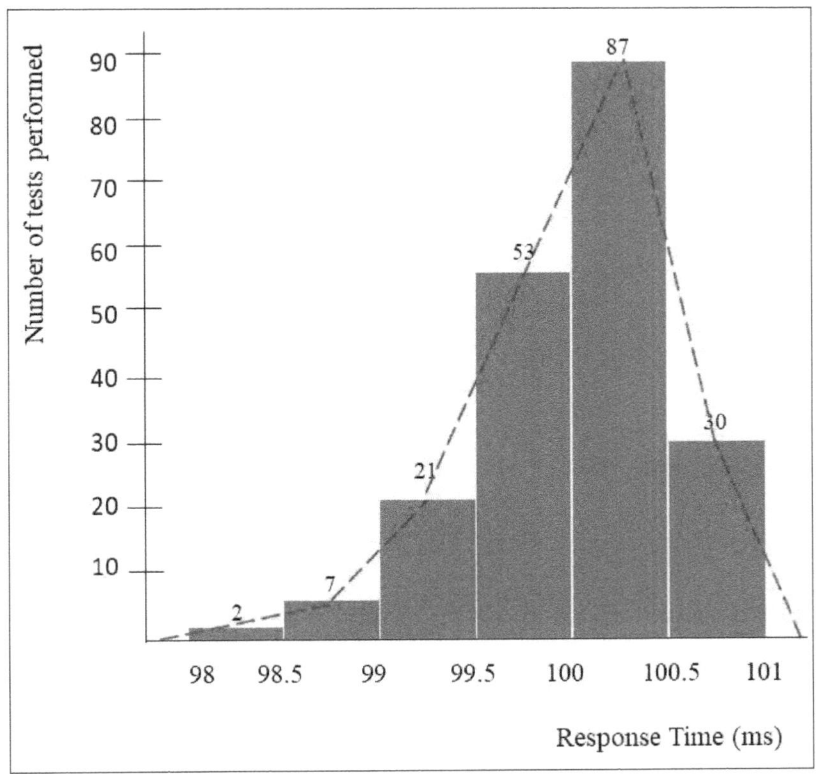

Solving this exercise requires the following table:

Software A							
Response Time (ms)	x_i	n_i	$n_i x_i$	$n_i x_i^2$	f_i	$f_i \uparrow$	$f_i \downarrow$
[98 , 98,5 [	98.25	2	196.50	19,306.13	1.00	1.00	100.00
[98,5 , 99 [	98.75	7	691.25	68,260.94	3.50	4.50	99.00
[99 , 99,5 [	99.25	21	2,084.25	206,861.81	10.50	15.00	95.50
[99,5 , 100 [	99.75	53	5,286.75	527,353.31	26.50	41.50	85.00
[100 , 100,5 [	100.25	87	8,721.75	874,355.44	43.50	85.00	58.50
[100,5 , 101 [	100.75	30	3,022.50	304,516.88	15.00	100.00	15.00
	Total	200	20,003.00	2,000,654.50	100.00		

A-2: The cumulative relative frequency curve of software **A** is:

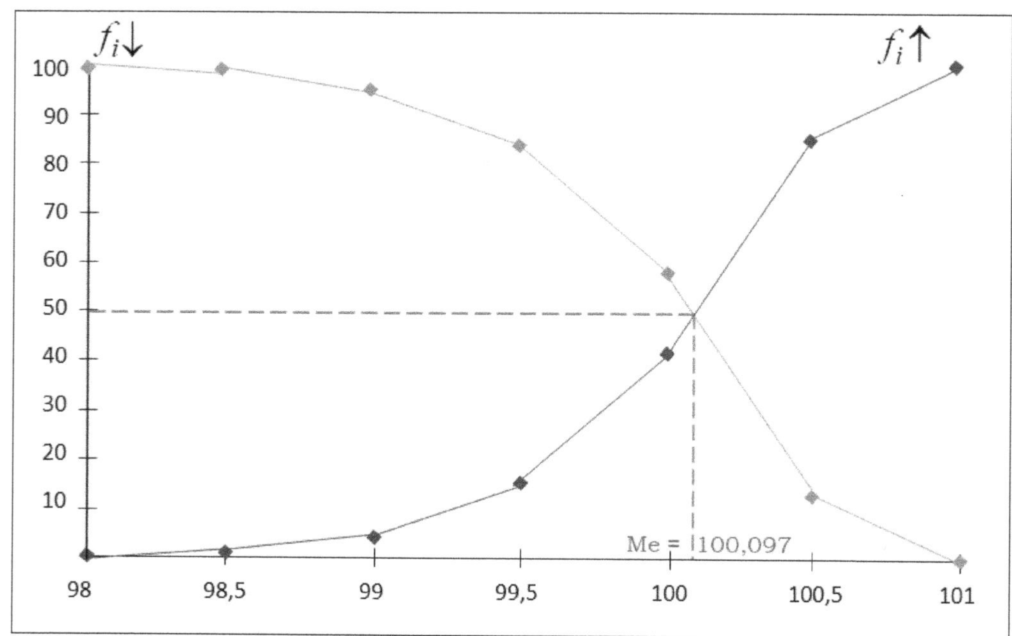

A-3: Determining the characteristics of central tendency:

A-3-a: The mode of software **A** is 100.25 ms.

A-3-b: The average of software **A** is:

$$\bar{x} = \frac{1}{n}\sum_{i=1}^{i=6} n_i x_i = \frac{20,003}{200}$$

$$= 100.015 \text{ ms}$$

A-3-c: The median of software **A** is:

$$Me = a_i + (a_{i+1} - a_i)\frac{(50 - F_i)}{(F_{i+1} - F_i)}$$

$$= 100 + (100.5 - 100)\frac{(50 - 41.50)}{(85 - 41.50)}$$

$$= 100.097701149425 \text{ ms}$$

B: Software B

B-1: The histogram of software **B** is:

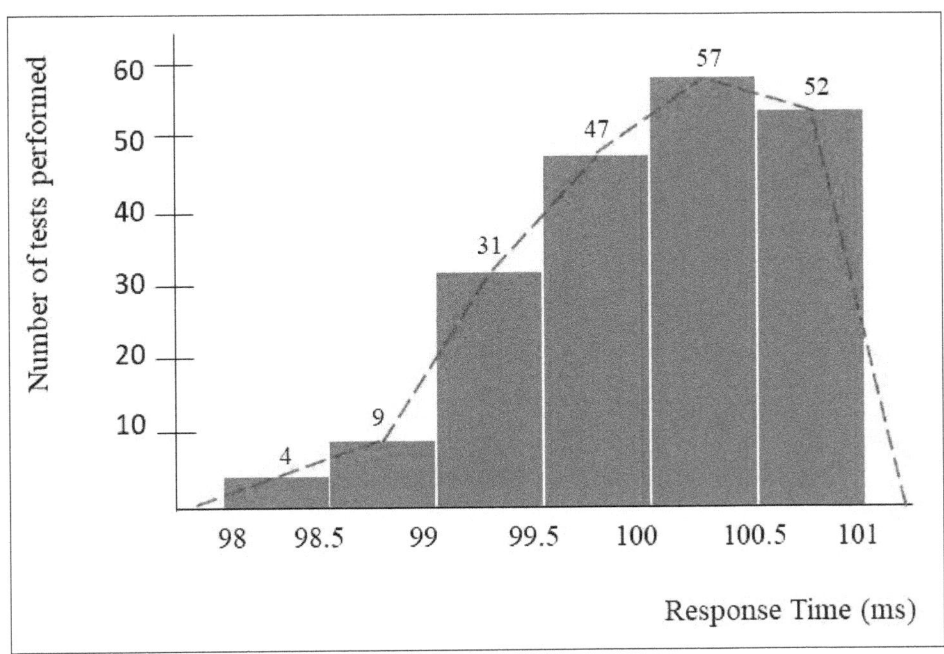

Solving this exercise requires the following table:

Software B							
Response Time (ms)	x_i	n_i	$n_i x_i$	$n_i x_i^2$	f_i	$f_i \uparrow$	$f_i \downarrow$
[98 , 98,5 [	98.25	4	393.00	38,612.25	2.00	2.00	100.00
[98,5 , 99 [	98.75	9	888.75	87,764.06	4.50	6.50	98.00
[99 , 99,5 [	99.25	31	3,076.75	305,367.44	15.50	22.00	93.50
[99,5 , 100 [	99.75	47	4,688.25	467,652.94	23.50	45.50	78.00
[100 , 100,5 [	100.25	57	5,714.25	572,853.56	28.50	74.00	54.50
[100,5 , 101 [	100.75	52	5,239.00	527,829.25	26.00	100.00	26.00
	Total	200	20,000.00	2,000,079.50	100.00		

B-2: The cumulative relative frequency curve is:

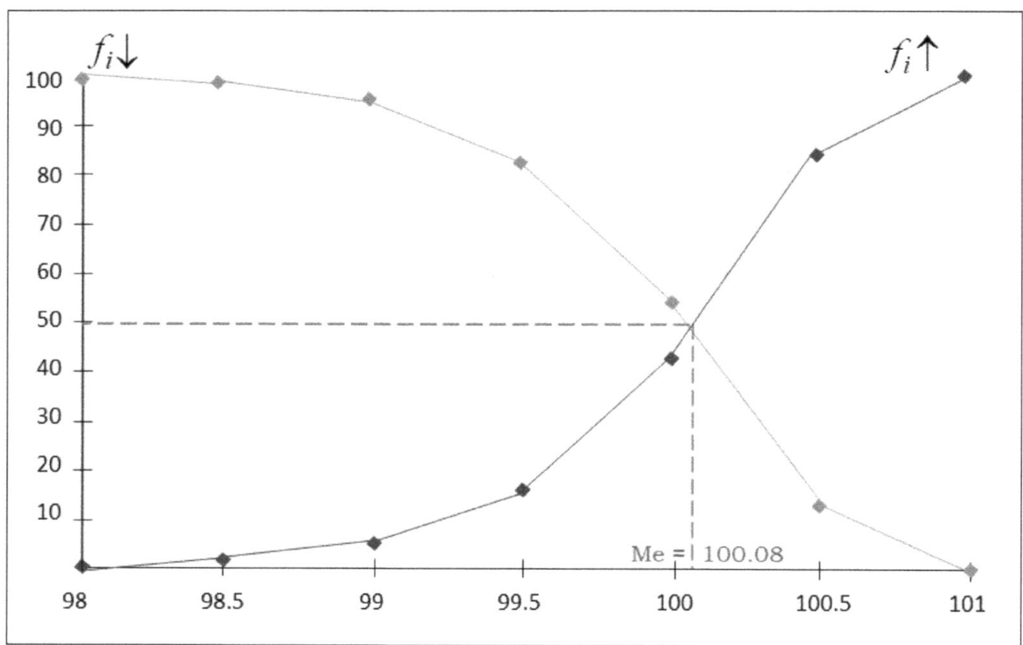

B-3: Determining the characteristics of central tendency:

B-3-a: The mode of software **B** is 100.25 ms.

B-3-b: The average of software **B** is:

$$\bar{x} = \frac{1}{n}\sum_{i=1}^{i=6} n_i x_i = \frac{20{,}000}{200}$$

$$= 100 \text{ ms}$$

B-3-c: The median of software **B** is:

$$Me = a_i + (a_{i+1} - a_i)\frac{(50 - F_i)}{(F_{i+1} - F_i)}$$

$$= 100 + (100.5 - 100)\frac{(50 - 45.5)}{(74 - 45.5)}$$

$$= 100.078947368421 \text{ ms}$$

C: Software C

C-1: The histogram of software **C** is:

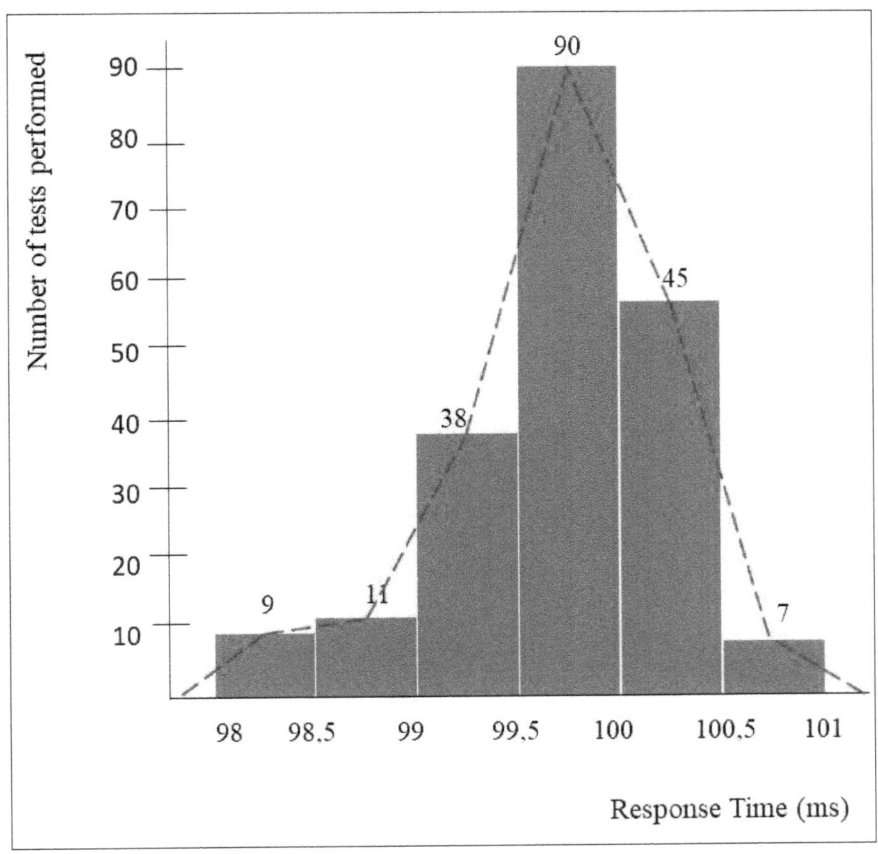

Solving this exercise requires the following table:

Software C							
Response Time (ms)	x_i	n_i	$n_i x_i$	$n_i x_i^2$	f_i	$f_i \uparrow$	$f_i \downarrow$
[98 , 98,5)	98.25	9	884.25	86,877.56	4.50	4.50	100.00
[98,5 , 99)	98.75	11	1,086.25	107,267.19	5.50	10.00	95.50
[99 , 99,5)	99.25	38	3,771.50	374,321.38	19.00	29.00	90.00
[99,5 , 100)	99.75	90	8,977.50	895,505.63	45.00	74.00	71.00
[100 , 100,5)	100.25	45	4,511.25	452,252.81	22.50	96.50	26.00
[100,5 , 101)	100.75	7	705.25	71,053.94	3.50	100.00	3.50
	Total	200	19,936.00	1,987,278.50	100.00		

C-2: The cumulative relative frequency curve of software **C** is:

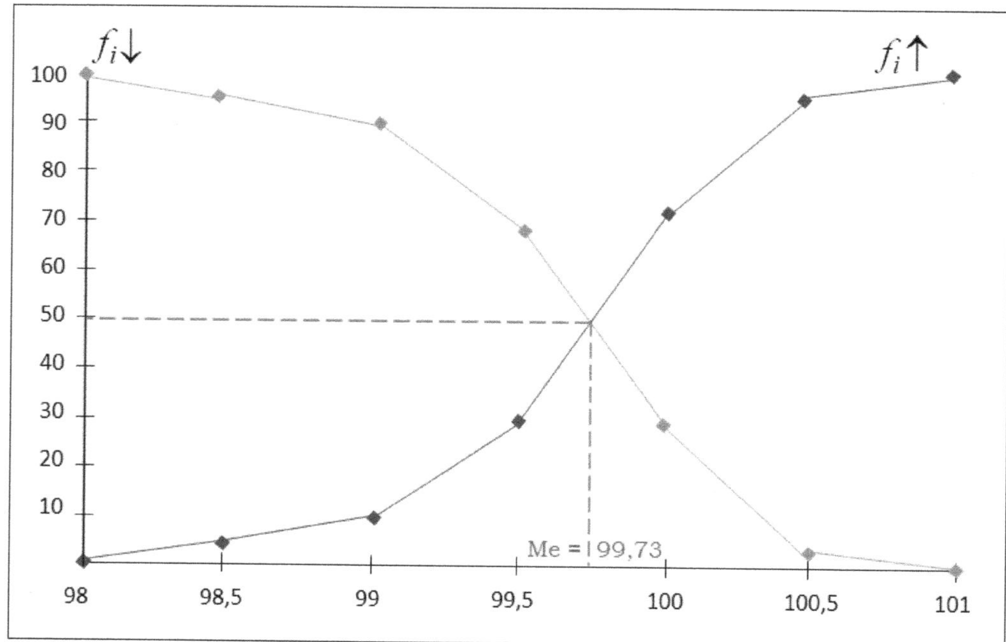

C-3: Determining the characteristics of central tendency:

C-3-a: The mode of software **C** is 99.75 ms.

C-3-b: The average of software **C** is:

$$\bar{x} = \frac{1}{n}\sum_{i=1}^{i=6} n_i x_i = \frac{19{,}936}{200}$$

$$= 99.68 \text{ ms}$$

C-3-c: The median of software **C** is:

$$Me = a_i + (a_{i+1} - a_i)\frac{(50 - F_i)}{(F_{i+1} - F_i)}$$

$$= 99.5 + (100 - 99.5)\frac{(50 - 29)}{(74 - 29)}$$

$$= 99.733 \text{ ms}$$

4: Interpretation

<u>INTERPRETATION OF RESULTS, ANALYSIS AND DECISION</u>

By comparing the three parameters, we observe that:

	Software A	Software B	Software C
	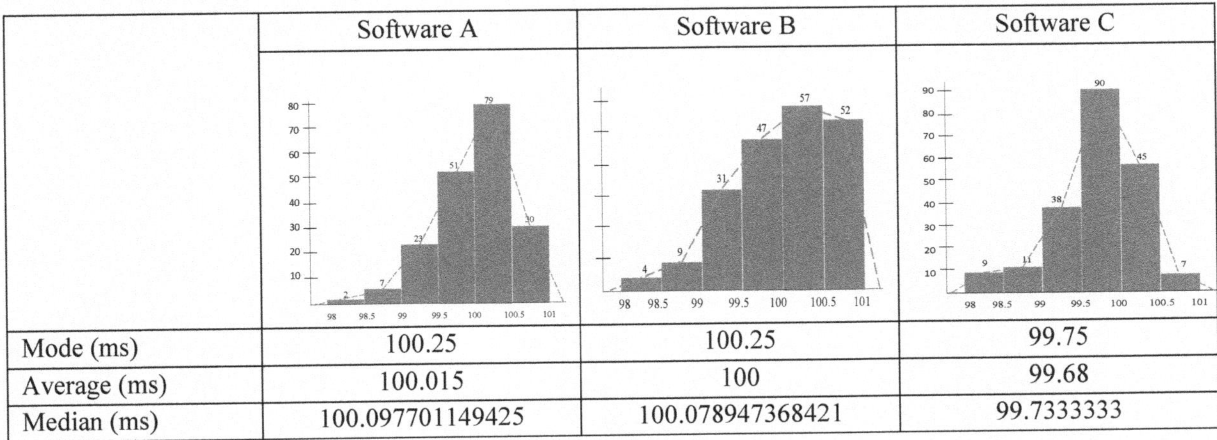		
Mode (ms)	100.25	100.25	99.75
Average (ms)	100.015	100	99.68
Median (ms)	100.097701149425	100.078947368421	99.7333333

Analysis

The quality control of three software revealed the following information:

Software **A**

- The response time exceeds the allotted time of 100 ms in the majority of the tests conducted.
- Moreover, even the average response time is above 100 ms, indicating a general inefficiency of the software.
- Furthermore, half of the tests showed a response time above 100 ms, proving that the software cannot be used to meet the foundry's needs under optimal conditions.

Software **B**

- The response time of the majority of tests does not meet the technical requirements as it exceeds 100 ms.
- The average response time is at the limit of the technical requirement.
- The response time of half of the tests does not meet the technical requirements as it exceeds 100 ms.
- Software B has obtained mixed results.
- Although the response time of the majority of tests does not meet the technical requirements as it exceeds 100 ms, the average response time is very close to the technical requirement of 100 ms, which could make it a viable option if the test results were more stable.
- However, the response time of half of the tests is also above 100 ms, suggesting that Software B may not be able to maintain stable and reliable performance under varying working conditions.

Software **C**

- The majority of response times in the tests meet the technical requirements for response time as they take less than 100 ms to respond. This means that most users should not encounter latency issues when using the software.
- In addition, the average response time of the tests also meets the technical requirements as it is less than 100 ms. This ensures a smooth and fast user experience for most operations performed with the software.
- It is also important to note that half of the tests have a response time below 100 ms, which is an additional indication of the software's speed and responsiveness.

Conclusion

Software A does not meet the requirements of the metal foundry in terms of response time.

Therefore, it is difficult for software **A** to enable the foundry to achieve its objectives within the allotted time and improve its production process.

Although software **B** has some advantages, it may not be the best choice for the highly specialized metal foundry that is looking for a reliable and fast solution to detect metal melting and perform necessary operations in less than 100 ms.

For software **C**, the majority of tests were performed in less than 100 ms, the average response time is also less than 100 ms, and half of the tests were performed in less than 100 ms. This performance ensures the fast detection of total metal melting, accurate calculation of the net weight of molten metal, exact amount of additive components to be added based on the weight of the metal, and quick activation of the mixture before cooling and hardening of the alloy. Therefore, the metal foundry can opt for software **C** to meet its needs.

Overall, software **A** and **B** do not meet the technical requirements in terms of response time, which can lead to negative consequences on the quality of the final product. Software **C** seems to be the only one that meets the technical requirements in terms of response time. Its performance in terms of response time in tests is in line with the expectations of the metal foundry.

11.5.3.2 Analysis of measures of dispersion for decision-making

A: Software A

A-5: Characteristics of dispersion:

The following table shows the data required to solve this exercise:

Software A							
Response Time (ms)	x_i	n_i	$n_i x_i$	$n_i x_i^2$	f_i	$f_i \uparrow$	$f_i \downarrow$
[98 , 98,5 [	98.25	2	196.50	19,306.13	1.00	1.00	100.00
[98,5 , 99 [	98.75	7	691.25	68,260.94	3.50	4.50	99.00
[99 , 99,5 [	99.25	21	2,084.25	206,861.81	10.50	15.00	95.50
[99,5 , 100 [	99.75	53	5,286.75	527,353.31	26.50	41.50	85.00
[100 , 100,5 [	100.25	87	8,721.75	874,355.44	43.50	85.00	58.50
[100,5 , 101 [	100.75	30	3,022.50	304,516.88	15.00	100.00	15.00
	Total	200	20,003.00	2,000,654.50	100.00		

A-5-1: The interquartile deviation of software **A** is:

$$Q_1 = a_i + (a_{i+1} - a_i)\frac{(25 - F_i)}{(F_{i+1} - F_i)}$$

$$= 99.5 + (100 - 99.5)\frac{(25 - 15)}{(41.5 - 15)}$$

$$= 99.688679245283 \text{ ms}$$

$$Q_3 = a_i + (a_{i+1} - a_i)\frac{(75 - F_i)}{(F_{i+1} - F_i)}$$

$$= 100 + (100.5 - 100)\frac{(75 - 41.5)}{(85 - 41.5)}$$

$$= 100.385057471264 \text{ ms}$$

$$\text{Interquartile Range} = Q_3 - Q_1$$

$$= 100.385057471264 - 99.688679245283$$

$$= 0.696378225981363 \text{ ms}$$

A-5-2: The variance of software **A** is:

$$V(x) = \frac{1}{n}\sum_{i=1}^{i=6} n_i x_i^2 - \bar{x}^2$$

$$= \frac{2{,}000{,}654.50}{200} - (100.015)^2$$

$$= 0.272275$$

A-5-3: The standard deviation of software **A** is:

$$\sigma = \sqrt{V(x)}$$

$$= \sqrt{0.272275}$$

$$= 0.521799770026 \text{ ms}$$

A-5-4: The coefficient of variation of software **A** is:

$$\text{CV} = \frac{\sigma}{|\bar{x}|}$$

$$= \frac{0.521799770026}{100.015} \text{x}100$$

$$= 0.52\%$$

A-6: Let's calculate the percentage of response times below the average:

$$\bar{x} = 100.015 \text{ ms}$$

Let $P_{<\bar{x}}$ be this percentage, we use the linear interpolation formula using the data pair $(a_i, f_i \uparrow)$:

$$\bar{x} = a_i + (a_{i+1} - a_i)\frac{(P_{<\bar{x}} - F_i)}{(F_{i+1} - F_i)}$$

$$100.015 = 100 + (100.5 - 100)\frac{(P_{<\bar{x}} - 41.5)}{(85 - 41.5)}$$

$$P_{<\bar{x}} = 42.805\%$$

A-7: Let's calculate the percentage of response times below $\bar{x} - \sigma$:

$$\bar{x} - \sigma = 100.015 - 0.521799770026$$
$$= 99.4932002299738 \text{ ms}$$

Let $P_{<\bar{x}-\sigma}$ be this percentage, we use the linear interpolation formula using the data pair $(a_i, f_i \uparrow$ $)$:

$$\bar{x} - \sigma = a_i + (a_{i+1} - a_i)\frac{(P_{<\bar{x}-\sigma} - F_i)}{(F_{i+1} - F_i)}$$

$$99.4932002299738 = 99 + (99.5 - 99)\frac{(P_{<\bar{x}-\sigma} - 4.5)}{(15 - 4.5)}$$

$$P_{<\bar{x}-\sigma} = 14.8572\%$$

A-8: Let's calculate the percentage of response times below $\bar{x} + \sigma$:

$$\bar{x} + \sigma = 100.015 + 0.52179977002616$$
$$= 100.53679977002616 \text{ ms}$$

Let $P_{<\bar{x}+\sigma}$ be this percentage, we use the linear interpolation formula using the data pair $(a_i, f_i \uparrow)$ as proceeded above:

$$\bar{x} + \sigma = a_i + (a_{i+1} - a_i)\frac{(P_{<\bar{x}+\sigma} - F_i)}{(F_{i+1} - F_i)}$$

$$100.53679977002616 = 100.5 + (101 - 100.5)\frac{(P_{<\bar{x}+\sigma} - 85)}{(100 - 85)}$$

$$P_{<\bar{x}+\sigma} = 86.1037\%$$

A-9: Let's calculate the percentage of response times between the mean, i.e., 100.015 cm and the mean minus the standard deviation, i.e., 99.4932002299738 ms.

This percentage is:

$$42.805\% - 14.8572\% = 27.9478\%$$

A-10: Let's calculate the percentage of response times between the mean, i.e., 100.015 cm and the mean plus the standard deviation, i.e., 100.53679977002616 ms.

This percentage is:

$$86.1037\% \; - \; 42.805\% = 43.2987\%$$

A-11: Let's calculate the percentage of response times between the mean minus the standard deviation, which is 99.4932002299738 ms and the mean plus the standard deviation, which is 100.53679977002616 ms.

This percentage is:

$$86.1037\% \; - \; 14.8572\% = 71.2465\%$$

The same response can be obtained by:

$$86.1037\% \; - \; 14.8572\% = 71.2465\%$$

A-12: Let's calculate the percentage of response times less than 100 ms.

This percentage can be obtained directly from the table, it is:

41.5%.

B: Software B

B-5: Characteristics of dispersion

The following table shows the data required to solve this exercise:

Software B							
Response Time (ms)	x_i	n_i	$n_i x_i$	$n_i x_i^2$	f_i	$f_i \uparrow$	$f_i \downarrow$
[98 , 98,5 [	98.25	4	393.00	38,612.25	2.00	2.00	100.00
[98,5 , 99 [	98.75	9	888.75	87,764.06	4.50	6.50	98.00
[99 , 99,5 [	99.25	31	3,076.75	305,367.44	15.50	22.00	93.50
[99,5 , 100 [	99.75	47	4,688.25	467,652.94	23.50	45.50	78.00
[100 , 100,5 [	100.25	57	5,714.25	572,853.56	28.50	74.00	54.50
[100,5 , 101 [	100.75	52	5,239.00	527,829.25	26.00	100.00	26.00
	Total	200	20,000.00	2,000,079.50	100.00		

B-5-1: The interquartile range of software **B** is:

$$Q_1 = a_i + (a_{i+1} - a_i)\frac{(25 - F_i)}{(F_{i+1} - F_i)}$$
$$= 99{,}5 + (100 - 99{,}5)\frac{(25 - 22)}{(45{,}5 - 22)}$$
$$= 99.563829787234 \text{ ms}$$
$$Q_3 = a_i + (a_{i+1} - a_i)\frac{(75 - F_i)}{(F_{i+1} - F_i)}$$
$$= 100.5 + (101 - 100.5)\frac{(75 - 74)}{(100 - 74)}$$
$$= 100.51923076923 \text{ ms}$$
$$\text{The interquartile range } = Q_3 - Q_1$$
$$= 100.51923076923 - 99.563829787234$$
$$= 0.955400981996732 \text{ ms}$$

B-5-2: The variance of software **B** is:

$$V(x) = \frac{1}{n}\sum_{i=1}^{i=6} n_i x_i^2 - \bar{x}^2$$
$$= \frac{2{,}000{,}079.5}{200} - (100)^2$$
$$= 0.397499999999127$$

B-5-3: The standard deviation of software **B** is:

$$\sigma = \sqrt{V(x)}$$

$$= \sqrt{0.397499999999127}$$
$$= 0.630476010645232 \text{ ms}$$

B-5-4: The coefficient of variation of software **B** is:

$$\text{CV} = \frac{\sigma}{|\bar{x}|}$$
$$= \frac{0.630476010645232}{100} \text{x} 100$$
$$= 0.630476010645232\%$$

B-6: Let's calculate the percentage of the response times less than the average.

This percentage can be obtained directly from the table, it is:

$$P_{<\bar{x}} = 41.5\%$$

B-7: Let's calculate the percentage of response times whose less than $\bar{x} - \sigma$

$$\bar{x} - \sigma = 100 - 0.630476010645232$$
$$= 99.369523989354768 \text{ ms}$$

Let $P_{<\bar{x}-\sigma}$ be this percentage, we use the linear interpolation formula using the data pair $(a_i, f_i \uparrow)$:

$$\bar{x} - \sigma = a_i + (a_{i+1} - a_i)\frac{(P_{<\bar{x}-\sigma} - F_i)}{(F_{i+1} - F_i)}$$
$$99.369523989354768 = 99 + (99.5 - 99)\frac{(P_{<\bar{x}-\sigma} - 6.5)}{(22 - 6.5)}$$
$$P_{<\bar{x}-\sigma} = 17.95512\%$$

B-8: Let's calculate the percentage of response times less than $\bar{x} + \sigma$

$$\bar{x} + \sigma = 100 + 0.630476010645232$$
$$= 100.630476010645232 \text{ ms}$$

Let $P_{<\bar{x}+\sigma}$ be this percentage, we use the linear interpolation formula using the data pair $(a_i, f_i \uparrow)$ as proceeded above:

$$\bar{x} + \sigma = a_i + (a_{i+1} - a_i)\frac{(P_{<\bar{x}+\sigma} - F_i)}{(F_{i+1} - F_i)}$$
$$100.630476010645232 = 100.5 + (101 - 100.5)\frac{(P_{<\bar{x}+\sigma} - 74)}{(100 - 74)}$$

$$P_{<\bar{x}+\sigma} = 80.78444\%$$

B-9: Let's calculate the percentage of response times between the mean, which is 100 ms and the mean minus the standard deviation, which is 99.369523989354768 ms.

This percentage is:

$$41.5\% - 17.95512\% = 23.54488\%$$

B-10: Let's calculate the percentage of response times between the mean, i.e., 100 ms and the mean plus the standard deviation, i.e.,100.630476010645232 ms.

This percentage is:

$$80.78444\% \ - \ 41.5\% = 39.28444\%$$

B-11: Let's calculate the percentage of response times between the mean minus the standard deviation, i.e., 99.369523989354768 ms and the mean plus the standard deviation, i.e., 100.630476010645232 ms.

This percentage is:

$$80.78444\ \% - 17.95512\% = 62.82932\%$$

B-12: Let's calculate the percentage of response times less than 100 ms.

This percentage can be obtained directly from the table, it is:

$$P_{<100ms} = 45.5\%$$

C: Software C

C-5: Characteristics of dispersion

The following table shows the data required to solve this exercise:

Software C							
Response Time (ms)	x_i	n_i	$n_i x_i$	$n_i x_i^2$	f_i	$f_i \uparrow$	$f_i \downarrow$
[98 , 98,5)	98.25	9	884.25	86,877.56	4.50	4.50	100.00
[98,5 , 99)	98.75	11	1,086.25	107,267.19	5.50	10.00	95.50
[99 , 99,5)	99.25	38	3,771.50	374,321.38	19.00	29.00	90.00
[99,5 , 100)	99.75	90	8,977.50	895,505.63	45.00	74.00	71.00
[100 , 100,5)	100.25	45	4,511.25	452,252.81	22.50	96.50	26.00
[100,5 , 101)	100.75	7	705.25	71,053.94	3.50	100.00	3.50
	Total	200	19,936.00	1,987,278.50	100.00		

C-5-1 The interquartile range of software **C** is:

$$Q_1 = a_i + (a_{i+1} - a_i)\frac{(25 - F_i)}{(F_{i+1} - F_i)}$$

$$= 99 + (99.5 - 99)\frac{(25 - 10)}{(29 - 10)}$$

$$= 99.3947368421053 \text{ ms}$$

$$Q_3 = a_i + (a_{i+1} - a_i)\frac{(75 - F_i)}{(F_{i+1} - F_i)}$$

$$= 100 + (100.5 - 100)\frac{(75 - 74)}{(96.5 - 74)}$$

$$= 100.022222222222 \text{ ms}$$

$$\text{The interquartile range} = Q_3 - Q_1$$

$$= 100.022222222222 - 99.3947368421053$$

$$= 0.627485380116966 \text{ ms}$$

C-5-2: The variance of software **C** is:

$$V(x) = \frac{1}{n}\sum_{i=1}^{i=6} n_i x_i^2 - \bar{x}^2$$

$$= \frac{1\,987\,278.5}{200} - (99.68)^2$$

$$= 0.290099999998347$$

C-5-3 The standard deviation of software **C** is:

$$\sigma = \sqrt{V(x)}$$

$$= \sqrt{0.290099999998347}$$

$$= 0.538609320378275 \text{ ms}$$

C-5-4 The coefficient of variation of software **C** is:

$$CV = \frac{\sigma}{|\bar{x}|}$$

$$= \frac{0.538609320378275}{99.68} x100$$

$$= 0.5403384032687\%$$

C-6: Let's calculate the percentage of response times less than average.

$$\bar{x} = 99.68 \text{ ms}$$

Let $P_{<\bar{x}}$ be this percentage, we use the linear interpolation formula on the couple $(a_i, f_i \uparrow)$, in the same way that we used for the median:

$$\bar{x} = a_i + (a_{i+1} - a_i)\frac{(P_{<-\sigma} - F_i)}{(F_{i+1} - F_i)}$$

$$99.68 = 99.5 + (100 - 99.5)\frac{(P_{<\bar{x}} - 29)}{(74 - 29)}$$

$$P_{<\bar{x}} = 45.2\%$$

C-7: Let's calculate the percentage of response times less than $\bar{x} - \sigma$

$$\bar{x} - \sigma = 99.68 - 0.538609320378275$$
$$= 99.141390679621725 \text{ ms}$$

Let $P_{<\bar{x}-\sigma}$ be this percentage, we use the linear interpolation formula using the data pair $(a_i, f_i \uparrow)$ as proceeded above:

$$\bar{x} - \sigma = a_i + (a_{i+1} - a_i)\frac{(P_{<\bar{x}-\sigma} - F_i)}{(F_{i+1} - F_i)}$$

$$99.141390679621725 = 99 + (99.5 - 99)\frac{(P_{<\bar{x}-\sigma} - 10)}{(29 - 10)}$$

$$P_{<\bar{x}-\sigma} = 15.37282\%$$

C-8: Let's calculate the percentage of response times less than $\bar{x} + \sigma$

$$\bar{x} + \sigma = 99.68 + 0.538609320378275$$
$$= 100.218609320378275 \text{ ms}$$

Let $P_{<\bar{x}+\sigma}$ be this percentage, we use the linear interpolation formula using the data pair $(a_i, f_i \uparrow)$ as proceeded above:

$$\bar{x} + \sigma = a_i + (a_{i+1} - a_i)\frac{(P_{<\bar{x}-\sigma} - F_i)}{(F_{i+1} - F_i)}$$

$$100.218609320378275 = 100 + (100.5 - 100)\frac{(P_{<\bar{x}+\sigma} - 74)}{(96.5 - 74)}$$

$$P_{<\bar{x}+\sigma} = 83.837\%$$

C-9: Let's calculate the percentage of response times between the mean, i.e., 99,68 ms and the mean minus the standard deviation, i.e., 99.141390679621725 ms.

This percentage is:

$$45.2\% - 15.37282\% = 29.82718\%$$

C-10: Let's calculate the percentage of response times between the mean, i.e., 99.68 ms and the mean plus the standard deviation, or.100.218609320378275 ms.

This percentage is:

$$83.837\ \% \ - \ 45.2\% = 38.637\%$$

C-11: Let's calculate the percentage of response times between the mean minus the standard deviation, i.e., 99.141390679621725 ms, and the mean plus the standard deviation, i.e., 100.218609320378275 ms.

This percentage is:

$$29.82718\% + 38.637\% = 68.46418\%$$

The same result can be obtained by:

$$83.837\% \ - \ 15.37282\% = 68.46418\%$$

C-12: Let's calculate the percentage of response times less than 100 ms.

This percentage can be obtained directly from the table, it is:

$$P_{<100ms} = 74\%$$

INTERPRETATION OF RESULTS, ANALYSIS AND DECISION

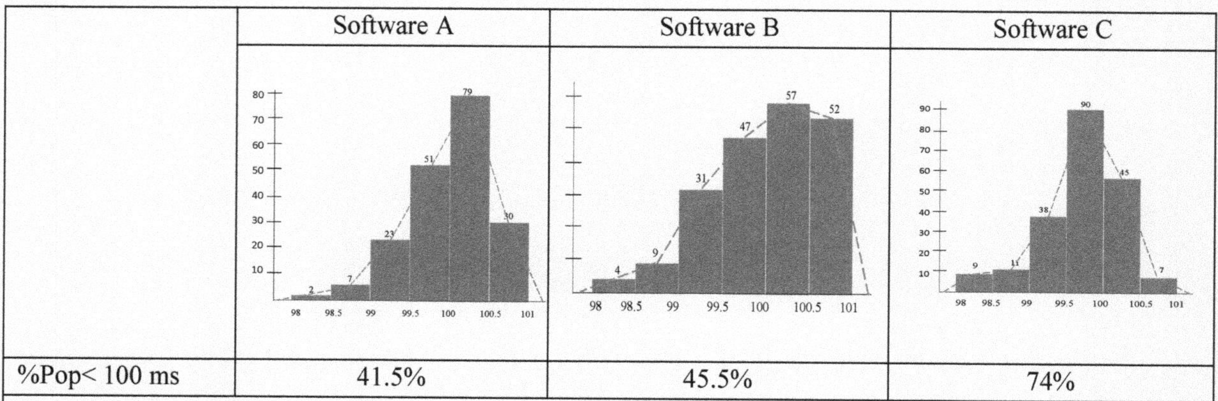

	Software A	Software B	Software C
%Pop< 100 ms	41.5%	45.5%	74%

Analysis

- Software **A** has 41.5% of the tests with a response time of less than 100 ms; therefore, has a success rate of 41.5% and a failure rate of 58.5%.
- Software **B** has 45.5% of the tests with a response time of less than 100 ms; therefore, has a success rate of 45.5% and a failure rate of 58.5%.
- Software **C** has 41.5% of the tests with a response time of less than 100 ms; therefore, has a success rate of 74% and a failure rate of 74%.

The three software programs have different results in terms of test success and response time under 100 ms. Software **C** has the highest success rate, with 74% of tests succeeded, followed by Software **B** with 45.5% and Software A with 41.5%.

However, test success is not the only criterion to consider. It is also important to take into account the failure rate of each software, which indicates the percentage of tests where the software did not respond in less than 100 ms. Software A has the highest failure rate with 58.5%, followed by Software B with 54.5% and Software **C** with only 26%.

Taking both criteria into account, Software C seems to be the most efficient and reliable, with a high success rate and a relatively low failure rate.

CV	0.52%	0.630%	0.5403384032%

The three software have a coefficient of variation of less than 1%, indicating that the dispersion of the results obtained is low and, therefore, the distributions are quite homogeneous.
Indeed, the fact that the distributions of the results of the three software are quasi-homogeneous means that there is little difference between the results obtained by each of them. Thus, even if the success and failure rates vary from one software to another, these variations are relatively small and, therefore, should not have a significant impact on the decision to choose the most performant and reliable software.

% in $[\bar{x}-\sigma, \bar{x}+\sigma]$	71.2465%	62.82932%	68.46418%

The population within the interval $[\bar{x}-\sigma, \bar{x}+\sigma]$ does not have much influence on the decision because this population is considered to be relatively homogeneous.

Conclusion

Software **C** is positioned as the most performant and reliable among the three tested software, with a success rate of 74% for tests with a response time under 100 ms. This result confirms the previously stated assumptions and reinforces the idea that software **C** best meets the user's needs in terms of result reliability.

It is important to note that the success rate is not the only criterion to consider when evaluating a software's performance. The failure rate, i.e., the percentage of tests where the software did not respond in less than 100 ms, is also an important criterion. In this case, software **C** has a relatively low failure rate of 26%, confirming its reliability.

11.5.3.3 Analysis of shape measures for decision-making

Let's calculate the shape characteristics:

A: Software **A**

The following table shows the data required to solve this exercise:

x_i	n_i	$n_i x_i$	$n_i(x_i-\bar{x})^3$	$n_i(x_i-\bar{x})^4$
98.25	2	196.50	-11.00	19.41
98.75	7	691.25	-14.17	17.93
99.25	21	2084.25	-9.40	7.19
99.75	53	5286.75	-0.99	0.26
100.25	87	8721.75	1.13	0.27
100.75	30	3022.50	11.91	8.76
	200	20003.00	-22.51365000	53.80854463

A-13: Let's calculate the YULE's coefficient of skewness:

$$C_Y = \frac{Q_1 - 2Q_2 + Q_3}{Q_3 - Q_1}$$

$$= \frac{99.68867924528 - 2(100.09770114) + 100.3850574712}{100.3850574712 - 99.68867924528}$$

$$= -0.17473$$

A-14: Let's calculate the PEARSON's coefficient of skewness:

$$\beta_1 = \frac{(\bar{x} - \text{Mode})}{\sigma}$$

$$= \frac{(100.015 - 100.25)}{0.52179977}$$

$$= -0.450364323$$

A-15: Let's calculate the FISHER's coefficient of skewness:

$$\mu_3 = \frac{\sum_{i=1}^{i=6} n_i (x_i - \bar{x})^3}{\sum_{i=1}^{i=6} n_i} = \frac{-22.51365}{200}$$

$$= -0.1125682497$$

$$\gamma_1 = \frac{\mu_3}{\sigma^3} = \frac{-0.1125682497}{(0.52179977)^3}$$

$$= -0.792327$$

A-16: Let's calculate the FISHER's coefficient of kurtosis:

$$\mu_4 = \frac{\sum_{i=1}^{i=6} n_i (x_i - \bar{x})^4}{\sum_{i=1}^{i=6} n_i} = \frac{53.80854463}{200}$$

$$= 0.26904278994$$

$$\gamma_2 = \frac{\mu_4}{\sigma^4} - 3 = \frac{0.26904278994}{(0.52179977)^4} - 3$$

$$= -0.6291576$$

B: Software **B**

The following table shows the data required to solve this exercise:

x_i	n_i	$n_i x_i$	$n_i(x_i - \bar{x})^3$	$n_i(x_i - \bar{x})^4$
98.25	4	393.00	-21.44	37.52
98.75	9	888.75	-17.58	21.97
99.25	31	3076.75	-13.08	9.81
99.75	47	4688.25	-0.73	0.18
100.25	57	5714.25	0.89	0.22
100.75	52	5239.00	21.94	16.45
	200	20000.00	-30.00000000	86.15625000

B-13: Let's calculate the YULE's coefficient of skewness:

$$C_Y = \frac{Q_1 - 2Q_2 + Q_3}{Q_3 - Q_1}$$

$$= \frac{99.56382978723 - 2(100.07894736) + 100.5192307692}{100.5192307692 - 99.56382978723}$$

$$= -0.078327493$$

B-14: Let's calculate the PEARSON's coefficient of skewness:

$$\beta_1 = \frac{(\bar{x} - \text{Mode})}{\sigma}$$

$$= \frac{(100 - 100.25)}{0.630476}$$

$$= -0{,}3965257996$$

B-15: Let's calculate the FISHER's coefficient of skewness:

$$\mu_3 = \frac{\sum_{i=1}^{i=6} n_i (x_i - \bar{x})^3}{\sum_{i=1}^{i=6} n_i} = \frac{-30}{200}$$

$$= -0.15$$

$$\gamma_1 = \frac{\mu_3}{\sigma^3} = \frac{-0.15}{(0.630476)^3}$$

$$= -0.598529$$

B-16: Let's calculate the FISHER's coefficient of kurtosis:

$$\mu_4 = \frac{\sum_{i=1}^{i=6} n_i (x_i - \bar{x})^4}{\sum_{i=1}^{i=6} n_i} = \frac{86.15625}{200}$$

$$= -0.430781$$

$$\gamma_2 = \frac{\mu_4}{\sigma^4} - 3 = \frac{-0.430781}{(0.630476)^4} - 3$$

$$= -0.2736443$$

C: Software **C**

The following table shows the data required to solve this exercise:

x_i	n_i	$n_i x_i$	$n_i(x_i - \bar{x})^3$	$n_i(x_i - \bar{x})^4$
98.25	9	884.25	-26.32	37.63
98.75	11	1086.25	-8.85	8.23
99.25	38	3771.50	-3.02	1.30
99.75	90	8977.50	0.03	0.00
100.25	45	4511.25	8.33	4.75
100.75	7	705.25	8.58	9.18
	200	19936.00	-21.24720000	61.09019400

C-13: Let's calculate the YULE's coefficient of skewness:

$$C_Y = \frac{Q_1 - 2Q_2 + Q_3}{Q_3 - Q_1}$$

$$= \frac{99.39473684210 - 2(99.7333) + 100.0222}{100.0222 - 99.39473684210}$$

$$= -0.079149122$$

C-14: Let's calculate the PEARSON's coefficient of skewness:

$$\beta_1 = \frac{(\bar{x} - \text{Mode})}{\sigma}$$

$$= \frac{(99.78 - 99.75)}{0.5386}$$

$$= -0.129964331$$

C-15: Let's calculate the FISHER's coefficient of skewness:

$$\mu_3 = \frac{\sum_{i=1}^{i=6} n_i (x_i - \bar{x})^3}{\sum_{i=1}^{i=6} n_i} = \frac{-21.2472}{200}$$

$$= -0.1062359998$$

$$\gamma_1 = \frac{\mu_3}{\sigma^3} = \frac{-0.1062359998}{(0.5386)^3}$$

$$= -0.679908$$

C-16: Let's calculate the FISHER's coefficient of kurtosis:

$$\mu_4 = \frac{\sum_{i=1}^{i=6} n_i (x_i - \bar{x})^4}{\sum_{i=1}^{i=6} n_i} = \frac{61.090194}{200}$$

$$= 0.30545091629$$

$$\gamma_2 = \frac{\mu_4}{\sigma^4} - 3 = \frac{0.30545091629}{(0.5386)^4} - 3$$

$$= 0.6294931$$

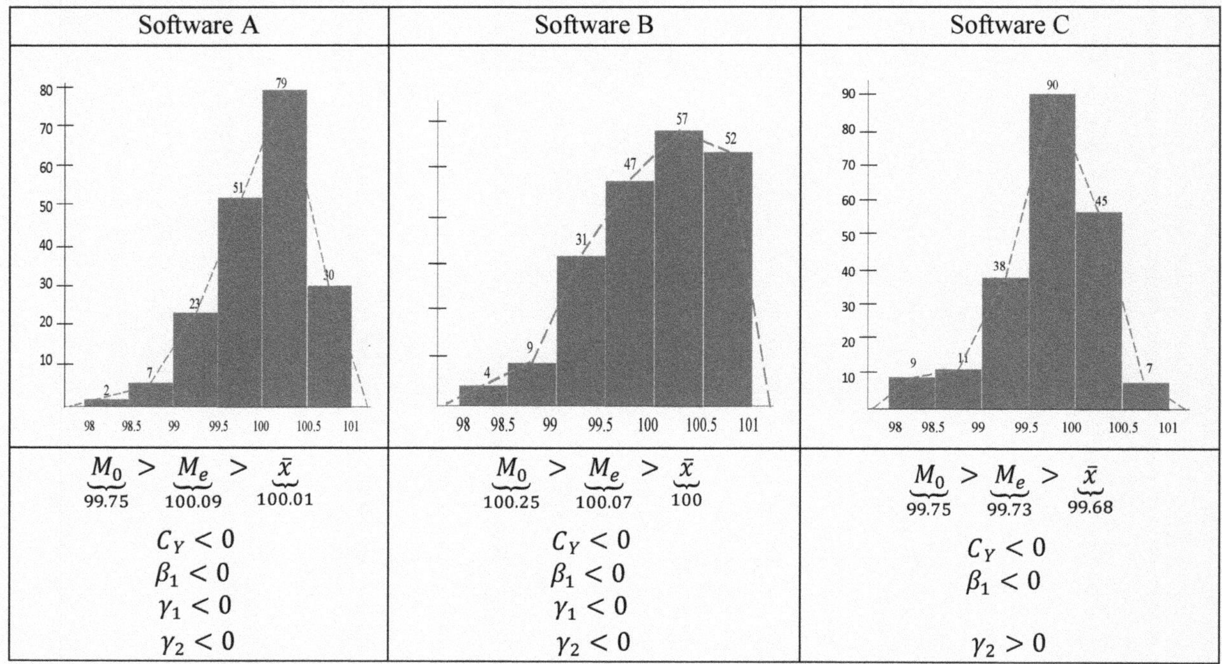

Analysis

The distribution curves of the response times for software **A**, **B**, and C all exhibit positive skewness, meaning their distribution tails are extended towards the left, indicating a preponderance of response times that are higher than the average. The distribution curves of **A** and **B** are flatter than the normal distribution curve, while **C**'s distribution curve is more pointed.

The **C** software stands out from the other two software in presenting a more concentrated distribution around the mean, with a more pronounced positive skewness. This allows it to achieve a success rate of 74%, the highest of the three software tested.

However, it is important to note that success rate is not the only criterion to consider when evaluating software performance. Other measures, such as failure rate or variability of response times, should also be taken into account. In this case, the **C** software also has advantages in terms of a relatively low failure rate and a more concentrated distribution, which reinforces its position as the best software among the three tested.

www.ingramcontent.com/pod-product-compliance
Ingram Content Group UK Ltd.
Pitfield, Milton Keynes, MK11 3LW, UK
UKHW061959290726
14090UKWH00021B/1294